HYPERSPACES OF SETS

PURE AND APPLIED MATHEMATICS

A Program of Monographs, Textbooks, and Lecture Notes

EXECUTIVE EDITORS—MONOGRAPHS, TEXTBOOKS, AND LECTURE NOTES

Earl J. Taft
Rutgers University
New Brunswick, New Jersey

Edwin Hewitt
University of Washington
Seattle, Washington

CHAIRMAN OF THE EDITORIAL BOARD

S. Kobayashi
University of California, Berkeley
Berkeley, California

EDITORIAL BOARD

Masanao Aoki
University of California, Los Angeles

W. S. Massey
Yale University

Glen E. Bredon
Rutgers University

Irving Reiner
University of Illinois at Urbana-Champaign

Sigurdur Helgason
Massachusetts Institute of Technology

Paul J. Sally, Jr.
University of Chicago

G. Leitman
University of California, Berkeley

Jane Cronin Scanlon
Rutgers University

Marvin Marcus
University of California, Santa Barbara

Martin Schechter
Yeshiva University

Julius L. Shaneson
Rutgers University

MONOGRAPHS AND TEXTBOOKS IN PURE AND APPLIED MATHEMATICS

1. *K. Yano*, Integral Formulas in Riemannian Geometry (1970)
2. *S. Kobayashi*, Hyperbolic Manifolds and Holomorphic Mappings (1970)
3. *V. S. Vladimirov*, Equations of Mathematical Physics (A. Jeffrey, editor; A. Littlewood, translator) (1970)
4. *B. N. Pshenichnyi*, Necessary Conditions for an Extremum (L. Neustadt, translation editor; K. Makowski, translator) (1971)
5. *L. Narici, E. Beckenstein, and G. Bachman*, Functional Analysis and Valuation Theory (1971)
6. *D. S. Passman*, Infinite Group Rings (1971)
7. *L. Dornhoff*, Group Representation Theory (in two parts). Part A: Ordinary Representation Theory. Part B: Modular Representation Theory (1971, 1972)
8. *W. Boothby and G. L. Weiss (eds.)*, Symmetric Spaces: Short Courses Presented at Washington University (1972)
9. *Y. Matsushima*, Differentiable Manifolds (E. T. Kobayashi, translator) (1972)
10. *L. E. Ward, Jr.*, Topology: An Outline for a First Course (1972) *out of print*
11. *A. Babakhanian*, Cohomological Methods in Group Theory (1972)
12. *R. Gilmer*, Multiplicative Ideal Theory (1972)
13. *J. Yeh*, Stochastic Processes and the Wiener Integral (1973) *out of print*
14. *J. Barros-Neto*, Introduction to the Theory of Distributions (1973) *out of print*
15. *R. Larsen*, Functional Analysis: An Introduction (1973)
16. *K. Yano and S. Ishihara*, Tangent and Cotangent Bundles: Differential Geometry (1973)
17. *C. Procesi*, Rings with Polynomial Identities (1973)
18. *R. Hermann*, Geometry, Physics, and Systems (1973)
19. *N. R. Wallach*, Harmonic Analysis on Homogeneous Spaces (1973)
20. *J. Dieudonné*, Introduction to the Theory of Formal Groups (1973)

21. *I. Vaisman*, Cohomology and Differential Forms (1973)
22. *B.-Y. Chen*, Geometry of Submanifolds (1973)
23. *M. Marcus*, Finite Dimensional Multilinear Algebra (in two parts) (1973, 1975)
24. *R. Larsen*, Banach Algebras: An Introduction (1973)
25. *R. O. Kujala and A. L. Vitter (eds.)*, Value Distribution Theory: Part A; Part B. Deficit and Bezout Estimates by Wilhelm Stoll (1973)
26. *K. B. Stolarsky*, Algebraic Numbers and Diophantine Approximation (1974)
27. *A. R. Magid*, The Separable Galois Theory of Commutative Rings (1974)
28. *B. R. McDonald*, Finite Rings with Identity (1974)
29. *J. Satake*, Linear Algebra (S. Koh, T. Akiba, and S. Ihara, translators) (1975)
30. *J. S. Golan*, Localization of Noncommutative Rings (1975)
31. *G. Klambauer*, Mathematical Analysis (1975)
32. *M. K. Agoston*, Algebraic Topology: A First Course (1976)
33. *K. R. Goodearl*, Ring Theory: Nonsingular Rings and Modules (1976)
34. *L. E. Mansfield*, Linear Algebra with Geometric Applications (1976)
35. *N. J. Pullman*, Matrix Theory and its Applications: Selected Topics (1976)
36. *B. R. McDonald*, Geometric Algebra Over Local Rings (1976)
37. *C. W. Groetsch*, Generalized Inverses of Linear Operators: Representation and Approximation (1977)
38. *J. E. Kuczkowski and J. L. Gersting*, Abstract Algebra: A First Look (1977)
39. *C. O. Christenson and W. L. Voxman*, Aspects of Topology (1977)
40. *M. Nagata*, Field Theory (1977)
41. *R. L. Long*, Algebraic Number Theory (1977)
42. *W. F. Pfeffer*, Integrals and Measures (1977)
43. *R. L. Wheeden and A. Zygmund*, Measure and Integral: An Introduction to Real Analysis (1977)
44. *J. H. Curtiss*, Introduction to Functions of a Complex Variable (1978)
45. *K. Hrbacek and T. Jech*, Introduction to Set Theory (1978)
46. *W. S. Massey*, Homology and Cohomology Theory (1978)
47. *M. Marcus*, Introduction to Modern Algebra (1978)
48. *E. C. Young*, Vector and Tensor Analysis (1978)
49. *S. B. Nadler, Jr.*, Hyperspaces of Sets (1978)

HYPERSPACES OF SETS

A Text with Research Questions

Sam B. Nadler, Jr.

MARCEL DEKKER, INC.   New York and Basel

Library of Congress Cataloging in Publication Data

Nadler, Sam B
  Hyperspaces of sets.

  (Monographs and textbooks in pure and applied mathematics ; v. 49)
    Bibliography: p.
    Includes indexes.
    1. Hyperspace.   I. Title.
QA691.N25   516'.182   78-9013
ISBN 0-8247-6768-3

COPYRIGHT © 1978 by MARCEL DEKKER, INC.   ALL RIGHTS RESERVED

Neither this book nor any part may be reproduced or transmitted in any form or by any means, electronic or mechanical, including photocopying, microfilming, and recording, or by any information storage and retrieval system, without permission in writing from the publisher.

MARCEL DEKKER, INC.
270 Madison Avenue, New York, New York 10016

Current printing (last digit):
10 9 8 7 6 5 4 3 2 1

PRINTED IN THE UNITED STATES OF AMERICA

DEDICATED

TO

PROFESSOR K. KURATOWSKI

who was responsible for my initial interest in hyperspaces and whose desire and ability to impart his wisdom and knowledge to others has been so deeply felt by all who know him

AND TO

THE MATHEMATICS DEPARTMENT

OF THE

UNIVERSITY OF GEORGIA

to whose faculty during my graduate student days (1961-1965) I owe a great debt of gratitude for their interest in the graduate program and their unselfish giving of themselves and their ideas to all the students.

CONTENTS

ABOUT THE BOOK		vi
USING THE BOOK IN A BEGINNING TOPOLOGY COURSE		ix
A VERY BRIEF HISTORY OF HYPERSPACE THEORY		xii
ACKNOWLEDGMENTS		xv
0.	SOME BASIC FACTS AND NOTATION	1
	Exercises	34
I.	THE GENERAL STRUCTURE OF HYPERSPACES	55
	A. Segments and Order-arcs-- Their Existence and Relationships	56
	B. Hyperspaces as Continuous Images and Preimages of the Cone Over the Cantor Set	81
	C. The Structure of $C(X)$ when X Is Indecomposable	99
	D. The Structure of 2^X and $C(X)$ when X Is Locally Connected	132
	E. The Hyperspace Operations and Inverse Limits	159
	Exercises	194
II.	DIMENSION OF $C(X)$	211
III.	EMBEDDING $C(X)$ IN R^n	229
IV.	MAPPINGS BETWEEN X, 2^X, AND $C(X)$	241
V.	SELECTIONS	253
VI.	RETRACTIONS	269
VII.	THE FIXED POINT PROPERTY	291
VIII.	RECOGNIZING HYPERSPACES AS CONES	301
IX.	RECOGNIZING HYPERSPACES AS SUSPENSIONS	335
X.	RECOGNIZING HYPERSPACES AS CARTESIAN PRODUCTS	339
XI.	ARCWISE DISCONNECTING HYPERSPACES	357
XII.	ARCWISE AND SEGMENTWISE ACCESSIBILITY	373
XIII.	CONTINUUMWISE ACCESSIBILITY	395

XIV.	WHITNEY PROPERTIES	399
	A. Basic Whitney Properties	399
	B. Some Non-Whitney Properties, Structure Results, and Questions	424
	C. Whitney Stable Continua	433
	D. Essentially Different Whitney Maps for $C(X)$	441
	E. Whitney-Reversible Properties	453
	F. Mappings between the Levels of $\mu^{-1}(t)$	465
	G. Properties of Whitney Maps for 2^X	466
	H. Spaces of Whitney Maps	473
	I. Cut Points of $\mu^{-1}(t)$	475
	J. The Covering Property (Revisited)	477
	K. Whitney Hyperspaces	502
	L. Two Hyperspace Functions	505
XV.	CONTINUITY PROPERTIES OF THE HYPERSPACE OPERATIONS	513
XVI.	CONTRACTIBILITY OF HYPERSPACES AND A PROPERTY OF KELLEY	531
XVII.	HOMOGENEITY OF HYPERSPACES	563
XVIII.	SPACES OF COMPACT CONVEX SETS	567
XIX.	OTHER SPACES OF SPECIAL SETS	601
XX.	APPENDIX--THE BOUNDARY BUMPING THEOREMS	625
BIBLIOGRAPHY		635
SUPPLEMENTARY BIBLIOGRAPHY, I		656
SUPPLEMENTARY BIBLIOGRAPHY, II		661
SUPPLEMENTARY BIBLIOGRAPHY, III		662
INDEX OF SPECIAL SYMBOLS		665
SUBJECT INDEX		667

ABOUT THE BOOK

This is the first book devoted to hyperspace theory. Primarily, the book is concerned with the structure and embedding properties of the two hyperspaces 2^X and $C(X)$ when X is a metric continuum.

I have attempted to treat much of the material from a unified point of view by using Whitney maps. These types of maps are defined and constructed in the first chapter of the book. They are used to obtain a systematic treatment of much of the theory of hyperspaces, including new proofs of some classical theorems. The idea of using Whitney maps in hyperspaces is due to Kelley [165]-- I have expanded on his idea.

The book may be considered as being essentially in three parts. Chapter 0 and Chapter I contain the basic classical results about hyperspaces and, thus, they comprise the first part. Many results center around the arc structure of hyperspaces. This arc structure is studied in depth throughout Chapter I. Because the material in Chapters 0 and I is so fundamental to hyperspace theory, many detailed proofs are included. Chapters II through XVII, which cover a variety of topics about 2^X and $C(X)$, may be considered as being the second part of the book. Chapters XVIII and XIX deal with hyperspaces other than 2^X and $C(X)$, and thus these two chapters comprise the third part. In Chapter XVIII, spaces of compact convex sets are studied. Results and techniques in infinite-dimensional topology, functional analysis, and convexity theory are used and some applications to these areas are presented. In Chapter XIX some other spaces of special sets are discussed.

The various topics in the book are, for the most part, covered thoroughly. Of notable exception is the theory of selections. Selections are discussed from the viewpoint of selections from subspaces of 2^X and $C(X)$ rather than from the viewpoint of selections for set-valued mappings. This is in keeping with the emphasis of the book. The general theory and numerous applications of selections would be too lengthy to discuss here. Many references are given for the interested reader.

Theorems are not always stated in the order in which they were originally proved, and sometimes results and ideas many years apart are combined. For example: See (1.8) and (1.25) where results and ideas of Borsuk-Mazurkiewicz and of Kelley are combined, or see the proof of (1.61) where the notion of terminal continua is used. However, though the material is not always treated from an historical perspective, there are a number of historical comments throughout the book.

Another comment about the theorems in the book is appropriate. The theorems are usually stated so as to be "true to the literature." By this I mean that they are usually stated with exactly the same information as is in the referenced paper. Sometimes, because of the exposition, this necessitates a minor generalization of a particular theorem or a change in its emphasis. When generalizations or modifications are combined *directly* with a theorem in the literature, statements are made to that effect. Hopefully, this results in the book being a convenient and accurate reference book to the literature.

The literary style in the book is in general governed by the following factors. One rarely reads a mathematics book, especially one of this size, from beginning to end--frequently one picks a topic and opens the book to read about that topic. There is almost no standard terminology and notation in hyperspace theory, and many fundamental ideas and theorems have only recently been discovered. Thus, as an aid to the reader, the exposition often contains many references to related material, and the proofs reference even the most basic facts and definitions so that the reader can easily find them in the book. By necessity, such references are given using the numbering system employed in the book. As a consequence, smoothness of style is often sacrificed--the overriding objective being to enable the reader to easily pinpoint all the information in the book about any particular topic. With this in mind, I have also included a comprehensive subject index.

Throughout the book there are a number of questions. When the word "Question" appears (to be differentiated from the "Exercises" at the ends of Chapters 0 and I), this means that the answer to the question is not known. Thus, such questions are "research questions." Many of these questions are my own and originate with the book. The same holds true for a number of theorems, examples, etc. Also, a number of new concepts are introduced in the book such as C-determined, C*-smooth, touching set, and Whitney hyperspace. In addition, some parts of the book are devoted almost entirely to topics which are original with the book--for example, Chapter XV. Conventions for credit are in Footnotes 1 and 5 on p.52.

In summary, the book is a text; a research monograph containing many original results and unanswered questions; and a survey of and a guide to the literature.

USING THE BOOK IN A BEGINNING
TOPOLOGY COURSE

The theory of hyperspaces is intimately connected with the structure of continua. Thus, once the ideas of connectedness and compactness are studied, hyperspace theory is an appropriate setting in which to broaden one's understanding of these concepts.

The book can be used as a text in a beginning course in topology. The background needed by students is roughly a semester of basic topology which emphasizes metric spaces. The major concepts and knowledge needed to read the parts of the book which are most fundamental to hyperspace theory are minimal. A general understanding of compactness, connectedness, and continuous functions is most of what is essential. Some understanding of limits of sets is desirable, but the connection between this and Hausdorff metric convergence is stated and proved in Chapter 0. Thus, the book starts at the very beginning of the subject. The only specific "specialized" theorem that is needed as background, and which is sometimes not covered in a one-semester basic topology course, is the following Boundary Bumping Theorem: If X is a metric continuum and if U is a nonempty proper open subset of X, then each component of the closure of U intersects the boundary of U. This result and its variants are proved in the Appendix--Chapter XX--for the convenience of the reader.

When one looks at the material covered in the book, it may seem surprising that so little pre-knowledge is necessary. However, let us take some examples. Dimension is discussed in Chapter I [see, for example, (1.107)]. But a careful examination reveals that what is really done is to locate certain n-cells in hyperspaces. Thus, if one accepts the intuitively obvious fact that an n-cell is of dimension n, no understanding of dimension theory is needed. Moreover, the existence of certain n-cells and the fact that they can be specifically located is more important, in many cases, than the dimension-theoretic facts. Thus, many theorems are stated in terms of where n-cells are. For another example, some algebraic concepts are used in Chapter I. However, a knowledge of unicoherence suffices in order to understand many of the applications of algebraic results. Furthermore, the fact that hyperspaces are nested intersections of absolute retracts [see (1.171)] is stronger than the algebraic results (and is used to obtain them). This intersection theorem is proved using only the most basic facts about inverse limits--these basic facts are stated and proved [except for (1.164)] in the beginning of Section E of Chapter I. Thus, the ins and outs of algebraic concepts can be avoided for the purpose of using the book in a beginning graduate course.

I suggest the outline in (1) through (4) below for using the book in a one-semester course.

(1) The notions of Hausdorff metric, convergence of sets, and their relationships [especially, (0.7)] should be covered first. Then the compactness of 2^X and C(X), (0.8), can be done. The Vietoris topology and its equivalence to the Hausdorff metric topology [(0.10) through (0.13)] provide additional understanding of the basic topology of hyperspaces. The examples in (0.54) through (0.56) should also be done.

(2) The notion of Whitney map in (0.50) and the construction of a specific Whitney map should be covered after the material in (1). Whitney maps are constructed in (0.50.1) through (0.50.3). The construction in (0.50.3) is the simplest and most appropriate in that it is related to ideas usually discussed in a first semester of topology. Whitney maps were not used in hyperspace theory until Kelley did so in 1942. The reason they are introduced so early in the book is that they are a unifying and convenient tool for the subject. Thus, I felt the reader should become familiar with their properties and their uses in basic situations. For

USING THE BOOK IN A TOPOLOGY COURSE

these reasons, they are used in a number of proofs in Chapter I. Thus, many of the proofs in Chapter I are quite different than those in the literature. In Chapter XIV, Whitney maps themselves are studied in depth. However, Chapter XIV is not an appropriate chapter for a one-semester introductory course.

(3) Order arcs are essential for an understanding of hyperspaces. The basic information about them is in (1.1) through (1.14) [Chapter XX should be covered first]. This material should be emphasized, as should the notion of segments and their relationships to order arcs [covered in (1.15) through (1.32)]. The result in (1.33) provides an interesting use of segments.

(4) After the material in (3) is covered, it would be appropriate to present results dealing with the location of cells in hyperspaces [especially (1.95), (1.100) - (1.107), and (1.145)]. In addition, the material leading to a proof of (1.61) can be presented. With time permitting Chapter IV can be done; Chapter IV contains some simple applications of results in Section B of Chapter I.

The material on arcs and cells, mentioned in (3) and (4) above, is especially suitable for a beginning topology course in that it has the effect of expanding, to a topological viewpoint, the student's understanding of these basic geometric objects.

There are many exercises at the ends of Chapters 0 and I. The purpose of these exercises is to give relatively easy problems which hopefully will increase one's familiarity with the basic ideas covered in these chapters. In addition, some exercises give a unified treatment and new proofs of some classical theorems in topology [for example, see (0.69)].

For a full year course, a variety of material augmenting Chapter I can be covered. An emphasis on Chapters 0 through III and Chapter XVI would probably be advantageous.

A VERY BRIEF HISTORY
OF
HYPERSPACE THEORY

Hyperspace theory had its beginnings in the early 1900's with the work of Hausdorff and Vietoris. During the 1920's and 1930's much of the fundamental structure of hyperspaces was determined. Of particular importance is the 1931 Borsuk-Mazurkiewicz paper [43] in which it is shown that 2^X and $C(X)$ are arcwise connected. This paper, together with the 1932 paper of Mazurkiewicz [213], laid the foundation for understanding the arc structure of hyperspaces. Also of importance is the 1939 paper of Wojdyslawski [342] in which it was proved that 2^X and $C(X)$ are absolute retracts when X is a locally connected continuum.

In 1942, Kelley's important paper [165] appeared. It discussed a variety of topics in hyperspace theory. Many previous results about hyperspaces were given a systematic treatment; also, a number of new results were obtained. Kelley's paper has been a significant contribution for the following major reasons. It was the first paper to study hyperspaces of hereditarily indecomposable continua. The role of these continua in hyperspace theory was not apparent until the 1970's, but Kelley determined the basic facts which have since been of such importance. Kelley's paper was also the first paper to use "Whitney maps" in investigating hyperspaces. Using Whitney maps, Kelley defined, studied, and made use of special mappings which he called "segments". The notion of segment is closely related to the work in [43] and [213]. Both Whitney maps and segments have become standard tools and have since been used in

almost all papers about hyperspaces. Kelley's paper was the first
hyperspace paper to give applications of hyperspace to other areas.
Finally, one of the main contributions of Kelley's 1942 paper was
that it brought the subject of hyperspaces to the attention of many
mathematicians, especially those in the United States.

In the 1950's, Michael's basic papers on selections and his
paper "Topologies on spaces of subsets" appeared. These papers
have been used in many areas of mathematics--differential equations,
functional analysis, topology, etc. In 1959, Segal [294] related
hyperspaces and inverse limits by showing that the hyperspace operation C commutes with inverse limits. Although Segal proved this result in order to show that $C(X)$ is acyclic (thus answering a question in Kelley's paper), Segal's inverse limit theorem has also been
of great importance.

Many important papers on hyperspaces were published in the
1960's. Among them are Transue's paper [315] on embedding hyperspaces in 3-space and Duda's incredibly detailed study of hyperspaces of finite graphs ([85] through [87]).

In 1969 the first conference on hyperspaces was held. It took
place at SUNY at Buffalo. There weren't very many people working
directly on hyperspace theory at that time, and the forty or so participants were mostly mathematicians who worked in areas which
"used hyperspaces," i.e., areas such as fixed point theory of multivalued mappings, theory of selections, and geometry of Banach
spaces. There was much enthusiasm on the part of the participants
about the diverse interest in hyperspace theory. A number of papers
grew out of the conference and simultaneously, it seemed, a lot of
interest in hyperspaces began to grow. Many papers on the subject
were written in the years 1970-1976. Among them, and of great interest, were the Curtis-Schori-West affirmative answers to the following famous question: If X is a locally connected continuum,
then is 2^X homeomorphic to the Hilbert cube? This question had
been of interest since the 1920's. In obtaining its solution, the

machinery and techniques of infinite-dimensional topology were introduced into the study of hyperspaces. In addition to the Curtis-Schori-West results, many papers on dimension, convex sets, embedding, etc. have recently appeared.

ACKNOWLEDGMENTS

I am especially indebted to Professor K. Kuratowski who devoted much time and effort to teaching me the fundamentals of hyperspace theory while I was a visitor at SUNY, Buffalo, January - June 1969. I also express my thanks to K. D. Magill, Jr., who arranged my visit to Buffalo.

The book was written during the years 1976-1977 while I was a visitor at the University of Delaware (Newark, Delaware), the University of Kentucky (Lexington, Kentucky), and the University of Saskatchewan (Saskatoon, Saskatchewan). I am grateful to Bob Gilbert and Margaret Waid of the University of Delaware for their encouragement during the time I worked on the original manuscript. I also express my gratitude to Carl Eberhart, J. B. Fugate, Bruce Hughes, William Nowell, Don Telage, and Calvin Van Niewaal of the University of Kentucky for their discussions and suggestions concerning the book. I am grateful to J. Grispolakis and E. D. Tymchatyn of the University of Saskatchewan for their valuable suggestions concerning the final version of the book. I also thank J. Grispolakis for doing the proof-reading of the final camera-ready manuscript.

I express my thanks to Joe Quinn who had encouraged me for several years to undertake the writing of the book.

I am most grateful to my wife, Elsa, who typed the manuscript.

Special thanks go to the following organizations and persons for obtaining funds used in the preparation of the book: The Institute for Mathematical Sciences at Delaware (Bob Gilbert, Director), Ivar Stakgold (Chairman of the Mathematics Department, University of Delaware), John Mack (Chairman of the Mathematics Department,

University of Kentucky), W. C. Royster (Dean of the Graduate School, University of Kentucky), and National Research Council of Canada [grants held by J. R. Martin (#A8205) and E. D. Tymchatyn (#A5616)].

S. B. N., Jr.

0.
SOME BASIC FACTS
AND NOTATION

Throughout this book[1] the word *continuum* means a compact connected metric space consisting of more than one point. The capital letter X will always denote a continuum with metric d, unless otherwise stated. A *subcontinuum* is a nonempty compact connected subspace of a space [a subcontinuum can be a one-point space]. Let[2]

$$2^X = \{A \subset X : A \text{ is nonempty and compact}\}$$

and let

$$C(X) = \{A \in 2^X : A \text{ is connected}\}.$$

A metric is defined for 2^X as follows:

(0.1) DEFINITION[3] [see (0.14)]. If $\varepsilon > 0$ and $A \in 2^X$, then

$$N_d(\varepsilon, A) = \{x \in X : d(x, a) < \varepsilon \text{ for some } a \in A\}.$$

If $A, B \in 2^X$, then define

$$H_d(A, B) = \inf\{\varepsilon > 0 : A \subset N_d(\varepsilon, B) \text{ and } B \subset N_d(\varepsilon, A)\}.$$

It is easy to see that H_d is a function from the cartesian product $2^X \times 2^X$ into the non-negative real numbers. In fact, as will be proved in (0.2), H_d is a metric for 2^X. It will be called the *Hausdorff metric induced by* d. Most of the time [except see, for example, (3.12)] the relationship between H_d and d is not important. When this is so or when there is no confusion, we will write H instead of H_d and N(,) instead of N_d(,); then, H will simply be called the *Hausdorff metric*. Throughout this book the symbols H and H_d will always denote the Hausdorff metric [sometimes in a more general setting than has been considered here--see, for example, (0.63.6) and (19.1)].

Some interesting results about the boundaries of the generalized balls N(ε, A) are in [54], [105], [121], [122], etc.

(0.2) THEOREM [see (0.14)]. *The function* H: $2^X \times 2^X \to [0, \infty)$ *is a metric for* 2^X.

Proof. We will prove the triangle inequality; the other simple verifications are left to the reader. Let A, B, C ε 2^X. We will show that

(*)   H(A, C) $\leq$ H(A, B) + H(B, C).

To prove (*), let $\eta > 0$ and let $\delta = \eta/2$. From the definition of H we see that:

(1)   A $\subset$ N(H(A, B) + δ, B),

(2)   B $\subset$ N(H(B, C) + δ, C).

Let a ε A. By (1), there exists b ε B such that

(3)   d(a, b) < H(A, B) + δ.

By (2), there exists c ε C such that

(4)   d(b, c) < H(B, C) + δ.

Using (3), (4), and the definition of δ, it follows easily that

d(a, c) < H(A, B) + H(B, C) + η.

Therefore, since a was an arbitrary point of A, we have proved that

(5) $A \subset N(H(A, B) + H(B, C) + \eta, C)$.

A similar argument shows that

(6) $C \subset N(H(A, B) + H(B, C) + \eta, A)$.

Since η was an arbitrary positive number, it follows from (5), (6), and the definition of $H(A, C)$ that (*) holds. This completes the proof of (0.2)

(0.3) **CONVENTION.** From now on, unless otherwise stated, when we write 2^X we mean the space 2^X with the topology obtained from the Hausdorff metric. Also, unless otherwise stated, $C(X)$ means the space $C(X)$ with the relative topology from 2^X. The spaces 2^X and $C(X)$ are called *hyperspaces of* X.

(0.4) **REMARK.** For any $x \in X$ and any $K \in 2^X$, let $d(x, K) = \inf\{d(x, y): y \in K\}$. Sometimes in the literature the following formula for a metric for 2^X appears: For $A, B \in 2^X$,

$$D(A, B) = \max\{\sup_{a \in A} d(a, B), \sup_{b \in B} d(b, A)\}.$$

It is easy to verify, using the definitions, that

$$D(A, B) = H_d(A, B)$$

for each $A, B \in 2^X$ [see (0.63.1)].

The following definition gives the classical notion of convergence of sets. As we will see in (0.7), this convergence is metrized by the Hausdorff metric.

(0.5) DEFINITION. Let $A_i \in 2^X$ for each $i = 1, 2, \ldots$. Then define:

$\lim \inf A_i = \{x \in X:$ if U is an open subset of X with $x \in U$, then $U \cap A_i \neq \emptyset$ for all but finitely many $i\}$;

$\lim \sup A_i = \{x \in X:$ if U is an open subset of X with $x \in U$, then $U \cap A_i \neq \emptyset$ for infinitely many $i\}$.

If $\lim \inf A_i = A = \lim \sup A_i$, then we say that the sequence $\{A_i\}_{i=1}^{\infty}$ *converges to* A, written $\lim A_i = A$ or $A_i \to A$.

(0.6) REMARK. Let $A_i \in 2^X$ for each $i = 1, 2, \ldots$. As is easy to verify:

(0.6.1) $\lim \inf A_i \subset \lim \sup A_i$;

(0.6.2) $\lim \inf A_i$ and $\lim \sup A_i$ are each closed subsets of X;

(0.6.3) If $\{A_{i(j)}\}_{j=1}^{\infty}$ is a subsequence of $\{A_i\}_{i=1}^{\infty}$, then $\lim \inf A_i \subset \lim \inf A_{i(j)}$ and $\lim \sup A_{i(j)} \subset \lim \sup A_i$.

(0.7) THEOREM [see (0.14)]. *Let* $A_i \in 2^X$ *for each* $i = 1, 2, \ldots$. *If the sequence* $\{A_i\}_{i=1}^{\infty}$ *converges to* A *in the sense of (0.5), then* $A \in 2^X$ *and the sequence* $\{A_i\}_{i=1}^{\infty}$ *converges with respect to the Hausdorff metric to* A. *Conversely, if the se-*

quence $\{A_i\}_{i=1}^{\infty}$ converges with respect to the Hausdorff metric to an $A \in 2^X$, then the sequence $\{A_i\}_{i=1}^{\infty}$ converges to A in the sense of (0.5).

Proof. First assume $\{A_i\}_{i=1}^{\infty}$ converges to A in the sense of (0.5). Since X is compact and each $A_i \neq \emptyset$, it follows easily that $\limsup A_i \neq \emptyset$. Thus, since $A = \limsup A_i$, we have that $A \neq \emptyset$. Also, by (0.6.2), A is a compact subset of X. Hence, $A \in 2^X$. Now we show that $\{A_i\}_{i=1}^{\infty}$ converges with respect to the Hausdorff metric to A. Let $\varepsilon > 0$. Note that

(1) $\limsup A_i = A \subset N(\varepsilon, A)$

and that, since $N(\varepsilon, A)$ is an open subset of X,

(2) the complement of $N(\varepsilon, A)$ is a compact subset
     of X.

Using (1), (2), and the second part of (0.6.3), it follows that there exists a natural number N_1 such that

(3) $A_i \subset N(\varepsilon, A)$ for each $i \geq N_1$.

Since A is a nonempty compact subset of X, there exist a finite number of open subsets $U_1, U_2, \ldots, U_k$ of X such that $A \subset [\cup_{j=1}^{k} U_j]$, the diameter of each U_j ($i \leq j \leq k$) is less than ε, and $U_j \cap A \neq \emptyset$ for each $j = 1, 2, \ldots, k$. Then, since $A = \liminf A_i$, there exists for each $j = 1, 2, \ldots, k$ a natural number M_j such that $A_i \cap U_j \neq \emptyset$ whenever $i \geq M_j$. Let $N_2 = \max\{M_1, M_2, \ldots, M_k\}$. It is easy to verify that

(4) $A \subset N(\varepsilon, A_i)$ for each $i \geq N_2$.

Let $N = \max\{N_1, N_2\}$. Then, by (3), (4), and the definition of H we see that $H(A, A_i) < \varepsilon$ for each $i \geq N$. Therefore, we have proved that $\{A_i\}_{i=1}^{\infty}$ converges with respect to the Hausdorff metric to A. This proves half of (0.7). To prove the other half, assume $\{A_i\}_{i=1}^{\infty}$ converges with respect to the Hausdorff metric to an $A \in 2^X$. We first show that

(5) $\lim \sup A_i \subset A$.

To verify (5), let $\varepsilon > 0$. Since $\{A_i\}_{i=1}^{\infty}$ converges with respect to H to A, there exists a natural number N such that $H(A, A_i) < \varepsilon$ for each $i \geq N$. Hence, by definition of H, we have that $A_i \subset N(\varepsilon, A)$ for each $i \geq N$. This implies that no point of $\lim \sup A_i$ can be more than ε from every point of A. Therefore, since $\varepsilon > 0$ was arbitrary, we have proved (5). Next we show that

(6) $A \subset \lim \inf A_i$.

To verify (6), let $\varepsilon > 0$. Let $a_o \in A$ and let

$U = \{x \in X: d(a_o, x) < \varepsilon\}$.

Since $\{A_i\}_{i=1}^{\infty}$ converges with respect to H to A, there exists a natural number N such that $H(A, A_i) < \varepsilon$ for each $i \geq N$. Hence, by the definition of H, we have that $A \subset N(\varepsilon, A_i)$ for each $i \geq N$. Thus, $U \cap A_i \neq \emptyset$ for each $i \geq N$. Therefore, since $\varepsilon > 0$ was arbitrary, $a_o \in \lim \inf A_i$. This proves (6). Combining (5) and (6) and using (0.6.1), we see that $\{A_i\}_{i=1}^{\infty}$ converges to A in the sense of (0.5). This completes the proof of (0.7).

NOTE: From now on when we say a sequence $\{A_i\}_{i=1}^{\infty}$ *converges* in 2^X, we will mean in the sense of (0.5) or with respect to the Hausdorff metric, interchangeably. Sometimes, when the viewpoint is important, we will specify which.

(0.8) THEOREM [see (0.14)]. *The spaces 2^X and $C(X)$ are each compact.*

Proof. First we show that

(*) 2^X is compact.

To prove (*), it suffices by (0.7) to show that every sequence in 2^X has a convergent subsequence in the sense of (0.5). To do this, let $\{A_i\}_{i=1}^{\infty}$ be a sequence of sets $A_i \in 2^X$. We define sequences

$\{A_i^1\}_{i=1}^{\infty}$ -- $A_1^1 \quad A_2^1 \quad \cdots \quad A_n^1 \cdots$

$\{A_i^2\}_{i=1}^{\infty}$ -- $A_1^2 \quad A_2^2 \quad \cdots \quad A_n^2 \cdots$

....

$\{A_i^n\}_{i=1}^{\infty}$ -- $A_1^n \quad A_2^n \quad \cdots \quad A_n^n \cdots$

....

inductively as follows. Let $U = \{U_1, U_2, \ldots, U_n, \ldots\}$ be a countable base for the topology for X. Define $\{A_i^1\}_{i=1}^{\infty}$ by $A_i^1 = A_i$ for each $i = 1, 2, \ldots$. Assume inductively that we have defined the sequence $\{A_i^n\}_{i=1}^{\infty}$. We define $\{A_i^{n+1}\}_{i=1}^{\infty}$ in one of the following two ways:

(1) If $\{A_i^n\}_{i=1}^{\infty}$ has a subsequence $\{A_{i(j)}^n\}_{j=1}^{\infty}$ such that $[\limsup A_{i(j)}^n] \cap U_n = \emptyset$, then let $\{A_i^{n+1}\}_{i=1}^{\infty}$ be one such subsequence of $\{A_i^n\}_{i=1}^{\infty}$;

(2) If every subsequence of $\{A_i^n\}_{i=1}^{\infty}$ has a point of its lim sup in U_n, then let $\{A_i^{n+1}\}_{i=1}^{\infty}$ be given by $A_i^{n+1} = A_i^n$ for each $i = 1, 2, \ldots$.

Now, having defined the sequence $\{A_i^n\}_{i=1}^{\infty}$ for each $n = 1, 2, \ldots$, consider the "diagonal sequence" $\{A_n^n\}_{n=1}^{\infty}$. Clearly $\{A_n^n\}_{n=1}^{\infty}$ is a subsequence of $\{A_i\}_{i=1}^{\infty}$, and we will show it converges. Suppose

$\{A_n^n\}_{n=1}^\infty$ does not converge. Then, by (0.6.1), there exists a point $p \in \limsup A_n^n$ such that $p \notin \liminf A_n^n$. Hence, there exists $U_m \in U$ such that $p \in U_m$ and such that $U_m \cap A_{n(i)}^{n(i)} = \emptyset$ for some subsequence $\{A_{n(i)}^{n(i)}\}_{i=1}^\infty$ of $\{A_n^n\}_{n=1}^\infty$. Clearly $\{A_{n(i)}^{n(i)}\}_{i=m}^\infty$ is a subsequence of $\{A_i^m\}_{i=1}^\infty$. Thus, $\{A_i^m\}_{i=1}^\infty$ satisfies (1) above [with m replacing n in (1)]. Hence, $[\limsup A_i^{m+1}] \cap U_m = \emptyset$. Therefore, since $\{A_n^n\}_{n=m+1}^\infty$ is a subsequence of $\{A_i^{m+1}\}_{i=1}^\infty$, it follows using the second part of (0.6.3) that $[\limsup A_n^n] \cap U_m = \emptyset$. But, since $p \in [\limsup A_n^n] \cap U_m$, we have a contradiction. Therefore, $\{A_n^n\}_{n=1}^\infty$ converges. This completes the proof of (*). Next we show that

(**) $C(X)$ is compact.

To prove (**), it suffices by (*) to show that $C(X)$ is a closed subset of 2^X. Let $\{K_n\}_{n=1}^\infty$ be a sequence in $C(X)$ such that $\{K_n\}_{n=1}^\infty$ converges to $K \in 2^X$. We will show that $K \in C(X)$. Suppose, to the contrary, K is not a connected subset of X. Then there exist disjoint open subsets U and V of X such that $K \subset [U \cup V]$, $K \cap U \neq \emptyset$, and $K \cap V \neq \emptyset$. Since $\{K_n\}_{n=1}^\infty$ converges to K in the sense of (0.5) and since $U \cup V$ is an open subset of X containing K, there exists a natural number N such that $K_n \subset [U \cup V]$ for each $n \geq N$. Thus: Since each K_n is connected and since U and V are disjoint open subsets of X we have that for any given $n \geq N$, $K_n \subset U$ or $K_n \subset V$. It follows easily that $\liminf K_n \neq K$. This contradicts the fact that $\{K_n\}_{n=1}^\infty$ converges to K. Hence, $K \in C(X)$ and we have proved (**).

(0.9) REMARK. In fact, 2^X and $C(X)$ are each continua. We delay the proof of this fact until Chapter I, where an even stronger result is proved--see (1.13).

Sometimes it is convenient to use the so-called Vietoris topology instead of the Hausdorff metric [see (0.13)]. The Vietoris topology is the one authors usually use for the nonempty closed subsets of a general topological space. Though almost all the results in the book are in the metric setting, we take this opportunity to introduce the Vietoris topology for the general setting. In (0.13), we will show that the Vietoris topology and the Hausdorff metric topology are the same when the underlying space is a (metric) continuum.

(0.10) NOTATION. Let S be a set with a topology T. Then:

(0.10.1) $CL(S) = \{A: A$ is a nonempty closed subset of $S\}$.

(0.10.2) For any finite collection of sets $U_1, U_2, \ldots, U_n \in T$, let

$$\langle U_1, U_2, \ldots, U_n \rangle = \{A \in CL(S): A \subset [\bigcup_{i=1}^{n} U_i] \text{ and } A \cap U_i \neq \emptyset \text{ for each } i = 1, 2, \ldots, n\}.$$

(0.11) THEOREM [see (0.14)]. *Let S be a set with a topology T. Then the collection of all subsets of $CL(S)$ of the form $\langle U_1, U_2, \ldots, U_n \rangle$, as defined in (0.10.2), is a base for a topology for $CL(S)$.*

Proof. Let

$$\beta = \{<U_1, U_2, \ldots, U_n>: U_i \in T \text{ for each } i = 1,2,\ldots,n\}.$$

First observe that, since $<S> \in \beta$ and $<S> = CL(S)$, $\cup \beta = CL(S)$. Next let, $B_1, B_2 \in \beta$, $B_1 = <U_1, U_2, \ldots, U_n>$ and $B_2 = <V_1, V_2, \ldots, V_m>$. Let $U = \cup_{i=1}^{n} U_i$ and let $V = \cup_{i=1}^{m} V_i$. Simple set-theoretic arguments give that

$$B_1 \cap B_2 = <U_1 \cap V, \ldots, U_n \cap V, V_1 \cap U, \ldots, V_m \cap U>.$$

It now follows that β is a base for a topology for $CL(S)$.

(0.12) DEFINITION. The topology for $CL(S)$ obtained from (0.11) is called the *Vietoris topology* or the *finite topology*.

(0.13) THEOREM [see (0.14)]. *The Vietoris topology and the Hausdorff metric topology for 2^X are the same.*

Proof. For the purpose of the proof, let T_V denote the Vietoris topology for 2^X and let T_H denote the Hausdorff metric topology for 2^X. By (0.8),

(1) 2^X with T_H is compact.

As is easy to prove [using the regularity of X],

(2) 2^X with T_V is a Hausdorff space.

Hence, by (1) and (2), it suffices to prove

(*) $T_V \subset T_H$.

To prove (*), let $A \in <U_1, U_2, \ldots, U_n> \in T_V$. For each $i = 1,2,\ldots,n$, let $a_i \in [U_i \cap A]$. Then, for each $i = 1,2,\ldots,n$, let $\varepsilon_i > 0$ such that

(3) $\{x \in X: d(x, a_i) < \varepsilon_i\} \subset U_i$.

Also choose $\varepsilon_o > 0$ such that

(4) $N(\varepsilon_o, A) \subset [\bigcup_{i=1}^{n} U_i]$.

Let $\varepsilon = \min\{\varepsilon_o, \varepsilon_1, \varepsilon_2, \ldots, \varepsilon_n\}$. It follows easily using (3) and (4) [and (0.1)] that

$\{K \in 2^X: H(A, K) < \varepsilon\} \subset <U_1, U_2, \ldots, U_n>$.

Hence, we have proved (*). This completes the proof of (0.13).

(0.14) REMARK. It does not seem to be clear where (0.1), (0.2), (0.7), (0.8), (0.11), and (0.13) each appeared first. Each may be found, some in more generality than stated above, in one or more of the following references: [138], [185], [186], [225], and [319]. In relation to (0.7), the reader may be interested in [419], [421], [428], and [436].

We now give a list of symbols, notation, and definitions which are used in the book. Symbols, etc. which are not in this list will be defined as they come up, or may be found in one of several standard references (for example, [185], [186], or [338]). Beginning with (0.49.1), some more results on hyperspaces are presented.

(0.15) A space is said to be *nondegenerate* provided it consists of more than one point.

(0.16) As mentioned above a *continuum* is a nondegenerate compact connected metric space, and X will always denote a continuum with metric d (unless otherwise specified). Sometimes the terminology "metric continuum" will be used when it is felt that the additional emphasis is appropriate. By a *Hausdorff continuum* we will mean a nondegenerate compact connected Hausdorff space; thus, a Hausdorff continuum is not necessarily (but could be) metric.

(0.17) The symbol [a, b], where $a \leq b$ are real numbers, denotes the closed interval $\{x: a \leq x \leq b\}$.

(0.18) By an *arc* we will mean a homeomorphism h of a nondegenerate closed interval [a, b], or we will mean the range h([a, b]) of such a homeomorphism. When it is necessary to know which is meant, the context will make it clear. Now let α be an arc in a space M and let h be a homeomorphism of [0, 1] onto α. Then: By an *end point of* α we will mean one of the two points h(0) and h(1). The arc α is said to be *free in* M provided that α without its end points [i.e., h((0, 1))] is an open subset of M. By a *generalized arc* we will mean a Hausdorff continuum with exactly two noncut points.

(0.19) For each $n = 1, 2, \ldots$, R^n will denote Euclidean n-dimensional space with the usual Euclidean metric obtained from the norm $||\ ||$:

$$||(x_1, x_2, \ldots, x_n)|| = [x_1^2 + x_2^2 + \cdots + x_n^2]^{\frac{1}{2}}$$

for each $(x_1, x_2, \ldots, x_n) \in R^n$. We will refer to R^n as *n-space*. For each $n = 1, 2, \ldots,$ the symbols B^n and S^{n-1} are defined by:

$$B^n = \{x \in R^n : ||x|| \leq 1\},$$
$$S^{n-1} = \{x \in R^n : ||x|| = 1\}.$$

A continuum which is homeomorphic to B^n [resp., S^n, S^1] will be called an *n-cell* [resp., *n-sphere*, *circle*].

(0.20) A *simple triod* is a continuum homeomorphic to

$$\{(0, y) \in R^2 : 0 \leq y \leq 1\} \cup \{(x, 1) \in R^2 : -1 \leq x \leq 1\}.$$

(0.21) A *triod* (see [32, p.653]) is a continuum M which contains a subcontinuum N such that the complement of N in M is the union of three nonempty mutually separated sets. An *n-odd* ($n \geq 2$) is a continuum M which contains a subcontinuum N such that the complement of N in M is the union of n nonempty mutually separated sets. Thus, for example, every n-odd (for $n \geq 3$) is a triod; also, B^2 is an n-odd for every $n \geq 2$. A continuum K is said to be *a-triodic* provided K contains no triod.

(0.22) By a *chain in a space* we will mean a finite collection $U = \{U_1, U_2, \ldots, U_n\}$ of open subsets $U_1, U_2, \ldots, U_n$ of the space such that $U_i \cap U_j \neq \emptyset$ if and only if $|i - j| \leq 1$. A member of U will be called a *link of* U.

(0.23) A continuum X is said to be *chainable* (some authors say *snake-like* or *arc-like*) provided that for each $\varepsilon > 0$, there is a chain in X covering X such that each link has diameter less than ε. It is well-known (see [186, p.226]) and not difficult to prove that a continuum is chainable if and only if it is an inverse limit of arcs with onto bonding maps. I refer the reader to [32] for basic information about chainable continua. We mention that, for the purposes of this book, the term *pseudo-arc* may be taken to mean the only hereditarily indecomposable chainable continuum (see [26, p.44] or (1) in the proof of (19.30)). The following theorem will be used several times in the book:

(0.23.1) THEOREM [see below]. *An arcwise connected chainable continuum is an arc.*

The result in (0.23.1) was noted in [117, Lemma 2], where it was mentioned that (0.23.1) was "probably well-known" and was a consequence of [306, Theorem 3.2]. The locally-connected case of (0.23.1) was proved by a different method in [364]. Later [245, Theorem 5], being aware (only) of [364], I proved (0.23.1) using a simple inverse limit argument (thus by-passing [306, Theorem 3.2], which I was unaware of at the time). Since Theorem 3.2 of [306] is difficult to prove, perhaps the proof in [245] is the easiest path to (0.23.1) for the person who has not read [306].

(0.24) A continuum X is said to be *circle-like* provided that for each $\varepsilon > 0$, there exists a finite collection V = $\{V_1, V_2, \ldots, V_n\}$ of open subsets of X covering X such that:

(1) $V_i \cap V_j \neq \emptyset$ if and only if $|i-j| \leq 1$ or $i,j \in \{1,n\}$,

(2) V_i has diameter less than ε for each $i = 1,2,\ldots,n$.

It is well-known and not difficult to prove that a continuum is circle-like if and only if it is an inverse limit of circles with onto bonding maps. Of special importance in various parts of the book is the structure of arcwise connected circle-like continua. This structure was determined in [245, Theorem 6]. For general facts about circle-like continua, I refer the reader to [28] and [56]. Let us note that there are continua which are both chainable and circle-like; Burgess [56] studied these continua. In this book, a continuum which is circle-like and not chainable will be called *proper circle-like* (this terminology apparently originated in [177, p.158]).

(0.25) By a *weakly chainable continuum* we will mean a continuum which is a continuous image of the pseudo-arc. This is not the definition of weakly chainable given in [193, p.272], but rather it is the Theorem in [193, p.274]. Continuous images of the pseudo-arc were also investigated in [102].

(0.26) A continuum is said to be a *dendrite* provided that it is locally connected and contains no circle. For some information about dendrites, see [40, p.138] and [338].

(0.27) A continuum is said to be *unicoherent* provided that $A \cap B$ is connected whenever A and B are subcontinua of X such that $A \cup B = X$. A continuum is said to be *hereditarily unicoherent* provided that each of its subcontinua is unicoherent.

(0.28) A *dendroid* is an arcwise connected hereditarily unicoherent continuum. The reader is referred to [39], [65], [66; Theorem 1 is false--see (1.66)], etc. for facts about dendroids. Note that every dendrite is a dendroid and every locally connected dendroid is a dendrite. Also [71], every dendroid is tree-like (where *tree-like* is as defined in [32, p.653].

(0.29) A continuum X is said to be *aposyndetic at $p \in X$* provided that if $q \in X$ such that $q \neq p$, then there exists a subcontinuum K of X and an open subset U of X such that $p \in U \subset K$ and $q \notin K$. A continuum is said to be *aposyndetic* provided that it is aposyndetic at each of its points. These concepts are originally due to Jones [156]; an excellent survey of basic facts is in [157]. We mention that (see [156, pp.546-547]) a continuum is aposyndetic if and only if it is semi-locally-connected (in the sense of [338, p.19]).

(0.30) A continuum is said to be *decomposable* provided that it is the union of two proper subcontinua. It is said to be *hereditarily decomposable* provided that each of its nondegenerate subcontinua is decomposable. A continuum X is said to be λ *connected* provided that if $x, y \in X$, then there exists an hereditarily decomposable subcontinuum K of X such that $x, y \in K$ (see [134]).

A continuum X is said to be *arcwise decomposable* provided there exist arcwise connected proper subcontinua A and B of X such that X = A ∪ B. Clearly, an arcwise decomposable continuum is arcwise connected. The Warsaw circle, $S^1_{\frac{1}{2}}$ in (8.22), is an arcwise connected circle-like continuum which is not arcwise decomposable; in fact, though there are uncountably many topologically different arcwise connected circle-like continuum [see (0.70.3)], the only arcwise decomposable circle-like continuum is a circle. This follows easily from [245, Theorem 6]. For an application, see (4.9).

(0.31) A continuum is said to be *indecomposable* provided that it is not decomposable. It is said to be *hereditarily indecomposable* (in [164] the term *Knaster continuum* is used) provided that each of its subcontinua is indecomposable. It is easy to see that a continuum X is hereditarily indecomposable if and only if whenever A and B are subcontinua of X such that A ∩ B ≠ ∅, then A ⊂ B or B ⊂ A. I refer the reader to [186, pp.204-214] for basic information about indecomposable continua. Also, see the first paragraph in section C of Chapter I.

(0.32) Let X be a continuum and let p ε X. By *the composant in X of p* we will mean the set κ given by

κ = {x ε X: there exists a proper subcontinuum A of

X such that p, x ε A}.

By a *composant of X* we will mean a composant in X of some point of X. For example, the composant in [0, 1] of 0 is [0, 1);

note that though [0, 1) is a composant of [0, 1] containing 1/2, [0, 1) is not *the* composant in [0, 1] of 1/2. This type of thing does not happen for indecomposable continua since their composants are mutually disjoint. See [186, pp.208-214] for basic information about composants.

(0.33) Let X be a continuum. A point $p \in X$ is said to be a *cut point* of X provided that $\{x \in X: x \neq p\}$ is not connected. A point of X which is not a cut point of X will be called a *noncut point of X*.

(0.34) Let X be a continuum. A point $p \in X$ is said to *continuumwise disconnect* X provided that there exist two points $x, y \in X$, with $x \neq p$ and $y \neq p$, such that if A is any subcontinuum of X such that $x, y \in A$, then $p \in A$.

(0.35) The symbol cl denotes closure, not to be confused with the symbol CL defined in (0.10.1).

(0.36) The slash \ denotes complementation for sets, i.e., $A \setminus B = \{x \in A: x \notin B\}$.

(0.37) The symbol × denotes (finite) cartesian product; $\pi_Y: Y \times Z \to Y$ denotes the *projection of* $Y \times Z$ *onto* Y defined by $\pi_Y((y, z)) = y$ for each $(y, z) \in Y \times Z$. Countable infinite cartesian products will be denoted with $\prod_{j=1}^{\infty}$. The topology for a finite or infinite cartesian product is assumed to be the usual product topology (as defined in [185, p.147]).

(0.38) The *cone over* X is the decomposition space of the upper semi-continuous decomposition

$$(X \times [0, 1]) / (X \times \{1\})$$

of $X \times [0, 1]$ obtained by "shrinking $X \times \{1\}$ to a point." The cone over X will be denoted by *Cone(X)*, its *base* $X \times \{0\}$ by $B(X)$, and its *vertex* $X \times \{1\} \in \text{Cone}(X)$ by v. If $Y \subset X$, then when we write Cone(Y) we mean the "natural" subset of Cone(X) given by

$$\text{Cone}(Y) = \{(x, t) \in [\text{Cone}(X) \setminus \{v\}] : x \in Y\} \cup \{v\},$$

and when we write $B(Y)$ we mean

$$B(Y) = \{(x, 0) \in B(X) : x \in Y\}.$$

The symbol π will denote the *projection of* $Cone(X) \setminus \{v\}$ *onto* $B(X)$ defined by $\pi((x, t)) = (x, 0)$ for each $(x, t) \in [\text{Cone}(X) \setminus \{v\}]$.

(0.39) The *suspension of* X is the decomposition space of the upper semi-continuous decomposition

$$(X \times [-1, +1]) / (X \times \{-1\}) \cup (X \times \{+1\})$$

of $X \times [-1, +1]$ obtained by "shrinking each of $X \times \{-1\}$ and $X \times \{+1\}$ to different points." The suspension of X will be denoted by *Sus(X)*. The two nondegenerate elements $X \times \{-1\}$ and $X \times \{+1\}$, as points of Sus(X), will be called the *vertices of* *Sus(X)*.

(0.40) The symbol ℓ_2 denotes the Hilbert space of all square-summable sequences of real numbers. The symbol I_∞ is defined by

$$I_\infty = \{(x_1, x_2, \ldots, x_j, \ldots) \in \ell_2 : 0 \le x_j \le 2^{-j} \text{ for each } j = 1, 2, \ldots\}.$$

Thus, I_∞ is the copy of $\prod_{j=1}^{\infty}[0, 2^{-j}]$ obtained by embedding $\prod_{j=1}^{\infty}[0, 2^{-j}]$ in ℓ_2 by inclusion. A continuum is said to be a *Hilbert cube* provided that it is homeomorphic to I_∞.

(0.41) The symbol *diam* denotes diameter, i.e., if (M, d) is a metric space and if G is a bounded subset of M, then

diam[G] = l.u.b.$\{d(x, y): x, y \in G\}$.

Obviously, the values of diam depend on the metric for X. We will not denote this dependence, since the context will make it clear what metric is being used. Some basic facts about diameter as a function are in (1.211).

(0.42) The symbol *Bd* denotes boundary defined as follows: If G is a subset of a space S, then

Bd[G] = cl[G] ∩ cl[S\G]

is called the *boundary of G in S* or, when there will be no confusion, simply the *boundary of G*. It is easy to prove

Bd[U] = cl[U]\U, for any open subset U of S.

(0.43) The symbol $\cong$ means "is homeomorphic to."

(0.44) In this book, dimension means inductive dimension as defined in [150]. The symbol dim will be used to denote (global) dimension, and dim_p will be used to denote dimension at a particular point p. Recall that the empty set (denoted in this book by ∅), and only the empty set, has dimension equal to -1 by definition [150, p.24]. Also, if M is a nonempty compact metric space, then [150, p.20] $dim[M] = 0$ if and only if M is totally-disconnected (a metric space is, by definition [150, p.15], *totally-disconnected* provided that no connected subset of it consists of more than one point). Though being zero-dimensional implies being totally-disconnected [150, p.15], the converse is false [150, pp.22-23].

(0.45) The word *mapping* is synonymous with continuous function. A mapping $f: X \xrightarrow{onto} Y$ is said to be

(0.45.1) *monotone* provided that $f^{-1}(y)$ is connected for each $y \in Y$;

(0.45.2) *open* provided that $f[U]$ is an open subset of Y for each open subset U of X;

(0.45.3) *confluent* provided that for each subcontinuum K of Y and each component A of $f^{-1}[K]$, we have $f[A] = K$ (this concept first appeared in [63]--note that, as observed in [63], monotone mappings and open mappings between continua are confluent);

(0.45.4) *weakly confluent* provided that for each subcontinuum K of Y, there exists a component A of $f^{-1}[K]$ such that $f[A] = K$ (this concept first appeared in [192]--also see [196]);

(0.45.5) *non-alternating* provided that for no two points $y, z \in Y$ do there exist disjoint nonempty open subsets U_1 and U_2 of X such that $X \setminus f^{-1}(y) = U_1 \cup U_2$ and $z \in (f[U_1] \cap f[U_2])$.

(0.46) By an *embedding* of a space Q in a space S, we will mean a homeomorphism of Q into (not necessarily onto) S, or we will mean a topological copy of Q contained in S. When it is necessary to know which we mean, the context will make it clear.

(0.47) Let (Y, d) and (Z, ρ) be nonempty compact metric spaces. A function $F: Y \to 2^Z$ is said to be *upper semi-continuous at a point* $y_0 \in Y$ provided that for each $\varepsilon > 0$ there exists a $\delta > 0$ such that if $y \in Y$ and $d(y_0, y) < \delta$, then $F(y) \subset N_\rho[\varepsilon, F(y_0)]$. It is easy to see that F is upper semi-continuous at y_0 if and only if $\lim \sup F(y_n) \subset F(y_0)$ for every sequence $\{y_n\}_{n=1}^{\infty}$ in Y converging to y_0. The function $F: Y \to 2^Z$ is said to be *upper semi-continuous* provided that it is upper semi-continuous at each point of Y. These definitions are equivalent to the ones in [185] and [186]. Some applications of upper semi-continuous functions to proving well-known theorems are in (0.69).

(0.48) For each $n = 1, 2, \ldots$, define $F_n(X)$ and $C_n(X)$ by:

$F_n(X) = \{A \in 2^X : A \text{ has at most } n \text{ points}\}$,

$C_n(X) = \{A \in 2^X : A \text{ has at most } n \text{ components}\}$.

Note that:

$F_1(X) \subset F_2(X) \subset \cdots \subset F_n(X) \subset \cdots \subset 2^X$

and

$C_1(X) \subset C_2(X) \subset \cdots \subset C_n(X) \subset \cdots \subset 2^X$.

Also note that $F_1(X) \cong X$ but $F_1(X) \neq X$. However, $C_1(X) = C(X)$; the symbol $C_1(X)$ will be used when it is appropriate to emphasize tha particular embedding, *by inclusion*, of $C(X)$ in $C_n(X)$ or 2^X. For each $n = 1, 2, \ldots$, $F_n(X)$ is called the *n-fold symmetric product of* X; $F_1(X)$ sometimes denoted $\hat{X}$ in the literature, is called the *space of singletons*. Symmetric products were introduced in [44] and subsequently studied in [38], [47], [120], [225], [229], [228], [344], and [345] (an error in [38] was noted and corrected in [47]). A survey of some basic results was done in [198]. A different type of symmetric product, called a permutation product in [198], was studied in [82], [197], [231], [277], [278], [301], [302], and [307]. Some generalizations appear in [201].

(0.49) Let $f: X \to Y$ be a mapping into a continuum Y. Then $\hat{f}: C(X) \to C(Y)$ defined by

$\hat{f}(A) = \{f(a): a \in A\}$, for each $A \in C(X)$

will be called the *map induced by* f, or simply the *induced map* (it follows easily, from the continuity of f, that $\hat{f}$ is indeed a mapping). It is easy to see that $\hat{f}$ need not always map $C(X)$ onto $C(Y)$ even when f maps X onto Y; in fact, the following theorem is easy to prove:

(0.49.1) THEOREM. *Let* $f: X \xrightarrow{onto} Y$ *be a mapping onto a continuum* Y. *Then:* $\hat{f}$ *maps* $C(X)$ *onto* $C(Y)$ *if and only if* f *is weakly confluent.*

Some basic facts about induced maps are in (0.67) and (1.212).

(0.50) Let $\Lambda = 2^X$ or $C(X)$. By a *Whitney map for* Λ we will mean any mapping $g: \Lambda \to [0, +\infty)$ satisfying

(0.50.a) if $A, B \in \Lambda$ such that $A \subset B$ and $A \neq B$,
then $g(A) < g(B)$

and

(0.50.b) $g(\{x\}) = 0$ for each $x \in X$.

Throughout this book, ω will denote a Whitney map for 2^X and μ will denote a Whitney map for $C(X)$. In a context different than hyperspaces, Whitney first constructed a mapping satisfying (0.50.a) and (0.50.b) in [336, p.275]; he constructed another such mapping in [337, pp.245-246]. We will give both constructions below [see (0.51.1) and (0.50.2)]. In (0.50.3) we will give a construction of a Whitney map due to Krasinkiewicz [178]. The construction in (0.50.3) [which is similar to the one in (0.50.2)] seems to be the

(0.50.1)   BASIC FACTS AND NOTATION   25

"simplest" and the "most natural" of all three. It is also the easiest for which to verify that the resulting map is indeed a Whitney map. Since we will mainly be using the fact that Whitney maps exist (rather than using a particular one), the verifications in (0.50.3) suffice from the standpoint of completeness for most of this book. We will not verify that the maps constructed in (0.50.1) and (0.50.2) are Whitney maps. The verifications for the map in (0.50.2) are not much more difficult than those given for the map in (0.50.3); however, the verifications for the map in (0.50.1) are somewhat harder. We make two comments about the map in (0.50.1): First, it seems to be the one most often used when a particular Whitney map is used [see (14.43) and (14.61)]; second, it has some properties that the others do not (necessarily) have. Finally, we mention that Kelley [165] was the first person to introduce Whitney maps into the study of hyperspaces. More will be said about this in Chapter I.

(0.50.1) CONSTRUCTION OF A WHITNEY MAP AS DONE IN [336, p.275].
Let $A \in 2^X$ be fixed. Let $n \geq 2$ be a fixed natural number. Define $\lambda_n : F_n(A) \to [0, +\infty)$ by: If $K = \{a_1, a_2, \ldots, a_n\} \in F_n(A)$ [where the enumeration of K may not be one-to-one], then

$\lambda_n(K) = \min\{d(a_i, a_j) : i \neq j\}$.

Note that $\lambda_n(K) \leq \operatorname{diam}[A]$ for each $K \in F_n(A)$; hence, $\omega_n(A)$ given by

$\omega_n(A) = \text{l.u.b.}\{\lambda_n(K) : K \in F_n(A)\}$

is a real number. This defines $\omega_n(A)$ for each natural number $n \geq 2$. Since $\omega_n(A) \leq \operatorname{diam}[A]$ for each $n = 2, 3, \ldots$, the series

$\sum_{n=2}^{\infty} 2^{1-n} \cdot \omega_n(A)$ converges; define $\omega(A)$ by

$$\omega(A) = \sum_{n=2}^{\infty} 2^{1-n} \cdot \omega_n(A).$$

Since A was an arbitrary member of 2^X, we have defined a function $\omega: 2^X \to [0, +\infty)$. I refer the reader to [336, pp.275-276] for details which verify that ω is a Whitney map. Let us note that ω has some other properties. For example [337, p.248]: Two members of 2^X which are isometric as subsets of X have the same value under ω [see (0.68.1)].

(0.50.2) CONSTRUCTION OF A WHITNEY MAP AS DONE IN [337, pp.245-246]. Let $\{x_1, x_2, \ldots, x_n, \ldots\}$ be a countable dense subset of X. For each $n = 1, 2, \ldots$, define $f_n: X \to [0, 1]$ by

$$f_n(x) = \frac{1}{1 + d(x_n, x)}, \quad \text{for each } x \in X.$$

For each $n = 1, 2, \ldots$, define $\omega_n: 2^X \to [0, 1]$ by

$$\omega_n(A) = \text{diam}[f_n(A)], \quad \text{for each } A \in 2^X.$$

Define $\omega: 2^X \to [0, 1]$ by

$$\omega(A) = \sum_{n=1}^{\infty} 2^{-n} \cdot \omega_n(A), \quad \text{for each } A \in 2^X.$$

It is not difficult to verify that ω is a Whitney map; the details are in [337, p.246].

(0.50.3) KRASINKIEWICZ'S CONSTRUCTION OF A WHITNEY MAP [178]. Let $U = \{U_1, U_2, \ldots, U_m, \ldots\}$ be a countable base for the topology for X. For each pair $\alpha = (U_i, U_j)$ of members of U such that $U_i \neq \emptyset$ and $\text{cl}[U_i] \subset U_j \neq X$, let $f_\alpha: X \to [0, 1]$ be a mapping such that $f_\alpha(\text{cl}[U_i]) = 0$ and $f_\alpha(X \setminus U_j) = 1$. This defines countably many functions f_α; enumerate them in a sequence $f_1, f_2, \ldots, f_n, \ldots$. For each $n = 1, 2, \ldots$, define $\omega_n : 2^X \to [0, 1]$ by

$$\omega_n(A) = \text{diam}[f_n(A)], \quad \text{for each } A \in 2^X.$$

Define $\omega : 2^X \to [0, 1]$ by

$$\omega(A) = \sum_{n=1}^{\infty} 2^{-n} \cdot \omega_n(A), \quad \text{for each } A \in 2^X.$$

It is clear that ω is continuous and satisfies (0.50.b). To see that ω satisfies (0.50.a), let $A, B \in 2^X$ such that $A \subset B$ and $A \neq B$. Let $x_0 \in [B \setminus A]$. It follows easily that there exists f_k such that $f_k(x_0) = 0$ and $f_k(A) = 1$. Hence, $\omega_k(A)$ and, since $f_k(B) \subset [0, 1]$ and $0, 1 \in f_k(B)$ [because $x_0 \in B$ and $A \subset B$], $\omega_k(B) = 1$. Thus, we have that $\omega_k(A) < \omega_k(B)$ and, since $A \subset B$, $\omega_n(A) \leq \omega_n(B)$ for each $n = 1, 2, \ldots$. Therefore, $\omega(A) < \omega(B)$. This proves that ω satisfies (0.50.a) and completes the proof that ω is a Whitney map.

Let us note the following result about induced maps and Whitney maps for $C(X)$. Some related information is in (0.67.6), (0.68.6), and (1.213.6).

(0.51) THEOREM [179, 2.1]. *Let* $f : X \to Y$ *be a mapping of a continuum* X *into a continuum* Y *such that*

$$\dim\{x \in X : \{x\} \neq f^{-1}[f(x)]\} \leq 0.$$

Then, the following hold:

(0.51.1) $\hat{f}$ *embeds* $C(X) \setminus F_1(X)$ *into* $C(Y) \setminus F_1(Y)$ *as a closed* subset of $C(Y)\setminus F_1(Y)$;

(0.51.2) If $\mu: C(Y) \to [0, +\infty)$ is a Whitney map for $C(Y)$, then $\mu \circ \hat{f}: C(X) \to [0, +\infty)$ is a Whitney map for $C(X)$.

Proof. To prove (0.51.1), let h denote the restriction of $\hat{f}$ to $C(X)\setminus F_1(X)$. By the dimension assumption, h sends $C(X)\setminus F_1(X)$ into $C(Y)\setminus F_1(Y)$. To see that h is one-to-one, let $A, B \in [C(X)\setminus F_1(X)]$ such that $A \neq B$. For the purpose of proof and without loss of generality assume $A \not\subset B$. Then, it follows easily using (20.1) that there is a nondegenerate subcontinuum K of $A\setminus B$. Since $K \cap B = \emptyset$ and $\dim[K] > 0$, it follows using the dimension assumption that $f[K] \not\subset f[B]$. Hence, since $K \subset A$, $f[A] \not\subset f[B]$ which implies $h(A) \neq h(B)$. This proves h is one-to-one. To see that h is a closed mapping of $C(X)\setminus F_1(X)$ onto $h[C(X)\setminus F_1(X)]$, we first prove the following:

(*) If Λ is a closed subset of $C(X)\setminus F_1(X)$, then
$h[\Lambda] = \hat{f}[cl(\Lambda)]\setminus F_1(Y)$ where, of course, cl denotes closure in $C(X)$.

To prove (*) first note that, since $\Lambda \subset [C(X)\setminus F_1(X)]$,
$h[\Lambda] = \hat{f}[\Lambda] \subset \hat{f}[cl(\Lambda)]$

and

$h[\Lambda] \subset h[C(X)\setminus F_1(X)] \subset [C(Y)\setminus F_1(Y)]$.

Hence,

(1) $h[\Lambda] \subset [\hat{f}[cl(\Lambda)]\setminus F_1(Y)]$.

Next, let $B \in [\hat{f}[cl(\Lambda)]\setminus F_1(Y)]$. Then, $B = \hat{f}(A)$ for some $A \in cl(\Lambda)$. Since $\hat{f}[F_1(X)] \subset F_1(Y)$ and $B \not\in F_1(Y)$, we have that $A \not\in F_1(X)$. Hence, since Λ is a closed subset of $C(X)\setminus F_1(X)$ and since $A \in [cl(\Lambda)\backslash F_1(X)]$, we have $A \in \Lambda$. Thus $B = \hat{f}(A) \in \hat{f}[\Lambda] = h[\Lambda]$, the last equality being valid because $\Lambda \subset [C(X)\backslash F_1(X)]$. Hence, $B \in h[\Lambda]$ and we have proved

(2) $[\hat{f}[cl(\Lambda)]\backslash F_1(Y)] \subset h[\Lambda]$.

Therefore, by combining (1) and (2), we have proved (*). Now, let Λ be a closed subset of $C(X)\backslash F_1(X)$. By (0.8), $cl(\Lambda)$ is a compact subset of $C(X)$. Hence, by the continuity of $\hat{f}$, we have that $\hat{f}[cl(\Lambda)]$ is a compact subset of $C(Y)$. Thus, by (*), $h[\Lambda]$ is a closed subset of $C(Y)\backslash F_1(Y)$. This completes the proof of (0.51.1). To prove (0.51.2), let $A, B \in C(X)$ such that $A \subset B$ and $A \neq B$. First assume $A, B \in [C(X)\backslash F_1(X)]$. Then, by (0.51.1), $\hat{f}(A) \neq \hat{f}(B)$. Hence, $f[A] \subset f[B]$ and $f[A] \neq f[B]$. Thus, $\mu(\hat{f}(A)) < \mu(\hat{f}(B))$. Next assume $A \in F_1(X)$ or $B \in F_1(X)$. Then, since $A \subset B$ and $A \neq B$, it must be that $B \notin F_1(X)$ and $A \in F_1(X)$. Thus, by (0.51.1), $\hat{f}(B) \notin F_1(Y)$, and, since $\hat{f}[F_1(X)] \subset F_1(Y)$, $\hat{f}(A) \in F_1(Y)$. Hence, again we have that $f[A] \subset f[B]$ and $f[A] \neq f[B]$. Thus, $\mu(\hat{f}(A)) < \mu(\hat{f}(B))$. This completes the proof that (0.50.a) holds for $\mu \circ \hat{f}$. The proof that (0.50.b) holds for $\mu \circ \hat{f}$ is done easily using the fact that $\hat{f}[F_1(X)] \subset F_1(Y)$. This completes the proof of (0.51.2).

The following basic result is easy to prove [the converse is false--see (0.58)].

(0.52) THEOREM [see (0.53)]. *If* $X \cong Y$, *then* $2^X \cong 2^Y$ *and* $C(X) \cong C(Y)$. *In fact if* $h: X \xrightarrow{onto} Y$ *is a homeomorphism, then*

$h*: 2^X \to 2^Y$ *defined by*

$h*(A) = \{h(a): a \in A\}$, for each $A \in 2^X$

is a homeomorphism of 2^X *onto* 2^Y *taking* $C(X)$ *onto* $C(Y)$.

(0.53) REMARK. The result in (0.52) has been in the folklore for years. It is not possible to pinpoint who first observed it, and it does not seem to be explicitly stated in the literature. It is alluded to in the third sentence of [165, p.22].

Sometimes it is possible to recognize a hyperspace as a homeomorph of some well-known space. Recognizing hyperspaces in this way is a subject of intense investigation and interest. It occurs throughout much of this book. At this time, we give two simple but fundamental examples.

(0.54) EXAMPLE [see (0.57)]. If X is an arc, then $C(X)$ is a 2-cell. By (0.52), it suffices to show this for $X = [0, 1]$. Define $f: C([0, 1]) \to R^2$ by

$f([a, b]) = (a, b)$, for each $[a, b] \in C([0, 1])$.

It is easy to verify that f is a homeomorphism of $C([0, 1])$ onto the closed convex hull in R^2 of $\{(0, 0), (1, 1), (0, 1)\}$. In [85, p.26], another homeomorphism g of $C([0, 1])$ onto a 2-cell was given; it was defined by

$g([a, b]) = (\frac{a + b}{2}, b - a)$, for each $[a, b] \in C([0, 1])$.

(0.55) EXAMPLE [see (0.57)]. If X is a circle, then $C(X)$

is a 2-cell. By (0.52), it suffices to show this for

$X = \{(x, y, 0) \in R^3 : x^2 + y^2 = 1\}$.

For each $p = (x, y, 0) \in X$, let $J[p]$ denote the convex arc in R^3 with end points p and $(0, 0, 2\pi)$. Let

$Q = \cup\{J[p] : p \in X\}$.

Thus, Q is the geometric cone over X in R^3 with vertex $(0, 0, 2\pi)$. Define $\ell : C(X)\setminus\{X\} \to [0, 2\pi)$ by letting $\ell(A)$ denote the arc length of A for each $A \in [C(X)\setminus\{X\}]$. Define $m : C(X)\setminus\{X\} \to X$ by letting $m(A)$, for each $A \in [C(X)\setminus\{X\}]$, denote that unique point of A which "divides" A into two subarcs of equal length. Define $h : C(X) \xrightarrow{\text{onto}} Q$ as follows: Let $h(X) = (0, 0, 2\pi)$ and, for each $A \in [C(X)\setminus\{X\}]$, let $h(A)$ be that unique point on $J[m(A)]$ whose third coordinate is equal to $\ell(A)$. It is easy to see that h is a homeomorphism of $C(X)$ onto the 2-cell Q. This is the homeomorphism described in [259, proof of Lemma 1]. Earlier [85, pp.267-268], a similar homeomorphism had been defined from $C(S^1)$ onto B^2. I refer the reader to the proof of (1.193) for an illustration of how computing geometric models for hyperspaces can be used to prove theorems!

(0.56) REMARK. It would now be instructive for the reader to "compute," on his own, geometric models for $C(X)$ when X is a simple triod and when $X = S^1 \cup \{(x, 0) \in R^2 : 1 \le x \le 2\}$. The first of these is done in (10.2). Note that we have not "computed" 2^X for any X. It has only recently been shown that $2^{[0,1]} \cong I_\infty$ ([290] and [291]; also, see (1.214.2) here). We will have more to say about this in the next chapter.

(0.57) REMARK. The models for $C(X)$ constructed in (0.54), (0.55), and (10.2) were well-known and in the folklore for many years. They are probably the ones Duda had in mind when he made the statement which appears at the end of the second paragraph of [85, p.265]. Duda investigated the topological and geometric nature of $C(X)$, when X is any finite graph, in [85] through [87]. In particular, geometric models for $C(X)$ are given in [87, see especially section 10] for many acyclic finite graphs X.

(0.58) REMARK. By (0.54) and (0.55), $C([0, 1]) \cong C(S^1)$ and, by [292, Theorem 4.4] or (1.97), $2^{[0,1]} \cong I_\infty \cong 2^{S^1}$. Hence, the converse of (0.52) is false [see (0.68.8) for related material]. However, the converse is true for the two special classes of continua in the following two results.

First we note that the example in (0.58) is exceptional for the class of finite graphs [see (1.108) for definition]. Duda has proved the following result:

(0.59) THEOREM [85, 9.1]. *If X and Y are finite graphs, at least one of which is different from an arc and a circle, such that $C(X) \cong C(Y)$, then $X \cong Y$.*

Next we consider the class of hereditarily indecomposable

continua. For any such continuum Z it follows easily, using (1.8), (1.11), and (1.61), that $A \in C(Z)$ is a cut point of an arc in $C(Z)$ if and only if $A \notin F_1(Z)$. The following result is a simple consequence of this.

(0.60) THEOREM (comp., paragraph 3 of section 4 of [93, p.1032]). *If* X *and* Y *are hereditarily indecomposable continua such that* $C(X) \cong C(Y)$, *then* $X \cong Y$. *In fact, if* $h: C(X) \xrightarrow{onto} C(Y)$ *is a homeomorphism, then* $h[F_1(X)] = F_1(Y)$.

We introduce the following terminology.

(0.61) DEFINITION. The members of a class Λ of continua are said to be *C-determined* provided that if $X, Y \in \Lambda$ and $C(X) \cong C(Y)$, then $X \cong Y$.

Thus, in (0.59) and (0.60), we have given two "diametrically opposite" classes of continua whose members are C-determined. No other classes are known.

(0.62) QUESTIONS. What are some other classes of continua, besides those in (0.59) and (0.60), whose members are C-determined? In particular, what about the class of chainable continua? What about the class of circle-like continua? Recently [439], it has been shown that the members of the class of smooth fans are C-determined. What about the class of fans?

EXERCISES[4,5]

(0.63) EXERCISES ON THE HAUSDORFF METRIC

(0.63.1) Prove the statement in (0.4) which says that $D(A, B) = H_d(A, B)$ for each $A, B \in 2^X$.

(0.63.2) For the closed unit interval $[0, 1]$ with the usual absolute-value metric and for any $[a_1, b_1], [a_2, b_2] \in C([0, 1])$, find a formula for $H_d([a_1, b_1], [a_2, b_2])$ in terms of the end points $a_1, b_1, a_2,$ and b_2.

(0.63.3) Prove that if $A, B, C \in 2^X$ such that $C \subset B$, then $H(A, A \cup C) \leq H(A, B)$.

(0.63.4) Let (M_1, d_1) and (M_2, d_2) be metric spaces. A function f from (M_1, d_1) into (M_2, d_2) is said to be an *isometry* provided that

$$d_2(f(p), f(q)) = d_1(p, q)$$

for each $p, q \in M_1$. Two metric spaces are said to be *isometric* provided that there is an isometry from one onto the other. Prove that X is isometric to $F_1(X)$ for any given continuum X. Show that, for $[0, 1]$ with the usual absolute-value metric d and for R^2 with the usual Euclidean metric [see (0.19)], the homeomorphisms f and g in (0.54) are not isometries from $(C([0, 1]), H_d)$ into R^2 [comp., (3.12)].

(0.63.5) Let (M_1, d_1) and (M_2, d_2) be metric spaces. A function f from (M_1, d_1) into (M_2, d_2) is said to be *nonexpansive* provided that

$$d_2(f(p), f(q)) \le d_1(p, q)$$

for each $p, q \in M_1$. The symbol *conv* denotes closed convex hull, i.e., if A is a subset of a linear space L, then

conv(A) = $\cap\{K: K$ is a closed convex subset of L

and $K \supset A\}$.

Now, let L be a Banach space, let Z be a nondegenerate compact convex subset of L, and let d denote the metric for Z obtained from the norm on L. It is well-known that if A is a compact subset of L, so is conv(A) [89, p.416]. Using this fact, we (may) consider conv as a function from $(2^Z, H_d)$ into $(2^Z, H_d)$. Prove that it is nonexpansive [247, Lemma 2, p.478].

(0.63.6) Let (M, d) be a metric space. Recall from the second footnote for Chapter 0 (see p.52) that 2^M denotes the set of all nonempty *compact* subsets of M. The Hausdorff metric was defined in (0.1) when (M, d) is a continuum. Clearly, the "same" definition gives a metric for 2^M when (M, d) is any metric space [comp., proof of (0.2)]. Assume 2^M has this metric and denote the metric by H_d. Prove that (M, d) is (Cauchy) complete if and only if $(2^M, H_d)$ is complete. Let H_d^* denote the metric for $C(M) \subset 2^M$ obtained from H_d. Prove that (M, d) is complete if and only if $(C(M), H_d^*)$ is complete. These results imply 2^{R^n}, $C(R^n)$, 2^{ℓ_2}, and $C(\ell_2)$ are complete [for further information, see

(19.38) through (19.40)]. All these facts are well-known. Some applications are in Chapter XIX [for example (19.27)].

(0.63.7) Let (X, d') and (Y, d'') be continua. Prove that a mapping f from (X, d') into (Y, d'') is nonexpansive [resp., an isometry] if and only if $\hat{f}$ is nonexpansive [resp., an isometry] from $(C(X), H_{d'})$ into $(C(Y), H_{d''})$. State and prove analogues of this result for arbitrary metric spaces and, for example, their spaces of nonempty compact subsets [i.e., $(2^M, H_d)$ as defined in (0.63.6)]. Such results have not appeared in the literature.

(0.64) EXERCISES ON LIM INF AND LIM SUP

(0.64.1) Prove (0.6.1), (0.6.2), and (0.6.3).

(0.64.2) Prove that if $A_i \in C(X)$ for each $i = 1, 2, \ldots$ and if $\liminf A_i \neq \emptyset$, then $\limsup A_i$ is connected.

(0.64.3) Give an example of a continuum X and $A_i \in C(X)$, $i = 1, 2, \ldots$, such that $\limsup A_i$ is connected but $\liminf A_i$ is not connected.

(0.64.4) Prove that a continuum X is hereditarily unicoherent if and only if for any $A_i \in C(X)$, $i = 1, 2, \ldots$, $\liminf A_i$ is connected (comp., [64, Lemma 1, p.6]).

(0.64.5) Let X be a continuum with metric d. For each $j = 0,1,2,\ldots$, let $\varepsilon_j > 0$ and let $A_j \in 2^X$. Prove [395, Lemma 3.2] that if $\varepsilon_j \to \varepsilon_0$ and $A_j \to A_0$ as $j \to \infty$, then
$$cl[N_d(\varepsilon_0, A_0)] \subset \liminf cl[N_d(\varepsilon_j, A_j)].$$

(0.64.6) Prove (20.7).

(0.65) EXERCISES ON CONTINUITY OF BALLS

(0.65.1) Let X be a continuum with metric d. For each $a \in X$ and $t \geq 0$, let
$$B_d(t, a) = \{x \in X: d(x, a) < t\}$$
and
$$C_d(t, a) = \{x \in X: d(x, a) \leq t\}.$$
Let $K_d: [0, +\infty) \times 2^X \to 2^X$ be given by
$$K_d(t, A) = \{x \in X: d(x, y) \leq t \text{ for some } y \in A\}$$
for each $(t, A) \in [0, +\infty) \times 2^X$. Prove that (1) through (3) below are equivalent [395, 3.3]:

(1) K_d is continuous;

(2) $cl[B_d(t, a)] = C_d(t, a)$ for each $t > 0$ and each $a \in X$;

(3) $cl[N_d(t, A)] = K_d(t, A)$ for each $t > 0$ and each $A \in 2^X$.

The reader may wish to use (0.64.5) in the proof of the equivalence above. The equivalence is original with the book, but has subsequently been written up for journal publication [395]. The continuity of K_d is used in the proofs of (6.12) and (14.42.1).

(0.65.2) Let K_d be as in (0.65.1). Prove that if K_d is continuous, then X is locally connected [395, 3.5]. Give an example of a locally connected continuum (X, d) such that K_d is not continuous. However, the following result [395, 3.5] is easy to prove using (0.65.1) through (0.65.4): A continuum X is locally connected if and only if there is a metric ρ for X (preserving the topology) such that K_ρ is continuous.

(0.65.3) A metric ρ for a continuum X is said to be *convex* provided that given x, y ε X, there exists z ε X such that
$$\rho(x, z) = \frac{\rho(x, y)}{2} = \rho(y, z).$$
Let (X, ρ) be a continuum with a convex metric ρ. (a) Prove that [390] for any x, y ε X, there exists a subset γ of X such that x, y ε γ and γ is isometric to the closed interval [0, $\rho(x, y)$]. (b) Letting γ be as in (a), prove that $\rho(x, y) = \rho(x, z) + \rho(z, y)$ for each z ε γ. (c) Let K_ρ be defined as in (0.65.1); prove that for any L ε 2^X and any t_1, $t_2 \geq 0$, $K_\rho(t_1, K_\rho(t_2, L)) = K_\rho(t_1 + t_2, L)$. (d) Letting $B_\rho(t, a)$ and $C_\rho(t, a)$ be as defined in (0.65.1), prove that cl$[B_\rho(t, a)] = C_\rho(t, a)$ for each t > 0 and each a ε X [7, Theorem 2].
(e) Prove that (X, ρ) is locally connected [390]--see (0.65.4).
(f) Letting K_ρ be defined as in (0.65.1), prove that K_ρ is continuous [395, 3.4]. Applications of (f) are in (0.65.4), (1.214.2), and the proof of (6.12).

(0.65.4) A continuum X is said to *admit a convex metric* ρ

provided that ρ is a convex metric for X and the (original) topology on X is the same as the topology for X obtained from ρ. Menger [390] proved that if a continuum X admits a convex metric, then X is a locally connected continuum [see (0.65.3)]. Menger [390] asked if the converse is true. Working independently, Bing [31] and Moise [392] each gave an affirmative answer to Menger's question by proving: Every locally connected continuum admits a convex metric. Using this result, prove (1.93) [Hint: Use (f) of (0.65.3)].

(0.65.5) Give an example of a continuum (X, d) such that K_d as defined in (0.65.1) is continuous but d is not a convex metric.

(0.65.6) Prove that if (X, d) and (Y, δ) are isometric continua, then K_d is continuous if and only if K_δ is continuous [K_d and K_δ as defined in (0.65.1)].

(0.66) EXERCISES ON THE VIETORIS TOPOLOGY

(0.66.1) Let (S, T) be a topological space. Prove that if (S, T) is regular, then $CL(S)$ with the Vietoris topology is Hausdorff (see [185, Theorem 3, p.168]). Prove that if (S, T) is a T_1-space and if $CL(S)$ with the Vietoris topology is Hausdorff, then (S, T) is regular (see [185, Theorem 4, p.168]).

(0.66.2) Let S be any countable discrete topological space. Prove that $CL(S)$ with the Vietoris topology is not metrizable.

Now, let $S^* = \{1, 1/2, 1/3, \ldots, 1/n, \ldots\}$. Let d' be the usual absolute-value metric for S^* and let d'' be the metric for S^* given by

$$d''(x, y) = \begin{cases} 1, & x \neq y \\ 0, & x = y \end{cases}.$$

Since (S^*, d') and (S^*, d'') are bounded metric spaces, clearly the "same" definition as in (0.1) gives metrics $H_{d'}$ and $H_{d''}$ for $CL(S^*)$. Note that (S^*, d') and (S^*, d'') are homeomorphic. Prove that $(CL(S^*), H_{d'})$ and $(CL(S^*), H_{d''})$ are not homeomorphic. Recall from the second footnote for Chapter 0 (see p.52) that 2^{S^*} denotes the set of all nonempty *compact* subsets of S^*. Let T_V denote the Vietoris topology for 2^{S^*} and let T' [resp., T''] denote the topology for 2^{S^*} as a subspace of $(CL(S^*), H_{d'})$ [resp., $(CL(S^*), H_{d''})$]. Which of the spaces $(2^{S^*}, T_V)$, $(2^{S^*}, T')$, and $(2^{S^*}, T'')$ are mutually homeomorphic?

(0.66.3) Let S be a compact Hausdorff space and let $F_1(S) = \{\{z\}: z \in S\}$. Prove that if $F_1(S)$ is a G_δ subset of $CL(S)$ with the Vietoris topology, then S is metrizable. This metrization theorem does not seem to have been noted in the literature until now.

(0.66.4) Let S be a T_1-space and let $F_1(S) = \{\{z\}: z \in S\}$. Prove that $F_1(S)$, with the (relativized) Vietoris topology for $CL(S)$, is homeomorphic to S [due to William Nowell].

(0.66.5) Let S denote the set of real numbers with the

cofinite topology T, i.e., $U \in T$ if and only if $S \setminus U$ is finite or $U = \emptyset$. Let $F_1(S) = \{\{z\}: z \in S\}$. Prove that $F_1(S)$ is dense in $CL(S)$ with the Vietoris topology (due to William Nowell; this exercise shows that [225, (c), p.156] is false).

(0.66.6) Let S be a T_1-space and, for each $n = 1, 2, \ldots$, let $F_n(S) = \{A \subset S: A$ has at most n points$\}$. Prove that $\bigcup_{n=1}^{\infty} F_n(S)$ is a dense subset of $CL(S)$ with the Vietoris topology (comp., [225, 2.4.1, p.156] where T_1 was left out).

(0.66.7) Let S be a T_1-space and, for each $n = 1, 2, \ldots$, let $F_n(S)$ be as in (0.66.6). Prove that if $F_1(S)$ is a closed subset of $CL(S)$ with the Vietoris topology, then S is a Hausdorff space [due to William Nowell]. Prove that if S is a Hausdorff space then, for each $n = 1, 2, \ldots$, $F_n(S)$ is a closed subset of $CL(S)$ with the Vietoris topology [225, 2.4.2, p.156].

(0.67) EXERCISES ON THE INDUCED MAP[6]

(0.67.1) Prove that for any mapping $f: X \to Y$ from a continuum X into a continuum Y, $\hat{f}: C(X) \to C(Y)$ is continuous [see (0.49)].

(0.67.2) Let $f: [0, 1] \to S^1$ be given by $f(t) = e^{2\pi i t}$ for each $t \in [0, 1]$. Draw a picture of $\hat{f}[C([0, 1])]$. Give an example of a mapping from $[0, 1]$ onto S^1 such that the induced map sends $C([0, 1])$ *onto* $C(S^1)$.

(0.67.3) Prove that if f is any mapping from [0, 1] onto [0, 1], then $\hat{f}$ sends C([0, 1]) onto C([0, 1]) [see (1.191)]. Prove that if g is any mapping of a circle X onto [0, 1], then $\hat{g}$ sends C(X) onto C([0, 1]) [comp., the notion of Class[W] in the paragraph preceding (14.73.21)].

(0.67.4) A mapping is said to be *essential* provided that it is not homotopic to a constant mapping; otherwise, it is said to be *nonessential*. Give an example of a weakly confluent mapping f of a circle onto a circle such that f is nonessential. This shows that the converse of (1.193) for the case when X is a circle is false--see (1.202). Show that (1.193) can not be extended to finite graphs by giving an example of an essential mapping of a "figure eight" onto itself which is not weakly confluent.

(0.67.5) Let X and Y be continua. Prove that if two mappings f_1, f_2: X → Y are homotopic, then $\hat{f}_1$, $\hat{f}_2$: C(X) → C(Y) are homotopic. Prove that if a mapping f: X → Y is nonessential, then $\hat{f}$: C(X) → C(Y) is nonessential; give an example to show that the converse is false. Using these ideas, prove (16.8).

(0.67.6) Let X be a continuum and let a, b ε X. Let Y = $X/\{a, b\}$ be the continuum obtained from X by identifying a and b; let ν: X → Y be the quotient map. Prove that $\hat{\nu}$ embeds $C(X)\setminus\{\{a\}, \{b\}\}$ into C(Y) [comp., (0.51) and (14.34)].

(0.68) EXERCISES ON WHITNEY MAPS[7]

(0.68.1) Let ω be the Whitney map for 2^X constructed in (0.50.1). Prove the following fact [mentioned at the end of (0.50.1)]: If $A, B \in 2^X$ such that A and B are isometric as subsets of X, then $\omega(A) = \omega(B)$. Give an example of a Whitney map for 2^X which does not have this property.

(0.68.2) Let $X = [0, 1]$ and let d denote the usual absolute-value metric for X. Let ω denote the Whitney map constructed in (0.50.1) using d. Compute $\omega([0, 1])$. For a given t such that $0 \leq t \leq \omega([0, 1])$, what are the members of the set $\omega^{-1}(t) \cap C([0, 1])$? What is the set $\omega^{-1}(t) \cap C([0, 1])$ topologically [as a subspace of $C([0, 1])$]? See (14.6).

(0.68.3) Verify that ω as constructed in (0.50.2) is a Whitney map.

(0.68.4) Prove that if ω' and ω'' are Whitney maps for 2^X, then so is their pointwise product $\omega' \cdot \omega''$ [see (14.71)]. Prove that if ω' and ω'' are Whitney maps for 2^X and if $t', t'' \in (0, +\infty)$, then $t' \cdot \omega' + t'' \cdot \omega''$ is a Whitney map for 2^X [see (14.71.3)].

(0.68.5) Let X be a continuum and let μ be a Whitney map for $C(X)$. Let $A \in C(X)$ and let $t \in [0, \mu(X)]$. Define $X(A, \mu, t)$ by

$X(A, \mu, t) = \{K \in \mu^{-1}(t) : K \cap A \neq \emptyset\}$.

Prove that $X(A, \mu, t)$ is compact [much more information about the structure of $X(A, \mu, t)$ is in (14.11.1)]. Prove that if X is a decomposable continuum, then there exists a proper subcontinuum A of X such that $X(A, \mu, t_o) = \mu^{-1}(t_o)$ for some $t_o < \mu(X)$ [the converse is also true--see (1.213.4)].

(0.68.6) Let $f: X \to Y$ be a mapping of a continuum X into a continuum Y and let μ be a Whitney map for $C(Y)$. In (0.51) it was shown that if

$$\dim\{x \in X: \{x\} \neq f^{-1}[f(x)]\} \leq 0,$$

then $\mu \circ \hat{f}$ is a Whitney map for $C(X)$. Give an example to show that the converse is false. Prove that for $X = [0, 1] = Y$, $\mu \circ \hat{f}$ is a Whitney map for $C(X)$ if and only if f is a homeomorphism of X into Y [in (2.213.6), necessary and sufficient conditions for arbitrary continua X and Y are given].

(0.68.7) Let X be a (metric) continuum and let $\Lambda \subset 2^X$. By an *irreducible member of* Λ we mean an $L_o \in \Lambda$ such that if D is any subset of X satisfying $D \subset L_o \neq D$, then $D \notin \Lambda$ [186, p.54]. Prove[8], by using Whitney maps, that if Λ is a nonempty compact subset of 2^X, then there is an irreducible member of Λ [Hint: The proof is very, very short!]. For the purpose of defining Λ_1 through Λ_4 below, assume that K_1 and K_2 are nonempty compact subsets of X, assume that G is any subset of X, and assume that $f: X \to S^1$ is an essential mapping from X onto the unit circle S^1. Define Λ_i for $i = 1,2,3,$ and 4 by

$\Lambda_1 = \{L \in 2^X: L \cap K_1 \neq \emptyset \neq L \cap K_2\}$,

$\Lambda_2 = \{L \in C(X): L \cap K_1 \neq \emptyset \neq L \cap K_2\}$,

$\Lambda_3 = \{L \in C(X): L \supset G\}$,

$\Lambda_4 = \{L \in 2^X:$ the restriction of f to L is essential$\}$.

Prove that, for each $i = 1,2,3,$ and 4, Λ_i is a nonempty compact subset of 2^X and, hence, has an irreducible member. Give an example of a (nonempty) subset Λ of 2^X such that Λ is not compact but such that Λ has an irreducible member. Thus, the first part of this exercise does not cover all cases. Whitney [337, p.247] has given a proof of the Brouwer Reduction Theorem using the Whitney map as constructed here in (0.50.2)--note that the construction in (0.50.2) is valid for a separable metric space, which is the setting for the Brouwer Reduction Theorem.

(0.69) EXERCISES ON USC FUNCTIONS[9]

(0.69.1) Let K and M be nonempty compact metric spaces. Prove that if $F: K \to 2^M$ is upper semi-continuous at a point $p \in K$ and if $F(p) = \{q\}$ for some $q \in M$, then F is continuous at p.

(0.69.2) Let K and M be nonempty compact metric spaces. For each $n = 1,2,\ldots,$ let $F_n: K \to 2^M$ be an upper semi-continuous function. Assume that $F_n(x) \supset F_{n+1}(x)$ for each $x \in K$ and each $n = 1,2,\ldots$. For each $x \in K$, let

$G(x) = \bigcap_{n=1}^{\infty} F_n(x)$.

Prove that this formula defines a function G from K into 2^M and that G is upper semi-continuous.

(0.69.3) Let K, M and F_n ($n = 1,2,\ldots$) satisfy the conditions in (0.69.2). Also assume that, for each $n = 1,2,\ldots$, $\cup_{x \in K} F_n(x) = M$. Let G be defined as in (0.69.2). Prove that $\cup_{x \in K} G(x) = M$.

(0.69.4) Prove that every nonempty compact metric space is a continuous image of the Cantor middle-thirds set [186, p.23]. Hint: Write the compact metric space M as the union of finitely many nonempty compact subsets, M_1, M_2, $\ldots$, M_n, each of diameter less than one. Divide the Cantor middle-thirds set K into n mutually disjoint sub-Cantor sets, K_1, K_2, $\ldots$, K_n. Define $F_1 : K \to 2^M$ by $F_1(x) = M_j$ if $x \in K_j$. Next, divide each of the sets M_1, M_2, $\ldots$, M_n into finitely many nonempty compact subsets each of diameter less than one-half. Divide each K_i into the number of pieces M_i was divided into. Define $F_2 : K \to 2^M$ in a natural way. Continue the process and obtain G as in (0.69.2). Use (0.69.1), (0.69.2), and (0.69.3) to complete the proof. Kelley's elegant application of the theorem in this exercise to hyperspaces is in the proof of (1.33).

(0.69.5) Use (0.69.1), (0.69.2), and (0.69.3) to produce a continuous function from $[0, 1]$ onto B^2. More generally, prove the Hahn-Mazurkiewicz theorem [186, p.256]: Every locally connected continuum is a continuous image of [0, 1].

(0.69.6) A compact metric space M is said to be *perfect* provided that each point of M is a limit point of M. Prove that [147, p.100] any two nonempty, compact, totally-disconnected, perfect metric spaces are homeomorphic [Hint: Use (0.69.1), (0.69.2), and (0.69.3)]. Such metric spaces are called *Cantor sets*. The *Cantor middle-thirds set* is the subspace of [0, 1] consisting of all numbers in [0, 1] which can be written in the ternary system without using the digit one [185, p.25].

(0.69.7) Prove the following special case of a theorem due to Pelczynski [265, p.85]: Let Q_1 and Q_2 be any two metric compactifications of a countable discrete space with a Cantor set as the remainder (i.e., for $i = 1$ and 2, $Q_i = D_i \cup R_i$ is a compact metric space where D_i is a countable discrete space, R_i is a Cantor set, and $D_i \cap R_i = \emptyset$); then, Q_1 is homeomorphic to Q_2. Hint: Use (0.69.1), (0.69.2), (0.69.3), and (0.69.6). For an application of this theorem, see (0.69.8).

(0.69.8) Let M be a compact metric space. Prove that M is totally-disconnected [resp., perfect] if and only if 2^M is totally-disconnected [resp., perfect]. Prove that M is a Cantor set if and only if 2^M is a Cantor set [see (0.69.6)]. Now, let $p = 0$ and let $Z = \{p, 1, 1/2, 1/3, \ldots, 1/n, \ldots\}$ with the topology inherited from R^1. Let

$$2^Z_p = \{A \subset Z : p \in A\}.$$

Prove that 2^Z_p is a Cantor set [comp., (1.40)]. Prove (1.40.1) and (1.40.2) [Hint: Use (0.69.7)]. In (0.58) it was mentioned that 2^X and 2^Y can be homeomorphic for nonhomeomorphic continua X and Y. This fact about *continua* is due to the recent results in [292]. The first examples of non-homeomorphic compact metric spaces F_1 and F_2 such that 2^{F_1} is homeomorphic to 2^{F_2} are due to Pelczynski [265]. In particular, it follows from (1.40.1) that there are uncountably many such spaces [265, Corollary 2]. The examples in [265] answered negatively a question raised in [270].

(0.70) EXERCISES ON HOMEOMORPHISMS

(0.70.1) Let X and Y be circles and let h denote any homeomorphism from C(X) onto C(Y). Prove that $h[F_1(X)] = F_1(Y)$.

(0.70.2) It is known [see Chapter VIII, especially the paragraph preceding (8.30)] that $\text{Cone}(S_o)$ is homeomorphic to $C(S_o)$ where S_o is the familiar $\sin[1/x]$-continuum defined in (8.22). Let h denote any homeomorphism from $\text{Cone}(S_o)$ onto $C(S_o)$ and let v denote the vertex of $\text{Cone}(S_o)$. What must $h(v)$ be? Verify your answer.

(0.70.3) In Figure (0.70.3), some chainable continua with exactly two arc components have been drawn. Which of them are mutually homeomorphic to which others [233, p.184]? Which of their hyperspaces C(X) are mutually homeomorphic to which others? Prove that there are uncountably many topologically different chainable continua with exactly two arc components [233, p.184]. These types of continua are related to the work in [233], [239], [245, Theorem 6], [253], etc. Prove that there are uncountably many topologically different arcwise connected circle-like continua [245, p.233].

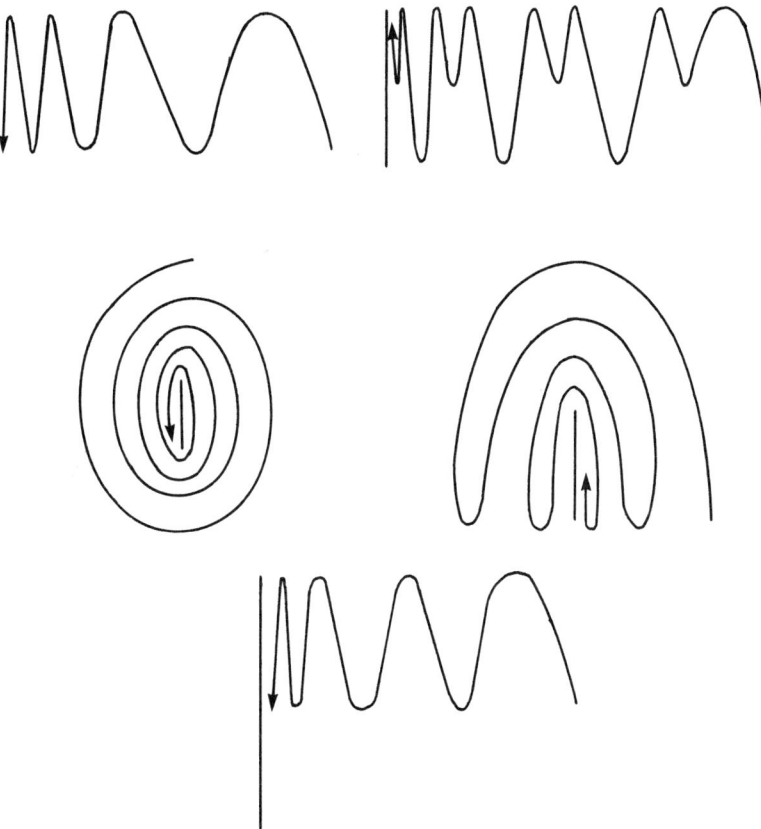

Figure (0.70.3)

(0.71) EXERCISES ON SYMMETRIC PRODUCTS

(0.71.1) Prove that $F_2(S^1)$ is a Moebius band [44, pp.876-877]. Prove that there is no retraction from $F_2(S^1)$ onto $F_1(S^1)$. By (6.17), this proves (6.18) for the special case of S^1; prove (6.17). This proof that S^1 does not admit a mean is quite different from those in the literature (see [94], [226, p.14], etc.).

(0.71.2) Let X be a continuum and let $C_2(X)$ be as defined in (0.48). For each $K \in C_2(X)$, let
$$f(K) = \{A \in C(X) : A \text{ is a component of } K\}.$$
Then, f is a function from $C_2(X)$ into $F_2(C(X))$. Prove that f is never onto, i.e., $f[C_2(X)] \neq F_2(C(X))$ for any continuum X. Also prove that f is never continuous on all of $C_2(X)$.

(0.72) EXERCISE ON SPACES OF ARCS

(0.72.1) Let $D = \bigcup_{n=0}^{\infty} A_n$ where
$$A_0 = \{(x, 0) \in R^2 : 0 \leq x \leq 1\}$$
and, for each $n = 1, 2, \ldots$,
$$A_n = \{(\lambda, \lambda/n) \in R^2 : 0 \leq \lambda \leq 1\}.$$
Let $p = (1, 0)$ and let [see (19.14)]
$$a(D, p) = \{B \in C(D) : B = \{p\} \text{ or } B \text{ is an arc with end point } p\}.$$
Draw a picture of the space $a(D, p)$ and find a homeomorphism between $a(D, p)$ and your picture.

(0.73) EXERCISES ON FIXED POINTS

(0.73.1) Prove that if $F: [0, 1] \to C([0, 1])$ is continuous, then there exists $p \in [0, 1]$ such that $p \in F(p)$ [this is a special case of (0.73.2)]. For each $\eta > 0$, give an example of a continuous function $G_\eta: B^2 \to C(B^2)$ such that, for each $p \in B^2$, $p \notin G_\eta(p)$ and
$$\inf\{d(p, q): q \in G_\eta(p)\} < \eta.$$
Such an example should be compared with [158].

(0.73.2) Prove that if X is a chainable continuum and if $F: X \to 2^X$ is continuous, then there exists $x \in X$ such that $x \in F(x)$ [325].

(0.73.3) In [246] and [247], the notion of a multi-valued contraction mapping was defined and studied for the first time. Let (M, d) be a metric space and let H_d be as in (0.63.6). A function F from (M, d) into $(2^M, H_d)$ is said to be a *multi-valued contraction mapping* provided that there exists $\lambda < 1$ such that $H_d[F(x), F(y)] \leq \lambda d(x, y)$ for all $x, y \in M$. Prove that if (M, d) is a complete metric space and if F is a multi-valued contraction mapping, then there exists $p \in M$ such that $p \in F(p)$. This result (and the definition above) is a special case of [247, Theorem 5]. These ideas were extended in [251] and in [72]. For more discussion, see the material about H shortly after (19.37.1).

(0.74) EXERCISE ON ABSOLUTE RETRACTS

(0.74.1) Use (1.48) and (1.96) to prove the following result: Let X be any continuum. If Λ is a locally connected subcontinuum of $C(X)$ such that $[\cup \Gamma] \in \Lambda$ for each subcontinuum Γ of Λ, then Λ is an absolute retract [178, 1.4]. For some applications of this result, see (1.208.5), (1.208.6), (1.214.3), and (14.73.32).

NOTES

1. Some results in the book were obtained by the author while writing the book--they are the ones for which the word "Theorem," "Corollary," etc. is not followed by a number [compare (6.12) with (1.9) and (1.12)]. For all other results, the word "Theorem," etc. is followed by a reference to the Bibliography [as in (1.9)] or to a statement of credit [as in (1.12)]. The same holds for questions, examples, etc. See Footnote 5 on this page.

2. More generally let S be any topological space; then 2^S will always denote the set of all nonempty *compact* subsets of S and $C(S)$ will always denote the members of 2^S which are connected. The symbol for the nonempty closed subsets of S is $CL(S)$--see (0.10).

3. Also, see (0.63.6) in relation to (0.1).

4. The exercises at the end of Chapter 0 are based on the material in Chapter 0. They are intended to be relatively straightforward, to illustrate the ideas in Chapter 0, and at times to connect these ideas with later ones.

5. Rather than "hiding" the answers to exercises, we have given references when the answers are in the literature or in the book. Some exercises contain results obtained while writing the book. When an exercise contains such a result, we have followed a procedure similar to the one outlined in Footnote 1 on this page, i.e., if there is no reference to the literature or to specific people, then the result in the exercise is (as far as I know) original with the book. However, some exercises are of interest only as exercises and are well-known--in such cases, the omission of a statement to that effect is a matter of convenience.

BASIC FACTS AND NOTATION 53

6. Other exercises on the induced map [besides those in (0.67)], based on the material in Chapter I, are in (1.212).

7. Other exercises on Whitney maps [besides those in (0.68)], based on material in Chapter I, are in (1.213).

8. The purpose of the exercise in (0.68.7) is to illustrate that for a very simple and "geometric" reason (via Whitney maps) compact families have irreducible members and that, for many cases of interest, the families are compact. In [186, Theorem 2, p.54] it is proved that for compact spaces (not necessarily metric), compact families have irreducible members--the proof is quite different (even for the case of compact metric spaces; see (186, Remark, p.54]) than the one asked for in (0.68.7). In other books, even when spaces are compact metric, the special case of compact families is not considered and either the Brouwer Reduction Theorem or Zorn's lemma is used. In this connection, the family Λ_4 of (0.68.7) may be of particular interest--comp., the proof in [338, 5.4, p.222].

9. The main purpose of the exercises in (0.69) is to present a unified treatment of some important theorems in topology. The first three exercises are easy-to-prove general facts and are used to prove the well-known results in (0.69.4) through (0.69.7). Many proofs have been given of, for example (0.69.4). The proof outlined here is by far the easiest of any I know. The first three exercises, and the use of them to prove (0.69.4) and (0.69.5), are due to M. K. Fort, Jr. They were part of his notes for a course in topology (at the University of Georgia), a theorem-proving course in which I was a student. Fort's treatment of these results has not appeared in print until now. The applications of (0.69.1) through (0.69.3) to proving (0.69.6) and (0.69.7) are my own.

I.
THE GENERAL STRUCTURE
OF HYPERSPACES

In Chapter 0, basic information about the topology of hyperspaces was given. Chapter I contains many of the fundamental structure results for hyperspaces. The material in it comprises a single unit in that, by and large, it presents the "classical" work on hyperspaces. However, the treatment is not for the most part historical. To the contrary, for example, Whitney maps are used to prove results which were published long before Kelley introduced Whitney maps into hyperspace theory [see, for instance, the proof of (1.4)]. Also, sometimes later ideas are combined, in the same theorem, with earlier ones. Some notions, never before used in a certain context, are used to obtain new results leading to new proofs of well-known results [for example, the use of terminal continua leading to a proof of (1.61)].

The two most important features of hyperspaces which facilitate their study are their arc structure and the continuity of the hyperspace operations with respect to inverse limits. We discuss the arc structure of hyperspaces in depth in the first section, (I.A). The material in the next three sections shows how the arc structure of hyperspaces may be used to obtain facts about the general structure. In the final section, (I.E), we discuss inverse limits and hyperspaces, giving a proof of the continuity referred to above and some applications. In various places in the chapter, some results original with the book have been included. Also, many unanswered questions are posed.

(I.A) Segments and Order-arcs--
Their Existence and Relationships

Let us begin with the following simple example which illustrates many important ideas about arcs in hyperspaces.

(1.1) EXAMPLE. Let X be the familiar $\sin[1/x]$-continuum defined in (1) of (8.22). For each $t \in [-1, +1]$, define $A_t \in C(X)$ by

$$A_t = \{(0, y) \in X: -1 \leq y \leq t\}.$$

For each $t \in (0, 1]$, define $B_t \in C(X)$ by

$$B_t = A_1 \cup \{(x, \sin[1/x]) \in X: 0 < x \leq t\}.$$

It is easy to see that the collection of sets just defined gives an arc in $C(X)$ with end points $\{(0, -1)\}$ and X; in fact, $f: [0, 1] \to C(X)$ defined by

$$f(t) = \begin{cases} A_{4t-1}, & 0 \leq t \leq 1/2 \\ B_{2t-1}, & 1/2 < t \leq 1 \end{cases}$$

is a homeomorphism. Note that this arc has the following special property: For any two points of it, at least one of them is contained in the other. Such an arc will be called an order arc [see

(1.2)]. As we will see, order arcs are extremely important in studying hyperspaces. Now, for each $t \in (0, 1]$, define $G_t \in C(X)$ by

$$G_t = \{(x, \sin[1/x]) \in X: t \le x \le 1\}.$$

Define $g: [0, 1] \to C(X)$ by

$$g(t) = \begin{cases} X, & t = 0 \\ G_t, & 0 < t \le 1. \end{cases}$$

Then, g is an arc in $C(X)$ with end points $\{(1, \sin[1])\}$ and X. With a little more work the reader can easily see that $C(X)$ is arcwise connected, even though X is not. This illustrates, for a special case, one of the most important results about hyperspaces [see (1.12) and more generally, (1.13)].

We begin the formal discussion of the arc structure of hyperspaces with the following definition, which was illustrated in (1.1).

(1.2) DEFINITION[1]. An *order arc* in 2^X is an arc α in 2^X such that if $A, B \in \alpha$, then $A \subset B$ or $B \subset A$.

The following simple lemma is a direct consequence of the definition of ω in (0.50).

(1.3) LEMMA. *Let* $\Lambda \subset 2^X$ *such that if* $A, B \in \Lambda$, *then* $A \subset B$ *or* $B \subset A$. *Then:* ω *is one-to-one on* Λ *and hence, if* Λ *is compact, the restriction of* ω *to* Λ *is a homeomorphism.*

The following theorem will be useful later on [see, for

example, the proof of (11.5)].

(1.4) THEOREM [213, Lemma 5]. *Let Λ be a nondegenerate subcontinuum of 2^X. Then Λ is an order arc if and only if $A, B \in \Lambda$ implies $A \subset B$ or $B \subset A$.*

Proof. Let Λ be a nondegenerate subcontinuum of 2^X such that $A, B \in \Lambda$ implies $A \subset B$ or $B \subset A$. Let ω_0 denote the restriction of ω to Λ. By the second part of (1.3), ω_0 is a homeomorphism. Thus, since Λ is a nondegenerate subcontinuum of 2^X, $\omega_0[\Lambda]$ is a nondegenerate compact subinterval of $[0, +\infty)$. It now follows that Λ is an arc, therefore an order arc. The other half of the theorem is trivial.

(1.5) LEMMA (comp., [213, proof of Lemma 2]). *If α is an order arc in 2^X, then $[\cap \alpha] \in \alpha$ and $[\cup \alpha] \in \alpha$.*

Proof. Let $t_0 = $ g.l.b.$(\omega[\alpha])$. Since α is compact and ω is continuous, there exists $A_0 \in \alpha$ such that $\omega(A_0) = t_0$. Now, suppose there exists $A \in \alpha$ such that $A_0 \not\subset A$. Then, since α is an order arc, $A \subset A_0$ and $A \neq A_0$. Hence, by (0.50.a),

$$\omega(A) < \omega(A_0) = t_0$$

which, since $A \in \alpha$, is a contradiction to the definition of t_0. Therefore, we have proved that $A_0 \subset A$ for all $A \in \alpha$. Thus, $A_0 \subset [\cap \alpha]$. But, since $A_0 \in \alpha$, we also have that $[\cap \alpha] \subset A_0$. Hence, $A_0 = \cap \alpha$ which, since $A_0 \in \alpha$, shows that $[\cap \alpha] \in \alpha$. Similarly, letting $t_1 = $ l.u.b.$(\omega[\alpha])$ and letting $A_1 \in \alpha$ such that $\omega(A_1) = t_1$, it can be shown that $A_1 = \cup \alpha$; hence, $[\cup \alpha] \in \alpha$.

(1.6) THEOREM (comp., [213, pp.172-173, esp. Lemma 5]). *If α is an order arc in 2^X, then the two end points of α are $\cap \alpha$ and $\cup \alpha$.*

Proof. By (1.5), $[\cap \alpha] \in \alpha$ and $[\cup \alpha] \in \alpha$. Therefore, since $[\cap \alpha] \subset A \subset [\cup \alpha]$ for all $A \in \alpha$, it follows easily using (0.50.a) and (1.3) that the restriction of ω to α is a homeomorphism of α onto the closed interval $[\omega(\cap \alpha), \omega(\cup \alpha)]$. This proves (1.6).

The next theorem is one of the main results [also, see (1.25)]. In order to state it as conveniently and as precisely as possible, we use (1.6) to introduce the following terminology.

(1.7) DEFINITION. If α is an order arc in 2^X, then α is said to be an *order arc from* [or, *beginning with*] $\cap \alpha$ *to* [or, *ending with*] $\cup \alpha$.

(1.8) THEOREM [see (1.10)]. *Let A_o, $A_1 \in 2^X$ such that $A_o \neq A_1$. Then, the following two statements are equivalent:*

(1.8.1) *There exists an order arc in 2^X from A_o to A_1;*

(1.8.2) $A_o \subset A_1$ *and each component of A_1 intersects A_o.*

Proof. Assume that (1.8.1) holds. Let α be an order arc in 2^X from A_o to A_1. By (1.7), $A_o = \cap \alpha$ and $A_1 = \cup \alpha$ so clearly $A_o \subset A_1$. Now, suppose that there exists a component L of A_1 such that $L \cap A_o = \emptyset$. Then, since A_o and L are closed subsets of the compact set A_1 such that no connected subset of A_1 intersects both A_o and L, there exist disjoint compact subsets B_o

and B_1 of A_1 such that $A_o \subset B_o$, $L \subset B_1$, and $A_1 = B_o \cup B_1$ [by (20.6)]. Let

$$\beta_o = \{A \in \alpha: A \subset B_o\}$$

and let

$$\beta_1 = \{A \in \alpha: A \cap B_1 \neq \emptyset\}.$$

Since $A_o \in \beta_o$ and $A_1 \in \beta_1$, $\beta_o \neq \emptyset$ and $\beta_1 \neq \emptyset$. Since $B_o \cap B_1 = \emptyset$, $\beta_o \cap \beta_1 = \emptyset$. It is easy to verify that β_o and β_1 are closed subsets of α [use (0.7)]. Finally, since $A \subset A_1 = B_o \cup B_1$ for all $A \in \alpha$, we have that $\alpha = \beta_o \cup \beta_1$. The facts proved about β_o and β_1 show that α is not connected, a contradiction. Hence, each component of A_1 intersects A_o. This completes the proof that (1.8.1) implies (1.8.2). Next, assume (1.8.2) holds [with $A_o \neq A_1$]. Let $\mathcal{F}$ denote the family of all subsets Γ of 2^X having the following three properties:

(1) If $G \in \Gamma$, then $A_o \subset G \subset A_1$;

(2) If $G \in \Gamma$, then each component of G intersects A_o;

(3) If $G', G'' \in \Gamma$, then $G' \subset G''$ or $G'' \subset G'$.

Since the union of any nest of members of $\mathcal{F}$ is again a member of $\mathcal{F}$, the Maximal Principle [164, p.33] may be applied to show there is a maximal member Γ_o of $\mathcal{F}$. We will show that Γ_o is an order arc in 2^X from A_o to A_1. Since $\Gamma_o \in \mathcal{F}$, we have:

(4) $\Gamma = \Gamma_o$ satisfies (1);

(5) $\Gamma = \Gamma_o$ satisfies (2);

(6) $\Gamma = \Gamma_o$ satisfies (3).

It is easy to verify that the closure of any member of $\mathcal{F}$ is a member of $\mathcal{F}$; thus, since Γ_o is a maximal member of $\mathcal{F}$, we

(1.8)   STRUCTURE OF HYPERSPACES   61

have that $\Gamma_o = cl[\Gamma_o]$. Hence, by (0.8),

(7) Γ_o is a compact subset of 2^X.

Let ω_o denote the restriction of ω to Γ_o. By (6), (7), and (1.3)

(8) ω_o is a homeomorphism.

Note that, by the maximality of Γ_o and by (1.8.2),

(9) $A_o \in \Gamma_o$ and $A_1 \in \Gamma_o$.

By (4) and (0.50.a),

(10) $\omega_o[\Gamma_o] \subset [\omega_o(A_o), \omega_o(A_1)]$.

To prove that Γ_o is an order arc in 2^X from A_o to A_1, it suffices by (6) through (10) [recall, $A_o \neq A_1$ so $\omega_o(A_o) < \omega_o(A_1)$] and (1.2) to show that

(*) $[\omega_o(A_o), \omega_o(A_1)] \subset \omega_o[\Gamma_o]$.

Suppose that (*) is false. Then, by (7) through (10), there exists a nondegenerate closed subinterval $[r_o, t_o]$ of $[\omega_o(A_o), \omega_o(A_1)]$ such that

(11) $r_o, t_o \in \omega_o[\Gamma_o]$

and

(12) $s \notin \omega[\Gamma_o]$ whenever $r_o < s < t_o$.

Using (11), let $R_o, T_o \in \Gamma_o$ such that $\omega_o(R_o) = r_o$ and $\omega_o(T_o) = t_o$. It follows easily using (6), (12) and (0.50.a) that

(13)   if $G \in \Gamma_o$, then $G \subset R_o$ or $G \supset T_o$.

Also, since $r_o < t_o$, it follows using (6) and (0.50.a) that

(14) $R_o \subset T_o$ and $R_o \neq T_o$.

By (14), there exists $\varepsilon > 0$ such that $cl[N(\varepsilon, R_o)] \not\subset T_o$. Let $p \in (T_o \setminus cl[N(\varepsilon, R_o)])$ and let K denote the component of T_o such that $p \in K$. By (5), $K \cap A_o \neq \emptyset$. Hence, since $A_o \subset R_o$ by (4), $K \cap R_o \neq \emptyset$. Let $q \in [K \cap R_o]$. Let $U = N(\varepsilon, R_o) \cap K$. Then U is a nonempty [because $q \in U$], proper [because $p \notin U$], open subset of the continuum K. Thus, letting M denote the component of $cl[U]$ such that $q \in M$, we have by (20.1) that

(15) $M \cap Bd_K[U] \neq \emptyset$ where Bd_K denotes boundary in K.

Let $G_o = R_o \cup M$. Let $\Gamma_1 = \Gamma_o \cup \{G_o\}$. We will show that $\Gamma = \Gamma_1$ satisfies (1) through (3). To do this, first note the following four facts: Clearly

(16) $R_o \subset G_o$

and it follows using (15) that

(17) $R_o \neq G_o$.

Also, since $M \subset K \subset T_o$ and $R_o \subset T_o$ [by (14)], we have that

(18) $G_o \subset T_o$

and, since $p \in [T_o \setminus G_o]$,

(19) $G_o \neq T_o$.

Now we show that $\Gamma = \Gamma_1$ satisfies (1). Since $R_o, T_o \in \Gamma_o$ and $R_o \subset T_o$ [by (14)], we have, using (4), that

(20) $A_o \subset R_o \subset T_o \subset A_1$.

By (16), (18), and (20),

(21) $A_o \subset G_o \subset A_1$.

Hence, since $\Gamma_1 = \Gamma_o \cup \{G_o\}$, it follows from (4) and (21) that $\Gamma = \Gamma_1$ satisfies (1). Next we show that $\Gamma = \Gamma_1$ satisfies (2). Since $G_o = R_o \cup M$ and since M is connected and $M \cap R_o \neq \emptyset$ [because $q \in (M \cap R_o)$], we see that each component of G_o contains a component of R_o. Thus, since each component of R_o

intersects A_o by (5),

(22) each component of G_o intersects A_o.

Hence, since $\Gamma_1 = \Gamma_o \cup \{G_o\}$, it follows from (5) and (22) that $\Gamma = \Gamma_1$ satisfies (2). Finally we show that $\Gamma = \Gamma_1$ satisfies (3). From (13), (16), and (18) we see that

(23) if $G \in \Gamma_o$, then $G \subset G_o$ or $G_o \subset G$.

Hence, since $\Gamma_1 = \Gamma_o \cup \{G_o\}$, it follows from (6) and (23) that $\Gamma = \Gamma_1$ satisfies (3). Therefore, we have shown that $\Gamma = \Gamma_1$ satisfies (1) through (3). Hence $\Gamma_1 \in \mathcal{F}$. Thus, since $\Gamma_1 \supset \Gamma_o$ and since Γ_o is a maximal member of $\mathcal{F}$, $\Gamma_1 = \Gamma_o$. Therefore,

(24) $G_o \in \Gamma_o$

and, by (16) through (19) and (0.50.a),

(25) $r_o < \omega(G_o) < t_o$.

But, (24) and (25) contradict (12). Hence, (*) holds. This completes the proof that (1.8.2) implies (1.8.1).

(1.9) **THEOREM** [43]. *The space 2^X is arcwise connected*.

Proof. [See (1.10)]. Let $A_o \in 2^X$ such that $A_o \neq X$. Let $A_1 = X$. By (1.8), there is an order arc in 2^X from A_o to A_1. Thus, each member of $2^X \setminus \{X\}$ can be joined to X by an arc in 2^X. The result follows.

(1.10) **REMARK**. Borsuk and Mazurkiewicz proved (1.9) in [43]. Their proof was somewhat different than the one given above. They first produced a subcontinuum $\Psi(C, D)$ of 2^X, where $C \subset D \neq C$ were two (arbitrary) subcontinua of X; they proved $\Psi(C, D)$ was locally connected [also $C, D \in \Psi(C, D)$]. Then, for $A \in 2^X$ and $p \in A$, they mapped $[0, 1]$ onto $\Psi(\{p\}, X)$. Denote such a mapping by f and let g be given by $g(t) = A \cup f(t)$ for each $t \in [0, 1]$. They observed that g is a mapping of $[0, 1]$ into 2^X such that $g(0) = A$ and $g(1) = X$. In the footnote in [43, p.152], Borsuk and Mazurkiewicz noted that $\Psi(C, D)$ satisfied: If $L, M \in \Psi(C, D)$, then $L \subset M$ or $M \subset L$. Thus, they were the first to discover order arcs (the terminology "order arc" comes from [234]). However, they did not seem to recognize that $\Psi(C, D)$ was an arc--this was proved later in [213, Lemma 5]. Also, since Borsuk and Mazurkiewicz [43] were only interested in proving that 2^X is arcwise connected, they did not formulate condition (1.8.2) which is why their proof of (1.9) differs from the one we gave. Condition (1.8.2) was first formulated by Kelley [165, 2.3] who proved that (1.8.2) was necessary and sufficient for there to be a segment from A_0 to A_1, segment being defined in [165, p.24] as a special type of mapping from $[0, 1]$ into 2^X [see (1.15)]. Kelley's proof of this fact and the proof above of (1.8) are essentially the same. However, Kelley did not seem to recognize how his segments and the type of set Borsuk and Mazurkiewicz [43] produced are related. The relationships are given here in (1.19) through (1.30). We will have more to say about the relationships of the results in [43], [213], and [165] later in this chapter."

(1.11) LEMMA. *If α is an order arc in 2^X beginning with $A_0 \in C(X)$, then $\alpha \subset C(X)$ [comp., (1.26)].*

Proof. Let $B \in \alpha$ such that $B \neq A_o$. Let β be the subarc of α with end points A_o and B. Clearly, by (1.2) and (1.7), β is an order arc from A_o to B. Hence, by (1.8),

(1) $A_o \subset B$

and

(2) each component of B intersects A_o.

Using (1) we see that

(3) $B = \cup \{A_o \cup K \colon K$ is a component of $B\}$.

By (2) and the fact that A_o is connected, we have that

(4) $A_o \cup K$ is connected for each component K of B.

It now follows from (3) and (4) that $B \in C(X)$. Therefore, $\alpha \subset C(X)$.

(1.12) THEOREM [see (1.14)]. *The space* $C(X)$ *is arcwise connected*.

Proof. Let $A_o \in C(X)$ such that $A_o \neq X$. By (1.8), there is an order arc α in 2^X from A_o to $A_1 = X$. Since $A_o \in C(X)$, $\alpha \subset C(X)$ by (1.11). Thus, each member of $C(X) \setminus \{X\}$ can be joined to X by an arc in $C(X)$. The result follows.

The next theorem, which summarizes some previous results, gives one of the most fundamental facts about 2^X and $C(X)$. A stronger result will be given in (1.33).

(1.13) THEOREM [see (1.14)]. *Each of* 2^X *and* $C(X)$ *is an arcwise connected continuum*.

Proof. The theorem is an immediate consequence of (0.8),

(1.9), (1.12), and the fact that 2^X and $C(X)$ are each nondegenerate [because each contains $F_1(X)$].

(1.14) REMARK. Borsuk and Mazurkiewicz's set $\Psi(C, D)$, produced in the proof in [43] and mentioned in (1.10) above, actually consisted only of subcontinua of X, i.e., $\Psi(C, D) \subset C(X)$. Hence, they proved (1.12), even though they did not mention it, before they proved (1.9). I also mention that McWaters [203, p.1209] has proved (1.13) for Hausdorff continua with the Vietoris topology [see (0.12)] and generalized arcs [see (0.18)]. Also, Ward [323] has some results for the case when the underlying space is not compact.

Next, Kelley's notion of segment is introduced. Then, the relationships between segments and order arcs are given.

(1.15) DEFINITION ([165, p.24]; see (1.16) below). Let ω be a given fixed Whitney map for 2^X. Let $A_0, A_1 \in 2^X$. A function $\sigma: [0, 1] \to 2^X$ is said to be a *segment with respect to* ω *from* A_0 *to* A_1 provided that

(1.15.1) σ is continuous on $[0, 1]$;

(1.15.2) $\sigma(0) = A_0$ and $\sigma(1) = A_1$;

(1.15.3) $\omega[\sigma(t)] = (1 - t)\cdot\omega[\sigma(0)] + t\cdot\omega[\sigma(1)]$ for each $t \in [0, 1]$;

(1.15.4) if $0 \leq t_1 \leq t_2 \leq 1$, then $\sigma(t_1) \subset \sigma(t_2)$.

A segment with respect to one Whitney map may not be a segment with respect to another Whitney map [see (1.24)]. However, most of the time it is not important to emphasize the dependence of segments on Whitney maps. Thus, for convenience, we adopt the following terminology: A function $\sigma: [0, 1] \to 2^X$ is said to be a *segment from* A_0 *to* A_1 provided that there is a Whitney map ω' for 2^X such that σ is a segment with respect to ω' from A_0 to A_1. Sometimes we will simply say σ is a *segment*, meaning σ is a segment from some member of 2^X to some member of 2^X. When σ is a segment and we want to indicate that $\sigma([0, 1]) \subset \Lambda$, some $\Lambda \subset 2^X$, we will say σ is a *segment in* Λ (in symbols, $\sigma: [0, 1] \to \Lambda$).

(1.16) REMARK. In [165], Kelley required ω (denoted μ in [165]) to satisfy $\omega(X) = 1$ in addition to (0.50.a) and (0.50.b) above; call such a Whitney map a *normalized Whitney map*. Hence, segments as defined in [165, p.24] are segments with respect to normalized Whitney maps. However, any segment as defined in (1.15) is also a segment with respect to some normalized Whitney map. For if σ is a segment with respect to a Whitney map ω, then σ is also a segment with respect to the normalized Whitney map $\eta = [1/\omega(X)] \cdot \omega$. This is easy to prove. It is only mentioned as a frame of reference for the reader who is familiar with segments as defined in the literature (they have always been defined as in [165, p.24], i.e., with respect to normalized Whitney maps). Also, the reader should compare this with (1.23) through (1.24).

(1.17) REMARK. Any constant function from $[0, 1]$ into 2^X is a segment with respect to any Whitney map for 2^X. As the following lemma shows, any non-constant segment is an arc.

(1.18) LEMMA. *If $\sigma: [0, 1] \to 2^X$ is a non-constant segment, then σ is a homeomorphism.*

Proof. Assume $\sigma: [0, 1] \to 2^X$ is a non-constant segment. We first show $\sigma(0) \neq \sigma(1)$. Suppose $\sigma(0) = \sigma(1)$. Denote the common value by B, i.e., $\sigma(0) = B = \sigma(1)$. Then: If $0 \leq t \leq 1$, it follows easily using (1.15.4) that

$B = \sigma(0) \subset \sigma(t) \subset \sigma(1) = B$.

Thus, $\sigma(t) = B$ for each $t \in [0, 1]$. Hence, σ is a constant segment, a contradiction. Therefore, we have proved

(1) $\sigma(0) \neq \sigma(1)$.

Now we show σ is one-to-one. Assume $s, t \in [0, 1]$ such that $\sigma(s) = \sigma(t)$. Then, $\omega[\sigma(s)] = \omega[\sigma(t)]$ where ω denotes a Whitney map for 2^X such that σ is a segment with respect to ω. Hence, by (1.15.3),

$(1 - s) \cdot \omega[\sigma(0)] + s \cdot \omega[\sigma(1)] = (1 - t) \cdot \omega[\sigma(0)] + t \cdot \omega[\sigma(1)]$.

Thus, by elementary arithmetic,

(2) $(t - s) \cdot \omega[\sigma(0)] = (t - s) \cdot \omega[\sigma(1)]$.

Note that, by (1.15.4),

(3) $\sigma(0) \subset \sigma(1)$.

Hence, by (1), (3), and (0.50.a),

(4) $\omega[\sigma(0)] < \omega[\sigma(1)]$.

Therefore, by (2) and (4), it must be that $(t - s) = 0$, i.e., $s = t$. This shows σ is one-to-one. Since σ is continuous by (1.15.1), σ is a homeomorphism.

(1.19)   STRUCTURE OF HYPERSPACES   69

Now, some results which relate segments and order arcs are given. The results are of two general types--compare (1.19), (1.22), and (1.25) with (1.30).

(1.19) THEOREM [special case of (1.22)]. *Let ω be a given fixed Whitney map for 2^X. A nondegenerate subset of 2^X is the range of a segment with respect to ω if and only if it is an order arc.*

Proof. Let $\Lambda \subset 2^X$ be nondegenerate. First, assume there is a segment $\sigma: [0, 1] \to 2^X$ with respect to ω such that $\sigma([0, 1]) = \Lambda$. Then, since Λ is nondegenerate, σ is a non-constant segment. Hence, by (1.18), σ is a homeomorphism. Thus, Λ is an arc. Therefore, using (1.15.4), we see that Λ is an order arc [by (1.2)]. This proves half of (1.19). Next, to prove the other half of (1.19), assume Λ is an order arc. Let ω_o denote the restriction of ω to Λ. Since Λ is an order arc, we have by (1.3) that

(1) ω_o is a homeomorphism.

Let $a, b \in [0, +\infty)$ such that $\omega_o[\Lambda] = [a, b]$. Define $\rho: [0, 1] \xrightarrow{\text{onto}} [a, b]$ by

$\rho(t) = (1 - t) \cdot a + t \cdot b$ for each $t \in [0, 1]$.

Define $\sigma: [0, 1] \to 2^X$ by

$\sigma = \omega_o^{-1} \circ \rho$.

Note that, since ρ maps $[0, 1]$ onto $[a, b]$ and ω_o^{-1} maps $[a, b]$ onto Λ,

(2) $\sigma([0, 1]) = \Lambda$.

Now we show σ is a segment with respect to ω. By (1), σ is continuous on [0, 1]. Hence, σ satisfies (1.15.1). Note that

(3) $\omega[\sigma(t)] = \rho(t) = (1 - t) \cdot a + t \cdot b$ for each $t \in [0, 1]$.

and

(4) $\omega[\sigma(0)] = \rho(0) = a$ and $\omega[\sigma(1)] = \rho(1) = b$.

Thus, combining (3) and (4), we have

(5) $\omega[\sigma(t)] = (1 - t) \cdot \omega[\sigma(0)] + t \cdot \omega[\sigma(1)]$ for each $t \in [0, 1]$.

Hence, by (5), σ satisfies (1.15.3). To show σ satisfies (1.15.4), let $0 \leq t_1 \leq t_2 \leq 1$. Clearly, since $a \leq b$,

(6) $(1 - t_1) \cdot a + t_1 \cdot b \leq (1 - t_2) \cdot a + t_2 \cdot b$.

Thus, by (3) and (6)

(7) $\omega[\sigma(t_1)] \leq \omega[\sigma(t_2)]$.

Suppose $\sigma(t_1) \not\subset \sigma(t_2)$. Then, since Λ is an order arc and $\sigma(t_1), \sigma(t_2) \in \Lambda$, $\sigma(t_2) \subset \sigma(t_1)$ and $\sigma(t_2) \neq \sigma(t_1)$. Hence, by (0.50.a), $\omega[\sigma(t_2)] < \omega[\sigma(t_1)]$. This contradicts (7). Therefore,

(8) $\sigma(t_1) \subset \sigma(t_2)$.

This proves σ satisfies (1.15.4). Hence, we have shown that σ satisfies (1.15.1), (1.15.3), and (1.15.4). Thus, by (1.15), σ is a segment with respect to ω from σ(0) to σ(1). Also, by (2), Λ is the range of σ. This completes the proof of (1.19).

(1.20) REMARK. In the proof of (1.19), we had

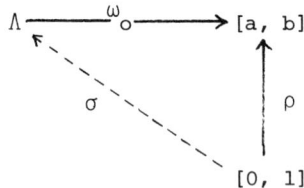

By (1.6), $\cap \Lambda$ and $\cup \Lambda$ are the two end points of Λ. Hence, since ω_o is a homeomorphism, it follows using (0.50.a) that $\omega_o(\cap \Lambda) = a$ and $\omega_o(\cup \Lambda) = b$. Thus, from the definition of σ and ρ, $\sigma(0) = \cap \Lambda$ and $\sigma(1) = \cup \Lambda$. In general, we have the following result.

(1.21) THEOREM. *If* $\sigma: [0, 1] \to 2^X$ *is a segment and* $\alpha = \sigma([0, 1])$, *then* $\sigma(0) = \cap \alpha$ *and* $\sigma(1) = \cup \alpha$.

Proof. Since the theorem is trivial if σ is constant, assume for the purpose of proof that σ is non-constant, i.e., α is nondegenerate. Then, by (1.19), α is an order arc. Hence, by (1.6), $\cap \alpha$ and $\cup \alpha$ are the two end points of α. Also, by (1.18), σ is a homeomorphism onto α and, thus, $\sigma(0)$ and $\sigma(1)$ are the two end points of α. Therefore,

(1) $\{\sigma(0), \sigma(1)\} = \{\cap \alpha, \cup \alpha\}$.

By (1.15.4),

(2) $\sigma(0) \subset \sigma(1)$.

Thus, by (1) and (2), $\sigma(0) = \cap \alpha$ and $\sigma(1) = \cup \alpha$.

(1.22) THEOREM [234, 1.4]. *Let* ω *be a given fixed Whitney map for* 2^X. *A subset* Λ *of* 2^X *is the range of a segment with respect to* ω *if and only if* Λ *is an order arc or* $\Lambda = \{A\}$ *for some* $A \in 2^X$.

Proof. If $\Lambda \subset 2^X$ is the range of a segment with respect to ω and if Λ is nondegenerate, then Λ is an order arc by (1.19). The other half of (1.22) is a simple consequence of (1.17) and (1.19).

(1.23) REMARK. Thus, (1.22) shows exactly what types of subsets of 2^X are images of [0, 1] under segments. Also, the proof of (1.19) shows how to define a segment whose range is a given order arc [see (1.31.2)]. Note that the notion of segment depends on a given Whitney map, whereas the notion of order arc is completely independent even of the existence of Whitney maps. Hence, by (1.22), any subset of 2^X which is the range of a segment with respect to one Whitney map is the range of a segment with respect to any other Whitney map. However, the segments may have to be different [for fixed ω, they are unique--see (1.30)]. The following example shows even more.

(1.24) EXAMPLE [comp., (1.32)]. Let ω_1 be a Whitney map for 2^X and let $\omega_2 = \omega_1^2$, i.e.,

$\omega_2(A) = \omega_1(A) \cdot \omega_1(A)$ for each $A \in 2^X$.

It is easy to verify that ω_2 is a Whitney map for 2^X [see (0.68.4)]. Let α_o be an order arc in 2^X [α_o exists by (1.8)]. By (1.19), there are segments $\sigma_i \colon [0, 1] \to 2^X$ with respect to ω_i such that $\sigma_i([0, 1]) = \alpha_o$, each $i = 1$ and 2. I will show that

(*) $\sigma_1(t) \neq \sigma_2(t)$ whenever $0 < t < 1$.

To prove (*), let $t \in [0, 1]$ such that $\sigma_1(t) = \sigma_2(t)$. Note that, by (1.21), $\sigma_i(0) = \cap \alpha_o$ and $\sigma_i(1) = \cup \alpha_o$. Let $a = \omega_1(\cap \alpha_o)$ and let $b = \omega_1(\cup \alpha_o)$. Then

(1) $\omega_1[\sigma_1(0)] = a$ and $\omega_1[\sigma_1(1)] = b$

and, using that $\omega_2 = \omega_1^2$,

(2) $\omega_2[\sigma_2(0)] = a^2$ and $\omega_2[\sigma_2(1)] = b^2$.

Hence, by (1) and (1.15.3),

(3) $\omega_1[\sigma_1(t)] = (1 - t)\cdot a + t\cdot b$

and, by (2) and (1.15.3),

(4) $\omega_2[\sigma_2(t)] = (1 - t)\cdot a^2 + t\cdot b^2$.

Since $\sigma_1(t) = \sigma_2(t)$, $\omega_2[\sigma_2(t)] = \omega_2[\sigma_1(t)]$ and therefore, since $\omega_2 = \omega_1^2$,

(5) $\omega_2[\sigma_2(t)] = (\omega_1[\sigma_1(t)])^2$.

Hence, by (3) through (5),

(6) $(1 - t)\cdot a^2 + t\cdot b^2 = [(1 - t)\cdot a + t\cdot b]^2$.

Since $a \neq b$ [because $(\cap \alpha_o) \subset (\cup \alpha_o) \neq \cap \alpha_o$ and thus, by (0.50.a), $a < b$], it follows after simple computations with the equation in (6) that $t = 0$ or $t = 1$. This proves (*). For each $i = 1$ and 2, let

$\Omega_i = \{\sigma : \sigma$ is a non-constant segment with respect to $\omega_i\}$

and let

$\rho_i = \{\sigma([0, 1]): \sigma \in \Omega_i\}$.

By (1.8),

$\Gamma_o(X) = \{\alpha: \alpha$ is an order arc in $2^X\} \neq \emptyset$

and, by (1.19),

$\rho_1 = \Gamma_o(X) = \rho_2$.

But, since α_o was an arbitrary member of $\Gamma_o(X)$ and since σ_i was an arbitrary member of Ω_i such that $\sigma_i([0, 1]) = \alpha_o$ for each $i = 1$ and 2, we have by (*) that $\Omega_1 \cap \Omega_2 = \emptyset$.

The following theorem is very important and will be used frequently throughout the book. It includes (1.8), and it shows that

(1.8.2) is equivalent to the existence of a segment from A_0 to A_1.

(1.25) THEOREM (the equivalence of (1.25.1) and (1.25.2) is due to Kelley [165, 2.3]). *Let* A_0, $A_1 \in 2^X$. *The following two statements are equivalent:*

(1.25.1) *There is a segment (with respect to any given Whitney map for* 2^X*) in* 2^X *from* A_0 *to* A_1;

(1.25.2) $A_0 \subset A_1$ *and each component of* A_1 *intersects* A_0.

Furthermore, if $A_0 \neq A_1$, *then (1.25.1) and (1.25.2) are each equivalent to the following statement:*

(1.25.3) *There exists an order arc in* 2^X *from* A_0 *to* A_1.

Proof. If $A_0 = A_1$, then clearly (1.25.1) holds [see (1.17)] and (1.25.2) holds. So, assume

(1) $A_0 \neq A_1$.

Then, by (1.8), (1.25.2) is equivalent to (1.25.3). Now, assume (1.25.1) holds. Let $\sigma: [0, 1] \to 2^X$ be a segment with respect to a given Whitney map for 2^X from A_0 to A_1. Let $\alpha = \sigma([0, 1])$. By (1), α is a nondegenerate subset of 2^X and hence, by (1.19), α is an order arc. By (1.21), $\sigma(0) = \cap \alpha$ and $\sigma(1) = \cup \alpha$. By (1.15.2), $\sigma(0) = A_0$ and $\sigma(1) = A_1$. Therefore, $A_0 = \cap \alpha$ and $A_1 = \cup \alpha$. Hence, α is an order arc from A_0 to A_1 by (1.7). Thus, (1.25.3) holds. Conversely, assume (1.25.3) holds. Let α be an order arc in 2^X from A_0 to A_1. By (1.7),

(2) $A_0 = \cap \alpha$ and $A_1 = \cup \alpha$.

By (1.19), there is a segment $\sigma: [0, 1] \to 2^X$ with respect to any given Whitney map for 2^X such that $\sigma([0, 1]) = \alpha$. Thus, by

(1.21), $\sigma(0) = \cap \alpha$ and $\sigma(1) = \cup \alpha$. Hence, by (2), $\sigma(0) = A_o$ and $\sigma(1) = A_1$. Therefore, by (1.15), σ is a segment from A_o to A_1. Thus, (1.25.1) holds. This completes the proof of (1.25).

Let us note the following analogue of (1.11) for segments.

(1.26) THEOREM [165, 2.6]. *If $\sigma: [0, 1] \to 2^X$ is a segment such that $\sigma(t_o) \in C(X)$ for some $t_o \in [0, 1]$, then $\sigma(t) \in C(X)$ for all $t \in [t_o, 1]$. Hence, if $\sigma(0) \in C(X)$, then $\sigma([0, 1]) \subset C(X)$.*

Proof. For the purpose of proof, assume that σ is a non-constant segment and that $t_o < 1$. Hence, by (1.18),

(1) σ is a homeomorphism.

Let $\alpha = \sigma([0, 1])$ and let $\beta = \sigma([t_o, 1])$. By (1) and (1.22),

(2) α is an order arc.

Since, as is easy to see, a subarc of an order arc is an order arc and since $t_o < 1$, we have by (1) and (2) that

(3) β is an order arc.

Since $\sigma(t_o) \subset \sigma(t)$ for all $t \in [t_o, 1]$ by (1.15.4), $\sigma(t_o) \subset [\cap \beta]$. Also, since $\sigma(t_o) \in \beta$, $[\cap \beta] \subset \sigma(t_o)$. Hence,

(4) $\cap \beta = \sigma(t_o)$.

Thus, by (3) and (4), β is an order arc beginning with $\sigma(t_o)$ [see (1.7)] and, by hypothesis, $\sigma(t_o) \in C(X)$. Therefore, by (1.11), $\beta \subset C(X)$. This completes the proof of (1.26).

In (1.22), ranges of segments were characterized. In (1.30), we will show that this characterization defines a homeomorphism. To aid in the statement and proof of (1.30), we adopt the following notation and prove the following two lemmas.

(1.27) NOTATION AND QUESTION.

(1.27.1) Let $\Gamma(X) = \{\alpha \in 2^{2^X} : \alpha$ is an order arc in $2^X\} \cup \{\{A\}: A \in 2^X\}$ with the metric obtained from the Hausdorff metric for 2^{2^X}.

(1.27.2) For any given fixed Whitney map ω for 2^X, let $S(\omega) = \{\sigma: [0, 1] \to 2^X : \sigma$ is a segment with the "sup metric" ρ given by

$$\rho(\sigma_1, \sigma_2) = \text{l.u.b.} \{H[\sigma_1(t), \sigma_2(t)] : 0 \le t \le 1\}$$

for each $\sigma_1, \sigma_2 \in S(\omega)$. Thus, we consider $S(\omega)$ as a subspace of the space of all mappings of $[0, 1]$ into 2^X with the topology of uniform convergence.

(1.27.3) QUESTION. The spaces $S(\omega)$ and $\Gamma(X)$ have never been investigated before [except for (1.29)]. It would be interesting to obtain more information about them. Some results in this direction are in (1.30), (1.32), and (1.203.3). The last of these implies $\Gamma(X)$ and 2^X have the same homotopy type. If X is a locally connected continuum, then is $\Gamma(X)$ an absolute retract[2] [see (1.96)] and, more specifically, is $\Gamma(X) \cong I_\infty$ [see (1.97)]? More generally [see (1.97)], is $\Gamma(X) \cong 2^X$ for any continuum X?

(1.28) LEMMA [165, 1.5]. *Let ω be a given fixed Whitney map for 2^X. Then: Given any $\varepsilon > 0$, there exists $\eta(\varepsilon) > 0$*

such that if $A, B \in 2^X$, $A \subset B$, and $[\omega(B) - \omega(A)] < \eta(\varepsilon)$, then $H(A, B) < \varepsilon$.

Proof. Suppose that the lemma is false for some $\varepsilon > 0$. Then there are two sequences $\{A_n\}_{n=1}^\infty$ and $\{B_n\}_{n=1}^\infty$ in 2^X such that, for each $n = 1, 2, \ldots$,

(1) $A_n \subset B_n$,

(2) $[\omega(B_n) - \omega(A_n)] < \dfrac{1}{n}$,

and

(3) $H(A_n, B_n) \geq \varepsilon$.

We assume, without loss of generality [since 2^X is compact by (0.8), we may go to subsequences if necessary], that

(4) $A_n \to A$ as $n \to \infty$, some $A \in 2^X$

and

(5) $B_n \to B$ as $n \to \infty$, some $B \in 2^X$.

By (1), (4), and (5),

(6) $A \subset B$.

By (2), (4), (5), and the continuity of ω [see (0.50)],

(7) $\omega(A) = \omega(B)$.

Hence, by (6), (7), and (0.50.a),

(8) $A = B$.

But, by (3) through (5),

(9) $H(A, B) \geq \varepsilon$.

Since (8) and (9) are incompatible, we have proved (1.28).

The following important lemma will be used in the proof of (1.30) and (1.33), the latter being a very interesting strengthening of (1.13). The reader should see (1.34), where the relationship between (1.29) and a result of Mazurkiewicz [213, Lemma 4] is discussed.

(1.29) LEMMA [165, first paragraph of the proof of 2.7]. *For any given Whitney map ω for 2^X, the space $S(\omega)$ as defined in (1.27.2) is a compact metric space.*

Proof. We first show that

(*) $S(\omega)$ is equicontinuous.

To prove (*), let $\varepsilon > 0$. Choose $\delta = \min\{\eta(\varepsilon), [\eta(\varepsilon)/\omega(X)]\}$, where $\eta(\varepsilon)$ is as guaranteed by (1.28). Now, let $\sigma \in S(\omega)$ and let $t_1, t_2 \in [0, 1]$ such that $|t_1 - t_2| < \delta$. By (1.15.3),

(1) $\omega[\sigma(t_i)] = (1 - t_i) \cdot \omega[\sigma(0)] + t_i \cdot \omega[\sigma(1)]$

for each $i = 1$ and 2.

Simple computations using (1) show that

(2) $\omega[\sigma(t_1)] - \omega[\sigma(t_2)] = (t_1 - t_2) \cdot (\omega[\sigma(1)] - \omega[\sigma(0)])$.

Thus, since $|t_1 - t_2| < \delta \leq \eta(\varepsilon)/\omega(X)$ and $|\omega[\sigma(1)] - \omega[\sigma(0)]| \leq \omega(X)$, we have using (2) that

(3) $|\omega[\sigma(t_1)] - \omega[\sigma(t_2)]| < \eta(\varepsilon)$.

Hence, by (3) and (1.28),

(4) $H[\sigma(t_1), \sigma(t_2)] < \varepsilon$.

Therefore, we have proved (*). It is easy to see that if a sequence in $S(\omega)$ converges uniformly to a function $f: [0, 1] \to 2^X$, then $f \in S(\omega)$ [use continuity of ω to see that f satisfies (1.15.3)]. From this fact and (*), we have that $(S(\omega), \rho)$ is compact by the Arzela-Ascoli Theorem (see [88, p.267] or the proof in [89, p.266]).

For some applications of the following theorem, see (1.31.2) and (1.32).

(1.30) THEOREM. *For any given Whitney map ω for 2^X, the spaces $S(\omega)$ and $\Gamma(X)$, as defined in (1.27), are homeomorphic. In fact, the function f_ω defined by*

(1.30.1) $f_\omega(\sigma) = \sigma([0, 1])$ *for each $\alpha \in S(\omega)$*

is a homeomorphism of $S(\omega)$ onto $\Gamma(X)$.

Proof. By (1.22),

(1) $f_\omega[S(\omega)] = \Gamma(X)$.

Using the fact that convergence in $S(\omega)$ is uniform convergence, it is easy to prove that

(2) f_ω is continuous [for a stronger fact, see (1.31.1)].

By (1), (2), and (1.29), it suffices to prove that

(*) f_ω is one-to-one.

To prove (*), assume that σ_1, $\sigma_2 \in S(\omega)$ such that $f_\omega(\sigma_1) = f_\omega(\sigma_2)$; denote the common value by α, i.e.,

$\sigma_1([0, 1]) = \alpha = \sigma_2([0, 1])$.

By (1.21), $\sigma_i(0) = \cap \alpha$ and $\sigma_i(1) = \cup \alpha$ for each $i = 1$ and 2. Thus, letting $t \in [0, 1]$, $a = \omega(\cap \alpha)$ and $b = \omega(\cup \alpha)$, we see from (1.15.3) that

$\omega[\sigma_1(t)] = (1 - t)\cdot a + t\cdot b = \omega[\sigma_2(t)]$.

Also, since $\alpha \in \Gamma(X)$, $\sigma_1(t) \subset \sigma_2(t)$ or $\sigma_2(t) \subset \sigma_1(t)$. Hence, using (0.50.a), we see that $\sigma_1(t) = \sigma_2(t)$. Therefore, since

$t \in [0, 1]$ was arbitrary, $\sigma_1 = \sigma_2$. This proves (*) and completes the proof of (1.30).

(1.31) REMARKS.

(1.31.1) Let f_ω be as in (1.30.1). Let H^2 denote the Hausdorff metric for 2^{2^X} induced by the Hausdorff metric H for 2^X, i.e. [comp., (0.1)], for each Λ_1, $\Lambda_2 \in 2^{2^X}$,

$$H^2(\Lambda_1, \Lambda_2) = \inf\{\varepsilon > 0: \Lambda_1 \subset N_H(\varepsilon, \Lambda_2) \text{ and }$$
$$\Lambda_2 \subset N_H(\varepsilon, \Lambda_1)\}.$$

Let ρ denote the "sup metric" for $S(\omega)$ defined in (1.27.2). Using the formulas for H^2 and ρ, one sees easily that f_ω is nonexpansive, i.e. [see (0.63.5)],

$$H^2[f_\omega(\sigma_1), f_\omega(\sigma_2)] \leq \rho(\sigma_1, \sigma_2)$$

for each σ_1, $\sigma_2 \in S(\omega)$. In general, f_ω is not an isometry [an isometry is a distance-preserving map--see (0.63.4)]. For example, let

$$Y = \{(x, y) \in R^2: \max\{|x|, |y|\} = 1 \text{ and such that}$$
$$\text{if } x = 1 \text{ then } y = -1 \text{ or } y = 1\}$$

with the usual Euclidean metric for R^2 [see (0.19)] restricted to $Y \times Y$. Let ω be any given Whitney map for 2^Y. Let σ_1, $\sigma_2 \in S(\omega)$ such that $\sigma_1(0) = \{(1, -1)\}$, $\sigma_1(1) = \{(x, -1): -1 \leq x \leq 1\}$, $\sigma_2(0) = \{(1, 1)\}$, and $\sigma_2(1) = Y$ [σ_1 and σ_2 exist by (1.25)]. Then, it can be seen that

$$H^2[f_\omega(\sigma_1), f_\omega(\sigma_2)] = 2 < \rho(\sigma_1, \sigma_2).$$

(1.31.2) Let X be a continuum, let Λ be an order arc in 2^X, and let ω be a given fixed Whitney map for 2^X. Define

$\sigma: [0, 1] \to \Lambda$ by

$$\sigma(t) = (\omega|\Lambda)^{-1}([1 - t]\cdot\omega(\cap\Lambda) + t\cdot\omega(\cup\Lambda))$$

for each $t \in [0, 1]$, where $\omega|\Lambda$ denotes the restriction of ω to Λ. It follows from previous results [see the proof of (1.19)] that $\sigma \in S(\omega)$ and that $\sigma([0, 1]) = \Lambda$, i.e., $\sigma \in S(\omega)$ and σ is a parameterization of Λ. *Furthermore, by (1.30), σ is the only segment with respect to ω parameterizing Λ.*

Recall that, by (1.24), $S(\omega_1)$ may not be equal (as a set) to $S(\omega_2)$ for different Whitney maps ω_1 and ω_2 for 2^X. However, let us note the following simple consequence of (1.30):

(1.32) COROLLARY. *For any two Whitney maps ω_1 and ω_2 for 2^X, the spaces $S(\omega_1)$ and $S(\omega_2)$ are homeomorphic.*

(I.B) Hyperspaces as Continuous Images and Preimages of the Cone Over the Cantor Set

In (1.13) we showed that each of 2^X and $C(X)$ is an arcwise connected continuum. Now, we prove an even stronger result.

(1.33) THEOREM ([165, 2.7] and [213]; see (1.35) below). *Each of 2^X and $C(X)$ is a continuous image of the cone over the Cantor middle-thirds set.*

Proof. We first prove (1.33) for the hyperspace 2^X. Let M denote the Cantor middle-thirds set. Let ω be a given fixed Whitney map for 2^X. Let

$$S_X(\omega) = \{\sigma \in S(\omega) : \sigma(1) = X\}.$$

Clearly, $S_X(\omega)$ is a closed subset of $S(\omega)$. Hence, by (1.29), $S_X(\omega)$ is a compact metric space. Thus, by (0.69.4), we have that there is a mapping $g: M \xrightarrow{\text{onto}} S_X(\omega)$. Define $f: M \times [0, 1] \to 2^X$ by

$$f(z, t) = [g(z)](t) \quad \text{for each} \quad (z, t) \in M \times [0, 1].$$

Since convergence in $S_X(\omega)$ is uniform convergence [see (1.27.2)] and since g is continuous, a simple sequence argument shows that

(1) f is continuous.

Next we show that f maps $M \times [0, 1]$ onto 2^X. Let $A \in 2^X$. By (1.25), there is a $\sigma \in S_X(\omega)$ such that $\sigma(0) = A$. Since g maps M onto $S_X(\omega)$, there exists $z_o \in M$ such that $g(z_o) = \sigma$. Now,

$$f(z_o, 0) = [g(z_o)](0) = \sigma(0) = A$$

and so we have proved that

(2) f sends $M \times [0, 1]$ onto 2^X.

Since $g(z) \in S_X(\omega)$ for each $z \in M$, we have using the definition of f that

(3) $f(z, 1) = X$ for each $(z, 1) \in M \times [0, 1]$.

Letting $\nu: M \times [0, 1] \xrightarrow{\text{onto}} \text{Cone}(M)$ [see (0.38)] be the natural map, it is easy to see using (1) through (3) that $\Psi: \text{Cone}(M) \to 2^X$ given by

$$\Psi(p) = f[\nu^{-1}(p)] \quad \text{for each} \quad p \in \text{Cone}(M)$$

is a mapping of $\text{Cone}(M)$ onto 2^X (see [88, pp.123-124] for a general theorem which implies the continuity of Ψ). Thus, we have

(1.35)   STRUCTURE OF HYPERSPACES   83

proved (1.33) for the hyperspace 2^X. To prove (1.33) for the case of $C(X)$, let

$$S_X^0(\omega) = \{\sigma \in S(\omega): \sigma(0) \in C(X) \text{ and } \sigma(1) = X\}.$$

Using the fact that $C(X)$ is a closed subset of 2^X [see (0.8)], it follows easily that $S_X^0(\omega)$ is a closed subset of $S(\omega)$. Hence, by (1.29), $S_X^0(\omega)$ is a compact metric space. Note that, by (1.26),

(4) $\sigma(t) \in C(X)$ for each $\sigma \in S_X^0(\omega)$ and each $t \in [0, 1]$.

The proof may now be completed as before [comp., (1.36)]: Let g_o be a mapping of M onto $S_X^0(\omega)$ and define $f_o: M \times [0, 1] \to 2^X$ by

$$f_o(z, t) = [g_o(z)](t) \text{ for each } (z, t) \in M \times [0, 1].$$

Then, (4) may be used to see that f_o sends $M \times [0, 1]$ into $C(X)$ and (1.25) may be used to see that f_o is onto $C(X)$. The desired mapping $\Psi_o: \text{Cone}(M) \xrightarrow{\text{onto}} C(X)$ is defined by

$$\Psi_o(p) = f_o[\nu^{-1}(p)] \text{ for each } p \in \text{Cone}(M).$$

(1.34) REMARK. In [213, Lemma 4], Mazurkiewicz stated and proved [comp., (1.203.2)]

(1.34.1) $\{\alpha \in \Gamma(X): X \in \alpha\}$ is a closed (therefore compact) subset of 2^{2^X}.

His proof also showed

(1.34.2) $\Gamma(X)$ is compact.

By the result in (1.30), (1.29) and (1.34.2) are really the same.

(1.35) REMARK. The proof given above for (1.33) is Kelley's proof [165, p.25]. The proof given in [213] is similar. Using (1.34.1), Mazurkiewicz first mapped the Cantor set onto

$\{\alpha \in \Gamma(X): X \in \alpha\}$ (see [213, p.175]); thus, Kelley used $S_X(\omega)$ in the same way as Mazurkiewicz used (1.34.1). Though the ideas in [213] and in [165] for completing the proof of (1.33) are the same, the details are carried out much more elegantly in [165] (note: Kelley meant to use $S_X(\omega)$ and $S_X^0(\omega)$ instead of his Σ and Σ_1). Mazurkiewicz only stated (1.33) for 2^X but, using the fact that $C(X)$ is a closed subset of 2^X, his proof also works for $C(X)$.

(1.36) REMARK. The following argument not only avoids repeating the proof in order to obtain (1.33) for the case of $C(X)$, but it also gives a mapping onto $C(X)$ which is closely related to the mapping onto 2^X. Let M, g, Ψ, and $S_X^0(\omega)$ be as in the proof of (1.33) and let $M' = g^{-1}[S_X^0(\omega)]$. It follows using (4) of the proof of (1.33) and (1.25) that $\Psi[\text{Cone}(M')] = C(X)$. Since g is continuous and $S_X^0(\omega)$ is compact, M' is a closed subset of M. Hence, by Corollary 2 of [185, p.281], there is a retraction r: M $\xrightarrow{\text{onto}}$ M'. As is easy to see, r "lifts up" to a retraction $\tilde{r}$: Cone(M) $\xrightarrow{\text{onto}}$ Cone(M'). Then, $\Psi \circ \tilde{r}$ is the desired mapping of Cone(M) onto $C(X)$.

By (1.33), 2^X and $C(X)$ are always continuous images of the cone over the Cantor set. Next, we investigate when 2^X and $C(X)$ are continuous preimages of the cone over the Cantor set. This can never happen when X is a locally connected continuum [by (1.92)]. However, the next main result, (1.39), shows 2^X is always such a preimage when X is a non-locally connected continuum. First, we prove two lemmas.

The following lemma was essentially obtained by Mazurkiewicz. Later, without knowing of Mazurkiewicz's paper, Bellamy obtained it in the form in which it appears here.

(1.37) LEMMA ([21, Theorem II] and [214, pp.66-71]). *There is a mapping of a (metric) continuum* Y *onto the cone over the Cantor middle-thirds set if and only if* Y *contains an open set with uncountably many components* [*see the second paragraph following (4.9)*].

Proof. Let Y be a (metric) continuum and let M denote the Cantor middle-thirds set. If there is a mapping $f: Y \xrightarrow{\text{onto}} \text{Cone}(M)$, then $f^{-1}[\text{Cone}(M) \setminus \{v\}]$ is an open subset of Y with uncountably many components. This proves half of (1.37). To prove the other half, assume Y is a metric continuum containing an open set U with uncountably many components. First we will show:

(*) there is an open subset W of U and a nonempty subset A of W such that no point of A is an isolated point of A, cl[A] ⊂ W, and distinct points of A belong to distinct components of cl[W].

To prove (*), let A_o be a "choice set" consisting of exactly one point of each component of U. We produce a nonempty subset A' of A_o such that no point of A' is an isolated point of A' as follows: Let

G = {x ∈ A_o : there is an open subset W_x of A_o such that x ∈ W_x and W_x is countable}.

Clearly, $G = \cup_{x \in G} W_x$ so that, since G is Lindelöf, it follows that G is countable (use the fact that W_x is countable for each $x \in G$). Let $A' = A_o \setminus G$. Since A_o is uncountable, $A' \neq \emptyset$ (in fact, uncountable). To see that no point of A' is an isolated point of A', let $x \in A'$. Suppose x is an isolated point of A', i.e., $\{x\}$ is an open subset of A'. Thus, there exists an open subset V of A_o such that $V \cap A' = \{x\}$. Since $x \in A_o$ and $x \notin G$, V is uncountable. In particular then, since G is countable, $[V \setminus \{x\}] \not\subset G$; hence, $[V \setminus \{x\}] \cap A' \neq \emptyset$. This is a contradiction. Therefore, no point of A' is an isolated point of A'. Now we show how to choose A. If $cl[A'] \subset U$, then let $A = A'$. If $cl[A'] \not\subset U$, then let $a \in A'$, let U_a be an open subset of Y such that $a \in U_a$ and $cl[U_a] \subset U$, and let $A = U_a \cap A'$ [note that since no point of A' is an isolated point of A', the same is true of any open subset of A']. Next, having chosen A, we show how to choose W. Since $cl[A] \subset U$, there exists (by normality) an open subset W of Y such that $cl[A] \subset W$ and $cl[W] \subset U$. Then, since distinct points of A belong to distinct components of U, distinct points of A belong to distinct components of $cl[W]$. This completes the proof of (*). Now,

    (**) we will construct a mapping $g: cl[W] \rightarrow M$ such that $g(cl[A]) = M$.

To write some statements concisely, let

    $B = B_1 | B_2$ mean B_1 and B_2 form a separation of B.

We construct g of (**) as follows. By (*) and (20.6), there exist $Q(1, 1)$ and $Q(1, 2)$ such that $cl[W] = Q(1, 1) | Q(1, 2)$ and

$Q(1, j) \cap A \neq \emptyset$ for each $j = 1$ and 2. Assume inductively that we have defined $Q(i, k)$ for each i with $1 \leq i \leq n$ and each k with $1 \leq k \leq 2^i$ such that, for $i = 2, 3, \ldots, n$, $Q(i - 1, k) = Q(i, 2k - 1) | Q(i, 2k)$ and such that, for each i and k with $1 \leq i \leq n$ and $1 \leq k \leq 2^i$, $Q(i, k) \cap A \neq \emptyset$. Since $Q(n, k) \cap A$ is a nonempty open subset of A for each k such that $1 \leq k \leq 2^n$ and since [by (*)] no point of A is an isolated point of A, $Q(n, k)$ contains at least two distinct points of A for each k with $1 \leq k \leq 2^n$. Thus, since [by (*)] different points of $Q(n, k) \cap A$ belong to different components of $Q(n, k)$ for each k with $1 \leq k \leq 2^n$, there exist [by using (20.6)], for each k with $1 \leq k \leq 2^n$, $Q(n + 1, 2k - 1)$ and $Q(n + 1, 2k)$ such that

$$Q(n, k) = Q(n + 1, 2k - 1) | Q(n + 1, 2k)$$

and

$Q(n + 1, j) \cap A \neq \emptyset$ for each $j = 1, 2, \ldots, 2^{n+1}$.

Thus, we have defined

$$Q(i, 1), Q(i, 2), \ldots, Q(i, 2^i)$$

for each $i = 1, 2, \ldots$. Note that

(1) statements (a) through (c) are equivalent:

(a) $\bigcap_{i=1}^{\infty} Q(i, k_i) \neq \emptyset$

(b) $Q(i, k_i) \supset Q(i + 1, k_{i+1})$ for each $i = 1, 2, \ldots$;

(c) $k_{i+1} \in \{2k_i - 1, 2k_i\}$ for each $i = 1, 2, \ldots$.

Since, for each $i = 1, 2, \ldots$, $Q(i, j)$ is compact and $Q(i, j) \cap A \neq \emptyset$ for each $j = 1, 2, \ldots, 2^i$, it follows using (1) that

(2) if $k_{i+1} \in \{2k_i - 1, 2k_i\}$ for each $i = 1, 2, \ldots$, then
$$[\bigcap_{i=1}^{\infty} Q(i, k_i)] \cap cl[A] \neq \emptyset.$$

As is easy to see from the way the sets $Q(i, j)$ were defined,

(3) for each $y \in cl[W]$, there is a unique sequence of natural numbers, denoted from now on by $\{k_i(y)\}_{i=1}^{\infty}$, such that $y \in [\bigcap_{i=1}^{\infty} Q(i, k_i(y))]$.

For each $y \in cl[W]$, let

$$g(y) = \sum_{i=1}^{\infty} \frac{a_i(y)}{3^i}$$

where, for each $i = 1, 2, \ldots$,

$$a_i(y) = \begin{cases} 0, & \text{if } k_i(y) \text{ is odd} \\ 2, & \text{if } k_i(y) \text{ is even.} \end{cases}$$

Since M is the Cantor middle-thirds set, the formula above defines a function $g: cl[W] \to M$. We will show that g satisfies the conditions in (**). First we show that $g(cl[A]) = M$. To do this, let $z \in M$. Then $z = \sum_{i=1}^{\infty} b_i/3^i$, where $b_i \in \{0, 2\}$ for each $i = 1, 2, \ldots$. We define a sequence $\{k_i\}_{i=1}^{\infty}$ of natural numbers as follows: Let

$$k_1 = \begin{cases} 1, & \text{if } b_1 = 0 \\ 2, & \text{if } b_1 = 2 \end{cases}$$

and, assuming inductively that we have defined k_i, let

$$k_{i+1} = \begin{cases} 2k_i - 1, & \text{if } b_{i+1} = 0 \\ 2k_i, & \text{if } b_{i+1} = 2. \end{cases}$$

By (2), there is a point y satisfying

(4) $y \in [\bigcap_{i=1}^{\infty} Q(i, k_i)]$

(1.37)   STRUCTURE OF HYPERSPACES   89

and

(5) $y \in cl[A]$.

Noting that $y \in cl[W]$, let $\{k_i(y)\}_{i=1}^{\infty}$ be as in (3). Then, by (4) and the uniqueness guaranteed by (3),

(6) $k_i(y) = k_i$ for each $i = 1, 2, \ldots$.

It follows from (6) and the way each k_i was defined that

(7) $b_i = \begin{cases} 0, & \text{if } k_i(y) \text{ is odd} \\ 2, & \text{if } k_i(y) \text{ is even} \end{cases}$ for each $i = 1, 2, \ldots$.

By (7) and the definition of $a_i(y)$, we have that

(8) $a_i(y) = b_i$ for each $i = 1, 2, \ldots$.

By (8) and the definition of g, it follows that $g(y) = z$. Hence, by (5), $z \in g(cl[A])$. Therefore, since z was an arbitrarily chosen point of M, we have shown that

(9) $g(cl[A]) = M$.

Next, we show that g is continuous. To do this, let $p \in cl[W]$ and let $\{p_j\}_{j=1}^{\infty}$ be a sequence, with $p_j \in cl[W]$ for each $j = 1, 2, \ldots$, such that $\{p_j\}_{j=1}^{\infty}$ converges to p. Let $\varepsilon > 0$. Choose I such that

(10) $\sum_{i > I} \frac{2}{3^i} < \varepsilon$.

Let $\{k_i(p)\}_{i=1}^{\infty}$ be as in (3), i.e.,

$$p \in [\bigcap_{i=1}^{\infty} Q(i, k_i(p))].$$

Since $Q(I, k_I(p))$ is an open subset of $cl[W]$ and since $\{p_j\}_{j=1}^{\infty}$ converges to $p \in Q(I, k_I(p))$, there exists a natural number J such that

$p_j \in Q(I, k_I(p))$ for each $j \geq J$.

Thus, since $Q(I, k_I(p)) \subset Q(i, k_i(p))$ for each $i \le I$,

$p_j \in Q(i, k_i(p))$ for each $i \le I$ and $j \ge J$.

Hence, from the definition and uniqueness of $\{k_i(p_j)\}_{i=1}^{\infty}$ for each $j \ge J$ [see (3)], we see that

$k_i(p_j) = k_i(p)$ for each $i \le I$ and $j \ge J$.

Hence,

(11) $a_i(p_j) = a_i(p)$ for each $i \le I$ and $j \ge J$.

Simple computations using the definition of g and (11) show that

(12) $|g(p_j) - g(p)| \le \left| \sum_{i>I} \dfrac{a_i(p_j) - a_i(p)}{3^i} \right|$ for each $j \ge J$.

Since $a_i(y) \in \{0, 2\}$ for each $y \in cl[W]$ and each $i = 1, 2, \ldots$,

(13) $|a_i(p_j) - a_i(p)| \le 2$ for each $i, j = 1, 2, \ldots$.

Hence, by (10), (12), and (13),

(14) $|g(p_j) - g(p)| < \varepsilon$ for each $j \ge J$.

Since $\varepsilon > 0$ was arbitrary, it follows from (14) that $\{g(p_j)\}_{j=1}^{\infty}$ converges to $g(p)$. Thus, we have shown that

(15) g is continuous.

By (9) and (15), g satisfies the conditions in (**). As the final step in the proof, we use (*) and g to show

(***)   there exists a mapping $f: Y \xrightarrow{\text{onto}} \text{Cone}(M)$.

By (*), there is a continuous "Urysohn" function $\psi: Y \to [0, 1]$ such that $\psi(cl[A]) = \{0\}$ and $\psi(Y\backslash W) = \{1\}$. Define $f: Y \to \text{Cone}(M)$ by

$$f(y) = \begin{cases} \nu((g(y), \psi(y))) & , \ y \in cl[W] \\ v & , \ y \in [Y\backslash W] \end{cases}$$

where $\nu: M \times [0, 1] \xrightarrow{\text{onto}} \text{Cone}(M)$ is the natural map and v is is the vertex of Cone(M) [see (0.38)]. Indeed, as is easy to show, f is a function. Also, since f is continuous on each of the closed sets cl[W] and Y\W, f is continuous on Y. It remains to show that

(#) $f[Y] = \text{Cone}(M)$.

To prove (#), let $z \in \text{Cone}(M)$. Then, $z = \nu((m, t))$ for some $m \in M$ and some $t \in [0, 1]$. By (9), there exists $y_0 \in \text{cl}[A]$ such that $g(y_0) = m$. Note that, since $\text{cl}[A] \subset W$ by (*), $y_0 \in W$. Let Γ denote the component of cl[W] such that $y_0 \in \Gamma$. By (20.1), there is a point $y_1 \in (\Gamma \cap \text{Bd}[W])$. Since $y_0 \in W$, we have from the formula for f that

$$f(y_0) = \nu((g(y_0), \psi(y_0))).$$

Since $y_0 \in \text{cl}[A]$, $\psi(y_0) = 0$. Thus, since $g(y_0) = m$,

(16) $f(y_0) = \nu((m, 0))$.

Since $y_1 \in \text{Bd}[W] = \text{cl}[W] \backslash W$ [because W is an open subset of Y; see (0.42)], $y_1 \in [Y \backslash W]$. Thus, by the formula for f,

(17) $f(y_1) = v$.

Now, since f is continuous and Γ is a subcontinuum of Y such that $y_0, y_1 \in \Gamma$, we have using (16) and (17) that

(18) $f[\Gamma]$ is a subcontinuum of Cone(M) such that

$$\nu((m, 0)), v \in f[\Gamma].$$

As is easy to see,

(19) every subcontinuum of Cone(M) is arcwise connected

and

(20) if α is an arc in Cone(M) such that $\nu((m, 0))$, $v \in \alpha$, then $\nu(\{m\} \times [0, 1]) \subset \alpha$.

Using (18) through (20), it follows easily that
$z = \nu((m, t)) \in f[\Gamma]$. Hence, $z \in f[Y]$. Therefore, since z was
an arbitrarily chosen point of Cone(M), we have proved (#). This
completes the proof of (1.37).

(1.38) LEMMA. *If* X *is a non-locally connected continuum,
then* 2^X *contains an open set with uncountably many components
[and conversely, by (1.92)].*

Proof. Since X is a non-locally connected continuum, there
exists [by (1.91)] a point $p \in X$ such that X is not connected
im kleinen at p. Hence, by (20.7), there exists $\varepsilon > 0$ such that
there is a sequence $\{K_n\}_{n=1}^{\infty}$ of distinct components K_n of cl[B],
where

$B = \{x \in X: d(p, x) < \varepsilon\}$,

such that $\{K_n\}_{n=1}^{\infty}$ converges to a nondegenerate subcontinuum K of
X satisfying $p \in K$ and $K \cap [\cup_{n=1}^{\infty} K_n] = \emptyset$. Without loss of generality, assume $K_n \cap B \neq \emptyset$ for each $n = 1, 2, \ldots$. Then, since
$p \in K = \lim K_n$, there is a sequence $\{x_n\}_{n=1}^{\infty}$ converging to p
such that $x_n \in [K_n \cap B]$ for each $n = 1, 2, \ldots$. Let

$Z = \{p, x_1, x_2, \ldots, x_n, \ldots\}$

and let

$2^Z_p = \{A \subset Z: p \in A\}$.

Clearly [for a more specific fact, see (1.40)],

(1) 2^Z_p is uncountable.

Now, let

$\beta = \{L \in 2^X: H(\{p\}, L) < \varepsilon\}$.

(1.38)

Note that, since $\beta = \langle B \rangle$ [see (0.10.2)] and since $Z \subset B$, we have that

(2) $2^Z_p \subset \beta$.

We will show

(*) each two different members of 2^Z_p belong to different components of $cl[\beta]$.

To prove (*), first note that since $K_i \cap K_j = \emptyset$ whenever $i \neq j$

(3) $Z \cap K_n = \{x_n\}$ for any $n = 1, 2, \ldots$.

Now, let A_1, $A_2 \in 2^Z_p$ with $A_1 \neq A_2$. Assume, without loss of generality, that $A_1 \not\subset A_2$. Then, from the definition of 2^Z_p and of Z, we see that there exists $n(1)$ such that $x_{n(1)} \in [A_1 \setminus A_2]$. Since $x_{n(1)} \notin A_2 \subset Z$, we have by (3) that

(4) $A_2 \cap K_{n(1)} = \emptyset$.

Hence, since $K_{n(1)}$ is a component of $cl[B]$, we have by (4) and (20.6) that there exist disjoint compact subsets M_1 and M_2 of $cl[B]$ such that $K_{n(1)} \subset M_1$, $A_2 \subset M_2$, and $cl[B] = M_1 \cup M_2$. Suppose that the component of $cl[\beta]$ determined by A_1 is equal to the component of $cl[\beta]$ determined by A_2; denote this one component by Γ (thus, Γ is a subcontinuum of $cl[\beta]$ and $A_1, A_2 \in \Gamma$). Let

$\Gamma_1 = \{G \in \Gamma : G \cap M_1 \neq \emptyset\}$

and let

$\Gamma_2 = \{G \in \Gamma : G \subset M_2\}$.

Since $x_{n(1)} \in [A_1 \cap K_{n(1)}]$ and $K_{n(1)} \subset M_1$, $x_{n(1)} \in [A_1 \cap M_1]$. Hence, $A_1 \cap M_1 \neq \emptyset$ and so, since $A_1 \in \Gamma$, $A_1 \in \Gamma_1$. Thus,

(5) $\Gamma_1 \neq \emptyset$.

Since $A_2 \in \Gamma$ and $A_2 \subset M_2$, $A_2 \in \Gamma_2$. Thus,

(6) $\Gamma_2 \neq \emptyset$.

It is easy to prove

(7) Γ_1 and Γ_2 are closed subsets of Γ.

Also, since $M_1 \cap M_2 = \emptyset$,

(8) $\Gamma_1 \cap \Gamma_2 = \emptyset$.

Simple sequence arguments, using the fact that $\beta = \langle B \rangle$ [as was mentioned just after the definition of β], show that $\cup cl[\beta] = cl[B]$. Hence, since $cl[B] = M_1 \cup M_2$, it follows that any given $D \in cl[\beta]$ satisfies $D \cap M_1 \neq \emptyset$ or $D \subset M_2$. Thus, since $\Gamma \subset cl[\beta]$, $\Gamma \subset [\Gamma_1 \cup \Gamma_2]$. Therefore, since $\Gamma_1 \subset \Gamma$ and $\Gamma_2 \subset \Gamma$,

(9) $\Gamma = \Gamma_1 \cup \Gamma_2$.

By (5) through (9), Γ is not connected which is a contradiction. This proves (*). By (1), (2), (*), and the fact that any component of β is contained in a component of $cl[\beta]$, we see that β has uncountably many components. This completes the proof of (1.38).

(1.39) THEOREM [215]. *If X is a non-locally connected continuum, then there is a mapping of 2^X onto the cone over the Cantor middle-thirds set [and conversely by (1.92)].*

Proof. By (1.38), $Y = 2^X$ satisfies the condition in the second part of (1.37). The converse part of (1.39) is a simple consequence of (1.92).

(1.40) REMARK. Let 2^Z_p be as in the proof of (1.38). It is not difficult to show that 2^Z_p is a Cantor set [see (0.69.8)].

Let us note the following two results [see (0.69.8) for historical remarks and indications of proofs]:

(1.40.1) THEOREM [265, p.88]. *If* F *is a compact totally-disconnected infinite metric space with a dense set of isolated points, then* $2^F \cong M \cup Q$ *where* M *is the Cantor middle-thirds set and* Q *is the set of midpoints of the maximal subintervals of* $[0, 1]\setminus M$.

(1.40.2) COROLLARY [265, p.89]. *If* F *is a compact countably infinite metric space, then* $2^F \cong M \cup Q$ *where* M *and* Q *are as in (1.40.1).*

Having determined in (1.39) when 2^X is a continuous preimage of the cone over the Cantor set, we now investigate when C(X) is such a preimage.

The following result is a trivial consequence of (1.37).

(1.41) THEOREM. *There is a mapping of* C(X) *onto the cone over the Cantor middle-thirds set if and only if* C(X) *contains an open set with uncountably many components.*

The characterization in (1.41) is not very satisfying in that it does not give necessary and sufficient conditions in terms of X itself. However, it does point the way towards the following example and results.

(1.42) EXAMPLE (comp., [250, p.194]). Let X be the familiar sin[1/x]-continuum defined in (1) of (8.22). Then, X is a non-locally connected continuum such that $C(X)$ is not a continuous preimage of the cone over the Cantor set. That $C(X)$ is not such a preimage can be seen using (1.41). It is not difficult to show directly that $C(X)$ does not contain an open set with uncountably many components (I leave the details to the reader); this can also be seen from the fact that $C(X) \cong \text{Cone}(X)$ ([280, p.284]; for a complete detailed proof, see [240, 4.11]). Also, I mention that a proof that $C(X)$ can not be mapped onto the cone over the Cantor set can be done from the following different point of view [not using (1.41)]: First, verify that

$$\{A \in C(X) : C(X) \text{ is not locally connected at } A\}$$

is contained in the 2-cell [see (0.54)]

$$C(\{(0, y) \in R^2 : -1 \leq y \leq +1\}).$$

Then use [99, 3, p.28] which says that, for a mapping $f: Q_1 \xrightarrow{\text{onto}} Q_2$ between compact metric spaces,

$$N(Q_2) \subset f[N(Q_1)]$$

where $N(Q_i)$ denotes the set of points of Q_i at which Q_i is not locally connected for each $i = 1$ and 2.

In (1.45) we give a sufficient condition in terms of X in order that $C(X)$ be a continuous preimage of the cone over the Cantor set; I do not know if the condition is also necessary [see (1.47)]. First, we prove two lemmas.

The following lemma is closely related to a result stated by

Kelley [(1.49) below].

(1.43) LEMMA (comp., proof of [165, 1.2]). *If Λ is a connected subset of 2^X such that $\Lambda \cap C(X) \neq \emptyset$, then $\cup \Lambda$ is a connected subset of* X.

Proof. Let $\Lambda \subset 2^X$ such that $\Lambda \cap C(X) \neq \emptyset$. Assume $\cup \Lambda$ is not connected. Then, $\cup \Lambda$ is the union of two nonempty mutually separated subsets M_1 and M_2. Let $A \in [\Lambda \cap C(X)]$. Then, $A \subset M_1$ or $A \subset M_2$. Assume without loss of generality, that $A \subset M_1$. Let

$$\Lambda_1 = \{L \in \Lambda : L \subset M_1\}$$

and let

$$\Lambda_2 = \{L \in \Lambda : L \cap M_2 \neq \emptyset\}.$$

Since $A \in \Lambda$ and $A \subset M_1$, $\Lambda_1 \neq \emptyset$. Since $M_2 \neq \emptyset$ and $M_2 \subset [\cup \Lambda]$, $\Lambda_2 \neq \emptyset$. Simple sequence arguments, using the fact that M_1 and M_2 are mutually separated, show that Λ_1 and Λ_2 are mutually separated. Since $[\cup \Lambda] \subset [M_1 \cup M_2]$, $\Lambda = \Lambda_1 \cup \Lambda_2$. Hence, Λ is not connected.

(1.44) LEMMA [250, 3.11]. *If* X *contains an open set with uncountably many components, then so does* C(X).[8]

Proof. Let U be an open subset of X with uncountably many components. Let $\Gamma = \{A \in C(X) : A \subset U\}$. Note that $\Gamma = \langle U \rangle \cap C(X)$ and so, by (0.13),

(1) Γ is an open subset of C(X).

Clearly,

(2) [∪Γ] ⊂ U

and thus, since {x} ∈ Γ for each x ∈ U,

(3) ∪Γ = U.

By (2) and (1.43) we have that

(4) if Λ is a component of Γ, then ∪Λ is contained in a component of U.

It follows easily from (3), (4), and the fact that U has uncountably many components that

(5) Γ has uncountably many components.

Therefore, combining (1) and (5), we have proved (1.44).

The following theorem is related to results in [250].

(1.45) THEOREM. *If* X *contains an open set with uncountably many components, then there is a mapping of* C(X) *onto the cone over the Cantor middle-thirds set.*[8]

Proof. The result is a simple consequence of (1.44) and (1.41).

I do not know conditions, in terms of X itself, which are both necessary and sufficient in order that C(X) be a continuous preimage of the cone over the Cantor set. In particular, I do not know the answers to the following two questions [as is easy to see using (1.41), the answers must be the same].

(1.46) QUESTION [250, 3.13]. Is the converse of (1.44) true? See (4.3) for a possible application.[8]

(1.47) QUESTION. Is the converse of (1.45) true?[8]

For some simple applications of results in (I.B), see Chapter IV.

(I.C) The Structure of C(X) when X Is Indecomposable

Some readers may not be familiar with indecomposable continua. For very simple proofs that certain continua are indecomposable, see the exercises in (1.209.2) through (1.209.4). These exercises use only the most basic facts about inverse limits [covered here in (1.149) through (1.160)]. Thus, the reader may wish to digress to the above-mentioned material before beginning this section. For the existence of hereditarily indecomposable continua, I refer the reader to (1.53) and (19.27). For basic structural information about indecomposable continua, see [186, pp.204-214].

One of the most valuable contributions to hyperspace theory was Kelley's study of C(X) when X is an indecomposable continuum [165, pp.34-35]. Throughout this book there will be many applications of the results in [165, pp.34-35], some outside the area of hyperspace theory [see, for example, the last part of (2.8)]. In this part of Chapter I, the results in [165, pp.34-35] and some closely related results are presented.

First, note the following two basic lemmas about the union function. They are valid for any continuum X, and will be used repeatedly here and throughout the book.

(1.48) LEMMA [165, 1.1(a)]. *The union function* $\cup$ *defined on* 2^{2^X} *is a mapping of* 2^{2^X} *onto* 2^X; *moreover*, $\cup: 2^{2^X} \to 2^X$ *is nonexpansive, i.e.*, [*see (0.63.5)*],

$$H(\cup \Lambda_1, \cup \Lambda_2) \leq H^2(\Lambda_1, \Lambda_2) \text{ for each } \Lambda_1, \Lambda_2 \in 2^{2^X}$$

where H^2 *denotes the Hausdorff metric for* 2^{2^X} *induced by the Hausdorff metric* H *for* 2^X [*see (1.31.1)*].

Proof. First we show $\cup$ sends 2^{2^X} onto 2^X. Let $\Lambda \in 2^{2^X}$. To show $\cup \Lambda$ is compact, it suffices by compactness of X to show $\cup \Lambda$ is a closed subset of X. Assume $z_0 \in X$ is a limit point of $\cup \Lambda$. Then there is a sequence $\{z_i\}_{i=1}^{\infty}$ converging to z_0 such that $z_i \in [\cup \Lambda]$ for each $i = 1, 2, \ldots$. For each $i = 1, 2, \ldots$, there exists $L_i \in \Lambda$ such that $z_i \in L_i$. Since Λ is compact, there is a subsequence $\{L_{i(j)}\}_{j=1}^{\infty}$ of $\{L_i\}_{i=1}^{\infty}$ such that $\{L_{i(j)}\}_{j=1}^{\infty}$ converges to a member L_0 of Λ. Hence, since $z_{i(j)} \in L_{i(j)}$ for each $j = 1, 2, \ldots$, and since $\{z_{i(j)}\}_{j=1}^{\infty}$ converges to z_0, it follows easily that $z_0 \in L_0$. Thus, since $L_0 \in \Lambda$, we have that $z_0 \in [\cup \Lambda]$. This proves $\cup \Lambda$ is a closed subset of X and, hence, compact. Since Λ is a nonempty collection of nonempty sets, $\cup \Lambda \neq \emptyset$. Hence, $\cup$ sends 2^{2^X} into 2^X. Therefore, noting that $\cup \{A\} = A$ for each $A \in 2^X$, we have that

(1) $\cup$ is a function from 2^{2^X} onto 2^X.

To show $\cup$ is a mapping (i.e., continuous), it clearly suffices to show $\cup$ is nonexpansive. First, note the following general fact which is easy to prove using the definition of H in (0.1) and

(1.48) STRUCTURE OF HYPERSPACES 101

compactness of A_1 and A_2:

(*) If $A_1, A_2 \in 2^X$ and $\eta = H(A_1, A_2)$, then
$A_1 \not\subset N(\eta, A_2)$ or $A_2 \not\subset N(\eta, A_1)$.

Now, to prove $\cup$ is nonexpansive, let $\Lambda_1, \Lambda_2 \in 2^{2^X}$. Let $A_1 = \cup \Lambda_1$ and let $A_2 = \cup \Lambda_2$. By (1), we have that $A_1, A_2 \in 2^X$. Let $\eta = H(A_1, A_2)$. Then, by (*), we assume without loss of generality that

(2) $A_1 \not\subset N(\eta, A_2)$.

By (2), there exists $p \in A_1$ such that

(3) $d(p, x) \geq \eta$ for all $x \in A_2$.

Since $p \in A_1 = \cup \Lambda_1$, there exists $L_1 \in \Lambda_1$ such that $p \in L_1$. Since $p \in L_1$ and $A_2 = \cup \Lambda_2$, it follows easily from (3) that

(4) $L_1 \not\subset N(\eta, L)$ for any $L \in \Lambda_2$.

By (4) and the definition of H in (0.1), it follows easily that $H(L_1, L) \geq \eta$ for any $L \in \Lambda_2$, i.e. [by definition of $N_H(\eta, \Lambda_2)$--comp., (0.1)],

(5) $L_1 \not\subset N_H(\eta, \Lambda_2)$.

Since $L_1 \in \Lambda_1$, we have by (5) that $\Lambda_1 \not\subset N_H(\eta, \Lambda_2)$. Hence, by the way H^2 is defined in terms of H [see (1.31.1)], it follows easily that $H^2(\Lambda_1, \Lambda_2) \geq \eta$. Therefore,

$$H(\cup \Lambda_1, \cup \Lambda_2) = H(A_1, A_2) = \eta \leq H^2(\Lambda_1, \Lambda_2).$$

This proves $\cup$ is nonexpansive and completes the proof of (1.48).

For an application of the nonexpansiveness of union, see the proof of (16.17).

(1.49) LEMMA [165, 1.2]. *If Λ is a subcontinuum of 2^X such that $\Lambda \cap C(X) \neq \emptyset$, then $\cup \Lambda$ is a subcontinuum of* X.

Proof. Assume Λ is as in the hypotheses of (1.49). Then, since $\Lambda \in 2^{2^X}$, $\cup \Lambda$ is a nonempty compact subset of X by (1.48). Since Λ is a connected subset of 2^X such that $\Lambda \cap C(X) \neq \emptyset$, $\cup \Lambda$ is a connected subset of X by (1.43). Therefore, we have proved (1.49).

Now, we begin presenting some of the results in and related to [165, pp.34-35].

The following theorem generalizes [165, 8.1] to arcwise connected subcontinua of 2^X [instead of arcs in C(X)]. It is also more general than [234, 3.1], where it was assumed Λ was a locally connected continuum. Another related result appears in [176, 2.2].

(1.50) THEOREM. *Let* Y *be an indecomposable continuum and let* $\Lambda \subset 2^Y$ *be an arcwise connected continuum. If* $\cup \Lambda = Y$ *and if* $\Lambda \cap C(Y) \neq \emptyset$, *then* $Y \in \Lambda$.

Proof. Let Y and Λ be as in the hypotheses of the theorem. Let $A \in [\Lambda \cap C(Y)]$. For the purpose of proof, assume $A \neq Y$. Then, since A is a proper subcontinuum of Y and since $\cup \Lambda = Y$, there exists $B \in \Lambda$ such that B intersects a composant of Y which does not contain A [186, p.212]. Since Λ is arcwise connected, there is a homeomorphism h from [0, 1] into Λ such that h(0) = A and h(1) = B. Since, for each $t \in [0, 1]$,

$h([0, t])$ is a subcontinuum of 2^Y such that
$A = h(0) \in [h([0, t]) \cap C(Y)]$, we have by (1.49) that

(1) $\cup h([0, t])$ is a subcontinuum of Y

for each $t \in [0, 1]$.

Thus, in particular, $\cup h([0, 1])$ is a subcontinuum of Y. Hence, since $[A \cup B] \subset [\cup h([0, 1])]$, $\cup h([0, 1])$ is a subcontinuum of Y which intersects two composants of Y. Therefore, by the indecomposability of Y and [186, p.212], we have $\cup h([0, 1]) = Y$. Hence, the number

$$t_o = g.l.b.\{t \in [0, 1]: \cup h([0, t]) = Y\}$$

exists. It follows from the continuity of h and of union [see (1.48)] that

(2) $\cup h([0, t_o]) = Y$.

Hence, since $A \neq Y$, $t_o > 0$. Now, choose any s_o such that $0 \leq s_o < t_o$. By (1), $\cup h([0, s_o])$ is a subcontinuum of Y and, by the definition of t_o and the choice of s_o, $\cup h([0, s_o]) \neq Y$. Hence, [186, p.207],

(3) $\cup h([0, s_o])$ is nowhere dense in Y.

Since $h([s_o, t_o]) \in 2^{2^Y}$ by the continuity of h and the compactness of $[s_o, t_o]$, we have by (1.48) that

(4) $\cup h([s_o, t_o])$ is a compact subset of Y.

Therefore, since [by (2)]

$$Y = [\cup h([0, s_o])] \cup [\cup h([s_o, t_o])],$$

it follows using (3) and (4) that

(5) $\cup h([s_o, t_o]) = Y$.

Since we have proved (5) for an arbitrary choice of $s_o \in [0, t_o]$, a simple argument using the continuity of h and of union [see (1.48)] shows that $h(t_o) = Y$. This proves $Y \in \Lambda$ and completes the proof of (1.50).

The next theorem is the first result in the literature about points of a hyperspace which arcwise disconnect the hyperspace. The first thorough and systematic study of such points was carried out in [234, section 4]. In Chapter XI, I discuss this topic in depth.

(1.51) THEOREM [165, 8.2]. *A continuum X is indecomposable if and only if* $C(X)\setminus\{X\}$ *is not arcwise connected* [*comp., (11.4)*].

Proof. First, assume that the continuum X is indecomposable. Let $x, y \in X$ such that x and y are points of different composants of X [186, p.212]. Let α be an arc in $C(X)$ such that $\{x\}, \{y\} \in \alpha$ [such an arc exists by (1.12)]. By (1.49), $\cup \alpha$ is a subcontinuum of X. Also, since $x, y \in \cup \alpha$, we have that $\cup \alpha$ intersects two different composants of X. Hence, since X is indecomposable, $\cup \alpha = X$ [186, p.212]. Therefore, by (1.50), $X \in \alpha$. This proves $C(X)\setminus\{X\}$ is not arcwise connected [see (1.52)]. Conversely, assume X is decomposable. Let A_1 and A_2 be proper subcontinua of X such that $X = A_1 \cup A_2$. Clearly,

(1) $[C(A_1) \cup C(A_2)] \subset [C(X)\setminus\{X\}]$

and, by (1.12),

(2) $C(A_1)$ and $C(A_2)$ are each arcwise connected.

Also, since $A_1 \cap A_2 \neq \emptyset$,

(3) $C(A_1) \cap C(A_2) \neq \emptyset$.

Thus, combining (1) through (3), it follows easily that

(4) $C(A_1) \cup C(A_2)$ is an arcwise connected subset of $C(X)\setminus\{X\}$.

Now, let $B \in [C(X)\setminus\{X\}]$. Clearly, since $B \neq \emptyset$ and $B \subset X = A_1 \cup A_2$, there is a point $b \in [(B \cap A_1) \cup (B \cap A_2)]$. By (1.25), there is a segment $\sigma: [0, 1] \to 2^X$ from $\{b\}$ to B. By (1.26), $\sigma([0, 1]) \subset C(X)$ and, by (1.15.4) and the fact that $\sigma(1) = B \neq X$, $X \notin \sigma([0, 1])$. Hence,

(5) $\sigma([0, 1])$ is an arcwise connected subset of $C(X)\setminus\{X\}$.

Also, since $\sigma(0) = \{b\} \in [C(A_1) \cup C(A_2)]$,

(6) $\sigma([0, 1]) \cap [C(A_1) \cup C(A_2)] \neq \emptyset$.

By (4) through (6) it follows easily that

(7) $\sigma([0, 1]) \cup C(A_1) \cup C(A_2)$ is an arcwise connected subset of $C(X)\setminus\{X\}$.

Since $\sigma(1) = B$,

(8) $B \in [\sigma([0, 1]) \cup C(A_1) \cup C(A_2)]$.

Using what we have proved about B [(7) and (8)], it is easy to see that $C(X)\setminus\{X\}$ is arcwise connected.

(1.52) REMARK. As regards the arc components of $C(X)\setminus\{X\}$ when X is indecomposable, the following theorem is easy to prove; its proof is left to the reader [see (1.206.1)].

(1.52.1) THEOREM (part (2) is [239, 5.1], less precisely stated in [281, p.286, lines 13-14 from the top]).

(1) If κ is a composant in a continuum X, then

$C(\kappa) = \{A \in C(X) : A \subset \kappa\}$ *is arcwise connected.*

(2) Let X *be an indecomposable continuum and let* $\Lambda \subset [C(X) \setminus \{X\}]$. *Then:* Λ *is an arc component of* $C(X) \setminus \{X\}$ *if and only if there is a composant* κ *in* X *such that* $\Lambda = C(\kappa)$.

Next our attention will be focused on the situation when X is an hereditarily indecomposable continuum. Such continua play a significant role in studying hyperspaces of general continua. One reason for this is the following result due to Bing.

(1.53) THEOREM [29, Theorem 5]. *Any (n+1)-dimensional continuum contains an n-dimensional hereditarily indecomposable continuum.*

For some applications of (1.53) to hyperspaces of general continua, see for example (1.86), (2.3) [and its proof preceding it], (8.18), and (10.8). For a proof that "most continua" in R^n or ℓ_2 are hereditarily indecomposable, see (19.27).

One of the main results about hereditarily indecomposable continua is (1.61). The proof here is somewhat different than the proof given in [165, pp.34-35]. I use the notion of terminal continua, which we now introduce and give some results about [some exercises are in (1.203.4) and (1.205)].

(1.54) DEFINITION [118, p.461]. Let A and B be subcontinua of a continuum Z such that $A \subset B$. Then, A is said to be a *terminal subcontinuum of* B, or *terminal in* B, provided that if K and L are subcontinua of B such that

$$K \cap A \neq \emptyset \neq L \cap A,$$

then $K \subset [A \cup L]$ or $L \subset [A \cup K]$. A point $p \in B$ such that $\{p\}$ is terminal in B is called a *terminal point of* B (see [32, p.660] and [228, p.190] for closely related concepts; also, see [128] for generalizations of the work in [228]).

Let us note the following convenient reformulation of (1.54); its simple proof is omitted.

(1.55) LEMMA. *Let* A *and* B *be subcontinua of a continuum* Z *such that* $A \subset B$. *Then,* A *is terminal in* B *if and only if given any two subcontinua* K *and* L *of* B *such that* $K \supset A$ *and* $L \supset A$, *we have* $K \subset L$ *or* $L \subset K$.

Terminal continua were used in hyperspace theory for the first time in [179]--see Chapter XIV here, for example (14.28) and (14.34).

Note the following two hyperspace characterizations of terminal continua, the first of which was observed in [179, section 2].

(1.56) THEOREM [179, section 2]. *Let* M *be a proper subcontinuum of a continuum* Z. *Then,* M *is terminal in* Z *if and only if*

$$C_M(Z) = \{K \in C(Z): M \subset K\}$$

is an order arc.

Proof. It is easy to prove that $C_M(Z)$ is a closed, hence [by (0.8)] compact subset of $C(Z)$. It follows easily using (1.8) and (1.11) that $C_M(Z)$ is (arcwise) connected [for a stronger result, see (1.208.5)]. Note that, since $M, Z \in C_M(Z)$ and $M \neq Z$, $C_M(Z)$ is nondegenerate. Thus, we have proved

(1) $C_M(Z)$ is a nondegenerate subcontinuum of $C(Z)$.

By (1.55),

(2) M is terminal in Z if and only if $K, L \in C_M(Z)$
    implies $K \subset L$ or $L \subset K$.

By (1) we may apply (1.4) to see that, by (2), M is terminal in Z if and only if $C_M(Z)$ is an order arc. This proves (1.56).

(1.57) THEOREM. *Let M be a proper subcontinuum of a continuum Z. Then, (1) there is (one and) only one order arc (in 2^Z) from M to Z if and only if (2) M is terminal in Z. Furthermore, if (1) or (2) holds, then the unique order arc is $C_M(Z)$ as defined in (1.56).*

Proof. Let $C_M(Z)$ be as defined in (1.56). First, assume there is one and only one order arc from M to Z; denote it by α. Clearly [use (1.11)], $\alpha \subset C_M(Z)$. Suppose there exists $K_o \in C_M(Z)$ such that $K_o \notin \alpha$. Then, (1.8) may be used twice to produce an order arc β from M to Z such that $K_o \in \beta$ [β is the union of two order arcs, one from M to K_o and the other from K_o to Z]. Since $K_o \in \beta$ and $K_o \notin \alpha$, we have $\alpha \neq \beta$

which contradicts the assumption that α is the only order arc from M to Z. Hence, $C_M(Z) \subset \alpha$ and thus, since $\alpha \subset C_M(Z)$, we have that $C_M(Z) = \alpha$. Therefore, in particular, $C_M(Z)$ is an order arc. Therefore, by (1.56), M is terminal in Z. This proves half of (1.57). Conversely, assume M is terminal in Z. Then, by (1.56), $C_M(Z)$ is an order arc. It is clear from the definition of $C_M(Z)$ that $M = \cap C_M(Z)$ and $Z = \cup C_M(Z)$. Hence, by (1.6), M and Z are the two end points of $C_M(Z)$ and [see (1.7)] $C_M(Z)$ is an order arc from M to Z. As is easy to see [use (1.11)], any order arc from M to Z must be contained in $C_M(Z)$ and, hence, must be equal to $C_M(Z)$ [because it has the same end points as $C_M(Z)$]. Therefore, $C_M(Z)$ is the one and only order arc from M to Z. This completes the proof of (1.57).

We have given some general facts about terminal continua. Now we apply these facts to the case when X is hereditarily indecomposable in order to prove (1.61). The following theorem sets the stage for doing this.

(1.58) THEOREM. *A continuum* Z *is hereditarily indecomposable if and only if each subcontinuum of* Z *is terminal in* Z.

Proof. Assume Z is hereditarily indecomposable. Let A be a subcontinuum of Z. Let K and L be any two subcontinua of Z such that $K \supset A$ and $L \supset A$. Then, $K \cap L \neq \emptyset$. Hence, $K \cup L$ is a subcontinuum of Z and, as such, is indecomposable. Therefore, $K \subset L$ or $L \subset K$. This proves [see (1.55)] that A is terminal in

Z, which completes the proof of half of (1.58). To prove the other half, assume Z contains a decomposable subcontinuum D. Let K and L be proper subcontinua of D such that $D = K \cup L$. Since $K \neq D \neq L$,

(1) $K \not\subset L$ and $L \not\subset K$.

Since $D = K \cup L$ is connected, there is a point $p \in [K \cap L]$. Then, $\{p\}$ is a subcontinuum of Z such that, by (1) and (1.55), $\{p\}$ is not terminal in Z. This completes the proof of the other half of (1.58).

(1.59) LEMMA (a special case of [165, 8.4], (1.61) below). *A continuum Z is hereditarily indecomposable if and only if for each proper subcontinuum A of Z there is (one and) only one order arc (in 2^X) from A to Z.*

Proof. The lemma is an immediate consequence of (1.57) and (1.58).

(1.60) LEMMA (lines 4-5 of the proof in [165, 8.4]). *If Y is an hereditarily indecomposable continuum and if Γ is an arc in C(Y), then $[\cup \Gamma] \in \Gamma$ [see (1.207.1)].*

Proof. By (1.49), $\cup \Gamma$ is a subcontinuum of Y and, as such, is indecomposable. The lemma now follows by taking Y and Λ in (1.50) to be $\cup \Gamma$ and Γ respectively.

The following theorem, concerning the arc structure of C(X) when X is hereditarily indecomposable, is one of the most impor-

(1.61) THEOREM [165, 8.4]. *A continuum* X *is hereditarily indecomposable if and only if* C(X) *is uniquely arcwise connected* [*i.e., given* A, B ∈ C(X) *with* A ≠ B, *there exists one and only one* α ⊂ C(X) *such that* α *is an arc with end points* A *and* B].

Proof. It is immediate from (1.8) and (1.59) that if C(X) is uniquely arcwise connected, then X is hereditarily indecomposable. To prove the converse, assume X is hereditarily indecomposable. Let A, B ∈ C(X) such that A ≠ B. By (1.12), there is an arc Λ ⊂ C(X) with end points A and B. I will show that Λ is the only such arc by considering two cases. Note that, by (1.60), [∪Λ] ∈ Λ.

Case 1: ∪Λ = A or ∪Λ = B. Then, assume without loss of generality that ∪Λ = B. Let h: [0, 1] $\xrightarrow{\text{onto}}$ Λ be a homeomorphism such that h(0) = A and h(1) = B = ∪Λ. Note that A is a proper subcontinuum of B and that, since B is a subcontinuum of X, B is an hereditarily indecomposable continuum. Hence, by taking Z = B in (1.59), we see that there is one and only one order arc from A to B; denote it by α_o. I will show Λ = α_o. To do this, let

Ω = {∪h([0, s]): 0 ≤ s ≤ 1}.

Note that, since A = h(0) and B = ∪Λ = ∪h([0, 1]),

(1) A ∈ Ω and B ∈ Ω.

Also, from the continuity of h and of union [see (1.48)], it follows that

(2) Ω is a continuum.

Since $h([0, s])$ is an arc in $C(X)$ for each $s > 0$, we have by (1.60) that $[\cup h([0, s])] \in h([0, s]) \subset \Lambda$ for each $s \in [0, 1]$. Hence,

(3) $\Omega \subset \Lambda$.

By (1) through (3) we see that Ω is a subarc of the arc Λ and that the end points of Λ belong to Ω. Thus,

(4) $\Omega = \Lambda$.

Since Ω is an arc [by (4)] and

$$[\cup h([0, s_1])] \subset [\cup h([0, s_2])] \text{ whenever } 0 \le s_1 \le s_2 \le 1,$$

we have by (1.2) that Ω is an order arc. Hence, from the way Ω was defined, it is clear that Ω is an order arc from A to B [see the definition in (1.7)]. Therefore,

(5) $\Omega = \alpha_o$.

By (4) and (5),

(6) $\Lambda = \alpha_o$.

Thus, we have proved under the assumptions of Case 1 that Λ is unique [by (6)].

Case 2: $A \ne \cup \Lambda \ne B$. Then, since $[\cup \Lambda] \in \Lambda$, there are subarcs Λ_1 and Λ_2 of Λ with A and $\cup \Lambda$ being the end points of Λ_1 and B and $\cup \Lambda$ being the end points of Λ_2. From what we showed in Case 1, Λ_1 [resp., Λ_2] is the unique order arc from A [resp., B] to $\cup \Lambda$. Hence, $\Lambda = \Lambda_1 \cup \Lambda_2$ is unique as an arc in $C(X)$ with end points A and B. This completes the proof of (1.61).

(1.63)   STRUCTURE OF HYPERSPACES   113

The following result is easily deducible from the proof of (1.61) or directly from (1.8), (1.11), (1.60), and (1.61). It is due to Kelley; though he did not explicitly state the result, it is a consequence of what he did in the proof of his 8.4 [165, pp.34-35]. The result was specifically stated and used for the first time in [176, 2.5].

(1.62) **COROLLARY** [see preceding paragraph]. *Let* X *be an hereditarily indecomposable continuum. If* α *is an arc in* C(X), *then either (1)* α *is an order arc or (2)* $\alpha = \alpha_1 \cup \alpha_2$ *where* α_1 *and* α_2 *are each order arcs such that* $\alpha_1 \cap \alpha_2 = \{\cup\alpha\}$.

Thus, by (1.62), we know everything about the arcs in C(X) when X is any hereditarily indecomposable continuum. For such X, some information is known about the arcwise connected subcontinua of C(X). In (1.63) through (1.75), we give some known and some new results about this.

(1.63) **THEOREM** [176, 2.3]. *If* X *is an hereditarily indecomposable continuum, then any arcwise connected subcontinuum of* C(X) *is contractible (in itself)*.

Proof. Let Λ be an arcwise connected subcontinuum of C(X). By (1.49), $Y = \cup\Lambda$ is a subcontinuum of X and, as such, is indecomposable. Thus, since Λ is an arcwise connected subcontinuum of C(Y) and $\cup\Lambda = Y$, we have by (1.50) that $Y \in \Lambda$. Thus, for each $L \in \Lambda$ such that $L \neq Y$, there is an arc, denoted by α_L, such that α_L has end points L and Y and $\alpha_L \subset \Lambda$. Since X is an hereditarily indecomposable continuum and each α_L is an arc in $C(X)$ with end points L and Y where $L \subset Y$, it follows from (1.61) and (1.62) that

    (1) each α_L is the unique order arc in $C(X)$ from L to Y.

Let ω denote a given fixed Whitney map for 2^X. By (1) and (1.30), we have that

    (2) for each $L \in \Lambda$, there is one and only one segment with respect to ω, denoted by σ_L, such that $\sigma_L([0, 1]) = \alpha_L$ if $L \neq Y$ and $\sigma_L([0, 1]) = \{Y\}$ if $L = Y$.

Define $\psi: \Lambda \times [0, 1] \to \Lambda$ by

$$\psi(L, t) = \sigma_L(t) \quad \text{for each } L \in \Lambda \text{ and } t \in [0, 1].$$

It is easy to prove using (1.15.4) that $\psi(L, 0) = \sigma_L(0) = L$ and $\psi(L, 1) = \sigma_L(1) = Y$ for each $L \in \Lambda$. Hence, it remains to show that ψ is continuous. To do this, let $(L, t) \in \Lambda \times [0, 1]$ and let $\{(L_n, t_n)\}_{n=1}^{\infty}$ be a sequence in $\Lambda \varepsilon [0, 1]$ converging to (L, t). For each $n = 1, 2, \ldots$, let σ_n denote σ_{L_n} where σ_{L_n} is as in (2). By (1.29), there is a subsequence $\{\sigma_{n(i)}\}_{i=1}^{\infty}$ of $\{\sigma_n\}_{n=1}^{\infty}$ such that $\{\sigma_{n(i)}\}_{i=1}^{\infty}$ converges to a segment σ with respect to ω. Since $L_{n(i)} \to L$ and $L_{n(i)} = [\sigma_{n(i)}](0) \to \sigma(0)$, we see that $\sigma(0) = L$. Since $[\sigma_{n(i)}](1) = Y$ for each $i = 1, 2, \ldots$ and since $[\sigma_{n(i)}](1) \to \sigma(1)$, we see that $\sigma(1) = Y$. Hence, by (1.22) and the uniqueness part of (1),

$$\sigma([0, 1]) = \alpha_L \quad \text{if } L \neq Y$$

and

$\sigma([0, 1]) = \{Y\}$ if $L = Y$.

Therefore, by (2),

(3) $\sigma = \sigma_L$.

Thus, since $\{\sigma_{n(i)}\}_{i=1}^{\infty}$ converges to $\sigma = \sigma_L$ and convergence is uniform convergence [see (1.27.2)],

(4) $[\sigma_{n(i)}](t_{n(i)}) \to \sigma_L(t)$.

Since $\psi(L_{n(i)}, t_{n(i)}) = [\sigma_{n(i)}](t_{n(i)})$ for each $i = 1,2,\ldots$ and $\psi(L, t) = \sigma_L(t)$, we have by (4) that

(5) $\psi(L_{n(i)}, t_{n(i)}) \to \psi(L, t)$.

It now follows from what we have shown in (5) that ψ is continuous. This completes the proof of (1.63).

The following corollary will be strengthened in (1.70).

(1.64) COROLLARY [176, 2.4]. *Let* X *be an hereditarily indecomposable continuum. If* Λ *is a one-dimensional arcwise connected subcontinuum of* C(X), *then* Λ *is a contractible dendroid.*

Proof. By (1.63), Λ is contractible. Therefore [see (1.65)], since Λ is one-dimensional, Λ is hereditarily unicoherent.

(1.65) REMARK. In the proof of (1.64), the following facts about the first Čech cohomology group over the integers were used: A contractible continuum is acyclic, each subcontinuum of a one-dimensional acyclic continuum is acyclic, and an acyclic continuum is unicoherent. For readers familiar with dimension theory and property (b) [see (16.2)], the final statement in the proof of (1.64) may be verified as follows (comp., [93, p.1029]). Assume K is a contractible one-dimensional continuum. Let L be a subcontinuum of K. Let $f: L \to S^1$ be a mapping. Since $\dim[K] = 1$, f can be extended to a mapping $g: K \to S^1$ [150, p.83]. As is easy to prove from the contractibility of K, g is homotopic to a constant mapping. Hence [338, p.226], there exists a mapping $\alpha: K \to R^1$ such that $g = e^{i\alpha}$ where $e: R^1 \to S^1$ denotes the exponential mapping. Clearly, $f = e^{i\beta}$ where β denotes the restriction of α to L. Since f was any given mapping of L into S^1, we have proved that L has property (b) [338, p.226]. Therefore, L is unicoherent [338, p.227]. This proves K is hereditarily unicoherent.

Contrary to [66, Theorem 1], there exist contractible dendroids with non-contractible subdendroids. An example due to F. B. Jones is in the Math. Reviews, vol.46, 1973, #8193. Charatonik showed me the following example.

(1.66) EXAMPLE [J. J. Charatonik]. Let
$A = \{(x, 0) \in R^2: -1 \le x \le +1\}$,
$B = \{(0, y) \in R^2: 0 \le y \le +1\}$,
and, for each $n = 1, 2, \ldots$, let
$C_n = \{(\frac{1}{n}, y) \in R^2: 0 \le y \le +1\}$,
$D_n = \{(\frac{-1}{n}, y) \in R^2: 0 \le y \le +1\}$,
$E_n = \{\lambda \cdot \left(\frac{2n+1}{2n(n+1)}, \frac{1}{2}\right) + [1 - \lambda] \cdot (\frac{1}{n}, 1): 0 \le \lambda \le 1\}$.

Let $Z = A \cup B \cup [\cup_{n=1}^{\infty} C_n] \cup [\cup_{n=1}^{\infty} D_n] \cup [\cup_{n=1}^{\infty} E_n]$. It is easy to see that Z is a contractible dendroid. For each $n = 1, 2, \ldots$, let

$$D_n' = \{(\frac{-1}{n}, y) \varepsilon R^2 \colon 0 \leq y \leq +\frac{1}{2}\}.$$

Then $Z' = A \cup B \cup [\cup_{n=1}^{\infty} C_n] \cup [\cup_{n=1}^{\infty} D_n'] \cup [\cup_{n=1}^{\infty} E_n]$ is a non-contractible subdendroid of Z.

In view of (1.66), let us note the following simple consequence of (1.64). A stronger result is given in (1.70).

(1.67) COROLLARY. *Let* X *be an hereditarily indecomposable continuum. If a dendroid* D *is embeddable in* $C(X)$, *then every nondegenerate subcontinuum of* D *is a contractible dendroid.*

Proof. As is well-known (see, for example, [65, pp.239-240]), every dendroid is one-dimensional and any subcontinuum of a dendroid is arcwise connected. Hence, (1.67) follows from (1.64).

In order to state (1.70), we give the following definition.

(1.68) DEFINITION [66, p.298]. A dendroid D is said to be *smooth at a point* $p \varepsilon D$ [also, see (19.14)] provided that if $\{a_n\}_{n=1}^{\infty}$ is any convergent sequence of points a_n of D, converging to say $a_0 \varepsilon D$, then

$$A_n \to A_0 \text{ as } n \to \infty$$

where, for each $n = 0, 1, 2, \ldots$, A_n denotes the unique arc in D with end points a_n and p if $a_n \neq p$ and $A_n = \{p\}$ if $a_n = p$. A dendroid is said to be *smooth* provided that it is smooth at *some* point. Clearly, a dendroid which is smooth at every point is locally connected and, hence, a dendrite [66, Corollary 5]. The cone over $\{0, 1, 1/2, 1/3,\ldots, 1/n,\ldots\}$ is a smooth non-locally connected dendroid. For examples of dendroids which are not smooth (i.e., not smooth at any point), see (1.66) and (1.69). Let us note the following fact about smooth dendroids:

(1.68.1) Every nondegenerate subcontinuum of a smooth dendroid is a smooth dendroid [67, Corollary 6, p.299] and every smooth dendroid is contractible [67, Corollary 12, p.311].

The following example shows that the converse of (1.67) is false.

(1.69) EXAMPLE. Let
$$A = \{(x, 0) \in R^2 : -2 \leq x \leq +2\}$$
and, for each $n = 1, 2, \ldots$, let
$$B_n = \{\lambda \cdot (-2, 0) + [1 - \lambda] \cdot (-1, \tfrac{1}{n}) : 0 \leq \lambda \leq 1\}$$
and
$$C_n = \{\lambda \cdot (2, 0) + [1 - \lambda] \cdot (1, \tfrac{1}{n}) : 0 \leq \lambda \leq 1\}.$$
Then $D = A \cup [\cup_{n=1}^{\infty} B_n] \cup [\cup_{n=1}^{\infty} C_n]$ is a dendroid, every subcontinuum of D is contractible, and D can not be embedded in $C(X)$ for any hereditarily indecomposable continuum X. The last fact follows from (1.70) and the fact that D is not smooth. However, a direct geometrical argument verifying the last fact is instructive.

By (1.68.1), the following result strengthens (1.64) and (1.67).

(1.70) THEOREM. *Let* X *be an hereditarily indecomposable continuum. If a dendroid* D *is embeddable in* C(X), *then* D *is a smooth dendroid. More specifically, if* $\Lambda \subset C(X)$ *is a dendroid,* Λ *is smooth at* $\cup \Lambda$.

Proof. Assume $\Lambda \subset C(X)$ is a dendroid. By (1.49), $\cup \Lambda$ is a subcontinuum of X. Hence, by (1.50), $[\cup \Lambda] \in \Lambda$. By fixing a Whitney map ω and using (1.29) and (1.30), it can be shown with a straightforward argument that Λ is smooth at $\cup \Lambda$ [use the fact that if $\sigma_n \to \sigma_o$ where $\sigma_n \in S(\omega)$ for each $n = 0,1,2,...$, then $\sigma_n([0, 1]) \to \sigma_o([0, 1])$; this follows from the fact that convergence in $S(\omega)$ is uniform convergence--see (1.27.2)]. I leave the details to the reader.

For dendroids with only one ramification point (such dendroids are called *fans* [64, p.6], the converse of (1.70) is true [see (1.73)]. To prove this, first observe the following lemma.

(1.71) LEMMA. *If* X *is an hereditarily indecomposable continuum [see (1.72)], then the cone over the Cantor set is embeddable in* C(X).

Proof. Assume X is an hereditarily indecomposable continuum. Then [186, p.213], there exists a Cantor set $K \subset X$ such that any two distinct points of K are in different composants of X. For each point $q \in K$, let $\alpha(q)$ denote the unique order arc in C(X) from $\{q\}$ to X [see (1.59)]. It is not difficult to show that Cone(K) $\cong \cup_{q \in K} \alpha(q)$.

(1.72) REMARK. By using (1.145), we see that (1.71) is true for any continuum X.

(1.73) THEOREM. *Let* X *be an hereditarily indecomposable continuum. A fan* F *is embeddable in* C(X) *if and only if* F *is smooth.*

Proof. Let F be a fan. If F is embeddable in C(X), then F is smooth by (1.70) [smoothness for fans was defined in [64, p.7] to mean smooth at the ramification point; for fans this is equivalent to the definition in (1.68)]. To prove the converse, assume F is a smooth fan. Charatonik [64, Theorem 9, p.27] proved every folding fan (see [64, p.24] for definition) is embeddable in the cone over the Cantor set and Eberhart [91] showed every smooth fan is folding. Hence, the theorem now follows using (1.71).

The following question remains unanswered.

(1.74) QUESTION. Is the converse of the first part of (1.70) true?[3] In (1.73) we have shown it is true for fans. Let us also note the following theorem:

(1.74.1) THEOREM. *Any dendrite can be embedded in* C(X) *for any continuum* X. *If a locally connected continuum* L *is embed-*

dable in C(X), *where* X *is an hereditarily indecomposable continuum, then* L *is a dendrite*.

Proof. It is not difficult to show [using (b) of (1.210.2)] that the universal dendrite constructed in [411, p.137] can be embedded in C(X) for any continuum X [by (1.145)] and the planarity of all dendrites, this need only be shown when X is hereditarily indecomposable]. The second part of (1.74.1) follows easily from (1.61).

The following result, about C(X) itself, completes the discussion [begun in (1.63)] of the arcwise connected subcontinua of C(X) when X is hereditarily indecomposable. For an interesting application of the result, see the proof of (7.6) in the paragraph just preceding (7.6).

(1.75) COROLLARY [see (1.76)]. *If* X *is an hereditarily indecomposable continuum, then* C(X) *is contractible*.

Proof. By (1.12), C(X) is arcwise connected. Hence, by (1.63), C(X) is contractible.

(1.76) REMARK. In [276, 3.5], Rhee stated that C(X) is contractible when X is a pseudo-arc. His proof showed that C(X) is contractible when X is any hereditarily indecomposable continuum. Krasinkiewicz [176, 2.3] noted the contractibility of C(X) for such X (his reference to Kelley appears to be a misprint). In Chapter XVI, we discuss contractibility of hyperspaces. I point out, in relation to (1.75), that not all hyperspaces are contractible [see (16.22)]. A stronger result than (1.75) is in (16.27). Also, by (1.75) and (16.7), 2^X is contractible if X is hereditarily indecomposable.

Let us recall the following basic facts about relationships between the decomposition space topology and the topology of hyperspaces.

(1.77) LEMMA ([185, pp.185-187]; also, see [93, p.1030]). *Let* M *be any compact metric space and let* T *denote the topology for* M. *Let* Γ *be any collection of nonempty compact mutually disjoint subsets of* M *whose union is* M *(i.e.,* Γ *is a* **set-theoretic decomposition** *of* M *into [nonempty] compact sets). Consider the following diagram*

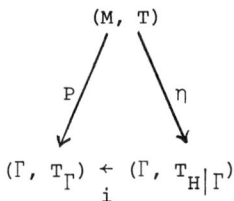

where T_Γ *denotes the decomposition topology for* Γ, P *denotes the natural map onto the decomposition space,* $T_{H|\Gamma}$ *denotes the topology (obtained from the Hausdorff metric) for* Γ *as contained in* 2^M *by inclusion,* η *denotes the natural "map" (not necessarily continuous), and* i *denotes the identity function. Then:*

(1.77.1) i *is continuous, i.e.,* $T_\Gamma \subset T_{H|\Gamma}$;

(1.77.2) *the following six statements are equivalent--*

(1.78)

(i) η *is continuous;*

(ii) $T_\Gamma = T_H|_\Gamma$;

(iii) $(\Gamma, T_H|_\Gamma)$ *is compact;*

(iv) (Γ, T_Γ) *is a continuous decomposition of* M;

(v) P *is both an open and a closed mapping;*

(vi) η *is both an open and a closed mapping.*

In (1.80) we will apply (1.77) to the case when M is an hereditarily indecomposable continuum and $\Gamma = \mu^{-1}(t)$ [see (0.50) for the definition of the symbol μ]. First, let us note the following lemma.

(1.78) LEMMA ([165, 8.3] and [165, part of the proof of 8.5]). *Let* X *be an hereditarily indecomposable continuum. Let* μ *be any given fixed Whitney map for* C(X). *Then, for each* $t \in [0, \mu(X)]$, $\mu^{-1}(t)$ *is a set-theoretic decomposition of* X *into subcontinua of* X.

Proof. Let $t \in [0, \mu(X)]$. Since $\mu^{-1}(t) \subset C(X)$ [see (0.50)], it suffices to show $\mu^{-1}(t)$ covers X and that the members of $\mu^{-1}(t)$ are mutually disjoint. By part (b) of (1.213.1), which is easily proved using results in Section (I.A) of this chapter, $\mu^{-1}(t)$ covers X. To prove the members of $\mu^{-1}(t)$ are mutually disjoint, let $A_1, A_2 \in \mu^{-1}(t)$ such that $A_1 \cap A_2 \neq \emptyset$. Then, $A_1 \cup A_2$ is a subcontinuum of X. Thus, since X is hereditarily indecomposable, $A_1 \subset A_2$ or $A_2 \subset A_1$. Hence, since $\mu(A_1) = \mu(A_2)$, we have using (0.50.a) that $A_1 = A_2$. This completes the proof of

(1.78).

(1.79) NOTATION. Let X be an hereditarily indecomposable continuum and let μ be a Whitney map for $C(X)$. For each $t \in [0, \mu(X)]$, let

$$\eta_t : X \xrightarrow{\text{onto}} \mu^{-1}(t)$$

denote the function η in (1.77) for the case when $M = X$ and $\Gamma = \mu^{-1}(t)$ [by (1.78), this choice for Γ in (1.77) is permissible]. In other words, for each $x \in X$, $\eta_t(x)$ is the unique [by (1.78)] member of $\mu^{-1}(t)$ to which x belongs.

The following result says what Kelley actually showed in his proof of [165, 8.5] rather than what he stated as 8.5. This precise formulation will be used in the proofs of (1.85), (2.2), and (14.1).

(1.80) THEOREM [165, proof of 8.5]. *Let X be an hereditarily indecomposable continuum. Let μ be any given fixed Whitney map for $C(X)$. Then [η_t as in (1.79)]:*

(1.80.1) *For each $t \in [0, \mu(X)]$, $\mu^{-1}(t)$ is a monotone continuous decomposition of X;*

(1.80.2) *For each $t \in [0, \mu(X)]$, η_t is a monotone open mapping of X onto $\mu^{-1}(t)$;*

(1.80.3) *Given any $\varepsilon > 0$, there exists $\delta > 0$ such that if $0 < t < \delta$ then $0 < \text{diam}[\eta_t^{-1}(A)] < \varepsilon$ for each $A \in \mu^{-1}(t)$.*

Proof. Let $t \in [0, \mu(X)]$. Since $\mu : C(X) \xrightarrow{\text{onto}} [0, \mu(X)]$

(1.80)   STRUCTURE OF HYPERSPACES   125

is a mapping and since $C(X)$ is compact [by (0.8)],

(1) $\mu^{-1}(t)$ is a compact subset of $C(X)$.

By (1.78), $\mu^{-1}(t)$ satisfies all the initial hypotheses on Γ in (1.77). Also, by (1), (iii) of (1.77.2) holds with $\Gamma = \mu^{-1}(t)$. Hence, by (1.77.2),

(2)   (iv) and (vi) of (1.77.2) hold with $\Gamma = \mu^{-1}(t)$

  and $\eta = \eta_t$.

By (2) and the fact that each member of $\mu^{-1}(t)$ is a subcontinuum of X we see that (1.80.1) and (1.80.2) each holds [by definition of η_t in (1.79),

(3) $\eta_t^{-1}(A) = A$ for each $A \in \mu^{-1}(t)$

from which we see that η_t is monotone as required in (1.80.2)].
To show (1.80.3) holds, first note that (1.80.3) may be restated as follows [by using (3)]:

(1.80.3')   *Given any* $\varepsilon > 0$, *there exists* $\delta > 0$ *such that if* $0 < t < \delta$ *then* $0 < \text{diam}[A] < \varepsilon$ *for each* $A \in \mu^{-1}(t)$.

We show (1.80.3) holds by showing (1.80.3') holds. Suppose (1.80.3') is false for some $\varepsilon_o > 0$. Then, for each $n = 1,2,\ldots,$ there exists $A_n \in C(X)$ such that

(4) $0 < \mu(A_n) < \frac{1}{n}$

and [note that since $\mu(A_n) \neq 0$, $A_n \notin F_1(X)$ by (0.50.b) and so $\text{diam}[A_n] \neq 0$]

(5) $\text{diam}[A_n] \geq \varepsilon_o$.

By (0.8), the sequence $\{A_n\}_{n=1}^{\infty}$ has a convergent subsequence $\{A_{n(i)}\}_{i=1}^{\infty}$ converging to say $A \in C(X)$. Since μ is continuous [by (0.50)], $\mu(A_{n(i)}) \to \mu(A)$ as $i \to \infty$. Hence, by (4), $\mu(A) = 0$. Therefore, it follows easily using (0.50.a) that $A \in F_1(X)$. But, since $\{A_{n(i)}\}_{i=1}^{\infty}$ converges to A, it follows from (5) that diam$[A] \geq \varepsilon_0 > 0$. Thus, we have a contradiction. This proves (1.80.3') holds and completes the proof of (1.80).

In (1.82) and (1.83) we give two of Kelley's most important results about hereditarily indecomposable continua. We will use the following lemma in the proof of (1.82).

(1.81) LEMMA [165, 7.8]. *Let* $f: M \xrightarrow{onto} Q$ *be a monotone open mapping from any compact metric space* M *onto a finite-dimensional compact metric space* Q. *Then: There exists a compact totally-disconnected subset* M_0 *of* M *such that* $f[M_0] = Q$ *if and only if* $\{q \in Q: f^{-1}(q) \in F_1(M)\}$ *is totally-disconnected.*

I refer the reader to [165, pp.33-34] for a proof of (1.81).

(1.82) THEOREM [165, 8.6]. *Let* X *be an hereditarily indecomposable continuum. Assume that given any* $\varepsilon > 0$, *there exists a monotone open mapping* $f: X \xrightarrow{onto} Y$ *from* X *onto a finite-dimensional continuum* Y *such that* $0 < $ diam$[f^{-1}(y)] < \varepsilon$ *for each* $y \in Y$. *Then,* dim$[X] = 1$.

Proof. Let $\varepsilon > 0$. Let f and Y be as guaranteed. Since $0 < $ diam$[f^{-1}(y)]$ for each $y \in Y$, clearly
$\{y \in Y: f^{-1}(y) \in F_1(X)\} = \emptyset$.

(1.82)

Hence, by (1.81), there exists

(1) a compact totally-disconnected subset X_o of X

such that

(2) $f[X_o] = Y$.

It follows easily from (1) that there exists an open subset U of X such that

(3) $X_o \subset U$

and

(4) the diameter of each component of $cl[U]$ is less than ε.

Now we prove

(*) the diameter of each component of $X \setminus U$ is less than ε.

To prove (*), assume $X \setminus U \neq \emptyset$ and let K be a component of $X \setminus U$. Then, since $K \neq \emptyset$, there is a point $p \in K$. Note that

(5) K is a subcontinuum of X.

Let $L = f^{-1}[f(p)]$. Since f is monotone,

(6) L is a subcontinuum of X.

Since $p \in L$,

(7) $K \cap L \neq \emptyset$.

Thus, by (5) through (7) and the fact that X is hereditarily indecomposable,

(8) $K \subset L$ or $L \subset K$.

Since $L = f^{-1}[f(p)]$, we have by (2) that $L \cap X_o \neq \emptyset$. Since $K \subset [X \setminus U]$, we have by (3) that $K \cap X_o = \emptyset$. Hence, $L \not\subset K$. Thus, by (8),

(9) $K \subset L$.

Since $diam[f^{-1}(y)] < \varepsilon$ for each $y \in Y$ and since $L = f^{-1}[f(p)]$,

(10) $\text{diam}[L] < \varepsilon$.

Therefore, by (9) and (10),

(11) $\text{diam}[K] < \varepsilon$

and we have proved (*) holds. Now, by (*), it follows that there exist a finite number of subsets $A_1, A_2, \ldots, A_m$ of $X\setminus U$ such that

(12) $X\setminus U = \bigcup_{i=1}^{m} A_i$, A_i is compact and $\text{diam}[A_i] < \varepsilon$
for each $i = 1, 2, \ldots, m$, and $A_i \cap A_j = \emptyset$
whenever $i \neq j$.

Similarly, by (4), there exist a finite number of subsets $B_1, B_2, \ldots, B_n$ of $\text{cl}[U]$ such that

(13) $\text{cl}[U] = \bigcup_{i=1}^{n} B_i$, B_i is compact and $\text{diam}[B_i] < \varepsilon$
for each $i = 1, 2, \ldots, n$, and $B_i \cap B_j = \emptyset$
whenever $i \neq j$.

Let $C = \{A_1, A_2, \ldots, A_m, B_1, B_2, \ldots, B_n\}$. From (12) and (13) we have that C is a finite cover of X by compact sets such that

(14) the diameter of each member of C is less than ε

and

(15) if $C' \subset C$ such that $\cap C' \neq \emptyset$, then the cardinality of C' is at most two.

It is easy to prove that there exists $\eta > 0$ such that

$$D = \{N(\eta, A_i) : 1 \leq i \leq m\} \cup \{N(\eta, B_i) : 1 \leq i \leq n\}$$

is a finite cover of X by open subsets of X such that D (in place of C) satisfies (14) and (15). Hence, since $\varepsilon > 0$ was arbitrary, we have by (150, Corollary, p.67] that $\dim[X] = 1$. This proves (1.82).

(1.83) COROLLARY [165, 8.7]. *Let* X *be an hereditarily indecomposable continuum such that* $\dim[X] \geq 2$ *[such continua exist by (1.53)]. Then:*

(1.83.1) *Given any* $\varepsilon > 0$, *there exists a monotone open mapping* $f: X \xrightarrow{onto} Y$ *such that* $0 < \operatorname{diam}[f^{-1}(y)] < \varepsilon$ *for each* $y \in Y$ *and* $\dim[Y] = \infty$;

(1.83.2) *There exists* $\varepsilon > 0$ *such that if* $f: X \xrightarrow{onto} Y$ *is any monotone open mapping satisfying* $0 < \operatorname{diam}[f^{-1}(y)] < \varepsilon$ *for each* $y \in Y$, *then* $\dim[Y] = \infty$;

(1.83.3) *There exist hereditarily indecomposable continua which are infinite-dimensional [see (1.84)].*

Proof. By (1.80),

(1) given any $\varepsilon > 0$ there exists a monotone open mapping $f_\varepsilon: X \xrightarrow{onto} Y_\varepsilon$ such that $0 < \operatorname{diam}[f_\varepsilon^{-1}(y)] < \varepsilon$ for $y \in Y_\varepsilon$.

If, for each $\varepsilon > 0$, f_ε and Y_ε could be chosen as in (1) so that Y_ε is finite-dimensional, then $\dim[X] = 1$ by (1.82). Therefore, since $\dim[X] \geq 2$,

(2) there exists $\varepsilon > 0$ such that for any choice of f_ε and Y_ε as in (1), $\dim[Y_\varepsilon] = \infty$.

By (1) and (2), we have proved (1.83.1) and, by (2), we have proved (1.83.2). By (1.83.1) and the easy-to-prove fact [in (1.207.4)] that the monotone image of an hereditarily indecomposable continuum is hereditarily indecomposable, we see that (1.83.3) holds.

(1.84) REMARK. We have proved the statement in (1.83.3) is true, but Kelley did not prove it as it stands (see the precise way [165, 8.7] reads). What Kelley stated and proved was that *if* there exist hereditarily indecomposable continua of dimension greater than one, *then* there exist such continua which are infinite-dimensional. At the time of Kelley's paper, the only known example of an hereditarily indecomposable continuum was the continuum we now call the pseudo-arc and was one-dimensional (see [165, footnote, p.34]). Ten years later, Bing [29, Theorem 4] showed for the first time that there exist infinite-dimensional hereditarily indecomposable continua. Bing constructed these continua in the Hilbert cube, whereas the proof given above for (1.83.3) used (1.53).

The following result is an application to (1.83) of what Kelley proved in [165, 8.5], stated above as (1.80). It gives information, as does (1.86), about $\mu^{-1}(t)$ which has not as yet been noted in the literature. The general theory of the "levels" $\mu^{-1}(t)$ is discussed in Chapter XIV.

Dimension of $C(X)$ will be discussed in Chapter II; in particular, in relation to the next result, see (2.2).

(1.85) THEOREM. *Let X be an hereditarily indecomposable continuum such that $\dim[X] \geq 2$. Let μ be any given fixed Whitney map for $C(X)$. Then, there exists $t_o > 0$ such that $\dim[\mu^{-1}(t)] = \infty$ whenever $0 < t < t_o$. Hence, $\dim[C(X)] = \infty$.*

Proof. Choose $\varepsilon > 0$ as in (1.83.2). Choose $\delta > 0$ so as to satisfy (1.80.3) for ε. Then, by (1.80.2) and (1.80.3),

$$\eta_t: X \xrightarrow{\text{onto}} \mu^{-1}(t) \text{ whenever } 0 < t < \delta$$

satisfies all the hypotheses on f in (1.83.2). Therefore, by (1.83.2), $\dim[\mu^{-1}(t)] = \infty$ whenever $0 < t < \delta$. By taking $t_0 = \delta$, we have proved (1.85).

The following corollary implies a result of Rogers [see (2.3)].

(1.86) COROLLARY. *Let* Y *be any continuum such that* $3 \leq \dim[Y] < \infty$. *Let* μ *be any given fixed Whitney map for* $C(Y)$. *Then, there exists* $t_0 > 0$ *such that* $\dim[\mu^{-1}(t)] = \infty$ *whenever* $0 < t < t_0$. *Hence*, $\dim[C(X)] = \infty$.

Proof. By (1.53), Y contains an hereditarily indecomposable continuum X such that $\dim[X] \geq 2$. Let μ_X denote the restriction of μ to $C(X)$. Clearly, μ_X is a Whitney map for $C(X)$ and therefore, by (1.85), there exists $t_0 > 0$ such that $\dim[\mu_X^{-1}(t)] = \infty$ whenever $0 < t < t_0$. Since $\mu_X^{-1}(t) \subset \mu^{-1}(t)$ for any $t < t_0$, the corollary is proved.

I do not know the answer to the following question.

(1.87) QUESTION. Does the first part of (1.86) remain valid if the assumption that $\dim[Y] < \infty$ is deleted?

This completes the discussion for now about the structure of $C(X)$ when X is an indecomposable continuum. There will be many more results in the book about hyperspaces of indecomposable continua. I call the reader's attention to three such results--(1.118), (1.122), and (1.124)--which are included in the next section of Chapter I for the purpose of contrast and motivation concerning some results about locally connected continua.

(I.D) The Structure of 2^X and $C(X)$ when X Is Locally Connected

The structure of the hyperspaces of locally connected continua is now fairly well understood. The global structure is completely known for 2^X [see (1.97)] and, to a large extent, for $C(X)$ [see (1.98)]. The dimension of $C(X)$ when X is locally connected [comp., Chapter II] has been completely determined [see (1.109)]. Also, a lot of work on the polyhedral structure of $C(X)$ when X is a finite graph has been done by Duda [two of his results mentioned below are in (1.115) and (1.116)]. In this part of Chapter I, we discuss these and related topics. Let us recall the following definitions.

(1.88) DEFINITIONS. Let X be a continuum and let $p \in X$. Then: X is said to be

(1.88.1) *locally connected at* p provided that given an open subset U of X such that $p \in U$, then there exists a connected open subset V of X such that $p \in V \subset U$;

(1.88.2) *connected im kleinen at* p provided that given an

open subset U of X such that p ε U, then there exists an open subset V of X such that p ε V ⊂ U and such that if x ε V then there is a connected subset Z of U such that p ε Z and x ε Z.

(1.89) REMARK. Clearly, a continuum X is connected im kleinen at a point p ε X if and only if given any open subset U of X such that p ε U, then there exists a subcontinuum K of X such that there is an open subset W of X with p ε W ⊂ K ⊂ U. Also, it is clear that if a continuum is locally connected at a point p, then it is connected im kleinen at p; the converse is false (see the example in [147, Figure 3-9, p.113]; also, see (1.91) here). Sometimes (see, for example, [338, p.18]) the condition in (1.88.2) is called "locally connected at p," but this is not the case in this book.

(1.90) DEFINITION. A continuum is said to be *locally connected* provided that it is locally connected at each of its points.

(1.91) REMARK. It is easy to prove that the conditions in (1.88.1) and (1.88.2) are equivalent globally, i.e., a continuum is locally connected if and only if it is connected im kleinen at each of its points.

The first results about hyperspaces of locally connected continua are due to Vietoris [320] and Wazewski [330]. We now state their results as one theorem [comp., (14.73.32)].

(1.92) THEOREM *(that (1.92.1) implies each of the other two in 320 ; the reverse implications are in [320]). The following three statements are equivalent:*

(1.92.1) *x is a locally connected continuum;*

(1.92.2) *C(X) is a locally connected continuum;*

(1.92.3) 2^X *is a locally connected continuum.*

The proof of (1.92) is not difficult and is left to the reader [see (1.208.1)]. A related result is in (1.208.2).

The next result obtained about the hyperspaces of locally connected continua is the following one. Its analogue for locally connected Hausdorff continua is false, even in simple cases--see (16.41).

(1.93) THEOREM [343]. *If X is a locally connected continuum, then 2^X and C(X) are each locally contractible and contractible (in themselves).*

Wojdyslawski [343] does not state (1.93) for the case of C(X), but his proof may be applied to show the C(X) part. It is appropriate to call to the reader's attention the result of Kelley in (16.7) which shows 2^X is contractible if and only if C(X) is contractible. In Chapter XVI, we give Kelley's proof [165] of (1.93). His proof is simpler than the one given by Wojdyslawski.

In (0.65.4), another proof of (1.93) was indicated.

(1.94) REMARK. One of the important aspects of [343] was that the following question appeared in print for the first time: If X is any locally connected continuum, then is 2^X homeomorphic to the Hilbert cube [343, p.248]? This question remained unanswered until very recently ([76, Theorem 1]; see (1.97) below). In fact--the special case, when $X = [0, 1]$, was only recently answered ([290]; for a complete proof, see [291]). Professor Kuratowski has told me that this special case, as well as the analogous question for $C(B^2)$, was worked on by Polish mathematicians in the 1920's.

In the light of Wojdyslawski's question, stated in (1.94), it is appropriate to note that Mazurkiewicz [216] had proved the following result in 1931 (later, unaware of [216], Kelley [165, 5.1] obtained the same result).

(1.95) THEOREM [216]. *For any continuum* X, 2^X *contains a Hilbert cube. Hence,* $\dim[2^X] = \infty$.

Proof. Let X be a continuum. Let $p \in X$. First we show

(*) there exist nondegenerate subcontinua

$A_1, A_2, \ldots, A_i, \ldots$ of X such that $A_j \cap A_k = \emptyset$ whenever $j \neq k$, $p \notin [\cup_{i=1}^{\infty} A_i]$, and $A_i \to \{p\}$.

To prove (*), let $\{x_i\}_{i=1}^{\infty}$ be a sequence of distinct points of $X \setminus \{p\}$. Then, a simple induction shows that there exist open subsets $U_1, U_2, \ldots, U_i, \ldots$ such that $x_i \in U_i$ for each

$i = 1,2,\ldots,$ $\text{cl}[U_j] \cap \text{cl}[U_k] = \emptyset$ whenever $j \neq k$, $p \notin \text{cl}[U_i]$ for any $i = 1,2,\ldots,$ and $\text{diam}[U_i] \leq 1/i$ for each $i = 1,2,\ldots$. For each $i = 1,2,\ldots,$ let A_i be the component of $\text{cl}[U_i]$ such that $x_i \in A_i$. Note that, by [186, Theorem 1, p.172], A_i is nondegenerate for each $i = 1,2,\ldots$. It is easy to prove that the sets $A_1, A_2, \ldots, A_i, \ldots$ just defined satisfy the other properties in (*). Now, for each $i = 1,2,\ldots,$ let $\alpha_i \subset 2^{A_i}$ be an arc [such arcs exist by (1.9) since each A_i is a *nondegenerate* subcontinuum]. For each $i = 1,2,\ldots,$ let $h_i: [0, 2^{-i}] \xrightarrow{\text{onto}} \alpha_i$ be a homeomorphism. Define $h: I_\infty \to 2^X$ by [see (0.40) for definition of I_∞]

$$h((t_1, t_2, \ldots, t_i, \ldots)) = \{p\} \cup [\bigcup_{i=1}^{\infty} h_i(t_i)]$$

for each $(t_1, t_2, \ldots, t_i, \ldots) \in I_\infty$. From the fact that $A_i \to \{p\}$ it follows easily that $h((t_1, t_2, \ldots, t_i, \ldots)) \in 2^X$ for each $(t_1, t_2, \ldots, t_i, \ldots) \in I_\infty$. It is easy to prove, using the properties in (*), that h is an embedding of I_∞ into 2^X. This completes the proof of (1.95).

As a partial answer to his question stated above in (1.94), Wojdyslawski [342] obtained the following result.

(1.96) THEOREM [342, pp.190-191]. *The following three statements are equivalent:*

(1.96.1) X *is a locally connected continuum;*

(1.96.2) 2^X *is an absolute retract;*

(1.96.3) C(X) *is an absolute retract.*

Wojdyslawski [342, p.191] also noted some other results--for example, (1.96.1) is equivalent to $C_n(X)$ being an absolute retract ($C_n(X)$, defined here in (0.48), was denoted by $(2^X)_n$ in [342]). We also remark that Neil Gray [131] found and corrected an error in Wojdyslawski's proof of (1.96). Kelley [165, pp.27-29] gave another proof of (1.96).

After Wojdyslawski's result (1.96), no partial answers were obtained to the question in (1.94) until 1967 Then, Neil Gray showed that if X is a finite simplicial complex, C(X) consists entirely of unstable points if and only if X contains no free one-simplex [133]. In 1969, Gray proved that if X is a locally connected continuum, then every point of 2^X is unstable [132]. In 1972, Jim West [333] proved that $C(D) \times I_\infty \cong I_\infty$ if D is any dendrite and that $C(D) \cong I_\infty$ if D is a dendrite whose branch points are dense in D. Also in 1972, Schori and West [290] announced and sketched a proof for the theorem $2^{[0,1]} \cong I_\infty$; a complete proof of this result appeared in [291]. Some extensions and related results were obtained in [291] and [78]. Finally, in 1974, Curtis and Schori [76] announced and sketched proofs for the following two important theorems.

(1.97) THEOREM [76, Theorem 1]. *If* X *is a locally connected continuum, then* $2^X \cong I_\infty$ *(and conversely).*

Proof. See (1.214.2); also, see the paragraph following (1.99).

(1.98) **THEOREM** [76, Theorem 2]. *If* X *is a locally connected continuum, then* $C(X) \times I_\infty \cong I_\infty$ *(and conversely). If* X *is a locally connected continuum which contains no free arc [see (0.18) for definition], then* $C(X) \cong I_\infty$ *(and conversely).*

Proof. See (1.214.2); also, see the paragraph following (1.99).

Curtis and Schori [76] also obtained the following result about "relative hyperspaces."

(1.99) **THEOREM** [76, Theorem 3]. *Let* X *be a locally connected continuum.*

(1.99.1) *If* $X \neq A \in 2^X$, *then* $\{B \in 2^X : A \subset B\} \cong I_\infty$.

(1.99.2) *If* $A \in C(X)$, *then* $\{B \in C(X) : A \subset B\} \times I_\infty \cong I_\infty$.

(1.99.3) *If* $X \neq A \in C(X)$ *and if* $X \setminus A$ *contains no free arc, then* $\{B \in C(X) : A \subset B\} \cong I_\infty$.

Proof. See (1.214.3); also, see the paragraph below.

Obtaining (1.97) through (1.99) represents the culmination and use of much mathematics, both inside and outside hyperspace theory. The original (complete) proofs of (1.97) through (1.99) will soon appear in [77]. These proofs rest heavily on near-homeomorphism techniques, inverse limits [see (1.199)], and techniques from infinite-dimensional topology. Discussions and outlines of these proofs and techniques are in [76], [289], [290], [331], and [445]. Recently, Toruńczyk has obtained an easy to use characterization of the

Hilbert cube--see (1.214.2)--and has shown how to use his characterization to prove (1.97) and the second part of (1.98). These proofs, based on Toruńczyk's characterization, have been left as an exercise in (1.214.2). In (1.214.3), I have left as an exercise a proof of (1.99) based on Toruńczyk's characterization. Though Toruńczyk's proof of his characterization uses sophisticated results from infinite-dimensional topology, the proofs requested in (1.214.2) and (1.214.3) avoid computations with inverse limits as done in [77]. Thus, (1.97) through (1.99) can now be understood from essentially one general theorem in infinite-dimensional topology-- Toruńczyk's beautiful characterization of the Hilbert cube. Some other applications of Toruńczyk's theorem to hyperspaces are in [438] and [439].

I leave the discussion of (1.97) through (1.99) with the hope that I have given the reader enough interest so that he will pursue the details on his own.

Locating n-cells and Hilbert cubes in hyperspaces is an important ingredient in understanding the structure of hyperspaces. Even for the case of 1-cells [i.e., arcs], there is ample evidence in (I.A), (I.B), and (I.C) above to support this statement. For historical remarks concerning who first found arcs in hyperspaces, see (1.10) and (1.14). The first person to find cells of dimension greater than one in hyperspaces was Mazurkiewicz, who actually found a Hilbert cube in every 2^X [(1.95) above]. The first person to find any cells, other than arcs, in C(X) was Kelley [165, proof of 5.3]. He showed that under certain conditions imposed on a locally connected continuum X, C(X) contains an n-cell [for some n which depends on the conditions--see (1.107)]. Later, Duda made significant and detailed use of n-cells in C(X) when X is a finite graph ([85, esp. section 6] and [86]; also [87]; here, see (0.59), (1.113), (1.115), (1.116), and (10.1)).

We will now discuss some results of Kelley and Duda, and some related results alluded to in the previous paragraph.

We begin with three general theorems for C(X) when X is not necessarily locally connected. They are (1.100), (1.102), and (1.103). We then apply (1.102) and (1.103) to the locally connected case. All of Kelley's results in [165, section 5] about C(X) when X is locally connected are discussed, and one of them is strengthened [see (1.111)].

The proof of the following theorem of Rogers is similar to the proof Kelley gave for [165, 5.3]. It is not known if the converse of the theorem is true--see (1.147).

(1.100) THEOREM [283, Theorem 1]. *If* X *is a continuum which contains an n-odd* [*see (0.21) for definition*], *then* C(X) *contains an n-cell* [*comp., (1.102)*].

Proof. Let X be a continuum which contains an n-odd M. Then, by definition (0.21), there is a subcontinuum N of M such that M\N is the union of n nonempty mutually separated sets $S_1, S_2, \ldots, S_n$. Let $B_i = S_i \cup N$ for each $i = 1, 2, \ldots, n$. Each B_i is a subcontinuum of X. Hence, by (1.25), there exists a segment $\sigma_i : [0, 1] \to 2^X$ from N to B_i for each $i = 1, 2, \ldots, n$.

(1.102)   STRUCTURE OF HYPERSPACES   141

Note that, by (1.26), $\sigma_i([0, 1]) \subset C(X)$ for each $i = 1, 2, \ldots, n$. Thus, since $N \in C(X)$ and $N \subset \sigma_i(t)$ for each $i = 1, 2, \ldots$, and each $t \in [0, 1]$,

(1) $[\cup_{i=1}^{n} \sigma_i(t_i)] \in C(X)$ for any choices of $t_i \in [0, 1]$.

Let I^n denote the cartesian product of n copies of $[0, 1]$. Let

$$h((t_1, t_2, \ldots, t_n)) = \cup_{i=1}^{n} \sigma_i(t_i) \text{ for each}$$

$(t_1, t_2, \ldots, t_n) \in I^n$.

Then we see, by (1), that h is a function from the n-cell I^n into $C(X)$. By the continuity of union [see (1.48)] and of each σ_i, we see that h is continuous. It follows easily that h is one-to-one. This completes the proof of (1.100).

(1.101) REMARK. For results related to (1.100), see the next two results, (1.210.1), and (14.33).

(1.102) COROLLARY [283, proof of Theorem 1]. *If* M *is an n-odd with* N *as in (0.21), then there exists an n-cell* $\Delta^n \subset C(M)$ *such that* $N \in \Delta^n$ *and such that* $K \supset N$ *for each* $K \in \Delta^n$.

Proof. The corollary is a simple consequence of the proof of (1.100).

The next result is an infinite-dimensional analogue of (1.100) and (1.102). It is not known if the converse (of the first part) is true--see (1.148).

(1.103) THEOREM [244, Theorem 6]. *Let* X *be a continuum which contains subcontinua* $Y, Y_1, Y_2, \ldots, Y_i, \ldots$ *such that*

(1) $Y_i \cap Y \neq \emptyset$ *and* $Y_i \not\subset Y$ *for any* $i = 1,2,\ldots;$

(2) $[Y_j \backslash Y] \cap [Y_k \backslash Y] = \emptyset$ *whenever* $j \neq k;$

(3) $\text{diam}[Y_i] \to 0$ *as* $i \to \infty$.

Then, $C(X)$ *contains a Hilbert cube which can be "located" as follows: For each* $i = 1,2,\ldots,$ *let* $\sigma_i: I_i = [0,1] \to C(X)$ *be a segment from* Y *to* $Y \cup Y_i$ *[such segments exist by (1.25) and (1.26)]. Then, if* $h: \Pi_{i=1}^{\infty} I_i \to C(X)$ *is defined by*

$$h((t_1, t_2, \ldots, t_i, \ldots)) = \bigcup_{i=1}^{\infty} \sigma_i(t_i)$$

for each $(t_1, t_2, \ldots, t_i, \ldots) \in \Pi_{i=1}^{\infty} I_i$, h *is a homeomorphism.*

A proof of (1.103) may be found in [244, p.243].

We now apply (1.102) and (1.103) to locally connected continua. First, let us note the following definition and classical result of Menger.

(1.104) DEFINITION [see (1.105)]. Let X be a locally connected continuum and let A be a subcontinuum of X. Let

$I^*(A) = \{$integers n: given any $\varepsilon > 0$, there exists an open subset U of X such that $A \subset U \subset N(\varepsilon, A)$ and $\text{Bd}[U]$ consists of at most n points$\}$.

Then: The *order of* A *in* X, written $\text{ord}_A[X]$, is defined by

$$\text{ord}_A[X] = \begin{cases} \min.[I^*(A)] & \text{, if } I^*(A) \neq \emptyset \\ \infty & \text{, if } I^*(A) = \emptyset \end{cases}.$$

(1.105) REMARK. The notion of order of a point in X goes back to early papers of Menger and Urysohn. The definition in (1.104) is as in [165, pp.29-30]. Let us note the following simple fact:

(1.105.1) A subcontinuum A of a locally connected continuum X is of order n in X ($n = 0,1,2,\ldots,$) if and only if A, as a point of the decomposition space $X/_A$ obtained by "shrinking A to a point," is of order n in $X/_A$.

(1.106) LEMMA (Menger; see [186, p.277]). *Let Y be a locally connected continuum and let $p \in Y$. If $\text{ord}_{\{p\}}[Y] \geq n < \infty$, then there exist n arcs $A_1, A_2, \ldots A_n$ in Y such that p is an end point of A_i for each $i = 1,2,\ldots,n$, and $A_j \cap A_k = \{p\}$ whenever $j \neq k$.*

(1.107) THEOREM [165, 5.3 and its proof]. *Let X be a locally connected continuum and let $A \in C(X)$. Then:*

(1.107.1) $\text{ord}_A[X] \leq \dim_A[C(X)]$;

(1.107.2) *if $\text{ord}_A[X] \geq n < \infty$, then there exists an n-cell $\Delta^n \subset C(X)$ such that $A \in \Delta^n$ and such that $K \supset A$ for each $K \in \Delta^n$.*

Proof. The theorem follows easily by using (1.106) and (1.105.1), and then applying (1.102).

Recall the following definition.

(1.108) DEFINITION. A continuum X is said to be a *finite graph* [or, *linear graph*] provided X can be written as the union of finitely many arcs any two of which are either disjoint or intersect only in one or both of their end points.

The next result is the combination of two of Kelley's theorems [165, 5.4 and 5.5]. It completely determines the dimension of C(X) when X is any locally connected continuum. From an historical point of view it was important as the first result which completely determined the dimension of C(X) for a general class of continua [comp., (1.95) and Chapter II].

(1.109) THEOREM [165, 5.4 and 5.5]. *If X is a locally connected continuum, then* $\dim[C(X)] < \infty$ *if and only if X is a finite graph* [*comp., (1.111)*]. *Furthermore, if X is a finite graph,*
$$\dim[C(X)] = \sup\{\text{ord}_A[X] : A \in C(X)\} = 2 + \sum_v (\text{ord}_{\{v\}}[X] - 2)$$
where the summation $\sum_v$ *runs over all points* $v \in X$ *such that* $\text{ord}_{\{v\}}[X] \geq 3$.

Proof. We only outline the proof of the first part of the theorem [see (1.110)]. First, Kelley proved the following:

(1.109.1) Let X be a locally connected continuum. Then $\text{ord}_A[X] < \infty$ for each $A \in C(X)$ if and only if X is a finite graph [165, 5.2].

The reader is referred to [165, p.30] for a proof of (1.109.1). Now, assume X is a locally connected continuum such that $\dim[C(X)] < \infty$. Then, by (1.107.1), $\text{ord}_A[X] < \infty$ for each

(1.110)   STRUCTURE OF HYPERSPACES   145

A ε C(X). Hence, by (1.109.1), X is a finite graph. Conversely, a simple induction shows that if X is a finite graph, then dim[C(X)] < ∞. This last fact is also a consequence of the "furthermore part" of the theorem [see (1.110)].

(1.110) REMARK. Kelley's proof [165, pp.30-31] for the validity of the formula in (1.109) is not correct (see the Remarks in [85, p.278]). I refer the reader to [85] for a correct proof.

This completes the presentation of Kelley's results about C(X) when X is locally connected, except to mention that Kelley observed that C(X) is a polyhedron when X is a finite graph [165, p.31]; see (1.113) below). Also, Kelley [165, section 6] used some hyperspace techniques to prove a theorem, originally due to Whyburn, about sequences of nonlocal separating points of locally connected continua.

In (1.111) we give a result which strengthens the first part of (1.109). First, we make some observations. As a simple consequence of (1.107.1) and (1.109.1), we see that [165, p.31]: If X is a locally connected continuum such that $\dim_A [C(X)] < \infty$ for each A ε C(X), then X is a finite graph and hence dim[C(X)] < ∞. Stated another way: If X is a locally connected continuum such that dim[C(X)] = ∞ [equivalently, such that X is not a finite graph], then there exists A_o ε C(X) such that $\dim_{A_o} [C(X)] = \infty$ [comp., (1.112)]. However, this does not prove the following stronger fact.

(1.111) THEOREM [244, Corollary 1]. *Let* X *be a locally connected continuum. Then* C(X) *contains a Hilbert cube if and only if* X *is not a finite graph [equivalently, if and only if* $\dim[C(X)] = \infty$].

Proof. It can be shown (I refer the reader to [244, pp.245-246] for the details) that if X is a locally connected continuum which is not a finite graph, then X contains subcontinua Y, Y_1, Y_2,..., Y_i,... which satisfy (1) through (3) of (1.103). Hence, by (1.103), C(X) contains a Hilbert cube. The other half of (1.111) is the "easy part" of (1.109).

(1.112) REMARK. There are continua Y such that C(Y) has the following properties:

(1) C(Y) contains an n-cell for each n = 1,2,... and, hence, $\dim[C(Y)] = \infty$;

(2) $\dim_A[C(Y)] < \infty$ for each A ε C(Y) and, hence, C(Y) does not contain a Hilbert cube.

An example of such a continuum, due to B. J. Ball, is in [244, Example 3].

Now, we turn our attention briefly to Duda's work in [85], [86], and [87].

In [85], Duda investigates the polyhedral structure of C(X) when X is any finite graph. Here, I mention only three of the many interesting and detailed facts he obtains in [85]; two other results from [85] are stated in (0.59) and (10.1).

Duda ([85, section 6] and [86]) obtains a specific decomposition of C(X) into balls when X is any finite graph. This decomposition is used to prove many results. In particular, he verifies a statement in Kelley's paper [165, p.31] by proving the following basic fact:

(1.113) THEOREM (stated in [165, p.31] and proved in [85, 6.4]). *The hyperspace* C(X) *is a (connected) polyhedron if and only if* X *is a finite graph.*

(1.114) NOTATION [85, p.280]. For a polyhedron P, let
$E_P = cl[\{p \in P:$ there exists a closed neighborhood V of p in P such that $V \cong B^2$ and p is a point in the manifold boundary of V$\}]$.

Duda obtains the following result on recovering X from C(X).

(1.115) THEOREM [85, 8.4]. *If* X *is a finite graph which is not an arc, then* $E_{C(X)} \cong X$.

Furthermore, Duda proves the following theorem.

(1.116) THEOREM [85, 9.2]. *Let* P *be a (connected) finite-dimensional polyhedron. Then:* P *is a hyperspace if and only if* E_P *is homeomorphic to a finite graph and* $P \cong C(E_P)$.

As Duda remarked in [85, p.284], he did not determine in an intrinsic way the structure of those polyhedra which are hyperspaces [comp., (1.116)]. In [87], he does such a determination for the special case of polyhedra which are hyperspaces of *acyclic* finite graphs. I refer the reader to [87] for an understanding of and for the statement of the Characterization Theorem [87, p.247]. As far as I know, the following question (implicit in Duda's work) is as yet unanswered.

(1.117) QUESTION ([85] and [87]). What is the structure of those polyhedra which are homeomorphic to $C(X)$ for non-acyclic finite graphs?

We note in passing that Duda also has two papers about convexity of hyperspaces of continua with a convex metric ([83] and [84]; for related work, see [416], [422], and [430]).

The next two results I give about hyperspaces of locally connected continua are (1.120) and (1.127). They are best motivated by comparison with the situation for hereditarily indecomposable continua.

In a conversation, B. J. Ball asked me the following question: If Z is an hereditarily indecomposable continuum, then can Cone(Z) be embedded in $C(Z)$? I answered this question by proving the following result (later, a somewhat weaker result was obtained in [281, Theorem 9]--see (1.207.7)).

(1.118) THEOREM [244, Theorem 3]. *A continuum* Z *is hereditarily indecomposable if and only if* C(Z) *does not (topologically) contain the cone over any continuum [for a stronger result, see (1.122)]*.

Thus, I became interested in the following general question (which, until now, has not been stated in the literature).

(1.119) QUESTION. For what classes of continua Y is it true that Cone(Y) is embeddable in C(Y) [see (1.121)]? We remark that by letting Y be the join at a point of two (otherwise disjoint) hereditarily indecomposable continua, we obtain a decomposable continuum Y such that Cone(Y) is not embeddable in C(Y); for a proof, see [244, Example 1].

One of the reasons I proved (1.111) in [244] was to obtain the following partial answer to (1.119).

(1.120) THEOREM [244, Theorem 8]. *If* Y *is any locally connected continuum, then* Cone(Y) *is embeddable in* C(Y).

Proof. Let Y be a locally connected continuum. By (1.111), we assume for the purpose of proof that Y is a finite graph. Then, by [150, p.56], Y is embeddable in R^3. Hence, Cone(Y) is embeddable in R^4. Thus, if C(Y) contains a 4-cell, we are done. So, for the purpose of proof, assume C(Y) does not contain a 4-cell. Then, by (1.107.2), $\text{ord}_A[A] \leq 3$ for each $A \in C(Y)$.

Hence, by a direct argument [or, by using the formula in (1.109)], it follows easily that Y has at most one point of order greater than two and that point, if it exists, is of order three. Therefore, Y must be an arc Y_1, a circle Y_2, a simple triod Y_3, or a *circle-with-a-sticker* Y_4, i.e.,

$$Y_4 \cong S^1 \cup \{(x, 0) \in R^2 : 1 \leq x \leq 2\}.$$

For the case when $Y = Y_1$ or $Y = Y_2$, $C(Y) \cong \text{Cone}(Y)$ by (0.54) and (0.55). For the case when $Y = Y_3$ or $Y = Y_4$, $C(Y)$ contains a 3-cell and Cone(Y) is embeddable in it since Y is embeddable in the plane. This completes the proof of (1.120).

The proof given in [244, p.247] for (1.120) was a little different than the one just given. It used the Kuratowski Graph Theorem [186, pp.305-306] instead of (1.107.2).

(1.121) REMARK. The only results I know of which provide answers to (1.119) are (1.118), (1.120), [244, Example 1], and those results concerning continua X whose cone and hyperspace C(X) are actually homeomorphic. This last topic will be discussed in depth in Chapter VIII.

The proof in [244] of (1.118) did not show the following stronger result, which answers a question raised in [244, Problem 1, p.241].

(1.122) THEOREM [176, 3.4]. *A continuum* Z *is hereditarily indecomposable if and only if* C(Z) *does not (topologically) contain the cartesian product of any continuum with an arc.*

The following question was asked in [244, Problem 1, p.241].

(1.123) QUESTION [244, Problem 1]. Can C(Z) ever (topologically) contain the cartesian product of two continua when Z is an hereditarily indecomposable continuum [see (1.125)]?

In his dissertation, Norman Passmore gave the following partial answer to (1.123). The result is new and of interest when Z is the pseudo-arc itself, any pseudo-soleniod, or any of Cook's continua [69].

(1.124) THEOREM [263, Theorem 5, p.69]. *If Z is an hereditarily indecomposable continuum such that each nondegenerate proper subcontinuum of Z is a pseudo-arc, then* C(Z) *does not (topologically) contain the cartesian product of any two continua.*

(1.125) REMARK. As far as I know, (1.122) and (1.124) are the only answers that have been obtained to (1.123).

Shortly after [244] was written, I became interested in the following general question.

(1.126) QUESTION. For what classes of continua Y is it true that Y × Y is embeddable in C(Y) [see (1.128)]?

This brings us to the next result about locally connected continua. It determines the class of locally connected continua Y such that $Y \times Y$ is embeddable in $C(Y)$.

(1.127) THEOREM [263, Theorem 4, p.17]. *Let Y be a locally connected continuum. Then: $Y \times Y$ is embeddable in $C(Y)$ if and only if Y is not a circle, simple triod, or circle-with-a-sticker [for definition, see Y_4 in proof of (1.120)].*

(1.128) REMARK. The questions in (1.123) and (1.126) were the basis of Norman Passmore's doctoral dissertation [263]. Besides proving the results that have been stated in (1.124) and (1.127), Passmore [263] also showed $Y \times Y$ is not embeddable in $C(Y)$ when Y is the "Buckethandle" [defined in (1.209.3)], when Y is any solenoid, and when Y is any compactification of $[0, +\infty)$ with an arc as remainder. There are no results, except the ones just mentioned and results in Chapter X, which provide answers to (1.126). In particular, answers to the following two questions are not known.

(1.129) QUESTION. Is an arc the only chainable continuum Y such that $Y \times Y$ is embeddable in $C(Y)$ [see (1.128) and, for example, (10.10)].

(1.130) QUESTION. Is there a circle-like continuum Y such that $Y \times Y$ is embeddable in $C(Y)$ [see (1.128); also, by (10.10) $C(Y) \not\approx Y \times Y$ for any circle-like continuum Y].

For the purposes of this chapter, we now end the discussion of hyperspaces of locally connected continua. Other results about this topic appear throughout the book [for example: (5.9), (6.12), (10.1), (10.20), (14.9), and (15.11)].

Goodykoontz has investigated points where hyperspaces are locally connected or connected im kleinen [see (1.88)]. We state some of his results below, and refer the reader to [123], [124], [126], and [127] for the others.

(1.131) THEOREM [124, Theorem 1]. *Let* $M \in C(X)$. *Then:* 2^X *is connected im kleinen at* M *if and only if for each open subset* U *of* X *such that* $M \subset U$, *the component of* U *containing* M *contains* M *in its interior (relative to* X).

(1.132) THEOREM [124, Theorem 2]. *Let* $M \in C(X)$. *Then:* 2^X *is locally connected at* M *if and only if for each open subset* U *of* X *such that* $M \subset U$, *there exists a connected open subset* V *of* X *such that* $M \subset V \subset U$.

The following result improves [124, Theorem 3] and answers Question 1 of [124, p.391].

(1.133) THEOREM [126, Theorem 1]. *Let* $M \in C(X)$. *If* 2^X *is connected im kleinen at* M, *then* $C(X)$ *is locally arcwise connected at* M [*i.e.*, M *belongs to arbitrarily small arcwise connected open subsets of* $C(X)$].

(1.134) THEOREM [124, Theorem 4 and Theorem 5]. *Let* A ε C(X). *Then:* 2^X *is connected im kleinen [resp., locally connected] at* A *if and only if* 2^X *is connected im kleinen [resp., locally connected] at each component of* A.

(1.135) THEOREM [124, Corollary 6]. *Let* A ε 2^X. *If* X *is connected im kleinen [resp., locally connected] at each point of* A, *then* 2^X *is connected im kleinen [resp., locally connected] at* A.

In relation to Koch's Arc Theorem [171, p.726], McWaters proved [203, p.1210] that the zero has an arcwise connected base. This fact, together with what McWaters showed about hyperspaces of Hausdorff continua [203, p.1209], gives the next theorem. It was proved and used in the metric case in [93, p.1032]; later, also in the metric case, in [123, Lemma 2]. We state it here for Hausdorff continua since it will be used in that form in (17.1).

(1.136) THEOREM [see paragraph above]. *For any Hausdorff continuum* Y, 2^Y *and* C(Y) *are each locally arcwise connected at* Y *[with the Vietoris topology, (0.12), and generalized arcs, (0.18)]*.

(1.137) REMARK. By (1.136), the converses of (1.135) are false as is noted in [124, p.397].

Though the following result is proved in [123] for Hausdorff continua, it was new even for the case of metric continua. For an application of (1.138), see the proof of (3.9). For other results about aposyndesis, see [123] and (14.19).

(1.138) THEOREM [123, Theorem 1]. *For any Hausdorff continuum X, 2^X and C(X) are each aposyndetic [with the Vietoris topology (0.12)].*

(1.139) THEOREM [126, Theorem 4]. *If X is an indecomposable continuum, then X is the only point at which 2^X is connected im kleinen.*

(1.140) REMARK. The converse of (1.139) is false [126, Example 1]. However, the following result holds.

(1.141) THEOREM [126, Theorem 5]. *If X is the only point at which C(X) is connected im kleinen, then X is indecomposable.*

(1.142) REMARK. Goodykoontz [126, Example 2] gives an example of an indecomposable continuum such that C(X) is connected im kleinen at points other than X.

The following theorem characterizes points of C(X) at which C(X) is connected im kleinen. A strong sufficient condition for connectedness im kleinen at a point of C(X) was given in [280, Theorem 6].

(1.143) THEOREM [126, Theorem 2]. *Let* $M \in C(X)$. *Then:* $C(X)$ *is connected im kleinen at* M *if and only if for each open subset* U *of* X *such that* $M \subset U$, *each sequence of subcontinua of* U *which converges to* M *is eventually contained in the component of* U *containing* M.

Goodykoontz has told me of the following as yet unanswered question [it does not appear in print until now].

(1.144) QUESTION [Jack T. Goodykoontz, Jr.]. What are necessary and sufficient conditions for $C(X)$ to be locally connected at $M \in C(X)$?

As we have seen, it is advantageous in studying the structure of hyperspaces to try to locate n-cells or Hilbert cubes in them if at all possible. We have also seen [in (1.61)] that the only such cells which exist in $C(X)$, when X is an hereditarily indecomposable continuum, are arcs. The next result shows that this statement about $C(X)$ is *only* true when X is hereditarily indecomposable. It is a result which will be of some use throughout the book [for example, see the proof of (2.1)]. It is included here because of its connection with results and techniques in this section, and because of its contrast with some results in the previous section. A related result is in (1.210.1).

(1.145) THEOREM [244, Theorem 1]. *If* D *is a decomposable continuum, then* C(D) *contains a 2-cell. Hence: If a continuum* X *contains a decomposable subcontinuum, then* C(X) *contains a 2-cell.*

Proof. Assume D is a decomposable continuum. To prove C(D) contains a 2-cell, it suffices by (1.100) to prove that

(*) D contains a 2-odd [see (0.21) for definition].

To prove (*), let A and B be proper subcontinua of D such that D = A ∪ B. If A ∩ B is connected, then D itself is a 2-odd [with N = A ∩ B in (0.21)]. Thus, for the purpose of proof, assume that A ∩ B is not connected. Let K_1 and K_2 be two distinct components of A ∩ B. Let W_1 and W_2 satisfy (1) through (3) below:

(1) W_i is an open subset of A for each i = 1 and 2;

(2) $K_i \subset W_i$ for each i = 1 and 2;

(3) $cl[W_1] \cap cl[W_2] = \emptyset$.

For each i = 1 and 2, let F_i denote the boundary of W_i in A; note that by (1), (0.42), and compactness of A,

(4) $F_i = cl[W_i] \setminus W_i$ for each i = 1 and 2.

For each i = 1 and 2, let M_i be the component of $cl[W_i]$ such that $K_i \subset M_i$. By (20.1),

(5) $M_i \cap F_i \neq \emptyset$ for each i = 1 and 2.

By (2), (4), and (5), $M_i \not\subset K_i$ for any i = 1 and 2. Hence, since M_i is a subcontinuum of A such that $K_i \subset M_i$ and K_i is a component of A ∩ B for each i = 1 and 2, we have that

(6) $M_i \setminus B \neq \emptyset$ for each i = 1 and 2.

Also, by (3),

(7) $M_1 \cap M_2 = \emptyset$.

Now, by letting $M = B \cup M_1 \cup M_2$ and $N = B$, it follows easily from (6) and (7) that M is a 2-odd as defined in (0.21). This proves (*) and completes the proof of (1.145).

(1.146) COROLLARY. *A continuum* X *is hereditarily indecomposable if and only if* C(X) *does not contain a 2-cell.*

Proof. Half the corollary is (1.61), and the other half is (1.145).

Note that if C(X) contains a 2-cell, then X contains a decomposable subcontinuum; hence, as was shown in the proof of (1.145), X contains a 2-odd. However, I do not know answers to the following questions.

(1.147) QUESTION. If C(X) contains an n-cell for some $n \geq 3$, then does X contain an n-odd? B. J. Ball asked me this question for the case when $n = 3$. If the answer is "yes," then by (1.100) we would have a characterization of those continua X such that C(X) contains an n-cell.

(1.148) QUESTION. If C(X) contains a Hilbert cube, then does X contain continua $Y, Y_1, Y_2, \ldots, Y_i, \ldots$ satisfying (1) through (3) of (1.103)?

(I.E) The Hyperspace Operations and Inverse Limits

One of the most important and widely used properties of hyperspaces is that they behave "nicely" with respect to inverse limits--the hyperspace of the inverse limit is homeomorphic to the inverse limit of the hyperspaces [see (1.169) for a precise statement]. The purpose of this section of Chapter I is to present some basic information about inverse limits and about their relationships to hyperspaces; in this respect, (1.169) and its applications are of particular importance. In (1.149) through (1.164), we give some fundamental properties of inverse limits independent of hyperspaces. These properties will be used in this chapter and in other parts of the book. In (1.169), we state and prove the important theorem which was roughly described above. Then, in (1.171) through (1.184), we give a number of applications of (1.169); other applications occur throughout the book. Finally, we discuss hyper-onto and minimal representations.

For simplicity and convenience, we state all results in terms of inverse *sequences*. Many of the results generalize to the setting of Hausdorff spaces and inverse systems over directed sets; in some places we will reiterate this. When it is appropriate [for example, (1.178) and (1.180)], some results are specifically stated for Hausdorff continua. But, for the most part, the reader can assume that all spaces are metric. In most statements, we will not be explicit about the types of spaces being used. Finally, we remind the reader that a continuum is, by definition, a *nondegenerate metric* continuum.

(1.149) DEFINITIONS. An *inverse sequence* is a "double sequence" $\{Y_n, f_n\}_{n=1}^{\infty}$ of spaces Y_n and mappings $f_n: Y_{n+1} \to Y_n$. The spaces Y_n are called *coordinate spaces* and the mappings f_n are called *bonding maps*.

(1.150) DEFINITION. Let $\{Y_n, f_n\}_{n=1}^{\infty}$ be an inverse sequence. The *inverse limit of* $\{Y_n, f_n\}_{n=1}^{\infty}$, denoted by

$$\varprojlim \{Y_n, f_n\}_{n=1}^{\infty},$$

is the subspace of the countable cartesian product $\Pi_{n=1}^{\infty} Y_n$ (with the product topology) given by:

$$\{(y_1, y_2, \ldots, y_n, \ldots) \in [\Pi_{n=1}^{\infty} Y_n]: f_n(y_{n+1}) = y_n \text{ for each } n = 1, 2, \ldots\}.$$

(1.151) NOTATION. Let $\{Y_n, f_n\}_{n=1}^{\infty}$ be an inverse sequence. For each $i = 1, 2, \ldots$, let

$$Q_i[Y_n, f_n] = \{(y_1, y_2, \ldots, y_n, \ldots) \in [\Pi_{n=1}^{\infty} Y_n]: f_n(y_{n+1}) = y_n \text{ for each } n \leq i\}.$$

The following lemma is easy to prove but is important in that it focuses our attention on the point of view from which we will think of inverse limits--namely, as nested intersections. Another way of thinking of inverse limits will be discussed in (1.199).

(1.152) LEMMA [see (1.159)]. *If* $\{Y_n, f_n\}_{n=1}^{\infty}$ *is an inverse* sequence, then

$$Q_1[Y_n, f_n] \supset Q_2[Y_n, f_n] \supset \cdots \supset Q_i[Y_n, f_n] \supset \cdots$$

and

$$\varprojlim\{Y_n, f_n\}_{n=1}^{\infty} = \bigcap_{i=1}^{\infty} Q_i[Y_n, f_n].$$

The next lemma relates the topological nature of the spaces $Q_i[Y_n, f_n]$ to that of the coordinate spaces Y_n, $n > i$. It will be of significance in the proof that 2^X and $C(X)$ are each acyclic [see the proof of (1.171)].

(1.153) LEMMA [see (1.159)]. *If* $\{Y_n, f_n\}_{n=1}^{\infty}$ *is an inverse sequence, then*

$$Q_i[Y_n, f_n] \cong \prod_{n=i+1}^{\infty} Y_n$$

for each $i = 1, 2, \ldots$.

Proof. Fix i and define $h: Q_i[Y_n, f_n] \to \prod_{n=i+1}^{\infty} Y_n$ by

$$h((y_1, y_2, \ldots, y_n, \ldots)) = (y_{i+1}, y_{i+2}, \ldots)$$

for each $(y_1, y_2, \ldots, y_n, \ldots) \in Q_i[Y_n, f_n]$. It is easy to prove h is a homeomorphism of $Q_i[Y_n, f_n]$ onto $\prod_{n=i+1}^{\infty} Y_n$.

(1.154) NOTATION. Let $\{Y_n, f_n\}_{n=1}^{\infty}$ be an inverse sequence. For each $i = 1, 2, \ldots$, let

$$\pi_i: \varprojlim\{Y_n, f_n\}_{n=1}^{\infty} \to Y_i$$

denote the restriction to $\varprojlim\{Y_n, f_n\}_{n=1}^{\infty}$ of the $i^{\underline{\text{th}}}$ projection map of $\prod_{n=1}^{\infty} Y_n$ to Y_i, i.e.,

$$\pi_i((y_1, y_2, \ldots, y_n, \ldots)) = y_i$$

for each $(y_1, y_2, \ldots, y_n, \ldots) \in \varprojlim\{Y_n, f_n\}_{n=1}^{\infty}$. The mappings π_i will be called *projection maps*, not to be confused with the projection maps of $\prod_{n=1}^{\infty} Y_n$. In particular, π_i may not map onto Y_i.

Note the following lemma about commutativity of the bonding maps and the projections.

(1.155) **LEMMA** [see (1.159)]. *Let* $\{Y_n, f_n\}_{n=1}^{\infty}$ *be an inverse sequence. Then, for each* $n = 1, 2, \ldots,$ $f_n \circ \pi_{n+1} = \pi_n$. *Hence, if* $1 \leq i < j$, $f_i \circ \cdots \circ f_j \circ \pi_{j+1} = \pi_i$.

The following lemma shows what should *only* be a subbase for $\varprojlim\{Y_n, f_n\}_{n=1}^{\infty}$ is actually a base.

(1.156) **LEMMA** (see (1.159)]. *Let* $\{Y_n, f_n\}_{n=1}^{\infty}$ *be an inverse sequence. For each* $i = 1, 2, \ldots,$ *let*
$$B_i[Y_n, f_n] = \{\pi_i^{-1}[U] : U \text{ is an open subset of } Y_i\}.$$
Let
$$B[Y_n, f_n] = \bigcup_{i=1}^{\infty} B_i[Y_n, f_n].$$
Then, $B[Y_n, f_n]$ *is a base for the topology for* $\varprojlim\{Y_n, f_n\}_{n=1}^{\infty}$.

Proof. Let $Z_1, Z_2, \ldots, Z_n, \ldots$ satisfy: Z_n is an open subset of Y_n for each $n = 1, 2, \ldots$ and, for some N, $Z_n = Y_n$ whenever $n \geq N$. Let
$$Z = [\prod_{n=1}^{\infty} Z_n] \cap \varprojlim\{Y_n, f_n\}_{n=1}^{\infty}.$$
Note that if $N = 1$, then

(1) $Z = \varprojlim\{Y_n, f_n\}_{n=1}^{\infty} = \pi_1^{-1}[Y_1]$.

Assume $N > 1$. Let $G_{N-1} = f_{N-1}^{-1}[Z_{N-1}]$ and, for each i such that $1 \le i < N-1$, let $G_i = (f_i \circ \cdots \circ f_{N-1})^{-1}[Z_i]$. Let $G = \bigcap_{i=1}^{N-1} G_i$. It is not difficult to show that

(2) $Z = \pi_N^{-1}[G]$.

By continuity of the bonding maps $f_1, f_2, \ldots, f_{N-1}$, we have that G is an open subset of Y_N. Hence, by (2) [or (1) if $N = 1$], $Z \in B[Y_n, f_n]$. Thus, from the conditions the sets $Z_1, Z_2, \ldots, Z_n, \ldots$ satisfied and the definition of the usual base for the product topology [185, p.147], we have shown that $B[Y_n, f_n]$ contains the collection of usual basic open sets for a subspace of $\prod_{n=1}^{\infty} Y_n$. The reverse containment is clear. This completes the proof of (1.156).

(1.157) LEMMA [see (1.159)]. *If* $\{Y_n, f_n\}_{n=1}^{\infty}$ *is an inverse sequence of nonempty compact metric spaces* Y_n, *then* $\varprojlim\{Y_n, f_n\}_{n=1}^{\infty}$ *is a nonempty compact metric space.*

Proof. Recall that the countable cartesian product of nonempty compact metric spaces is a nonempty compact metric space ([185, pp.212-213] and [186, p.17]). Using this fact, the lemma follows: By (1.153), $Q_i[Y_n, f_n]$ is a nonempty compact metric space for each $i = 1, 2, \ldots$. Hence, by (1.152), $\varprojlim\{Y_n, f_n\}_{n=1}^{\infty}$ is a nonempty compact metric space.

(1.158) LEMMA [see (1.159)]. *If* $\{Y_n, f_n\}_{n=1}^{\infty}$ *is an inverse sequence of continua* Y_n, *then* $\varprojlim\{Y_n, f_n\}_{n=1}^{\infty}$ *is a continuum or a one-point space.*

Proof. Similar to the proof of (1.157)--use the additional fact that, since Y_n is connected for each $n = 1,2,\ldots$, $Q_i[Y_n, f_n]$ is connected for each $i = 1,2,\ldots$ by (1.153) and [186, Theorem 11, p.137].

(1.159) REMARK. I can not say with any certainty who first obtained any of the results in (1.152), (1.153), and (1.155) through (1.158). The first person to publish any definitive treatment of inverse limits as nested intersections seems to have been Capel [57]. Indeed, the second part of (1.152) is Capel's definition of $\varprojlim\{Y_n, f_n\}_{n=1}^{\infty}$ [57, p.234], and the first part of (1.152) is [57, 2.3]. Also: (1.155) is [57, 2.1]; (1.156) is [57, 2.2]; (1.157) is [57, 2.5]; and (1.158) is [57, 2.10]. The results just mentioned are in [57] for the general setting of Hausdorff spaces and inverse systems over directed sets. The result in (1.153), which can also be formulated in this setting, was not stated in [57]. It has been used in some proofs in the literature [see, for example, the proof of 1.4 of [173]; also, see (1.175)]. Capel stated [57, p.234] that many of the results in [57, section 2] were known. He included some for completeness and some to illustrate that several proofs in the literature could be simplified by considering inverse limits as nested intersections. For some generalizations and results related to Capel's work in [57, sections 2 and 4], see [245].

The next lemma will be used in the proof of (1.169). It has

applications elsewhere in the book. In particular, I refer the reader to (12.17) and to the discussion in the two paragraphs preceding (12.14).

(1.160) LEMMA [57, 2.8(i) and (ii)]. *Let* $\{Y_n, f_n\}_{n=1}^{\infty}$ *be an inverse sequence and let* Y *denote* $\varprojlim\{Y_n, f_n\}_{n=1}^{\infty}$. *Let* A *be a compact subset of* Y. *For each* $n = 1,2,\ldots$, *let* g_n *denote the restriction of* f_n *to* $\pi_{n+1}(A)$. *Then* $g_n[\pi_{n+1}(A)] = \pi_n(A)$ *for each* $n = 1,2,\ldots$ *and, hence,* $\{\pi_n(A), g_n\}_{n=1}^{\infty}$ *is an inverse sequence with onto bonding maps. Furthermore,*

(1.160.1) $\quad \varprojlim\{\pi_n(A), g_n\}_{n=1}^{\infty} = A = [\prod_{n=1}^{\infty} \pi_n(A)] \cap Y$

when, of course, we consider $\prod_{n=1}^{\infty} \pi_n(A)$ *as being contained in* $\prod_{n=1}^{\infty} Y_n$ *by inclusion.*

Proof. Since $f_n \circ \pi_{n+1} = \pi_n$ for each $n = 1,2,\ldots$ [by (1.155)], we have that

$g_n[\pi_{n+1}(A)] = \pi_n(A)$ for each $n = 1,2,\ldots$.

Now we prove (1.160.1). It is easy to see that

$\varprojlim\{\pi_n(A), g_n\}_{n=1}^{\infty} = [\prod_{n=1}^{\infty} \pi_n(A)] \cap Y.$

Thus, to prove (1.160.1), it suffices to prove that

(*) $\quad A = [\prod_{n=1}^{\infty} \pi_n(A)] \cap Y.$

Clearly, $A \subset [\prod_{n=1}^{\infty} \pi_n(A)] \cap Y$. So, to prove (*), let

$y = (y_1, y_2, \ldots, y_n, \ldots) \in [\prod_{n=1}^{\infty} \pi_n(A)] \cap Y.$

For each $n = 1,2,\ldots$, let

$$K_n = A \cap \pi_n^{-1}(y_n).$$

Fix n. Since $y \in [\Pi_{n=1}^{\infty} \pi_n(A)]$, $y_n \in \pi_n(A)$. Hence, there exists $a \in A$ such that $\pi_n(a) = y$. Clearly, then, $a \in K_n$. We have proved that

(1) $K_n \neq \emptyset$.

Next we show $K_n \supset K_{n+1}$. To do this, let $x \in K_{n+1}$. Then, by definition of K_{n+1}, $x \in A$ and $\pi_{n+1}(x) = y_{n+1}$. Since $x \in A$ and $A \subset Y$, $x \in Y$ and thus $f_n[\pi_{n+1}(x)] = \pi_n(x)$. Thus, since $\pi_{n+1}(x) = y_{n+1}$, $f_n(y_{n+1}) = \pi_n(x)$ and, since $y \in Y$, $f_n(y_{n+1}) = y_n$. Hence, $y_n = \pi_n(x)$, i.e., $x \in \pi_n^{-1}(y_n)$. Therefore, since $x \in A$, $x \in K_n$. We have proved that

(2) $K_n \supset K_{n+1}$.

Since the projections π_n are continuous and A is compact, it is clear from the definition of K_n that

(3) K_n is a compact subset of A.

Since (1) through (3) hold for each $n = 1, 2, \ldots$, there is a point $z \in [\cap_{n=1}^{\infty} K_n]$. Hence, $z \in A$ and $\pi_n(z) = y_n$ for each $n = 1, 2, \ldots$. Therefore, $z = y$ and (thus) $y \in A$. This completes the proof of (*) and of the lemma.

The next lemma will be used in the proof of (1.195) and in other parts of the book [see, for example, Chapter II and the proof of (10.10)]. Let us recall the following definition.

(1.161) DEFINITION. A mapping $f: M_1 \to M_2$ from a metric space M_1 into a space M_2 is said to be an ε-*mapping* provided

(1.163)

that $\operatorname{diam}[f^{-1}(f(p))] < \varepsilon$ for each $p \in M_1$.

(1.162) LEMMA [see (1.163)]. *Let* $\{Y_n, f_n\}_{n=1}^{\infty}$ *be an inverse sequence of compact metric spaces* Y_n. *Then: Given* $\varepsilon > 0$, *there exists* N *such that there is an* ε-*mapping of* $\varprojlim\{Y_n, f_n\}_{n=1}^{\infty}$ *into* Y_k *for each* $k \geq N$. *Furthermore, by a suitable choice of a (topologically equivalent) metric for* $\varprojlim\{Y_n, f_n\}_{n=1}^{\infty}$, N *may be chosen so that* π_k *is an* ε-*mapping for each* $k \geq N$.

Proof. Assume, as we may [185, pp.212-213], that the metric d_∞ for $\Pi_{n=1}^{\infty} Y_n$ is given by the formula

$$d_\infty(y, z) = \sum_{n=1}^{\infty} 2^{-n} \cdot \frac{d_n(y_n, z_n)}{1 + d_n(y_n, z_n)}$$

for each $y = (y_1, y_2, \ldots, y_n, \ldots)$ and $z = (z_1, z_2, \ldots, z_n, \ldots)$ in $\Pi_{n=1}^{\infty} Y_n$, where d_n denotes the metric for Y_n for each $n = 1, 2, \ldots$. Then, it is easy to see that π_i is a $[\sum_{n=i+1}^{\infty} 2^{-n}]$-mapping of $\varprojlim\{Y_n, f_n\}_{n=1}^{\infty}$ into Y_i for each $i = 1, 2, \ldots$. The result now follows easily.

(1.163) REMARK. The result in (1.162) is related to [205, Lemma 1], where the authors assume the spaces Y_n are compact polyhedra. The import of [205] is that the authors prove the converse of their Lemma 1.

The following lemma is the last of the preliminary facts--then, we begin applying inverse limits to hyperspaces.

(1.164) LEMMA [116, p.183]. *Any (metric) continuum is homeomorphic to an inverse limit* $\varprojlim \{P_n, f_n\}_{n=1}^{\infty}$ *where the spaces* P_n *are compact connected polyhedra.*

(1.165) REMARK. We refer the reader to [116] or [260, pp.305-306] for a proof of (1.164). Also, let us note that (1.164) may be formulated for Hausdorff continua and inverse systems of (metric) polyhedra over directed sets [260, pp.303-305].

We begin the discussion of inverse limits and hyperspaces with the following lemma. It gives a base which will be convenient to use in the proof of (1.169).

(1.166) LEMMA [see (1.167)]. *Let* X *be a continuum and assume* $X = \varprojlim \{X_n, f_n\}_{n=1}^{\infty}$. *For each* $i = 1, 2, \ldots$, *let*
$$\beta_i = \{<\pi_i^{-1}(U_1), \pi_i^{-1}(U_2), \ldots, \pi_i^{-1}(U_k)>:$$
$U_1, U_2, \ldots, U_k$ *is an open subset of* X_i\}.
Let $\beta = \cup_{i=1}^{\infty} \beta_i$. *Then,* β *is a base for the topology for* 2^X.

Proof. It follows easily from (0.11), (0.12), and (1.156) that
$$\beta^* = \{<\pi_{n(1)}^{-1}(U_{n(1)}), \pi_{n(2)}^{-1}(U_{n(2)}), \ldots, \pi_{n(m)}^{-1}(U_{n(m)})>: U_{n(j)}$$
is an open subset of $X_{n(j)}$ for each $j = 1, 2, \ldots, m$\}
is a base for the Vietoris topology for 2^X. We will show that $\beta = \beta^*$. To do this, let
$$<\pi_{n(1)}^{-1}(U_{n(1)}), \pi_{n(2)}^{-1}(U_{n(2)}), \ldots, \pi_{n(m)}^{-1}(U_{n(m)})> \in \beta^*.$$
Let $k = \max\{n_1, n_2, \ldots, n_m\}$. Then, by (1.155),

$$(1)\quad \pi^{-1}_{n(j)}(U_{n(j)}) = \begin{cases} \pi_k^{-1}[f^{-1}_{k-1}(U_{n(j)})] & \text{, if } n(j) = k-1 \\ \pi_k^{-1}[(f_{n(j)} \circ \cdots \circ f_{k-1})^{-1}(U_{n(j)})] & \text{, if } n(j) < k-1. \end{cases}$$

Let

$$V_{n(j)} = \begin{cases} U_{n(j)} & \text{, if } n(j) = k \\ f^{-1}_{k-1}(U_{n(j)}) & \text{, if } n(j) = k-1 \\ (f_{n(j)} \circ \cdots \circ f_{k-1})^{-1}(U_{n(j)}) & \text{, if } n(j) < k-1. \end{cases}$$

By the continuity of the bonding maps, $V_{n(j)}$ is an open subset of X_k for each $j = 1, 2, \ldots, m$. Hence,

(2) $\langle \pi_k^{-1}(V_{n(1)}), \pi_k^{-1}(V_{n(2)}), \ldots, \pi_k^{-1}(V_{n(m)}) \rangle \in \beta$.

By (1), $\pi^{-1}_{n(j)}(U_{n(j)}) = \pi_k^{-1}(V_{n(j)})$ for each $j = 1, 2, \ldots, m$; hence,

(3) $\langle \pi^{-1}_{n(1)}(U_{n(1)}), \pi^{-1}_{n(2)}(U_{n(2)}), \ldots, \pi^{-1}_{n(m)}(U_{n(m)}) \rangle =$
$\langle \pi_k^{-1}(V_{n(1)}), \pi_k^{-1}(V_{n(2)}), \ldots, \pi_k^{-1}(V_{n(m)}) \rangle$.

By (2) and (3), $\langle \pi^{-1}_{n(1)}(U_{n(1)}), \pi^{-1}_{n(2)}(U_{n(2)}), \ldots, \pi^{-1}_{n(m)}(U_{n(m)}) \rangle \in \beta$. Therefore, $\beta^* \subset \beta$. Clearly, $\beta \subset \beta^*$. Hence, we have proved $\beta = \beta^*$. The lemma now follows [using (0.13)].

(1.167) REMARK. By intersecting each member of β in (1.166) with $C(X)$, one obtains a result for $C(X)$ analogous to the one for 2^X. The result for $C(X)$ is [294, Lemma 1.3]. The proof in [294], which used the metric, is quite complicated and contains an error; the error was noted and corrected in [109, pp.132-

133]. The proof given above has two main advantages: First, it is simpler than the proof in [294] and [109]; secondly, it obviously works for Hausdorff continua.

In (1.169), we give the important theorem which relates the hyperspaces of an inverse limit to the hyperspaces of the coordinate spaces. To facilitate its statement and proof, we adopt the following notation.

(1.168) NOTATION. Let X be a continuum and assume $X = \varprojlim\{X_n, f_n\}_{n=1}^{\infty}$ where each of the spaces X_n is a continuum. For each $n = 1, 2, \ldots$, let

$$f_n^*: 2^{X_{n+1}} \to 2^{X_n}$$

be given by $f_n^*(A) = \{f_n(a) : a \in A\}$ for each $A \in 2^{X_{n+1}}$. Note that, for each $n = 1, 2, \ldots$, the restriction of f_n^* to $C(X_{n+1})$ is the induced map $\hat{f}_n: C(X_{n+1}) \to C(X_n)$ defined in (0.49). Also note that $\{2^{X_n}, f_n^*\}_{n=1}^{\infty}$ and $\{C(X_n), \hat{f}_n\}_{n=1}^{\infty}$ are inverse sequences. Denote $\varprojlim\{2^{X_n}, f_n^*\}_{n=1}^{\infty}$ by 2_{∞}^{X} and denote $\varprojlim\{C(X_n), \hat{f}_n\}_{n=1}^{\infty}$ by $C_{\infty}(X)$. Thus, *by definition*,

$$2_{\infty}^{X} = \varprojlim\{2^{X_n}, f_n^*\}_{n=1}^{\infty} \quad \text{and} \quad C_{\infty}(X) = \varprojlim\{C(X_n), \hat{f}_n\}_{n=1}^{\infty}.$$

I will consider $C_{\infty}(X)$ as being contained in 2_{∞}^{X} by inclusion: $C_{\infty}(X) \subset 2_{\infty}^{X}$.

We are ready to state and prove the main result, after which we will give some important applications. Roughly, the result says that the hyperspace operations are continuous with respect to [or, commute with] inverse limits. Here is its precise formulation.

(1.169) THEOREM [see (1.170)]. *Let* X *be a continuum and assume* $X = \varprojlim\{X_n, f_n\}_{n=1}^{\infty}$ *where each of the spaces* X_n *is a continuum. Let* 2_{∞}^X *and* $C_{\infty}(X)$ *be as in (1.168). Then,* $2_{\infty}^X \cong 2^X$ *and* $C_{\infty}(X) \cong C(X)$. *Furthermore, there is a homeomorphism* $h: 2_{\infty}^X \xrightarrow{\text{onto}} 2^X$ *such that* $h[C_{\infty}(X)] = C(X)$.

Proof. We will define [see (5)] the homeomorphism $h: 2_{\infty}^X \xrightarrow{\text{onto}} 2^X$ after making some observations about an arbitrary point of 2_{∞}^X. Let

$$A = (A_1, A_2, \ldots, A_n, \ldots) \in 2_{\infty}^X.$$

Then, by definition of 2_{∞}^X, $f_n^*(A_{n+1}) = A_n$ for each $n = 1, 2, \ldots$. Hence, by definition of f_n^* [see (1.168)],

(1) $f_n[A_{n+1}] = A_n$ for each $n = 1, 2, \ldots$.

Therefore, letting $f_n|A_{n+1}$ denote the restriction of f_n to A_{n+1} for each $n = 1, 2, \ldots$, we have that

$\{A_n, f_n|A_{n+1}\}_{n=1}^{\infty}$ is an inverse sequence.

Note that, since $A \in 2_{\infty}^X = \varprojlim\{2^{X_n}, f_n^*\}_{n=1}^{\infty}$, $A_n \in 2^{X_n}$ for each $n = 1, 2, \ldots$; this means

(2) $A_n \subset X_n$ for each $n = 1, 2, \ldots$

and

(3) A_n is nonempty and compact for each $n = 1, 2, \ldots$.

Since $X = \varprojlim\{X_n, f_n\}_{n=1}^{\infty}$, it follows using (2) that $[\varprojlim\{A_n, f_n|A_{n+1}\}_{n=1}^{\infty}] \subset X$. Hence, by (3) and (1.157),

(4) $[\varprojlim\{A_n, f_n|A_{n+1}\}_{n=1}^{\infty}] \in 2^X$.

Let

(5) $h(A) = \varprojlim \{A_n, f_n|A_{n+1}\}_{n=1}^{\infty}$.

Recalling that A was an arbitrarily chosen point of 2_{∞}^{X}, we see from (4) that (5) defines a function $h: 2_{\infty}^{X} \to 2^{X}$. We now show, in (a) through (d) below, that h has all the desired properties.

(a) Proof that h sends 2_{∞}^{X} onto 2^{X}. Let $K \in 2^{X}$. Then K is a compact subset of $X = \varprojlim \{X_n, f_n\}_{n=1}^{\infty}$. Hence, by (1.160),

(6) $f_n[\pi_{n+1}(K)] = \pi_n(K)$ for each $n = 1, 2, \ldots$

and

(7) $K = \varprojlim \{\pi_n(K), f_n|\pi_{n+1}(K)\}_{n=1}^{\infty}$

where the vertical line in (7) denotes destriction. Also, since the projections $\pi_n: X \to X_n$ are continuous and since K is a nonempty compact subset of X,

(8) $\pi_n(K) \in 2^{X_n}$ for each $n = 1, 2, \ldots$.

By (8), f_n^* is defined at $\pi_{n+1}(K)$ for each $n = 1, 2, \ldots$; thus, by (6) and the definition of f_n^*,

(9) $f_n^*(\pi_{n+1}(K)) = \pi_n(K)$ for each $n = 1, 2, \ldots$.

Hence, by (8) and (9), $(\pi_1(K), \pi_2(K), \ldots, \pi_n(K), \ldots) \in 2_{\infty}^{X}$. By (7) and the definition of h, $h((\pi_1(K), \pi_2(K), \ldots, \pi_n(K), \ldots)) = K$.

(b) Proof that h is one-to-one. Let

$L = (L_1, L_2, \ldots, L_n, \ldots) \in 2_{\infty}^{X}$

and

$M = (M_1, M_2, \ldots, M_n, \ldots) \in 2_{\infty}^{X}$

such that $h(L) = h(M)$. We will show that $L_n = M_n$ for each $n = 1, 2, \ldots$. Choose and fix $n = k$. Let $p \in L_k$. Using (1) and (5), it is easy to produce a point $(x_1, x_2, \ldots, x_n, \ldots) \in h(L)$ such that $x_k = p$. Since $h(L) = h(M)$,

(1.169) STRUCTURE OF HYPERSPACES 173

$(x_1, x_2, \ldots, x_n, \ldots) \in h(M)$. Hence, by (5), $x_k \in M_k$. Therefore, since $p = x_k$, we have $p \in M_k$. This proves $L_k \subset M_k$. A similar argument shows that $M_k \subset L_k$. Thus, $L_k = M_k$.

(c) Proof that h is continuous. As before, let $\pi_n : X \to X_n$ denote projection from X into X_n. Let $\pi'_n : 2^X_\infty \to 2^{X_n}$ denote projection from 2^X_∞ into 2^{X_n}. To prove h is continuous, it suffices by (1.166) to show

$$h^{-1}(<\pi_i^{-1}(U_1), \pi_i^{-1}(U_2), \ldots, \pi_i^{-1}(U_k)>)$$

is an open subset of 2^X_∞ for each

$$<\pi_i^{-1}(U_1), \pi_i^{-1}(U_2), \ldots, \pi_i^{-1}(U_k)> \in \beta,$$

where β is the base defined in (1.166). Let

$$<\pi_i^{-1}(U_1), \pi_i^{-1}(U_2), \ldots, \pi_i^{-1}(U_k)> \in \beta.$$

We will show (as is sufficient)

(#) $\quad h^{-1}(<\pi_i^{-1}(U_1), \pi_i^{-1}(U_2), \ldots, \pi_i^{-1}(U_k)>) =$

$$\pi_i'^{-1}(<U_1, U_2, \ldots, U_k>).$$

First note the following fact, which can be proved using (1) and (5):

(10) If $A = (A_1, A_2, \ldots, A_n, \ldots) \in 2^X_\infty$, then $\pi_i[h(A)] = A_i$.

Now, we prove (#). Except for the fourth equality below, each of the equalities is easy to verify using set-theoretic facts and the definition of <> in (0.10.2). The fourth equality is valid by (10).

$$h^{-1}(<\pi_i^{-1}(U_1), \pi_i^{-1}(U_2), \ldots, \pi_i^{-1}(U_k)>) =$$

$$\{A \in 2^X_\infty : h(A) \subset [\bigcup_{j=1}^{k} \pi_i^{-1}(U_j)] \text{ and } h(A) \cap \pi_i^{-1}(U_j) \neq \emptyset \text{ for each } j = 1, 2, \ldots, k\} =$$

$$\{A \in 2^X_\infty : h(A) \subset [\pi_i^{-1}(\bigcup_{j=1}^{k} U_j)] \text{ and } h(A) \cap \pi_i^{-1}(U_j) \neq \emptyset$$
$$\text{for each } j = 1,2,\ldots,k\} =$$
$$\{A \in 2^X_\infty : \pi_i[h(A)] \subset [\bigcup_{j=1}^{k} U_j] \text{ and } \pi_i[h(A)] \cap U_j \neq \emptyset \text{ for}$$
$$\text{each } j = 1,2,\ldots,k\} =$$
$$\{A \in 2^X_\infty : A_i \subset [\bigcup_{j=1}^{k} U_j] \text{ and } A_i \cap U_j \neq \emptyset \text{ for each}$$
$$j = 1,2,\ldots,k\} =$$
$$\{A \in 2^X_\infty : A_i \in \langle U_1, U_2, \ldots, U_k\rangle\} = \pi_i'^{-1}(\langle U_1, \ldots, U_k\rangle).$$

This proves (#).

(d) Proof that $h[C_\infty(X)] = C(X)$. The proof consists of some observations based on what we have already done. First, assume A at the beginning of the proof of the theorem is a point of $C_\infty(X)$. Then, in addition to the properties in (3), A_n is connected for each $n = 1,2,\ldots$. Hence, by (1.158) and (5), $h(A) \in C(X)$. Thus, we have that $h[C_\infty(X)] \subset C(X)$. On the other hand, assume K in (a) is also connected, i.e., $K \in C(X)$. Then, $\pi_n(K) \in C(X_n)$ for each $n = 1,2,\ldots$ [comp., (8)], and the rest of the proof of (a) shows

$$(\pi_1(K), \pi_2(K), \ldots, \pi_n(K), \ldots) \in C_\infty(X)$$

and

$$h((\pi_1(K), \pi_2(K), \ldots, \pi_n(K), \ldots)) = K.$$

Therefore, we have $h[C_\infty(X)] = C(X)$. This completes the proof of (1.169) [since each 2^{X_n} is compact by (0.8), 2^X_∞ is compact by (1.157); hence, h is a homeomorphism since h is one-to-one and continuous].

(1.170) REMARK. The result in (1.169) may be formulated in the general setting of Hausdorff continua and inverse systems over directed sets--the proof is a simple reworking of what we have done. The result appears in this general setting in [300, Theorem 1] for the case of 2^Y (only) where Y is a compact Hausdorff space; it appears in the metric setting in [294, Theorem 1.1] for the case of $C(X)$ where X is a (metric) continuum.

A number of applications of (1.169) occur throughout the book. The following one is of a general nature, and will be used to obtain the others in this chapter. In later chapters, the applications of (1.169) are somewhat different [see Chapter II, the proof of (10.10), etc.].

(1.171) THEOREM [see (1.175)]. *Let* X *be a continuum. Then, there is a continuum* M *and an embedding* $Z \subset M$ *of* 2^X *or* $C(X)$, *either one, such that*

$$Z = \bigcap_{i=1}^{\infty} M_i$$

where $M_1 \supset M_2 \supset \cdots \supset M_i \supset \cdots$ *are compact absolute retracts contained in* M. *Moreover,* M *and all the sets* M_i ($i = 1, 2, \ldots$) *can be chosen to be Hilbert cubes.*

Proof. Let X be a continuum. We first prove (1.171) for 2^X. By (1.164),

$$X \cong \varprojlim \{P_n, f_n\}_{n=1}^{\infty}$$

where the spaces P_n are compact connected polyhedra. By (1.169)

(1) $\quad 2^X \cong \varprojlim \{2^{P_n}, f_n^*\}_{n=1}^{\infty}$.

Choose M, Z, and M_i, $i = 1, 2, \ldots$, for (1.171) as follows:

$$M = \prod_{n=1}^{\infty} 2^{P_n},$$

$$Z = \varprojlim \{2^{P_n}, f_n^*\}_{n=1}^{\infty},$$

$M_i = Q_i[2^{P_n}, f_n^*]$ as defined in (1.151).

By (1), Z is an embedding of 2^X in M. By (1.97), 2^{P_n} is a Hilbert cube for each $n = 1,2,\ldots$. Hence, by (1.153) and the fact that the countable cartesian product of Hilbert cubes is a Hilbert cube, we have that M_i is a Hilbert cube for each $i = 1,2,\ldots$. Finally, by (1.152), $M_1 \supset M_2 \supset \cdots \supset M_i \supset \cdots$ and $Z = \cap_{i=1}^{\infty} M_i$. This completes the proof of (1.171) for 2^X. The proof for $C(X)$ follows easily from the proof given above for 2^X by using (1.98) in place of (1.97) and the following theorem due to Jim West [412, Theorem 6.2]: The countable infinite cartesian product of continua is a Hilbert cube if the cartesian product of each continuum with the Hilbert cube is a Hilbert cube.

Now, using (1.171), we derive some specific properties of hyperspaces, some of which have not been precisely stated in the literature. The treatment will be somewhat historical.

All homology and cohomology groups to be used are assumed to be over a coefficient group for which the theory is continuous--for example, the integers.

Kelley proved the following result.

(1.172) THEOREM [165, 3.4]. *For any continuum* X, 2^X *is acyclic in the sense of Vietoris homology.*

(1.175)   STRUCTURE OF HYPERSPACES   177

Kelley's approach to proving (1.172) did not work for $C(X)$, and he asked the following question [165, p.27]: Is $C(X)$ acyclic (in the sense of Vietoris homology) for any continuum X? This question was answered affirmatively by Segal [see (1.173)]. In fact, the reason Segal proved (1.169) for $C(X)$ was to use it to answer Kelley's question. Still, Segal did not use a result such as (1.171); instead, he used the contractibility of $C(P)$ when P is a polyhedron [see (1.93)], (1.164), and (1.169) for the case of $C(X)$, together with the continuity of Čech homology.

(1.173) THEOREM [294, Theorem 1.2]. *For any continuum* X, $C(X)$ *is acyclic in the sense of Vietoris homology (or Čech homology).*

Segal [294] has also proved that $C(X)$ is acyclic in the sense of singular homology.

The proof Segal gave in [294] for (1.173) can be easily modified to show acyclicness in the sense of Čech cohomology. Note the following corollary to (1.171).

(1.174) COROLLARY. *For any continuum* x, 2^X *and* $C(X)$ *are each acyclic in the sense of Čech cohomology.*

Proof [see (1.175)]. Use (1.171).

(1.175) REMARK. The approach used here to prove (1.174) is the same as the one used in [173, section 1] to give another proof

of (1.173) and to derive some other facts [see (1.181) and (1.182)].
It is also the approach I used to prove the result [see (1.176)] announced as Theorem 3 in [243]. I had proved the first part of
(1.171) in order to prove Theorem 3 of [243]. Krasinkiewicz [173,
1.5] states the first part of (1.171) for C(X).

(1.176) COROLLARY ([243, Theorem 3]; see (1.177) below). *For any continuum* X, *both* 2^X *and* C(X) *have property(b) and, hence* [*338, 7.3, p.227*], *are unicoherent.*

Proof. Use (1.171).

For an application of (1.176) of some importance, see the proof of (14.2) and the paragraph following it.

(1.177) REMARK. I obtained (1.176) in [243] without knowing that a continuum has property(b) if and only if its first Čech cohomology group (over the integers) is zero. This was pointed out to me by John Isbell (see the last sentence in [243, p.413]). Thus, (1.176) is a special case of (1.174).

The following result was specifically stated for the first time in [203], where it was observed to be a simple consequence of a semigroup theorem of Wallace's [321, Corollary 1]. The proof given here is different [see (1.179)].

(1.178) COROLLARY ([203, p.1209]; comp., [321, Corollary 1]).

For any Hausdorff continuum Y, 2^Y *is acyclic in the sense of Alexander-Kolmogoroff-Spanier cohomology.*

Proof [see (1.179)]. As has been mentioned, many of the results preceding (1.171) can be formulated for the setting of Hausdorff continua and inverse systems over directed sets. Thus, (1.171) can be reformulated so as to be used to prove (1.178).

(1.179) REMARK. Corollary 1 of [321], which McWaters used to prove (1.178), does not apply to C(Y). Nonetheless, the ideas we indicated in the proof of (1.178) can be used to prove the following result.

(1.180) COROLLARY [190]. *For any Hausdorff continuum* Y, C(Y) *is acyclic in the sense of Alexander-Kolmogoroff-Spanier cohomology.*

Proof. See (1.179).

Now we come to two results of Krasinkiewicz for C(X). The corresponding results for 2^X are in (1.183) and (1.184) respectively.

(1.181) COROLLARY [173, 1.6]. *If* X *is a continuum, then* C(X) *is contractible with respect to any absolute neighborhood retract.*

Proof. Use (1.171).

(1.182) COROLLARY [173, 1.9]. *If X is a continuum, then $C(X)$ has trivial shape. In particular, $C(X)$ is a fundamental retract and is movable.*

Proof. Use (1.171).

For some other results about shape and hyperspaces, see [178].

The following two results are proved using the 2^X part of (1.171). They are analogous to (1.181) and (1.182) respectively, but they are not stated in [173].

(1.183) COROLLARY. *If X is a continuum, then 2^X is contractible with respect to any absolute neighborhood retract.*

Proof. Use (1.171).

(1.184) COROLLARY. *If X is a continuum, then 2^X has trivial shape. In particular, 2^X is a fundamental retract and is movable.*

Proof. Use (1.171).

This completes the applications of (1.169) for this chapter. There are many more throughout the book.

We turn our attention to another aspect of the relationship between hyperspaces and inverse limits which has been of some interest. It is the subject of hyper-onto and minimal representations, which began with a paper of Fort and Segal [109]. First, some discussion seems appropriate.

Recall that if X is a continuum, then by (1.164) there exists an inverse sequence $\{P_n, f_n\}_{n=1}^{\infty}$, where the spaces P_n are compact connected polyhedra, such that

$$X \cong \varprojlim\{P_n, f_n\}_{n=1}^{\infty};$$

furthermore, by (1.169) [and (0.52)], $2^X \cong \varprojlim\{2^{P_n}, f_n^*\}_{n=1}^{\infty}$ and $C(X) \cong \varprojlim\{C(P_n), \hat{f}_n\}_{n=1}^{\infty}$. As is well-known (see the proof in [116] of (1.164) above), the bonding maps f_n may be chosen so that they are onto, i.e., $f_n[P_{n+1}] = P_n$ for each $n = 1, 2, \ldots$. Assume they are so chosen. Then, clearly, the bonding maps f_n^* are onto. However, and this is the point, *the bonding maps $\hat{f}_n$ need not be onto*, i.e. [by (0.49.1)], the bonding maps $f_n: P_{n+1} \xrightarrow{\text{onto}} P_n$ need not be weakly confluent. The following example is included for the purpose of illustrating and motivating the notion of a hyper-onto representation [to be defined in (1.186)].

(1.185) EXAMPLE. For each $n = 1, 2, \ldots$, let $P_n = S^1$ and let $f_n: P_{n+1} \xrightarrow{\text{onto}} P_n$ be given by

$$f_n(e^{i\theta}) = \begin{cases} e^{i2\theta}, & 0 \le \theta \le \pi \\ e^{-i2\theta}, & \pi \le \theta \le 2\pi. \end{cases}$$

Then, $\hat{f}_n$ does not map $C(P_{n+1})$ onto $C(P_n)$ for any $n = 1, 2, \ldots$. This is seen by taking $A \in C(P_n)$ to be

$$\{e^{i\theta} \in P_n : 0 \le \theta \le \frac{\pi}{2} \text{ or } \frac{3\pi}{2} \le \theta \le 2\pi\}.$$

On the other hand, as the reader who is familiar with inverse limits can easily see, the continuum $X = \varprojlim\{P_n, f_n\}_{n=1}^{\infty}$ is chainable, i.e., X can be represented as an inverse limit of arcs [see

(0.23)]. We now give such a representation for X. For each $n = 1, 2, \ldots$, let $A_n = [0, 2\pi]$ and let $g_n: A_{n+1} \xrightarrow{\text{onto}} A_n$ be given by

$$g_n(\theta) = \begin{cases} 2\theta & , \quad 0 \leq \theta \leq \pi \\ -2\theta + 4\pi & , \quad \pi \leq \theta \leq 2\pi. \end{cases}$$

Let $Y = \varprojlim \{A_n, g_n\}_{n=1}^{\infty}$. Note that Y is homeomorphic to the "Buckethandle" continuum defined in (1.209.3). We define a homeomorphism $h: Y \xrightarrow{\text{onto}} X$ as follows: For each $n = 1, 2, \ldots$, define k_n" $A_n \xrightarrow{\text{onto}} P_n$ by

$$k_n(\theta) = e^{i\theta} \quad \text{for each } \theta \in A_n.$$

It is easy to verify that

(1) $f_n \circ k_{n+1} = k_n \circ g_n$ for each $n = 1, 2, \ldots$.

From (1) it follows easily that if $(y_1, y_2, \ldots, y_n, \ldots) \in Y$, then $(k_1(y_1), k_2(y_2), \ldots, k_n(y_n), \ldots) \in X$. Define $h: Y \to X$ by

$$h((y_1, y_2, \ldots, y_n, \ldots)) = (k_1(y_1), k_2(y_2), \ldots, k_n(y_n), \ldots)$$

for each $(y_1, y_2, \ldots, y_n, \ldots) \in Y$. Since $X \subset [\Pi_{n=1}^{\infty} P_n]$ and each coordinate function k_n is continuous, h is continuous. As a general fact: When the coordinate spaces are compact and the maps k_n are onto and commute as in (1), then the map h defined as above is onto X. Using the formulas for the maps f_n, g_n, and k_n, it follows that h is one-to-one. Therefore, Y is homeomorphic to X and (thus) $Y = \varprojlim \{A_n, g_n\}_{n=1}^{\infty}$ is a representation of X as an inverse limit of arcs. Clearly, the maps g_n are weakly confluent [comp., proof of (1.191)]; hence, by (0.49.1), the induced maps

$$\hat{g}_n: C(A_{n+1}) \to C(A_n)$$

are onto. Therefore, we have represented X as an inverse limit of polyhedra in two ways--in the first representation, none of the induced maps

$$\hat{f}_n : C(P_{n+1}) \to C(P_n)$$

sends $C(P_{n+1})$ onto $C(P_n)$ but, in the second representation, all the induced maps

$$\hat{g}_n : C(A_{n+1}) \to C(A_n)$$

are onto.

Keeping in mind what has been done in (1.185), we introduce the following definition.

(1.186) DEFINITION [109, p.133]. A continuum X *is said to have a hyper-onto representation* provided that there is an inverse sequence $\{P_n, f_n\}_{n=1}^{\infty}$ such that

(i) $X \cong \varprojlim \{P_n, f_n\}_{n=1}^{\infty}$,

(ii) P_n is a compact connected polyhedron for each $n = 1, 2, \ldots$,

and

(iii) $\hat{f}_n : C(P_{n+1}) \to C(P_n)$ maps $C(P_{n+1})$ onto $C(P_n)$ for each $n = 1, 2, \ldots$.

An inverse sequence $\{P_n, f_n\}_{n=1}^{\infty}$ satisfying (i) through (iii) is called a *hyper-onto representation for* X.

Fort and Segal were interested in hyper-onto representations because such representations imply the existence of minimal representations [to be defined in (1.188)]. Some discussion of a general nature is in order.

One of the values of inverse limits is that they allow complicated continua to be "approximated by" relatively simple ones [comp., (1.164)]. One class of simple continua is the class of one-dimensional polyhedra. By (1.162) and [150, p.72], any continuum which is an inverse limit of one-dimensional polyhedra must be one-dimensional. However, as we see from Kelley's work [(1.107) or (1.109)], $\dim[C(P)]$ can be as large (but finite) as we like by making suitable choices of one-dimensional polyhedra P. Thus, it is conceivable that for any continuum X, there are one-dimensional polyhedra P_1, P_2,..., P_n,... such that $C(X)$ is homeomorphic to an inverse limit of the hyperspaces $C(P_n)$. In fact, it is true:

(1.187) THEOREM [109, Theorem 3]. *If X is any continuum, then there exists an inverse sequence $\{C(P_n), \psi_n\}_{n=1}^{\infty}$ such that $C(X) \cong \varprojlim\{C(P_n), \psi_n\}_{n=1}^{\infty}$, where the spaces P_n are one-dimensional compact connected polyhedra [but the bonding maps ψ_n are not necessarily onto, nor are they necessarily induced in the sense of (0.49) from maps between the polyhedra P_n].*

The result in (1.187) is especially nice in view of what we now know about the structure of $C(P)$ when P is a one-dimensional polyhedron (see [85], [86], and [87]; a few of these results were discussed in (I.D) of this chapter). However, (1.187) has one drawback--the bonding maps ψ_n are not necessarily onto [perhaps it is too much to hope that they can be induced maps]. Note the following definition:

(1.188) DEFINITION [109, p.129]. Let X be a continuum. We say that $C(X)$ *has a minimal representation* provided that there exist one-dimensional compact connected polyhedra $P_1, P_2, \ldots, P_n, \ldots$ such that
$$C(X) \cong \varprojlim \{C(P_n), \gamma_n\}_{n=1}^{\infty}$$
where γ_n maps $C(P_{n+1})$ onto $C(P_n)$ for each $n = 1, 2, \ldots$.

From the discussion which follows (1.186), we see why minimal representations are of interest. The following result expresses the reason Fort and Segal were interested in hyper-onto representations.

(1.189) THEOREM [109, Theorem 5]. *If a continuum* X *has a hyper-onto representation, then* $C(X)$ *has a minimal representation.*

Now, we discuss what is known about hyper-onto and minimal representations. Some of the treatment is different, and some is more general, than that in the literature. This is because Lelek's relatively new notion of weakly confluent mapping has not been used in this context before.

Note the following general theorem. It simply restates (1.186) in a more convenient form, one which avoids the induced maps and, thus, is intrinsic in terms of the inverse sequence representing X. Also, it focuses one's attention on the types of bonding maps one must use to have a hyper-onto representation.

(1.190) THEOREM. *A continuum* X *has a hyper-onto representation if and only if there is an inverse sequence* $\{P_n, f_n\}_{n=1}^{\infty}$ *satisfying (i) and (ii) of (1.186) and such that* $f_n: P_{n+1} \xrightarrow{onto} P_n$ *is weakly confluent for each* $n = 1, 2, \ldots$.

Proof. The result in (0.49.1) shows that (iii) of (1.185) is equivalent to $f_n: P_{n+1} \xrightarrow{onto} P_n$ being weakly confluent for each $n = 1, 2, \ldots$.

(1.191) COROLLARY [109, p.133]. *If* X *is a chainable continuum, then* X *has a hyper-onto representation. In particular, if* $\{P_n, f_n\}_{n=1}^{\infty}$ *is an inverse sequence of arcs* P_n *and onto bonding maps* $f_n: P_{n+1} \xrightarrow{onto} P_n$ *such that*

$$X \cong \varprojlim \{P_n, f_n\}_{n=1}^{\infty},$$

then $\{P_n, f_n\}_{n=1}^{\infty}$ *is a hyper-onto representation for* X.

Proof. By (0.67.3), any mapping from an arc onto an arc is weakly confluent. Hence, the result follows from (1.190) and (0.23)

(1.192) COROLLARY [109, Theorems 6 and 7]. *If* $X \cong \varprojlim \{P_n, f_n\}_{n=1}^{\infty}$ *where the spaces* P_n *are compact connected polyhedra and the bonding maps* $f_n: P_{n+1} \xrightarrow{onto} P_n$ *are monotone or open, then* $\{P_n, f_n\}_{n=1}^{\infty}$ *is a hyper-onto representation for* X. *Hence*, C(X) *has a minimal representation*.

Proof. As remarked in (0.45.3), any monotone or open map is confluent and, hence, weakly confluent. Thus, the first part of

(1.192) follows from (1.190). The second part is a consequence of the first part and (1.189).

Though the result in (1.192) is quite simple, it was used in [109, pp.136-137] to show that the Menger universal curve and the Sierpinski universal plane curve each have hyper-onto representations. As might be expected, however, the result in (1.192) is somewhat limited in its possible applications. Fort and Segal illustrate this by giving an example of a locally connected continuum which is not homeomorphic to any inverse limit of polyhedra with either monotone or open onto bonding maps [109, pp.134-136].

In (1.191) we saw that every chainable continuum has a hyper-onto representation. Moreover, we saw in the proof of (1.191) that this was a consequence of the fact that every mapping of an arc onto an arc is weakly confluent. In (1.185) we saw that mappings of a circle onto a circle need not be weakly confluent. So, the following two questions arise: Does every circle-like continuum have a hyper-onto representation? What conditions are necessary and sufficient for a mapping of a circle onto a circle to be weakly confluent [see (1.202)]? Rogers [286, Proposition 2.2] answered the first of these questions affirmatively [see (1.195)]. An examination of his proof reveals that he obtained the following partial answer to the second question: If a mapping of a circle onto a circle is essential, then it is weakly confluent. At about the same time as [286] was being written, Feuerbacher [106, Lemma 5] proved the more general result in (1.193). The proof of (1.193) given below is somewhat different than the one for circles in [286] and the one in [106].

(1.193) LEMMA [106, Lemma 5]. *If a mapping of a continuum* X *onto a circle is essential, then it is weakly confluent* [see (0.67.4) and (1.202)].

Proof. Let f: $X \xrightarrow{onto} S^1$ be a mapping. Assume f is not weakly confluent. Then, by (0.49.1),

(1) $\hat{f}: C(X) \to C(S^1)$ is not onto.

Note that, since f maps X onto S^1,

(2) $\hat{f}[F_1(X)] = F_1(S^1)$.

Recall from (0.55) that

(3) $C(S^1)$ is a 2-cell and $F_1(S^1)$ is the manifold boundary of $C(S^1)$.

By (1) through (3), there is a retraction

$r: \hat{f}[C(X)] \xrightarrow{onto} F_1(S^1)$.

Since $F_1(S^1) \cong S^1$ and since, by (1.176), C(X) has property (b), we have that

(4) $r \circ \hat{f}: C(X) \to F_1(S^1)$ is homotopic to a constant mapping.

Let i: $X \to C(X)$ be given by $i(x) = \{x\}$ for each $x \in X$ and let j: $F_1(S^1) \to S^1$ be given by $j(\{z\}) = z$ for each $\{z\} \in F_1(S^1)$. It is easy to verify the commutativity of the following diagram:

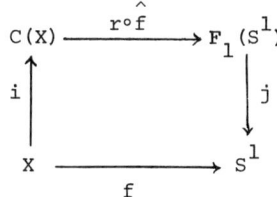

It follows easily from (4) and the commutativity of the above diagram that f is homotopic to a constant mapping. This completes the proof of (1.193).

It will be convenient to have the following lemma for the proof of (1.195).

(1.194) LEMMA [177, 3.1]. *If* X *is a continuum such that for each* $\varepsilon > 0$ *there exists an* ε-*mapping of* X *into* S^1 *which is homotopic to a constant mapping, then* X *is chainable.*

Proof. Assume X satisfies the hypotheses in (1.194). First we will show that

(*) for each $\varepsilon > 0$, there exists an
    ε-mapping of X into R^1.

To prove (*), let $\varepsilon > 0$. By the hypotheses in (1.194), there exists an ε-mapping $f: X \to S^1$ such that f is homotopic to a constant mapping. Hence, by [338, 6.2, p.226], there exists a mapping $\alpha: X \to R^1$ such that $f = e^{i\alpha}$. Clearly, since f is an ε-mapping and f is the composition of α followed by e^i, α must be an ε-mapping. This proves (*). It is well-known that (*) is equivalent to X being chainable; but, for completeness, we show why (*) implies X is chainable. Let $\varepsilon' > 0$. By (*), there is an ε'-mapping $g: X \to R^1$. Cover $g[X]$ by a chain [see (0.22)] of open intervals $U_1, U_2, \ldots, U_n$ such that
$\text{diam}[g^{-1}(U_j)] < \varepsilon'$ for each $j = 1, 2, \ldots, n$.
Then, $\{g^{-1}(U_1), g^{-1}(U_2), \ldots, g^{-1}(U_n)\}$ is a chain of open subsets

of X convering X such that each link has diameter less than ε'. Therefore, since $\varepsilon' > 0$ was arbitrary, we have proved that X is chainable [see (0.23)]. This completes the proof of (1.194).

(1.195) THEOREM [286, Proposition 2.2]. *If X is a circle-like continuum, then X has a hyper-onto representation.*

Proof. Let X be a circle-like continuum. If X is also chainable, then X has a hyper-onto representation as an inverse limit of arcs by (1.191). Thus, for the purpose of proof, assume X is not chainable. Then, by (1.194),

(1) there exists $\varepsilon > 0$ such that any
ε-mapping of X into S^1 is essential.

By (0.24), there exists an inverse sequence $\{X_n, f_n\}_{n=1}^{\infty}$, where each of the spaces X_n is a circle and each of the bonding maps f_n sends X_{n+1} onto X_n, such that $X \cong \varprojlim \{X_n, f_n\}_{n=1}^{\infty}$. By (1) and the second part of (1.162), there exists N such that

(2) $\pi_k: \varprojlim \{X_n, f_n\}_{n=1}^{\infty} \to X_k$ is essential for each $k \geq N$.

It follows easily from (2) and (1.155) that

(3) $f_k: X_{k+1} \xrightarrow{\text{onto}} X_k$ is essential for each $k \geq N$.

By (3) and (1.193),

(4) $f_k: X_{k+1} \xrightarrow{\text{onto}} X_k$ is weakly confluent for each $k \geq N$.

As is easy to prove, $\varprojlim \{X_n, f_n\}_{n=1}^{\infty} \cong \varprojlim \{X_k, f_k\}_{k=N}^{\infty}$, and hence

(5) $X \cong \varprojlim \{X_k, f_k\}_{k=N}^{\infty}$.

The result now follows from (4), (5), and (1.190).

(1.196) REMARK. Let X be a circle-like continuum. As might

have been expected, and as the proof of (1.195) shows, X has a hyper-onto representation as an inverse limit of arcs if X is chainable or as an inverse limit of circles if X is not chainable. Hence, in either case, C(X) has a minimal representation as an inverse limit of 2-cells by (0.54), (0.55), (0.67.3), (1.169), and the proof of (1.195).

We note the following theorem which shows that if P is a polyhedron, then C(P) has a special minimal representation.

(1.197) THEOREM [see (1.198)]. *Let* P *be a compact connected polyhedron. Then,* C(P) *has a minimal representation as follows: There exist one-dimensional polyhedra* P_n, $P_1 \subset P_2 \subset \cdots \subset P_n \subset \cdots \subset P$, *and retractions* Φ_n,

$$\Phi_n : C(P) \xrightarrow{onto} C(P_n) \quad \text{for each} \quad n = 1, 2, \ldots,$$

such that $C(P) \cong \varprojlim \{C(P_n), \psi_n\}_{n=1}^{\infty}$, *where* ψ_n *denotes the restriction of* Φ_n *to* $C(P_{n+1})$ *for each* $n = 1, 2, \ldots$.

(1.198) REMARK. The result in (1.197) is partly [109, Theorem 1] and partly its proof. The sets denoted E_n in [109, Theorem 1] are the hyperspaces $C(P_n)$ in (1.197). The statement in [109, Theorem 1] that E_n is acyclic is evident here by (1.173). Though the polyhedra P_n in (1.197) were not mentioned in [109, Theorem 1], it is clear from the proof in [109] that the sets E_n in [109, Theorem 1] were the hyperspaces of one-skeletons of certain subdivisions of P (see [109, Lemma 3]).

(1.199) REMARK. One of the lemmas used in [109] to prove (1.197) has become a fundamental tool in infinite-dimensional topology. In particular, it plays an important role in [77] in the proofs of (1.97) and (1.98). We state it below.

(1.199.1) LEMMA [109, Lemma 4]. *Let* Y *be a compact metric space and assume* $Y_1, Y_2, \ldots, Y_n, \ldots$ *are compact subsets of* Y. *For each* $n = 1, 2, \ldots,$ *assume there are mappings*

$$f_n : Y \xrightarrow{onto} Y_n \text{ and } g_n : Y_{n+1} \xrightarrow{onto} Y_n$$

such that $f_n = g_n \circ f_{n+1}$ *and such that the sequence* $\{f_n\}_{n=1}^{\infty}$ *converges uniformly to the identity mapping on* Y. *Then, the function* h *defined by*

$$h(y) = (f_1(y), f_2(y), \ldots, f_n(y), \ldots)$$

for each $y \in Y$ *is a homeomorphism of* Y *onto* $\varprojlim \{Y_n, g_n\}_{n=1}^{\infty}$.

Note that we can think of $Y \cong \varprojlim \{Y_n, g_n\}_{n=1}^{\infty}$ in (1.199.1) as being "approximated from within." Traditionally, inverse limits have been thought of as nested intersections; we have adopted the nested intersection viewpoint in this chapter. Recently, the "interior approximation" approach, exemplified in (1.199.1), has been of significant value in infinite-dimensional topology. Roughly speaking-- when the Y_n's are nice (usually, nested Hilbert cubes) and the g_n's are nice (usually, near-homeomorphisms which are retractions), then it can be shown that Y is a Hilbert cube. I refer the reader to [76, p.928] for more information. The first use of the "interior approximation" approach was the proof in [109] of (1.197).

Except for some details we have discussed all that is known about hyper-onto and minimal representations. Answers to the following questions are not known.

(1.200) QUESTION [109, p.134]. Does every continuum have a hyper-onto representation?

(1.201) QUESTION [109, p.134]. For any continuum X, does C(X) have a minimal representation?

If the answer to (1.200) is "yes," then by (1.189) the answer to (1.201) is "yes."

(1.202) QUESTION. What conditions characterize weakly confluent mappings of circles onto circles? The converse of (1.193), for X a circle, is false [see (0.67.4)].

Especially in view of how much is now known about C(P) when P is a one-dimensional polyhedron ([85], [86], and [87]), it would seem to be fruitful to know more about hyper-onto and minimal representations. Such information, together with the work of Duda referenced above, should be able to yield new insight into the structure of hyperspaces.

EXERCISES[4]

(1.203) EXERCISES ON ORDER ARCS

(1.203.1) Let X be any continuum and let A, B ε 2^X. Prove that there is exactly none, one, or an uncountable number of order arcs in 2^X from A to B. Find necessary and sufficient conditions in order that there be exactly one order arc from A to B [comp., (1.57)].

(1.203.2) Let $\Gamma(X)$ be as defined in (1.27.1). By (1.29) and (1.30), $\Gamma(X)$ is compact [see (1.34)]. Give a "direct" proof ["using the topology of 2^{2^X} and not using $s(\omega)$"] that $\Gamma(X)$ is compact. This type of proof will be similar to the proof of Lemma 4 in [213].

(1.203.3) Let $\Gamma(X)$ be as defined in (1.27.1). Define [see (1.7)] b: $\Gamma(X) \to 2^X$ and e: $\Gamma(X) \to 2^X$ by:

b(α) = $\cap \alpha$ for each $\alpha \varepsilon \Gamma(X)$,

e(α) = $\cup \alpha$ for each $\alpha \varepsilon \Gamma(X)$.

Prove that b and e are continuous functions from $\Gamma(X)$ onto 2^X and that b and e are homotopic. Also, prove that $F_1(2^X)$ is a strong deformation retract of $\Gamma(X)$ [see (1.27.3)]. These results have not appeared in the literature.

(1.203.4) Let X be a continuum and let A, B ε C(X) such that $A \subset B \neq A$. Prove that if there is a point $p \in A$ such that p is a terminal point of B, then there is one and only one order arc in 2^X from A to B. Give an example to show that the converse is false.

(1.203.5) Let X be a continuum lying in a Banach space E. Let $cc(X)$ be as defined in (18.1). Prove that $cc(X)$ is arcwise connected if and only if X is arcwise connected [Hint: To prove half the theorem, use ψ_p as defined in the proof of (18.13)]. This theorem has not appeared in the literature.

(1.204) EXERCISES ON NONDISCONNECTING HYPERSPACES

(1.204.1) Prove that no point of $C(X)$ [resp., 2^X] is a cut point of $C(X)$ [resp., 2^X]. The $C(X)$ part is a very special case of (2.15).

(1.204.2) For any continuum X, any Whitney map μ for $C(X)$, and any proper subset Λ of any $\mu^{-1}(t)$, prove that $C(X) \setminus \Lambda$ is connected.

(1.204.3) Prove that if Λ is any subset of $C_1(X)$, then $2^X \setminus \Lambda$ is connected. This result is original with the book and appears in [395, 2.11] for the case when $\Lambda = C_1(X)$. It will be used in the proof of (6.11).

(1.205) EXERCISES ON TERMINAL CONTINUA

(1.205.1) Let X and Y be continua and let $f: X \xrightarrow{onto} Y$ be a monotone mapping. Prove that if A is a terminal subcontinuum of X, then $f[A]$ is a terminal subcontinuum of Y.

(1.205.2) Let X be a continuum and let A be a subcontinuum of X. Let Y be the decomposition space of the upper semi-continuous decomposition $X/_A$ of X obtained by "shrinking A to a point"; denote the point by p. Prove that A is terminal in X if and only if p is a terminal point of Y.

(1.205.3) Prove that no proper subcontinuum of $C(X)$ is terminal in $C(X)$.

(1.206) EXERCISES ON COMPOSANTS

(1.206.1) Prove (1.52.1).

(1.206.2) Prove that a continuum has exactly one, three, or uncountably many composants.

(1.207) EXERCISES ON HEREDITARILY INDECOMPOSABLE CONTINUA

(1.270.1) Prove that $[\cup \Gamma] \in \Gamma$ for every arc Γ in $C(X)$ if and only if X is hereditarily indecomposable [half the result is in (1.60)].

(1.207.2) Prove that a continuum X is hereditarily indecomposable if and only if S_o can not be embedded in C(X), where S_o is the familiar sin[1/x]-continuum defined in (8.22). Hint: The proof is *easy*.

(1.207.3) Show that the three dendroids drawn in Figure (1.207.3) can be embedded in C(X) for any continuum X [see (1.74)].

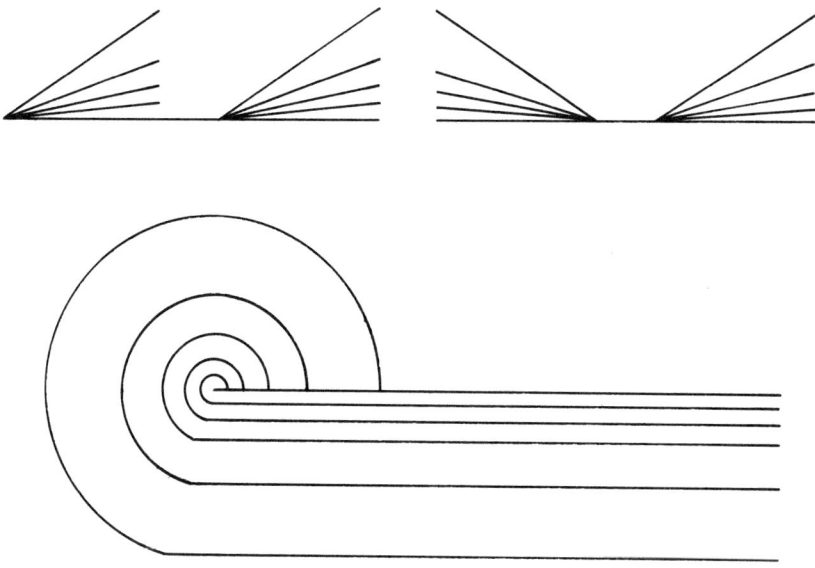

Figure (1.207.3)

(1.207.4) Prove that if f is a monotone mapping from an indecomposable [resp., hereditarily indecomposable] continuum X onto a continuum Y, then Y is indecomposable [resp., hereditarily indecomposable]. These facts are well-known. For an application, see the proof of (14.1).

(1.207.5) Prove that if f is an open mapping [or, more generally, a confluent mapping] from an hereditarily indecomposable continuum X onto a continuum Y, then Y is hereditarily indecomposable. However [comp., (1.207.4)], prove that there is an open mapping of the "Buckethandle" continuum [see (1.209.3)] onto an arc. Prove that there is an open mapping of any solenoid [see (1.209.4)] onto a circle. These facts are well-known. For an application of the last one, see (14.73.26).

(1.207.6) Prove that if X is any continuum and Y is any hereditarily indecomposable continuum, then any mapping from X onto Y is confluent [70, Theorem 4].

(1.207.7) Let X be any hereditarily indecomposable continuum. Prove that C(X) is not embeddable in Cone(X). It is known that Cone(X) is not embeddable in C(X); this follows directly from [244, Theorem 3], here (1.118), and was also obtained in [281, Theorem 9]. However, prove that there is a monotone open mapping from Cone(X) onto C(X). This is most of Theorem 9 of [280], where it is also claimed that such a mapping exists which is *arc-preserving* (= the image of an arc is an arc or a point [147, p.103]). It is easy to see that the mapping used in the proof in [280] is not arc-preserving. In fact, prove that no mapping from Cone(X) onto C(X) is arc-preserving.

(1.207.8) Let X be any hereditarily indecomposable continuum. Prove that if $\{A_n\}_{n=1}^{\infty}$ is a sequence of subcontinua of X such that $\{A_n\}_{n=1}^{\infty}$ converges to $A \in C(X)$, then the sequence $\{C(A_n)\}_{n=1}^{\infty}$ converges to $C(A)$ [Hint: Use the function $\alpha: X \to C(C(X))$ given by $\alpha(x)$ is the unique order arc in $C(X)$ from $\{x\}$ to X]. In the language of Chapter XV, this result says that any hereditarily indecomposable continuum is C*-smooth. The result was obtained independently by myself and by Grispolakis and Tymchatyn [443].

(1.208) EXERCISES ON LOCALLY CONNECTED CONTINUA

(1.208.1) Prove (1.92); prove it for Hausdorff continua with the Vietoris topology for the hyperspaces.

(1.208.2) Prove that if X is a continuum such that $C(X) \setminus F_1(X)$ is locally connected, then X is locally connected (and conversely). Prove the analogous result obtained by replacing $C(X)$ with 2^X. These results have not appeared in the literature.

(1.208.3) Can $C(X) \setminus F_1(X)$ be homeomorphic to R^n for some n and some continuum X? If so, then determine all such n and X.

(1.208.4) Prove that if X is a continuum such that $C(X)$ is homeomorphic to B^n for some $n = 1, 2, \ldots,$ then $n = 2$ and X is an arc or a circle.

(1.208.5) For any continuum X and any subcontinuum M of X, let [as in (1.56)] $C_M(X) = \{K \in C(X) : M \subset K\}$. Prove that $C_M(X)$ is a locally connected subcontinuum of $C(X)$ [418]. Hence, by (0.74.1), $C_M(X)$ is an absolute retract [438]. The proof of this last fact is somewhat different than the proof in [438]. For related results and applications, see (1.214.3), [418], [438], and [439]. For subcontinua A and B of X, prove that $\{K \in C(X) : A \subset K \subset B\}$ is a locally connected subcontinuum of $C(X)$ [418], hence an absolute retract.

(1.208.6) For any continuum X and any subcontinuum M of X, let $C^M(X) = \{K \in C(X) : K \cap M \neq \emptyset\}$. Prove that $C^M(X)$ is an arcwise connected subcontinuum of $C(X)$. Prove that if M is a locally connected subcontinuum of X, then $C^M(X)$ is an absolute retract [Hint: Use (0.74.1)]. Prove that X is locally connected if and only if $C^M(X)$ is locally connected for each subcontinuum M of X. These results have not appeared in the literature.

(1.208.7) Prove that if X is any locally connected continuum, then the union of any open subset of $C(X)$ is an open subset of X; give an example to show that the converse is false. Give an example of a continuum Y and a nonempty open subset of $C(Y)$ whose union is nowhere dense in Y. On the other hand prove that if Z is any continuum, then the union of any open subset of 2^Z is an open subset of Z.

(1.209) EXERCISES ON INVERSE LIMITS

(1.209.1) Define $f: [0, 1] \to [0, 1]$ by

$$f(t) = \begin{cases} 2t, & 0 \le t \le \frac{1}{2} \\ 1, & \frac{1}{2} \le t \le 1 \end{cases}$$

and let $\{X_n, f_n\}_{n=1}^{\infty}$ be the inverse sequence where, for each $n = 1, 2, \ldots$, $X_n = [0, 1]$ and $f_n = f$. Prove that $\varprojlim \{X_n, f_n\}_{n=1}^{\infty}$ is an arc.

(1.209.2) An inverse sequence $\{X_n, f_n\}_{n=1}^{\infty}$ of (nondegenerate) continua X_n will be called an *indecomposable inverse sequence* provided that, for each $n = 1, 2, \ldots$, whenever A_{n+1} and B_{n+1} are subcontinua of X_{n+1} such that $X_{n+1} = A_{n+1} \cup B_{n+1}$, we have $f_n[A_{n+1}] = X_n$ or $f_n[B_{n+1}] = X_n$. Prove that if $\{X_n, f_n\}_{n=1}^{\infty}$ is an indecomposable inverse sequence, then $\varprojlim \{X_n, f_n\}_{n=1}^{\infty}$ is an indecomposable continuum. The theorem is due to van Heemert [409], though the terminology originates here. Some applications are in (1.209.3) and (1.209.4). Related work is in [402] and, in another direction, in [373].

(1.209.3) Define $f: [0, 1] \to [0, 1]$ by

$$f(t) = \begin{cases} 2t, & 0 \le t \le \frac{1}{2} \\ -2t + 2, & \frac{1}{2} \le t \le 1 \end{cases}$$

and let $\{X_n, f_n\}_{n=1}^{\infty}$ be the inverse sequence where, for each $n = 1, 2, \ldots$, $X_n = [0, 1]$ and $f_n = f$. Prove that $\varprojlim \{X_n, f_n\}_{n=1}^{\infty}$ is an indecomposable continuum [Hint: Use (1.209.2)]. The continuum $\varprojlim\{X_n, f_n\}_{n=1}^{\infty}$ is called the *"Buckethandle,"* and is homeomorphic to the continuum described in [186, p.204] and pictured in Figure 4 of [186, p.205]. Prove that any nondegenerate proper subcontinuum of the "Buckethandle" continuum is an arc.

(1.209.4) For a given $p \in \{1, 2, 3,...\}$, define $f^p: S^1 \to S^1$ by $f^p(z) = z^p$ for each $z \in S^1$ (where z^p means the $p\underline{\text{th}}$ power of z, complex multiplication). Let $\{X_n, f_n\}_{n=1}^{\infty}$ be the inverse sequence where, for each $n = 1, 2, ...$, $X_n = S^1$ and $f_n^p = f^p$, p fixed. Prove that if $p \neq 1$, then $\varprojlim\{X_n, f_n^p\}_{n=1}^{\infty}$ is an indecomposable continuum [Hint: Use (1.209.2)]. Prove that if $p = 1$, then $\varprojlim\{X_n, f_n^p\}_{n=1}^{\infty}$ is a circle. The continuum $\varprojlim\{X_n, f_n^p\}_{n=1}^{\infty}$ is called the *p-adic solenoid*. When $p = 2$, it is called the *dyadic solenoid*. A *solenoid* is any continuum of the form $\varprojlim\{X_n, f_n^{p(n)}\}_{n=1}^{\infty}$ where $p(n) \in \{1, 2, 3,...\}$ for each $n = 1, 2, ...$. Prove that a solenoid is either a circle or an indecomposable continuum. Prove that any nondegenerate proper subcontinuum of a solenoid is an arc. Prove that a solenoid admits the structure of a topological group. All these results are well-known.

(1.209.5) Prove (12.17.1).

(1.210) EXERCISES ON LOCATING CELLS

(1.210.1) Prove that if X contains subcontinua A and B such that $A \cap B$ has at least n components, $n < \infty$ [resp., $n = \infty$], then $C(X)$ contains an n-cell [resp., a Hilbert cube].

This result is in [244, Theorems 4 and 14].

(1.210.2) Let $Y = \bigcup_{n=1}^{\infty} Y_n$ where, for each $n = 1, 2, \ldots$, $Y_n = \{(\frac{\lambda}{n}, \frac{\lambda}{2n}) \in R^2 : 0 \le \lambda \le 1\}$.

The continuum Y is the so-called "hairy point." (a) Describe $C(Y)$ [244, Example 2]. The hairy point is a particular smooth fan. Geometric models for $C(X)$ when X is *any* smooth fan have been determined in [439]. For an exercise related to (a) above, see (b) of (1.214.3). (b) Prove that if X is any continuum and $A \in [C(X) \setminus F_1(X)]$, then there is a hairy point in $C(X)$ with A as its "vertex." An application of (b) is in the proof of (1.74.1).

(1.210.3) Prove (2.5) [show how to fill in the details in the paragraph following (2.5)].

(1.211) EXERCISES ON DIAMETER

(1.211.1) Let (X, d) be a continuum. Considering diam [defined in (0.41)] as a function from 2^X into R^1, prove that diam is continuous [186, p.55].

(1.211.2) Let (X, d) be a continuum. Considering diam as a function from $C(X)$ onto $[0, \text{diam}(X)]$, prove that diam is a monotone mapping in the sense of (0.45.1) [comp., (1.211.3)]. Give an example to show that it is not necessarily an open mapping. These facts have not been noted in the literature.

(1.211.3) Let S^1 be the unit circle in the plane R^2 and let d denote the metric for S^1 inherited from the Euclidean metric for R^2 [see (0.19)]. Considering diam as a function from 2^{S^1} onto [0, 2], prove that diam is neither a monotone nor an open mapping [comp., (14.67) and (14.68)].

(1.211.4) Prove (16.16).

(1.212) MORE EXERCISES ON THE INDUCED MAP[5]

(1.212.1) Prove that if f is a monotone mapping of a continuum X onto a continuum Y, then $\hat{f}^{-1}(B)$ is an arcwise connected subcontinuum of C(X) for each $B \in C(Y)$.

(1.212.2) Prove that a mapping f of a continuum X onto a continuum Y is monotone if and only if $\hat{f}$ is monotone [381].

(1.212.3) A mapping g: X → Y is said to be *light* provided that $g^{-1}(g(x))$ is totally-disconnected for each $x \in X$. A mapping g: X → Y is said to be *hereditarily irreducible* provided that for any given subcontinuum A of X, no proper subcontinuum of A maps onto g[A]. Prove that, for a mapping f of a continuum X into a continuum Y, $\hat{f}$ is light if and only if f is hereditarily irreducible (due to J. B. Fugate and myself). See (1.213.6).

(1.213) MORE EXERCISES ON WHITNEY MAPS[6]

(1.213.1) Let X be a continuum and let μ be any Whitney map for $C(X)$. (a) Prove that if $0 \leq t \leq \mu(X)$, then $\mu^{-1}(t)$ is nondegenerate [see (0.15)]. (b) Prove that if $t \in [0, \mu(X)]$, then $\cup \mu^{-1}(t) = X$. (c) Prove that if $A \in \mu^{-1}([t, \mu(X)])$, then $\cup [C(A) \cap \mu^{-1}(t)] = A$. (d) Prove that μ is an open mapping from $C(X)$ onto $[0, \mu(X)]$. Part (d) was originally proved in [93, p.1032]--see (14.44) here.

(1.213.2) Let X be a circle and let μ be any Whitney map for $C(X)$. Show that $\mu^{-1}([t_o, \mu(X)])$ is a 2-cell for any $t_o < \mu(X)$ [comp., the paragraph after (14.73.17); also, see (14.73.31)]. More generally [see (14.7)], prove that if Y is any continuum and μ is any Whitney map for $C(Y)$, then $\mu^{-1}([t_o, \mu(Y)])$ is a 2-cell if $\mu^{-1}(t_o)$ is a circle [comp., (1) of (14.73.13)]. These results have not appeared in the literature. Give an example of a continuum Z and a Whitney map μ_o for $C(Z)$ such that, for some t_o, $\mu_o^{-1}([t_o, \mu_o(Z)])$ is a 3-cell.

(1.213.3) Let S_o be the familiar $\sin[1/x]$-continuum defined in (8.22). Let μ be any given Whitney map for $C(S_o)$. Determine the topological types of $\mu^{-1}(t)$ for all t [249, p.404].

(1.213.4) Prove that a continuum X is indecomposable if and only if given any proper subcontinuum A of X and any $t < \mu(X)$,

$X(A, \mu, t) \neq \mu^{-1}(t)$ [half of this result is in (0.68.5)].

(1.213.5) Let X be a continuum and let μ be a Whitney map for C(X). Prove that if $\mu^{-1}(t)$ contains an arc for some t, then C(X) contains a circle. Prove the converse using results in Chapter I and (14.8.1).

(1.213.6) Let $f: X \to Y$ be a mapping of a continuum X into a continuum Y and let μ be a Whitney map for C(Y). Prove that $\mu \circ \hat{f}$ is a Whitney map for C(X) if and only if $\hat{f}$ is light [Hint: Use (1.212.3)]. This result has not appeared in the literature; it extends part of (0.51)--see (0.68.6).

(1.213.7) A continuum X is said to be *planar* [resp., *nonplanar*] provided that X is [resp., is not] embeddable in R^2. Give an example of a planar continuum X and a Whitney map μ for C(X) such that $\mu^{-1}(t)$ is not planar for some t. Give an example of a nonplanar continuum X and a Whitney map μ for C(X) such that $\mu^{-1}(t)$ is planar for some t, $0 < t < \mu(X)$. These examples show, in the language of Chapter XIV, that planarity and nonplanarity are not Whitney properties [comp., (14.18)].

(1.213.8) Give an example of a continuum X and a Whitney map μ for C(X) such that $\mu^{-1}(t)$ is a simple triod for some $t_o > 0$ [267, Example 14(b)]. By using (14.47) and material in Chapter I, prove the following result: If X is a continuum for which

(1.214.1)   STRUCTURE OF HYPERSPACES   207

there is a Whitney map μ for $C(X)$ such that $\mu^{-1}(t)$ is a finite graph for each $0 < t < \mu(X)$, then X is an arc or a circle. See (14.50) and (14.51) for related results. Also, the hypotheses of the result here can be weakened to only assuming that, for some sequence $\{t_n\}_{n=1}^{\infty}$ such that $t_n \to 0$ as $n \to \infty$, $\mu^{-1}(t_n)$ is a finite graph for each $n = 1, 2, \ldots$.

(1.213.9) Give an example of a 3-odd M and a Whitney map μ for $C(M)$ such that for some $\delta > 0$, $\mu^{-1}(t)$ does not contain a 2-cell for any $t < \delta$ [comp., (14.33) and (14.55)].

(1.214) EXERCISES ON TORUŃCZYK'S THEOREM

(1.214.1) A closed subset A of a continuum X (with metric d) is said to be a *Z-set in* X^7 [354, p.2] provided that for each $\varepsilon > 0$ there exists a mapping $f_\varepsilon : X \to X \setminus A$ such that $d(f_\varepsilon(x), x) < \varepsilon$ for each $x \in X$. Prove that any closed subset of a Z-set in X is a Z-set in X and that the union of two Z-sets in X is a Z-set in X (these facts are well-known; see, for example, [354, p.2]). A mapping $f: X \to X$ is said to be a *Z-map* [431] provided that the range $f[X]$ of f is a Z-set in X. Prove that if X is the Hilbert cube, then the identity mapping of X onto X is a uniform limit of Z-maps. This well-known fact, together with the special case of a result of Toruńczyk [431] stated below in (1.214.2), gives a characterization of the Hilbert cube from among all compact absolute retracts.

(1.214.2) The following amazing result is a special case of a theorem due to Toruńczyk [431, Theorem 1]: If Y is a compact absolute retract such that the identity mapping of Y onto Y is a uniform limit of Z-maps, then Y is a Hilbert cube. Using this result, (1.96), the Bing-Moise result mentioned in (0.65.4), and (f) of (0.65.3), prove (1.97) and the second part of (1.98); proofs along these lines are due to Toruńczyk and are in the last part of his paper [431]. I remark that though the first part of (1.98) is a simple consequence of (1.96) and the special case above of Toruńczyk's theorem, it is not appropriate to prove the first part of (1.98) in this way. This is because Toruńczyk uses, in his proof [431], the recent result of R. D. Edwards that the cartesian product of any compact absolute retract with the Hilbert cube is the Hilbert cube. Eberhart [438] and Eberhart and I [439] have recently used the special case above of Toruńczyk's theorem to obtain some results about C(X) when X is not (necessarily) locally connected.

(1.214.3) (a) Use (1.208.5) and the special case of Toruńczyk's theorem in (1.214.2) to prove (1.99.3). (b) Also prove that if v is the vertex of the cone, Cone(E), over any infinite compact metric space E, then $C_{\{v\}}(\text{Cone}(E))$ is a Hilbert cube [439]; for applications and related results, see [439]. (c) Prove that for any continuum X and any $A \in 2^X$, 2^X_A defined by

$$2^X_A = \{B \in 2^X : A \subset B\}$$

is a retract of 2^X. (d) Use (c) in proving that the following three statements are equivalent:

(1) X is a locally connected continuum;

(2) 2^X_A is locally connected for each $A \in 2^X$;

(3) 2^X_A is an absolute retract for each $A \in 2^X$.

The results in (c) and (d) have not appeared in the literature. However, the fact that (1) of (d) implies (2) and (3) of (d) is implicit in [165, Remark, p.29]. Use (d) and the special case of Toruńczyk's theorem in (1.214.2) to prove (1.99.1). Note that (1.99.2) is a consequence of (1.208.5) and Edward's theorem mentioned in (1.214.2).

NOTES

1. The notion of an order arc has been used in the more general setting of partially ordered spaces. It is difficult to say when the notion was first formally introduced in the context of hyperspaces. The definition in (1.2) appears in [234, p.6].

2. In relation to (1.27.3), William Nowell and I have shown that X is a locally connected continuum if and only if $\Gamma(X)$ is a locally connected continuum.

3. This question has recently been answered affirmatively in [443].

4. The exercises at the end of Chapter I are based on the material in Chapters 0 and I. Also, see Footnote 5 on p.52.

5. Other exercises on the induced map [besides those in (1.212)] are in (0.67).

6. Other exercises on Whitney maps [besides those in (1.213)] are in (0.68).

7. The definition of Z-set in (1.214.1) is the definition used in [354], and is the definition usually used in hyperspace theory (for example, see [292, p.240]). Anderson [1] is responsible for

the notion of property Z, defined as follows: A closed subset A of a continuum X is said to have *property Z in X* [1, p.366] provided that for each homotopically trivial nonempty open subset U of X, U\A is homotopically trivial and nonempty. In all of our applications, the notion of property Z agrees with the notion of Z-set given above.

8. With respect to (1.44) through (1.47), see the footnote on p.251.

II.
DIMENSION OF C(X)

Complete determination of the dimension of $C(X)$ when X is any locally connected continuum was done by Kelley [see (1.109); also, (1.110)]. Mazurkiewicz [see (1.95)] showed that, for any continuum X, $\dim[2^X] = \infty$ by showing that 2^X contains a Hilbert cube (this result was also obtained by Kelley [165, 5.1]). As Kelley said [165, p.22]: "...the question of dimension is resolved except for the dimension of $C(X)$ when X is non-Peanian." This question remained untouched until 1959. Then, Segal observed [using his result that $C_\infty(X) \cong C(X)$; see (1.169)]:

(2.0) THEOREM [294, section 2]. *If* $X = \varprojlim\{X_n, f_n\}_{n=1}^\infty$ *where* X_n *is a continuum such that* $\dim[C(X_n)] \le k < \infty$ *for each* $n = 1, 2, \ldots$, *then* $\dim[C(X)] \le k$.

The result in (2.0) is what Segal proved in section 2 of [294]; the theorem he states is not a good representation of what he proved. Segal [294, Example 2.1] used (2.0) to show that $\dim[C(\Sigma_2)] \le 2$ when Σ_2 is the dyadic solenoid. The following argument shows how Segal proved this fact, and indicates how to prove the more general result in (2.0). Since Σ_2 is an inverse limit of circles X_n,

$C(\Sigma_2)$ is homeomorphic to an inverse limit of 2-cells $C(X_n)$ by (0.55) and (1.169); hence, $\dim[C(\Sigma_2)] \leq 2$ by (1.162) and [150, p.72]. *Note that, though Segal did not say so, the argument shows $\dim[C(X)] \leq 2$ for any circle-like or chainable continuum X* [for the chainable case, use (0.54) instead of (0.55)]. Indeed, it was a problem in the folklore to determine the exact dimension of $C(X)$ for all chainable and all circle-like continua X (the problem was rumored to be due to R. H. Bing). Let us note that if X is any continuum which contains an arc, then $C(X)$ contains a 2-cell by (0.54); hence, if X is chainable or circle-like and contains an arc, it follows from what we have just observed about Segal's work that $\dim[C(X)] = 2$. For the special case of the p-adic solenoids, this conclusion was reached by Rhee [276, 2.1]. In [93], the following two results [(2.1) and (2.2)] were proved. Because of the development of material in Chapter I, the proof of (2.1) given here is shorter and somewhat different than the proof in [93].

(2.1) THEOREM [93, Theorem 1]. *If X is any continuum, then $\dim[C(X)] \geq 2$. Hence, [93, Corollary 1], if X is a chainable or circle-like continuum, $\dim[C(X)] = 2$.*

Proof. To prove the first part of (2.1), let X be a continuum. Suppose $\dim[C(X)] = 1$. Then, by the second part of (1.145), X must be an hereditarily indecomposable continuum. Thus, since $C(X)$ is one-dimensional and an arcwise connected continuum [by (1.13)], we have by (1.64) that $C(X)$ is a dendroid [see (0.28) for definition]. Let x_1, $x_2 \in X$ such that $x_1 \neq x_2$. By (1.8) and

(1.11), there exist order arcs α_i in $C(X)$ from $\{x_i\}$ to X, $i = 1$ and 2. Clearly, $F_1(X) \cup \alpha_1 \cup \alpha_2$ is a non-unicoherent subcontinuum of $C(X)$. This contradicts that $C(X)$ is a dendroid. Hence, $\dim[C(X)] \neq 1$ and we have proved the first part of (2.1). The second part of (2.1) follows immediately from the first part and the italicized statement in the paragraph preceding (2.1).

(2.2) THEOREM [93, Theorem 2]. *If X is an hereditarily indecomposable continuum, then $\dim[C(X)] = 2$ or $\dim[C(X)] = \infty$. Furthermore [for X hereditarily indecomposable]:*

(2.2.1) *If $\dim[C(X)] = 2$, then each monotone open image of X is one-dimensional or a one-point space;*

(2.2.2) *If each monotone open image of X is finite-dimensional, then $\dim[C(X)] = 2$;*

(2.2.3) *Hence [93, p.1031], $\dim[C(X)] = \infty$ if $\dim[X] \geq 2$ [comp., (1.85)].*

Proof. We will prove four facts, (a) through (d) below. Then, we use these facts to prove (2.2). The reader who is familiar with decomposition spaces will note that (b) is a simple consequence of (1.77.2). However, for the benefit of persons not knowledgeable about decomposition spaces, we indicate a direct proof for (b) which does not use (1.77.2).

Throughout the proof, assume X is an hereditarily indecomposable continuum [by (1.207.4)],

(a) If there is a monotone open image X_o of X such that $\dim[X_o] \geq 2$, then there is a monotone open image Y of X such that $\dim[Y] = \infty$.

To prove (a), assume $g: X \xrightarrow{\text{onto}} X_o$ is a monotone open mapping such that $\dim[X_o] \geq 2$. Since the monotone image of an hereditarily indecomposable continuum is hereditarily indecomposable [this is easy to prove],

(1) X_o is an hereditarily indecomposable continuum such that $\dim[X_o] \geq 2$.

By (1), we may replace X in (1.83) with X_o and apply (1.83.1) to X_o to obtain a monotone open mapping f from X_o onto a continuum Y such that $\dim[Y] = \infty$. Therefore, since $f \circ g: X \xrightarrow{\text{onto}} Y$ is a monotone open mapping, we have proved (a). Next, we prove the following general fact [which is well-known, but whose proof we include for the reasons mentioned above]:

(b) If Q is a monotone open image of a continuum Z, then Q is embeddable in $C(Z)$.

To prove (b), let Z be a continuum and let $k: Z \xrightarrow{\text{onto}} Q$ be a monotone open mapping. Let $\Gamma = \{k^{-1}(q): q \in Q\}$. Since k is a monotone mapping, $k^{-1}(q)$ is a subcontinuum of Z for each $q \in Q$--so, as we may, consider Γ as a subspace Γ_o of $C(Z)$ by inclusion, i.e.,

$\Gamma_o = \{A \in C(Z): A = k^{-1}(q) \text{ for some } q \in Q\}$.

Define $h: Q \xrightarrow{\text{onto}} \Gamma_o$ by $h(q) = k^{-1}(q)$ for each $q \in Q$. Using convergence in $C(Z)$ in the sense of (0.5) [see (0.7)] and using the fact that k is an open mapping, it is easy to prove that h is continuous (the continuity of h is proved in this way in [338, 4.31, p.130]). Clearly, h is one-to-one. Therefore, since Q is

compact, h is a homeomorphism. This completes the proof of (b).
Now, let μ be a given fixed Whitney map for $C(X)$. We prove

(c) If $\dim[C(X)] < \infty$, then $\dim[\mu^{-1}(t)] \leq 1$

for each $t \in [0, \mu(X)]$.

To prove (c), assume there exists $t_0 \in [0, \mu(X)]$ such that $\dim[\mu^{-1}(t_0)] \geq 2$. Then, by (1.80.2), $\mu^{-1}(t_0)$ satisfies the conditions on X_0 in (a). Hence, there exists Y as in (a) and, by (b), Y is embeddable in $C(X)$. Thus, since $\dim[Y] = \infty$ [by (a)], $\dim[C(X)] = \infty$. This proves (c). Next we prove

(d) If $\dim[\mu^{-1}(t)] \leq 1$ for each $t \in [0, \mu(X)]$,

then $\dim[C(X)] = 2$.

To prove (d), first note the following general fact from dimension theory [150, pp.91-92; especially the restatement at the beginning of the proof on p.92]:

(#) If $\alpha: M_1 \to M_2$ is a closed mapping between separable
metric spaces M_1 and M_2 and if $\dim[\alpha^{-1}(y)] \leq m$
for each $y \in M_2$, then $\dim[M_1] \leq m + \dim[M_2]$.

Now, assume as in (d) that $\dim[\mu^{-1}(t)] \leq 1$ for each $t \in [0, \mu(X)]$. Then, using (#) with $M_1 = C(X)$, $M_2 = [0, \mu(X)]$, and $\alpha = \mu$, we see that

(2) $\dim[C(X)] \leq 1 + 1 = 2$.

By the first part of (2.1),

(3) $\dim[C(X)] \geq 2$.

By (2) and (3), $\dim[C(X)] = 2$ and, thus, we have proved (d). By (c) and (d) we have that: If $\dim[C(X)] < \infty$, then $\dim[C(X)] = 2$. This is the first part of (2.2). To prove (2.2.1), assume the con-

clusion of (2.2.1) is false. Then, using (a) and (b), we have that
$\dim[C(X)] = \infty$. Thus, (2.2.1) holds. To prove (2.2.2), assume the
hypothesis of (2.2.2). Then, by (a), each monotone open image of X
is of dimension less than or equal to one. Hence, by (1.80.2), the
hypothesis of (d) holds. Therefore, by (d), $\dim[C(X)] = 2$. This
proves (2.2.2). Since X is a monotone open image of itself,
(2.2.3) follows from (2.2.1) and the first part of (2.2). Let us
note that (2.2.3) also follows from (1.85). This completes the
proof of (2.2).

The results in (2.1) and (2.2) did much to answer Kelley's
question [in the first paragraph of this chapter]. First, the
exact dimension of $C(X)$ is determined by (2.1) when X is a
chainable or circle-like continuum. Secondly, (2.2) can be used to
determine $\dim[C(X)]$ for many hereditarily indecomposable continua
X. Of course, by (2.2.3) or (1.85), $\dim[C(X)] = \infty$ if X is an
hereditarily indecomposable continuum such that $\dim[X] \geq 2$. Also,
consider the following examples. For each $j = 1, 2, \ldots,$ Cook [69]
constructed an hereditarily indecomposable continuum X^j in the
plane such that $X^j = \underleftarrow{\lim}\{X_n^j, f_n\}_{n=1}^{\infty}$ where each X_n^j is the union
of j circles all meeting at a point (and otherwise disjoint). For
each $j = 1, 2, \ldots,$ X^j has j complementary domains in R^2 and
each nondegenerate proper subcontinuum of X^j is a pseudo-arc. By
(1.109), $\dim[C(X_n^j)] = 2j$ for each $n = 1, 2, \ldots$. Hence, by (2.0),
$\dim[C(X^j)] \leq 2j < \infty$ for each $j = 1, 2, \ldots$. Thus, by (2.2),

$\dim[C(X^j)] = 2$ [93, p.1031]. Therefore, the exact dimension of each $C(X^j)$ is determined by (2.2). We will return to Cook's continua later [see the paragraphs before and after (2.14)]. For now, let us note another application of (2.2) which shows that (2.2) almost completely answers Kelley's question! By (2.2.3), $\dim[C(X)] = \infty$ whenever X contains an hereditarily indecomposable continuum Z such that $\dim[Z] \geq 2$. Hence, by Bing's result [see (1.53)], we arrive at Rogers' application of (2.2):

(2.3) THEOREM [283, Theorem 5]. *If* X *is a continuum such that* $\dim[X] \geq 3$, *then* $\dim[C(X)] = \infty$.

Proof and comment. In the last two sentences of the paragraph preceding (2.3), we gave Rogers' proof of (2.3). In the process of writing the book, the result in (1.86) was obtained. It also implies (2.3).

Thus, Kelley's question is solved when $\dim[X] \geq 3$. No example is known of a two-dimensional continuum X such that $\dim[C(X)] < \infty$.

(2.4) QUESTION [250, 2.2]. If X is any two-dimensional continuum, then must $\dim[C(X)] = \infty$? Some partial answers and references to partial answers are given below.

In relation to (2.4), we mention the following partial results.

(2.5) THEOREM [see each part]. *The theorem is in four parts:*

(1) *If* X *is an arcwise connected continuum such that* dim[X] $\geq$ 2, *then* dim[C(X)] = ∞ [*283, Corollary 3"*].

(2) *If* X *contains a subcontinuum which is homeomorphic to a cartesian product of two (nondegenerate) continua, then* dim[C(X)] = ∞ [*283, Corollary 2*].

(1') *Under the same hypotheses as (1),* C(X) *contains a Hilbert cube.*[*244, Theorem 13*].

(2') *Under the same hypothesis as (2),* C(X) *contains a Hilbert cube* [*244, Theorem 9*].

The results in (1) and (2) of (2.5) are proved in [283] using (1.100); the ones in (1') and (2') of (2.5) are proved in [244] using (1.103). The proofs are straightforward except for mentioning that the proofs for (1) and (1') use the fact that any continuum of dimension greater than one contains a nondegenerate indecomposable subcontinuum [212]. Other results along the lines of (2.5) are in [287] and [244, section 5]. Also, see (1.210.1).

The following theorem gives an affirmative answer to Question 2 of [93, p.1031]. It also gives some information relevant to (2.4). In view of (2.1) and (2.3), the result is only of interest for determining information about dim[C(X)] when dim[X] = 2.

(2.6) THEOREM [173, 2.3]. *If* X *is a finite-dimensional continuum, then* dim[C(X)] > dim[X].

A generalization of (2.6) is in (2.10). Also, Lau [190] has

extended (2.6) to Hausdorff continua using cohomological dimension.

We now turn our attention from two-dimensional to one-dimensional continua.

(2.7) QUESTION [93, Question 3]. Is there a one-dimensional hereditarily indecomposable continuum X such that $\dim[C(X)] = \infty$ [see (2.8)]? By (2.2), this is equivalent to asking: Is there a one-dimensional hereditarily indecomposable continuum X such that X admits a monotone open image of dimension greater than one?

Krasinkiewicz [175] has the following partial answer to (2.7), for which he gives a very elegant and beautiful proof. Another partial answer due to Krasinkiewicz is given in (14.17.1).

(2.8) THEOREM [175]. *If* X *is an hereditarily indecomposable continuum which is embeddable in the plane, then* C(X) *is embeddable in* R^4 *[see (2.9)]. Hence, by (2.2),* $\dim[C(X)] = 2$ *and each monotone continuous decomposition of* X *is at most one-dimensional.*

Proof. Let X be as above and consider $X \subset R^2$. For each point $x = (x_1, x_2) \in R^2$, let $\pi_i(x) = x_i$ for each $i = 1$ and 2. We define three real-valued functions ρ_1, ρ_2, and ψ on C(X) as follows [recall that $X \subset R^2$ so, as we may, consider any member of C(X) as a subcontinuum of R^2]: For each $A \in C(X)$, let

$\rho_i(A) = \text{g.l.b.}(\pi_i[A])$ for each $i = 1$ and 2;

$\psi(A) = \text{diam}(\pi_1[A])$.

It is easy to prove that ρ_1, ρ_2, and ψ are continuous. Let μ

be a Whitney map for $C(X)$ [μ is continuous, see (0.50)]. Define $h: C(X) \to R^4$ by

$$h(A) = (\rho_1(A), \rho_2(A), \psi(A), \mu(A))\text{ for each } A \in C(X).$$

Clearly, h is continuous and thus, since $C(X)$ is compact [see (0.8)], it suffices to show that h is one-to-one. Let $A, B \in C(X)$ such that $A \neq B$. Suppose $h(A) = h(B)$. Then

(1) $\rho_1(A) = \rho_1(B)$,

(2) $\rho_2(A) = \rho_2(B)$,

(3) $\psi(A) = \psi(B)$,

(4) $\mu(A) = \mu(B)$.

Since $A \neq B$, we have by (4) and (1.78) that

(5) $A \cap B = \emptyset$.

Now, let L_1, L_2, and L_3 be the three lines in R^2 given by:

$$L_1 = \{(x_1, x_2) \in R^2: x_1 = \rho_1(A)\}$$
$$L_2 = \{(x_1, x_2) \in R^2: x_2 = \rho_2(A)\}$$
$$L_3 = \{(x_1, x_2) \in R^2: x_1 = \rho_1(A) + \psi(A)\}.$$

By (1),

(6) $A \cap L_1 \neq \emptyset \neq B \cap L_1$

and, by (2),

(7) $A \cap L_2 \neq \emptyset \neq B \cap L_2$.

By (1) and (3), $\rho_1(A) + \psi(A) = \rho_1(B) + \psi(B)$; hence [note, as may be helpful, $\rho_1(A) + \psi(A) = \text{l.u.b.}(\pi_1[A])$ and $\rho_1(B) + \psi(B) = \text{l.u.b.}(\pi_1[B])$],

(8) $A \cap L_3 \neq \emptyset \neq B \cap L_3$.

Now we prove that $L_1 \neq L_3$ (to be used when we define D). Suppose $L_1 = L_3$. Then, $\psi(A) = 0$ which implies $A \subset L_1$. Thus, since A

is indecomposable, $A \in F_1(X)$. Hence, by (4) [and (0.50)], $B \in F_1(X)$. Let $A = \{(a_1, a_2)\}$ and $B = \{(b_1, b_2)\}$. By (1), $a_1 = b_1$ and, by (2) $a_2 = b_2$. Therefore, $A = B$ which contradicts our assumption that $A \neq B$. Thus, we have proved

(9) $L_1 \neq L_3$.

By (6) there exist points $a \in [A \cap L_1]$ and $b \in [B \cap L_1]$. By (5), $a \neq b$. Thus, since $a, b \in L_1$, $\pi_2(a) \neq \pi_2(b)$. So, for the purpose of proof, assume $\pi_2(a) < \pi_2(b)$. By (8), there is a point $a' \in [A \cap L_3]$ and, by (7), there is a point $b' \in [B \cap L_2]$. For some real number r_2, $A \cup B$ is contained in the rectangle

$$D = \{(x_1, x_2) \in R^2 : \rho_1(A) \leq x_1 \leq \rho_1(A) + \psi(A)$$
$$\text{and } \rho_2(A) \leq x_2 \leq r_2\}.$$

Note that a, a', b, and b' are in the boundary S of the rectangle D and that the set $\{a, a'\}$ disconnects S between b and b'. Therefore, by facts about the topology of the plane, $A \cap B \neq \emptyset$ (if $A \cap B = \emptyset$, then one can obtain two disjoint arcs α and β in D such that $\alpha \cap S = \{a, a'\}$ and $\beta \cap S = \{b, b'\}$; then, the θ-curve Theorem [186, p.511] is contradicted). This contradicts (5). Therefore, $h(A) \neq h(B)$. We have proved h is one-to-one. This completes the proof of (2.8).

The last part of (2.8) would seem to be especially interesting to mathematicians working in decomposition theory.

(2.9) REMARK. Tymchatyn [316] has recently shown that if X is an hereditarily indecomposable continuum which is embeddable in the plane, then $C(X)$ is embeddable in R^3. This strengthens the conclusion to the first part of (2.8). Also, the hypotheses are less restrictive than those in (3.4) and [282, Theorem 3.2 and Corollaries 3.3 and 3.5]. Tymchatyn's result answers a question in [176, Problem 3, p.185].

The first paper to use Whitney maps μ to investigate the dimension of $C(X)$ was [93]--see the proof of (2.2). In (1.85) and and (1.86), we gave some results about $\dim[\mu^{-1}(t)]$. A result related to $\dim[\mu^{-1}(t)]$ is in (14.33). Rogers [282] has studied relationships between $\dim[C(X)]$ and $\dim[\mu^{-1}(t)]$. We state his results in (2.10), (2.11), and (2.13).

(2.10) THEOREM [282, Theorem 1.5]. *For any continuum* X *and any Whitney map* μ *for* $C(X)$, $\dim[\mu^{-1}(t)] + 1 \leq \dim[C(X)]$ *for each* $t \in [0, \mu(X)]$.

The result in (2.10) is a generalization of (2.6), as can be seen by taking $t = 0$.

(2.11) THEOREM [282, Theorem 1.6]. *For any continuum* X *such that* $\dim[C(X)] < \infty$ *[see (2.12)] and for any Whitney map* μ *for* $C(X)$, *there exists* $t_o \in [0, \mu(X)]$ *such that* $\dim[\mu^{-1}(t_o)] + 1 = \dim[C(X)]$.

(2.12) REMARK. Though it is not stated in [282], the result in

(2.11) is not necessarily true when $\dim[C(X)] = \infty$. In other words, the dimension of $C(X)$ may be infinite and yet the dimension of $\mu^{-1}(t)$ may be finite for each $t \in [0, \mu(X)]$. This last statement is true for the continuum X in [244, Example 3].

(2.13) THEOREM [282, Theorem 1.7]. *For any continuum* X *such that* $\dim[C(X)] < \infty$ *and for any Whitney map* μ *for* $C(X)$, *either* $\dim[\mu^{-1}(t_o)] > \dim[X]$ *for some* $t_o \in [0, \mu(X)]$ *or* $\dim[\mu^{-1}(t)] = \dim[X]$ *for infinitely many* $t \in [0, \mu(X)]$.

Many continua which are of interest have the property that each proper subcontinuum is a member of a well-studied class of continua. Such is the case, for example, for Cook's continua X^j mentioned in the paragraph following the proof of (2.2). There we saw that the dimension of $C(X^j)$ is equal to two for each $j = 1, 2, \ldots$ Each proper subcontinuum of each X^j is a pseudo-arc. Hence, using the second part of (2.1), the following general result of Rogers applies to these continua to give another proof that $\dim[C(X^j)] = 2$.

(2.14) THEOREM [282, Theorem 2.2]. *Let* X *be any continuum. If* $\dim[C(Y)] < n < \infty$ *for each proper subcontinuum* Y *of* X, *then* $\dim[C(X)] < n$.

Cook [69] also constructed an hereditarily indecomposable continuum X^∞ in the plane such that X^∞ has infinitely many complementary domains in R^2 and such that each nondegenerate proper subcontinuum of X^∞ is a pseudo-arc. Hence, by (2.1) and (2.14), $\dim[C(X)] = 2$. At the time Eberhart and I wrote [93], we could not determine $\dim[C(X^\infty)]$ because (2.0) did not apply to show $\dim[C(X^\infty)] < \infty$ [hence, equal to two by (2.2)]. I also mention that (2.8), which is more recent than [93] and [282], shows $\dim[C(X^\infty)] = 2$. However, the *first* paper to have a result which applies to $C(X^\infty)$ was [176]. The appropriate result is given in (3.4); it, together with (2.2), shows that $\dim[C(X^\infty)] = 2$.

We have finished our discussion of results and problems about the *global* dimension of $C(X)$. The question of the dimension of $C(X)$ at certain points of $C(X)$ is of interest. For locally connected continua, this question is fairly well settled by (1.107), (1.109), and related results of Duda ([85], [86], and [87]). The first result about this for non-locally connected continua was given in [93, pp.1032-1033] where it was proved: $\dim_A[C(P)] = 2$ when A is any nondegenerate proper subcontinuum of the pseudo-arc P. It was asked [93, Question 4] if $\dim_A[C(P)] = 2$ when A is a singleton or $A = P$. Also, it was conjectured [93, p.1033] that $C(P)$ is a 2-dimensional Cantor manifold [a continuum of dimension n ($1 \leq n < \infty$) is called an *n-dimensional Cantor manifold* provided that it can not be disconnected by a subset of dimension $\leq n-2$]. In [261, Theorem 4.1], Nishiura and Rhee verified the conjecture and, in [174], Krasinkiewicz proved the following more general result [some related results are in (1.204)].

(2.15) THEOREM [174]. *For any continuum* X, *no zero-dimen-*

sional subset of C(X) *can disconnect* C(X). *Hence, if* dim[C(X)] = 2, C(X) *is a two-dimensional Cantor manifold.*

Let us note [174, p.758] that if X is a simple triod, then dim[C(X)] = 3 but C(X) is not a Cantor manifold [see (10.2)].

In (173, 2.1], Krasinkiewicz generalized (2.15) by showing that if $f: X \to Y$ is a mapping from a continuum X into a continuum Y, then no zero-dimensional subset of $\hat{f}[C(X)]$ can disconnect $\hat{f}[C(X)]$ ($\hat{f}$ is the induced map defined in (0.49)).

For continua X which contain an n-odd, (1.102) can be used to determine a lower bound for the dimension of C(X) at certain points. The proofs of some of the results in [244, especially Theorem 4] can be used in the same way. On the other hand, the second part of (1.103) can be used to show that, for some continua X, $\dim_Y[C(X)] = \infty$ for certain $Y \in C(X)$. This can be observed in some special cases by examining how (1.103) is used to prove the results in [244, section 5].

A space is said to be *dimensionally homogeneous* provided that it has the same dimension at every point. Duda [85, 9.4] showed that if X is a finite graph which is not an arc or a circle, then C(X) is not dimensionally homogeneous. This result follows from the fact that every finite graph has a free arc and from the fact that any finite graph which is not an arc or a circle has a point of order greater than two [hence, (1.107.1) can be applied to prove the result]. We give the following somewhat more general result.

(2.16) THEOREM. *Let* X *be a locally connected continuum. Then,* C(X) *is dimensionally homogeneous if and only if* X *does not contain a free arc or* X *is an arc or a circle.*

Proof. Assume C(X) is dimensionally homogeneous. Assume X contains a free arc. Then, by using (0.54), we have that dim[C(X)] = 2. Hence, since X is a locally connected continuum, X must be an arc or a circle [any other locally connected continuum has a point of order greater than two]. Conversely, if X does not contain a free arc, C(X) $\cong$ I$_\infty$ by the second part of (1.98); if X is an arc or a circle, then C(X) is a 2-cell by (0.54) and (0.55).

(2.17) QUESTION. For what continua X is C(X) dimensionally homogeneous? The answer for locally connected continua is given in (2.16). Also, by (2.15), C(X) is dimensionally homogeneous if dim[C(X)] = 2 (as is the case, for example, when X is chainable or circle-like [by (2.1)] or when X satisfies the hypotheses of (2.8)). I conjecture that C(X) is dimensionally homogeneous if X is any cartesian product of (nondegenerate) continua-- the second part of (1.103) may be helpful in proving this.

We have covered almost everything known about dimension of C(X). As we saw in (2.8), embedding results can be used to determine the exact dimension of C(X). Embedding will be discussed in the next chapter.

Permit me to make the following subjective comment. Of all the questions in the book, I feel that (2.4) and (2.7), especially

(2.4), are among the most important and interesting. If an affirmative answer is obtained to (2.4), the methods used to obtain the answer will almost surely give new and significant information about the structure of two-dimensional continua [about which little is now known]. A complete answer to (2.7) would certainly be of interest not only to mathematicians in hyperspace theory, but also to those working in the theory of mappings between continua, dimension theory, etc.

III.
EMBEDDING C(X) IN R^n

The topic of this chapter was first investigated by Transue [315], whose elegant proof of the result in (3.1) stimulated much subsequent activity. Before stating and proving (3.1), we make several comments about the manner in which the proof will be presented.

Because of the importance to hyperspace theory of (3.1) and the techniques of its proof [315], I have tried to make the proof given here as self-contained as possible. Thus, though the statements in (7) and (8) of the proof are easily implied by well-known facts about the topology of the plane, they are proved for the special situation encompassed by (3.1) [I have also indicated (by reference) what specific general results about the topology of the plane can be used to obtain them]. To be more explicit, in the proof of (3.1) a mapping f onto C(X) is constructed. The mapping f is defined very naturally. Roughly speaking, the inverse image under f of a given member A of C(X) is a subcontinuum of R^3 obtained by multiplying each point of A by a given real number and then translating up in 3-space into a plane parallel to the x,y-plane [see the formula in (3) of the proof of (3.1)]. The fact that f behaves in this simple manner enables one to easily see [using (7)] that (8) holds. Thus, for example, we can avoid using the fact that separating the plane is a "positional invariant" [150, Corollary 1, p.101] in obtaining (8) from (7).

The Dyer-Hamstrom theorem [see (3.0)] will be used in the proof of (3.1). I state it without proof and refer the reader to [90] for the proof.

(3.0) THEOREM [90, Theorem 8, p.116]. *If* D *is an upper semi-continuous decomposition of* R^3 *such that each element of* D *is a subcontinuum of* R^3 *which is contained in a horizontal plane in* R^3 *and does not separate that plane, then the decomposition space of* D *is homeomorphic to* R^3.

One final comment about the proof given for (3.1) is appropriate. In Transue's original proof [315] he decomposed $X \times [0, 1]$. Instead, we will decompose the cone over X [see (2) in the proof of (3.1)]. This slight modification of Transue's proof is due to Krasinkiewicz [176, proof of 5.1]. The reason for doing it is that with no extra work a somewhat better result is obtained in (3.4). A stronger result [than (3.4)] is stated, without proof, in (2.9).

Now, we state and prove Transue's theorem.

(3.1) THEOREM [315]. *If* X *is an hereditarily indecomposable continuum in the plane which does not separate the plane, then* C(X) *is embeddable in* R^3.

Proof. Assume X satisfies the hypotheses of (3.1) and, without loss of generality, assume $X \subset R_0^3$ where, for each $t \in R^1$, R_t^3 is defined by

$$R_t^3 = \{(x, y, z) \in R^3 : z = t\}.$$

Let K(X) denote the geometric cone over X in R^3 with vertex $v = (0, 0, 1)$, i.e.,

$$K(X) = \{(1 - t) \cdot p + t \cdot v : p \in X \text{ and } t \in [0, 1]\}.$$

Let μ be a Whitney map for C(X) and assume, as we may [see (1.16)], that $\mu(X) = 1$. Then: For any given $p \in X$ and $t \in [0, 1]$, there exists by (1.78) one and only one $A \in C(X)$, denoted by A(p, t), such that $p \in A$ and $A \in \mu^{-1}(t)$. Thus, by letting

$$f[(1 - t) \cdot p + t \cdot v] = A(p, t)$$

for each $[(1 - t) \cdot p + t \cdot v] \in K(X)$, we see that we have defined a function f from K(X) onto C(X) [see (3.1.1) for historical comments]. Using the compactness of C(X) [see (0.8)] and the continuity of μ [see (0.50)], it follows easily that

(1) f is continuous.

By (1), [186, Theorem 3, p.65], and [185, Theorem 2, p.184]

(2) $D_1 = \{f^{-1}(A) : A \in C(X)\}$ is an upper semi-continuous decomposition of K(X) and the decomposition space of D_1 is homeomorphic to C(X).

Recall from the formula for f that:

$f[(1 - t) \cdot p + t \cdot v] = A$ means $A = A(p, t)$ which, in turn, means $p \in A$ and $\mu(A) = t$.

Hence, for each $A \in C(X)$,

$$f^{-1}(A) = \{[(1-t) \cdot p + t \cdot v] \in K(X) : p \in A \text{ and } \mu(A) = t\}.$$

In other words:

(3) For each $A \in C(X)$, $f^{-1}(A) =$

$$\{[(1 - \mu(A)) \cdot p + \mu(A) \cdot v] \in K(X) : p \in A\}.$$

It follows easily from (3) that

(4) $f^{-1}(A) \cong A$ for each $A \in [C(X) \setminus \{X\}]$

and [recall that $\mu(X) = 1$]

(5) $f^{-1}(X) = \{v\}$.

Also, since $v = (0, 0, 1)$ and $X \subset R_0^3$, it follows easily from (3) that

(6) $f^{-1}(A) \subset R_{\mu(A)}^3$ for each $A \in C(X)$.

For readers who are familiar with some general facts ([150, Theorem IV3, p.44], [150, Theorem VI4, p.83], [150, Theorem VI13, p.100], and [150, Corollary 1, p.101]), the statement in (8) below follows easily from (4), (5), and the conditions on X in (3.1). However, for reasons mentioned in the second paragraph of this chapter, we will give a direct proof of (8). First, we prove the statement in (7). Suppose some subcontinuum B of X separates R_0^3. Then, $R_0^3 \setminus B = U \cup V$ where U and V are two disjoint nonempty open subsets of R_0^3. Since X does not separate R_0^3, it follows easily that $U \subset X$ or $V \subset X$. Hence, since U and V are open subsets of the plane R_0^3, X must contain a 2-cell. This contradicts the assumption in (3.1) that X is hereditarily indecomposable. Thus, we have proved

(7) no subcontinuum of X separates R_0^3.

It is easy to see by using (7) and the formula in (3) [see the second paragraph of this chapter] that

(8) $f^{-1}(A) \subset R^3_{\mu(A)}$ [by (6)] does not separate $R^3_{\mu(A)}$ for any $A \in C(X)$.

Finally, it follows easily from the first part of (2) and from the definition of upper semi-continuous decomposition that

(9) $D = D_1 \cup \{\{(x, y, z)\}: (x, y, z) \in [R^3 \setminus K(X)]\}$ is an upper semi-continuous decomposition of R^3.

By (4) and (5), each element of D is a subcontinuum of R^3. Hence, by (8) and (9), (3.0) applies to show that the decomposition space of D is homeomorphic to R^3. Therefore, by the second part of (2), $C(X)$ is embeddable in R^3. This proves (3.1).

(3.1.1) COMMENTS ABOUT THE MAPPING f IN THE PROOF OF (3.1).

Let F_μ be the mapping defined in (16.12). It was first defined by Kelley in [165, proof of 3.3]. As the reader can easily see, when X is an hereditarily indecomposable continuum the mapping F_μ is the same as f (except that their domains and ranges are formally different). For the case when X is an hereditarily indecomposable continuum, F_μ has been rediscovered several times in the literature. As far as I know, what follows is the chronological rediscovery of F_μ: F_μ is the mapping described in the first part of the proof of the theorem in [315]; it is the mapping Φ used in [276, proof of 3.5] and in [280, proof of Theorem 9--see (1.207.7)], where it was attributed to Rhee [276]; it is the mapping λ in [176, 1.6], which is a statement of what was in [280, proof of Theorem 9].

It is perhaps worth noting that by using (3.1), (1.61), and [150, Theorem IV3, p.44], it is easy to see $\dim[C(X)] \leq 2$ for X as in (3.1). However, neither this fact nor the exact dimension of C(X) for such an X was discovered until [93, Corollary 3, p.1031]; see Chapter II.

Transue's paper [315] was actually written to give an affirmative answer to the following question raised by Andrew Conner [34, p.152]: "Can the space of all subcontinua of the pseudo-arc be embedded in E^3?" Thus, by (3.1), Transue answered Conner's question for a class of continua which included the pseudo-arc. Several years later, Henderson [139] proved the following theorem:

(3.2) THEOREM [139]. *If* X *is a chainable continuum, then* C(X) *is embeddable in 3-space.*

Thus, Henderson gave another class of continua [besides those in (3.1)] which included the pseudo-arc and whose hyperspaces are embeddable in 3-space. Another such class of continua was given by Rogers [284, Theorem 1] in the following result:

(3.3) THEOREM [284, Theorem 1]. *If* X *is a circle-like continuum in the plane, then* C(X) *is embeddable in* R^3 *[see (3.6)].*

The interesting and important feature of the proof in [284] of (3.3) is the use, for the first time, of techniques and results from differential topology. This novel approach should prove useful in the future.

It is worth mentioning that it follows from (1.169) and [152, Theorem 1, p.78] that if X is a circle-like continuum, then C(X) is embeddable in 4-space.

To show the necessity that X in (3.3) be embeddable in the plane, Rogers [284, Theorem 2] proved the result stated here in (8.1). Then he observed [284, p.168] that for solenoids X which are not embeddable in the plane, C(X) is not embeddable in 3-space by (8.1) and the result in [24] which says that the cone over any non-planar circle-like continuum is not embeddable in 3-space.

Let $X^1, X^2, \ldots, X^j, \ldots$ be Cook's continua mentioned in the paragraph following the proof of (2.2). As was mentioned there, $\dim[C(X^j)] = 2$ for each $j = 1, 2, \ldots$. Motivated somewhat by (3.1) and the fact that each nondegenerate proper subcontinuum of each X^j is a pseudo-arc, Eberhart and I asked the following question [93, Question 1, p.1031]: Can $C(X^j)$ be embedded in 3-space for each of Cook's continua X^j? In answering this question, Krasinkiewicz proved the following result (which was also obtained later [282, Corollary 3.3] by using a different method of proof).

(3.4) THEOREM [176, 5.1]. *If* X *is an hereditarily indecomposable continuum in the plane such that no proper subcontinuum of* X *separates the plane, then* C(X) *is embeddable in* R^3 *[see (2.9) for the recent result of Tymchatyn].*

Proof. See the paragraph between (3.0) and (3.1). In particular, the proof for (3.4) is the same as the one given for (3.1) except for one minor technical observation. Since X in (3.4) may separate R_0^3, (7) in the proof of (3.1) is not necessarily true with the hypotheses in (3.4). However, because of (5) in the proof of (3.1), (8) is still valid [by (3) through (5)] with the hypotheses in (3.4). Thus, with the modifications we have indicated, (3.4) is proved.

In [176, Problem 4, p.185], Krasinkiewicz asks the following question:

(3.5) QUESTION [176, Problem 4]. If X is a continuum such that C(X) is embeddable in R^3, then must X be embeddable in the plane [see (3.6)]?

Krasinkiewicz [178, 3.5] has proved the following result, which provides a converse to (3.3) and a partial answer to (3.5).

(3.6) THEOREM [178, 3.6]. *If* X *is a circle-like continuum, then* C(X) *is embeddable in* R^3 *if and only if* X *is embeddable in the plane.*

Earlier, Ball and Sher [16, 5.4] had proved that if X is a pseudo-solenoid, then C(X) is embeddable in 3-space if and only if X is embeddable in the plane (i.e., X is the pseudo-circle [16, p.791]). So (3.6) is an extension of Ball and Sher's result.

However, in general, (3.5) remains unsolved. In [250, p.190], the following related question was asked:

(3.7) QUESTION [250, 2.1]. If X is a continuum such that $C(X)$ is embeddable in R^3, then must X be one-dimensional? Let us note that if $C(X)$ is embeddable in R^3, then $\dim[X] \leq 2$ by (2.3). Also note that an affirmative answer to (3.5) would imply (by using [150, Theorem IV3, p.44]) an affirmative answer to (3.7). Finally, note that an affirmative answer to (2.4) would give an affirmative answer to (3.7); in fact, trying to obtain answers to (2.4) is in part the motivation for (3.7).

The following result is related to (3.7) in that it says X has a property in common with one-dimension subsets of R^3 --namely, no one-dimensional subset of R^3 can separate R^3 [150, Theorem IV4, p.48].

(3.8) THEOREM [250, 2.5]. *If* X *is a continuum such that* $C(X)$ *is embeddable in* R^n, *then no subcontinuum of* X *can separate* R^n *or* X *is a circle.*

Though the result in (3.8) is stated for general "n," it is only of interest for $n = 3$. The reasons are as follows. Assume X is as in (3.8). Then, $\dim[C(X)] < \infty$. Hence, by (2.3), $\dim[X] \leq 2$. Therefore: If $n > 3$, X can not separate R^n by [150, Theorem IV4, p.48]. Clearly [by (2.1)], $n = 1$ is vacuous. Finally, for the case when $n = 2$, we have proved the following theorem which completely determines what X must be.

(3.9) THEOREM [250, 2.3]. *If* X *is a continuum, then* C(X) *is embeddable in* R^2 *if and only if* X *is an arc or a circle.*

Proof. Assume X is a continuum and note the following two general facts:

(1) C(X) has property (b) [by (1.176)] or, equivalently [see (16.2)], every mapping of C(X) into S^1 is homotopic to a constant mapping;

(2) C(X) is aposyndetic [by (1.138)].

Now, assume C(X) ⊂ R^2. Then, by (1) and [150, Theorem VI13, p.100],

(3) C(X) does not separate R^2.

Jones [157, Theorem 1, p.139] has a result which, in particular, says that an aposyndetic continuum in the plane which does not separate the plane is locally connected. Hence, by (2) and (3), C(X) is locally connected. Thus, by (1.92), X is locally connected. It now follows easily using (1.107.1) [or using (1.109)] that X must be an arc or a circle. This proves half of (3.9). The other half is immediate from (0.54) and (0.55).

The short proof given above for (3.9) is the one given in [250]. Also, a longer but "less sophisticated" and "more geometric" proof of (3.9) was indicated in [250, p.191]. It used (15.13). By reading some of Chapter XV, the reader can get some idea of the geometry which is hidden by the proof given above.

Recently, Hagopian proved the following result related to (3.9).

(3.10) THEOREM [134, Theorem 4, p.234]. *Let* X *be a* λ *connected continuum. Then:* C(X) *can be ε-mapped into the plane for each* $\varepsilon > 0$ *if and only if* X *is chainable or circle-like.*

(3.11) QUESTION [134, p.234]. Does (3.10) remain valid without the assumption of λ connectedness? The "if" part of (3.10) is valid without λ connectedness as can be seen using (1.169), (1.162), (0.54), and (0.55).

We mention that in the proof of [134, Theorem 3] the author says that if X is chainable or circle-like, then C(X) is disklike by (1.169). Actually, the reason is (1.191) and (1.195).

Some authors have been concerned with special types of embeddings. For example, Ball and Sher [16, 5.1] give a "particularly nice" embedding of C(X) in 4-space for any pseudo-solenoid X [see the second paragraph after (3.3)--the emphasis in [16, 5.1] was on the *type* of embedding]. In [250, p.192], the following question was asked. Recall [from (0.63.4)] that an isometry is a distance-preserving map.

(3.12) QUESTION [250, 2.7]. When, if ever, does a continuum (X, d) have the property that $(C(X), H_d)$ or $(2^X, H_d)$ is isometric to a subset of ℓ_2?

To examine (3.12) somewhat closer, let us take the "simplest example": Let X be the closed unit interval [0, 1] with the

usual absolute-value metric d. Take $A_1 = \{0\}$, $A_2 = [0, 1]$,
$B_1 = \{1/2\}$, and $B_2 = [0, 1/2]$. Then, easy computations show that

$$H_d(A_1, B_i) = \frac{1}{2} = \frac{1}{2} \cdot H_d(A_1, A_2) = H_d(A_2, B_i)$$

for each i = 1 and 2. Hence, it follows easily that
$(C([0, 1]), H_d)$ can not be isometrically embedded in ℓ_2. However, I do not know the answer to (3.12) when (X, d) is a continuum which does not contain an isometric copy of any nondegenerate interval in R^1.

IV.
MAPPINGS BETWEEN X, 2^X, AND C(X)

The only study of the existence of mappings between X, 2^X, and C(X) was the one done in [250, section 3]. Let us note that these spaces are continuous images of one another whenever X is a locally connected continuum, as is easy to see using (1.92). Thus, we are interested in the situation when X is not locally connected. We will obtain some conditions under which mappings between X, 2^X, and C(X) exist, and we will show that they do not always exist. A number of questions are asked.

The only case in which there is always a mapping is given in the following result.

(4.1) THEOREM [250, 3.6]. *There is always a mapping from 2^X onto C(X).*

Proof. If X is locally connected, then (4.1) follows from (1.92). So, assume X is not locally connected. Then, by (1.39), there is a mapping f from 2^X onto the cone over the Cantor set. By (1.33), there is a mapping g from the cone over the Cantor set onto C(X). Hence, g∘f is a mapping of 2^X onto C(X).

The following result completely determines when there is a mapping from X or from C(X) [in fact, from any continuum] onto 2^X, and will be used to see that such mappings do not always exist.

(4.2) THEOREM [250, 3.5]. *Let* X *and* Y *be continua. Then: There is a mapping from* Y *onto* 2^X *if and only if*

(1) X *is locally connected*

   or

(2) Y *contains an open set with uncountably many components*.

Proof. Assume there is a mapping from Y onto 2^X. Assume (1) is false. Then, by (1.39), there is a mapping from 2^X onto the cone over the Cantor set. Hence, there is a mapping from Y onto the cone over the Cantor set. Therefore, by the "only if" part of (1.37), (2) holds. Conversely, if (1) holds, then it follows from (1.92) that there is a mapping from Y onto 2^X. Assume (2) holds. Then, by the "if" part of (1.37), there is a mapping from Y onto the cone over the Cantor set. By (1.33), there is a mapping from the cone over the Cantor set onto 2^X. Hence, there is a mapping from Y onto 2^X. We have proved (4.2).

Now, let us note that there is not always a mapping of X or of C(X) onto 2^X. For example, let X be the familiar sin[1/x]-continuum defined in (1) of (8.22). Then, neither X nor C(X) satisfies (2) of (4.2) [several proofs are given in (1.42) which show that C(X) does not satisfy (2); the "last proof" in (1.42) shows that $C(X)$ can not be mapped onto the cone over the Cantor set which, by (1.37), is equivalent to $C(X)$ not satisfying (2) of (4.2)].

(4.3) REMARK. In (1.44) it was proved that if X contains an open set with uncountably many components, then so does $C(X)$. In (1.46) we asked if the converse of (1.44) is true.[1] If the answer to (1.46) is "yes," then [using (4.2)] we would have the following result: There is a mapping from $C(X)$ onto 2^X if and only if X is locally connected or X contains an open set with uncountably many components. This result would be a characterization, in terms of X, of when $C(X)$ can be mapped onto 2^X. It would also show that being able to map $C(X)$ onto 2^X is equivalent to being able to map X onto 2^X.

Now, (4.1) and (4.2) leave open the following three questions, for which we will give some partial solutions.

(4.4) QUESTION. When is there a mapping from 2^X onto Y? Of special interest is the case when $Y = X$ [250, 3.1].

(4.5) QUESTION. When is there a mapping from $C(X)$ onto Y? Of special interest is the case when $Y = X$ [250, 3.1].

(4.6) QUESTION. When is there a mapping from X onto $C(X)$? By (4.1) and (4.2) if X is locally connected or if X contains an open set with uncountably many components, then there is a mapping from X onto C(X).

The next three results, (4.13), and (4.14) are partial solutions to (4.4) and (4.5).

(4.7) THEOREM [250, 3.2; stated in the special case when Y = X]. *If* X *and* Y *are continua such that there is a mapping from* 2^X *or from* C(X) *onto* Y, *then* Y *is weakly chainable [see (0.25) for definition] and* Y *is arcwise decomposable [see (0.30) for definition].*

Proof. We first show

(*) any indecomposable continuum (in particular, the pseudo-arc) contains an open set with uncountably many components.

To prove (*), let Z be an indecomposable continuum and let W be a nonempty open subset of Z such that cl[W] ≠ Z. We will show that W has uncountably many components. Let Q be a component of W. Then, since cl[Q] ⊂ cl[W] ≠ Z, cl[Q] is a proper subcontinuum of Z. Hence, cl[Q] is contained in a composant of Z. Thus, Q is contained in a composant of Z. This proves any component of W is contained in a composant of Z. Using this fact, the proof of (*) can be easily completed by recalling that each composant is dense in Z [186, Theorem 2, p.209] and, since Z is an indecomposable continuum, the composants of Z are mutually disjoint [186, Theorem 5, p.212] and there are uncountably many of them [186, (i), p.213]. Now, by (*) and (1.37) [also, see (0.25)], we have that

(1) the cone over the Cantor set is weakly chainable.

By (1) and (1.33),

(2) 2^X and $C(X)$ are each weakly chainable for any continuum X.

Assume, as in (4.7), that X and Y are continua such that Y is a continuous image of 2^X or of $C(X)$. Then, by (2), Y is weakly chainable. This proves the first part of (4.7). We now prove the second part. Since Y is a continuous image of 2^X or of $C(X)$, we have using (1.33) that there is a mapping f from the cone over the Cantor set onto Y. From now on the proof uses a technique Bellamy used in a specific example and for a different purpose [21, Example II]. Let A_o be irreducible with respect to the property of being a subcontinuum of the cone over the Cantor set which maps onto Y by f [A_o exists by Zorn's lemma or by using (0.68.7)]. If the vertex v of the cone over the Cantor set is a cut point of A_o, then A_o is the union of two arcwise connected proper subcontinua; if v is not a cut point of A_o, then A_o is an arc. In either case, there exist arcwise connected proper subcontinua A_1 and A_2 of A_o such that $A_o = A_1 \cup A_2$. Then: $f[A_1]$ and $f[A_2]$ are arcwise connected proper [by the minimality property of A_o] subcontinua of Y whose union is Y. This completes the proof of (4.7).

The next two theorems show exactly which X are continuous images of 2^X or of $C(X)$ for two special classes of continua.

(4.8) THEOREM [250, 3.3]. *If* X *is a chainable continuum, then* X *is a continuous image of* 2^X *or of* C(X) *if and only if* X *is an arc.*

Proof. See below.

The result in (4.8) is a simple consequence of (1.13) and the fact that an arc is the *only* arcwise connected chainable continuum [see (0.23.1)]. However, there are uncountably many topologically different arcwise connected circle-like continua [see (0.70.3)]. The structure of these continua as determined in [245, Theorem 6] has had some applications to hyperspace theory (see, for example, [280, proof of Theorem 7] and Chapter VIII of this book). We give one such application now; the result is analogous to (4.8) but is not a consequence of (1.13).

(4.9) THEOREM [250, 3.3]. *If* X *is a circle-like continuum, then* X *is a continuous image of* 2^X *or of* C(X) *if and only if* X *is a circle.*

Proof. Assume X is a circle-like continuum which is a continuous image of 2^X or of C(X). By (4.7), X is arcwise decomposable. It follows easily from [245, Theorem 6] that the circle is the only arcwise decomposable circle-like continuum. Hence, X is a circle. Since the other half of (4.9) is trivial, this completes the proof of (4.9).

A space is said to be *g-contractible* provided there is a mapping of the space onto itself which is homotopic to a constant map. Relationships between g-contractibility and the topic of this chapter will be given in (4.13) and (4.14). We will first investigate g-contractibility of hyperspaces.

The notion of g-contractibility originated with Bellamy [21, p.15], whose paper [21] I recommend highly to the reader as a beautiful and interesting work. In (1.37) and in this chapter we cover a large portion of [21]. We remark that Bellamy did not apply his ideas to hyperspace theory. A few other comments are appropriate. The result in (1.37) was essentially obtained by Mazurkiewicz [214] and later, independently, by Bellamy [21] whose proof we have given. Bellamy [22] has another characterization of continuous preimages of the cone over the Cantor set. Also, Bellamy has some results in [21] for Hausdorff continua. The characterization [21, Theorem 1] for a Hausdorff continuum to be a continuous preimage of the cone over the Cantor set is not quite as simple as the one given for metric continua in (1.37).

Let us note that any locally connected continuum is g-contractible [250, p.194], as is any continuum which is both a continuous image and a continuous preimage of a contractible space. Thus, in particular, the following result can be proved using (1.92) if X is locally connected and using (1.33) and (1.39) otherwise.

(4.10) THEOREM [250, 3.9]. *For any continuum* X, 2^X *is g-contractible*.

Proof. See paragraph above.

However, I do not know the answer to the following question.

(4.11) QUESTION [250, 3.10]. Is $C(X)$ g-contractible for any continuum X? A partial solution will be given in (4.12). Of course if X is a locally connected continuum, $C(X)$ is contractible [by (16.18)]. However, let us note that $C(X)$ and 2^X are not always contractible [see (16.22)].

(4.12) THEOREM [250, 3.12]. *Let* X *be a continuum. If* X *or if* $C(X)$ *[see (1.44) and (1.46)] contains an open set with uncountably many components, then* $C(X)$ *is g-contractible.*

Proof. Assume that X is a continuum which contains an open set with uncountably many components. Then, by (1.44), $C(X)$ contains an open set with uncountably many components. Hence, by (1.37), there is a mapping f from $C(X)$ onto the cone over the Cantor set. By (1.33), there is a mapping g from the cone over the Cantor set onto $C(X)$. Since the cone over the Cantor set is contractible, it follows that $g \circ f : C(X) \xrightarrow{\text{onto}} C(X)$ is homotopic to a constant map. This proves (4.12).

Thus, the question in (4.11) is unanswered (only) for those continua X which are "between" locally connected continua and "very non-locally connected" continua.

Now we come to the first of two results which show what g-contractibility has to do with the topic of this chapter.

The following result gives a partial answer to (4.4).

(4.13) THEOREM [250, 3.4]. *If* X *is a g-contractible continuum, then there is a mapping from* 2^X *onto* X.

Proof. Since (4.13) is trivial when X is locally connected, we assume for the purpose of proof that X is not locally connected. Then, by (1.39), there is a mapping from 2^X onto the cone over the Cantor set. Hence, we are done once we show

(*) there is a mapping from the cone over the
     Cantor set onto X.

Bellamy has the following general result:

(4.13.1) THEOREM [21, Lemma II]. *Any g-contractible continuum is a continuous image of the cone over the Cantor set.*

Since X in (4.13) is g-contractible, (4.13.1) implies (*). For completeness, we give the proof of (4.13.1). Let Z be a g-contractible continuum. Let M denote the Cantor set. Since every compact metric space is a continuous image of M [186, Corollary 3a, p.23], there is a mapping f: M $\xrightarrow{\text{onto}}$ Z. The mapping f "lifts up" to a mapping

F: Cone(M) $\xrightarrow{\text{onto}}$ Cone(Z).

We now show how to obtain a mapping from Cone(Z) onto Z (the existence of such a mapping is a generalization of the well-known fact that a contractible continuum is a continuous image, in fact an r-image, of the cone over it). Since Z is g-contractible, there is a mapping g: Z $\xrightarrow{\text{onto}}$ Z such that g is homotopic to a constant map. Let h: Z × [0, 1] → Z be a homotopy such that

(1) $h(z, 0) = g(z)$ for each $z \in Z$

and

(2) $h(z, 1) = p$ for each $z \in Z$

where p is some fixed point of Z. We have the following diagram, where ν is the natural map:

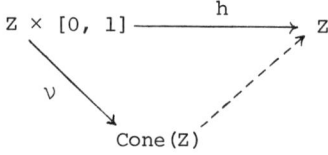

It follows from (2) that $h \circ \nu^{-1}$ is single-valued. Hence, by [88, 3.2, pp.123-124],

(3) $h \circ \nu^{-1}$: Cone(Z) → Z is continuous.

Since g sends Z onto Z, it follows from (1) that

(4) $h \circ \nu^{-1}$ sends Cone(Z) onto Z.

Thus, by (3) and (4), $h \circ \nu^{-1}$ is a mapping from Cone(Z) onto Z. Therefore, $h \circ \nu^{-1} \circ F$ is a mapping from Cone(M) onto Z and we have proved (4.13.1).

For the final result of this chapter, we give a theorem which completely determines when X [or C(X)] is both a continuous image and preimage of 2^X.

(4.14) THEOREM [250, 3.14]. *Let* X *and* Y *be continua. Then: There are mappings from* Y *onto* 2^X *and from* 2^X *onto* Y *if and only if*

(1) X *and* Y *are both locally connected*

or

(2) X is not locally connected and Y contains an open set with uncountably many components and is g-contractible.

Proof. The theorem can be proved using (1.39), (1.92), (4.2), (4.10), and (4.13.1). I leave the details to the reader.

So far we have not been concerned with the type of mapping which may or may not exist between two of the spaces X, 2^X, and $C(X)$. As far as I know, no study has been made of this, so the following question is virtually untouched (except, see the next two chapters).

(4.15) QUESTION. When does there exist a monotone, or open, or confluent, or non-alternating, etc. mapping between two of the spaces X, 2^X, and $C(X)$? See (6.8) for a related question.

We will discuss two specific aspects of (4.15) in the next two chapters. Also, see (14.70) for a question about mappings from 2^X onto $[0, 1]$.

NOTE

1. With respect to (4.3), E. D. Tymchatyn and I have recently shown that the answer to (1.46) and (1.47) is "yes". Our paper, entitled "Mappings between continua and hyperspaces", has been submitted for publication.

V.
SELECTIONS

The study of selections began with Michael in [225] and was continued by him in [217] through [219], etc. Let M be a metric space and let $\Gamma \subset 2^M$. A function $f: \Gamma \to M$ is called a *selection for* Γ provided f is continuous and for each $A \in \Gamma$, $f(A) \in A$ [225, p.154]. Two comments are in order. First, Michael [225] considers more general spaces than metric spaces. Second, our 2^M is denoted by $\mathcal{E}(M)$ in [225].

Some authors have considered non-continuous choice functions. For example, in [188, p.400] the following important result was proved.

(5.1) THEOREM [188, Corollary 2]. *If* Y *is a complete separable metric space, then there is a Baire class one choice function for* CL(Y) *with the Vietoris topology [see (0.10) through (0.12)]*.

This result was extended by Engelking [97].

Here we will only be concerned with selections, i.e., continuous choice functions. Also, the point of view will be that of considering selections from subspaces of hyperspaces rather than selections for set-valued mappings.

In [225, p.154], Michael raised and studied the following question: For what spaces Z is there a selection for 2^Z [recall: our 2^Z is Michael's (Z)]? His answer was the following:

(5.2) THEOREM [225, 1.9 and 7.6]. *Let* (Z, T) *be a Hausdorff space such that every component of* Z *is an open subset of* Z. *Then:*

(1) *There is a selection for* 2^Z *[with the Vietoris topology] if and only if there exists a linear ordering on* Z *such that the order topology is coarser than* T *[225, 1.9].*

(2) *If* Z *is connected, then having a selection for* $F_2(Z)$ *is equivalent to having one for* 2^Z *[225, 7.6].*

For some interesting results related to (5.2), see the notion of topological well-ordering and its applications to selections in [98]. Some of the results in [98] are stated here after (5.23).

(5.3) THEOREM [187, Theorem 1]. *A continuum* X *admits a selection for* $F_2(X)$ *if and only if* X *is an arc. The same result holds when* X *is a Hausdorff continuum and "arc" is replaced by "generalized arc"* [187, p.7].

Not realizing how close (1) of (5.2) came to proving the result in (5.3), the authors of [187] gave one proof of (5.3), the "first proof" [187, pp.6-7], using (2) of (5.2). The "first proof"

[187, pp.6-7] proves the result for 2^X independent of (2) of (5.2), but (2) of (5.2) was used in the "first proof" to obtain the result for $F_2(X)$. The "second proof" [187, p.7] shows the result for $F_2(X)$ independent of [225] and for Hausdorff continua. The "first proof" is very geometric and led to a proof of the following main result of [187, p.5]:

(5.4) THEOREM [187, Theorem 2]. *If* M *is a locally compact separable metric space such that there is a selection for* $F_2(M)$, *then* M *can be embedded in the real line.*

The proof in [187, pp.9-10] of Theorem 2 of [187] yields a result, outside of hyperspace theory, which is worth mentioning. In [227, Theorem 1] Miller determined conditions in order that a compact metric space be embeddable in the line. The following generalization of Miller's theorem can be deduced from part of the proof of Theorem 2 in [187].

(5.5) THEOREM [see paragraph above]. *Let* M *be a locally compact separable metric space satisfying the following two conditions:*
  (1) each component of M *is a point, arc, half-line, or line;*
  (2) if D *is a component of* M *and* E *denotes the set of end points [at most two] of* D, *then* D\E *is an open subset of* M.
Then, M *is embeddable in the real line.*

An application outside the context of hyperspaces, of (5.5) and of other ideas used in [187], appears in [242, section 2].

By (5.3), the selection question for 2^X when X is a continuum is completely solved. Therefore, since C(X) is an important subset of 2^X, the following question arises naturally:

(5.6) QUESTION. For what continua X is there a selection for C(X)? We will give a number of partial answers and related questions.

As far as I know the only places this question is investigated are [259] and [328]. The next three results are some of the results in [259].

(5.7) THEOREM [259, Lemma 3]. *If a continuum* X *admits a selection for* C(X), *then* X *is a dendroid* [see (0.28) for definition].

Proof. Assume X is a continuum such that there is a selection f: C(X) → X. Since f({x}) = x for each {x} ε C(X) and since f(K) ε K for each K ε C(X),

(1) f[C(Y)] = Y for each subcontinuum Y of X.

By (1), the continuity of f, and (1.12) we have that

(2) each subcontinuum of X is arcwise connected.

Now, suppose X is not hereditarily unicoherent. Then [see (0.27)], there exist subcontinua A and B of X such that A ∩ B is not connected. Hence, it follows easily using (2) that $A \cup B$ contains a circle Y_o. From (0.55) we see that $C(Y_o)$ is a 2-cell whose manifold boundary is the circle $F_1(Y_o)$. Let g denote the restriction of f to $C(Y_o)$ and let $i: Y_o \xrightarrow{\text{onto}} F_1(Y_o) \subset C(Y_o)$ be given by $i(y) = \{y\}$ for each $y \in Y_o$. By (1), $i \circ g$ is a continuous function from $C(Y_o)$ onto $F_1(Y_o)$. Furthermore,

$$i \circ g(\{y\}) = i(y) = \{y\} \text{ for each } \{y\} \in F_1(Y_o).$$

Hence, $i \circ g$ is a retraction of the 2-cell $C(Y_o)$ to its manifold boundary, a contradiction [185, Corollary 1a, p.314]. Thus, X is hereditarily unicoherent and, by (2), X is arcwise connected. This proves (5.7).

(5.8) THEOREM [259, Theorem 1]. *Let* X *be a continuum which is a directed space with partial order* Γ *such that* Γx *is connected for each* $x \in X$ *(see [259, p.370] for definitions). Then: There is a selection for* $C(X)$ *if and only if* X *is a generalized tree.*

The following theorem completely determines which locally connected continua X admit selections for $C(X)$.

(5.9) THEOREM [259, p.371]. *Let* X *be a locally connected continuum. Then: There is a selection for* $C(X)$ *if and only if* X *is a dendrite.*

The relationship between (5.3) and (5.9) is especially pleasant.

When the requirement of a selection for 2^X is weakened to the requirement of one for $C(X)$, then, in the presence of local connectivity, the class of continua changes from arcs to the one-dimensional absolute retracts, i.e., (by [40, 13.5, p.138]), the dendrites. Since a locally connected dendroid is a dendrite [and conversely], it would indeed have been pleasing if the converse to (5.7) had been true. However, as shown in [259, Theorem 2], this is not the case. We now give a different and shorter proof of this fact than is given in [259, pp.371-372].

(5.10) EXAMPLE [259, Theorem 2]. Let D_1 be the geometric cone in the plane with vertex $(-1, 0)$ and base consisting of the points $(1, 0)$ and $(1, 2^{-n})$ for each $n = 0,1,2,\ldots$. Let D_2 be the geometric cone in the plane with vertex $(1, 0)$ and base consisting of the points $(-1, 0)$ and $(-1, -2^{-n})$ for each $n = 0,1,2,\ldots$. Let $D = D_1 \cup D_2$ (see Figure 1 of [259, p.372]). Clearly, D is a dendroid and it is easy to see that D is not contractible. It is also easy to see that $D \in \text{prop}[\kappa]$ as defined in (16.10). Hence, by (16.15), $C(D)$ is contractible. Now, suppose there is a selection $f: C(D) \to D$ for $C(D)$. Let $i: D \to C(D)$ be the inclusion given by $i(x) = \{x\}$ for each $x \in D$. Then, $i \circ f$ is a retraction from $C(D)$ onto $F_1(D)$, the space of singletons. Therefore, since $C(D)$ is contractible and $F_1(D) \cong D$ is not, we have a contradiction. Thus, there is no selection for $C(D)$.

The proofs in (5.7) and (5.10) illustrate an important feature of selections--namely, they can be thought of as being very special types of retractions. Note that this last statement is not *literally* correct: When the domain of a selection does not contain the singletons, then the selection may not even be an r-map. However, in the context of selections as they are discussed in most of this chapter, the statement above is correct. By "special types" I mean, for example, the restriction to a subhyperspace maps that subhyperspace "back into itself." A discussion of r-maps between the spaces X, 2^X, and C(X) takes place in Chapter VI. Thus, Chapter VI may be interpreted as a continuation, for more general maps, of what is discussed here.

In the light of (5.7), (5.6) can be replaced by the following question.

(5.11) QUESTION. For what dendroids D is there a selection for C(D)? By (5.10), not all dendroids admit such a selection. Also, there exists a dendroid D which admits a selection for C(D) but which can not be partially ordered as a generalized tree [259, Theorem 3]. I do not know if every contractible dendroid D admits a selection for C(D). However, every smooth [see (1.68) for definition] dendroid D admits a selection for C(D). This last fact can be seen from (5.8) above and the equivalent formulation of smoothness given in [172, p.680]. Ward has the following result which characterizes those dendroids which are smooth in terms of the existence of a certain type of selection.

(5.12) THEOREM [328, Theorem 2]. *A dendroid* D *admits a rigid selection (see* [328] *for definition) for* C(D) *if and only if* D *is smooth.*

In [225, p.178] Michael defines an S_4 space to be a Hausdorff space which admits a selection for each covering of it by mutually disjoint nonempty compact subsets.

(5.13) QUESTION [225, p.155]. What spaces are S_4 spaces? See (5.14) for a partial answer and see (5.15) and (5.16) for related questions.

It is easy to see that being an S_4 space is a topological invariant and is hereditary by closed subsets. It is also easy to see that the unit circle in the plane is not an S_4 space--this follows from using the cover by all two-element sets of the form $\{z, -z\}$. Thus, no S_4 space can contain a circle [225, 8.4]. Michael [225, 8.6] also states that certain trees (see [225, p.178] for definition) are S_4 spaces. J. T. Rogers, Jr. and I have noticed the following fact about S_4 spaces.

(5.14) THEOREM [Sam B. Nadler, Jr. and J. T. Rogers, Jr., unpublished]. *No* S_4 *space can contain an hereditarily indecomposable metric continuum. Hence, a finite-dimensional metric continuum which is an* S_4 *space must be one-dimensional.*

Proof. Since being an S_4 space is hereditary by closed

subsets, it suffices to show that no hereditarily indecomposable metric continuum is an S_4 space; once this is done, the second part of (5.14) follows using (1.53). So, let Z be an hereditarily indecomposable metric continuum. Let μ be a Whitney map for $C(Z)$. Fix t_o such that $0 < t_o < \mu(X)$. Let us note the following properties [(1) through (4)] of $\mu^{-1}(t_o)$. By (1.78),

(1) $\mu^{-1}(t_o)$ is a covering of Z

and

(2) the members of $\mu^{-1}(t_o)$ are mutually disjoint subcontinua of Z.

Since $0 < t < \mu(Z)$, it follows from (0.50) that

(3) each member of $\mu^{-1}(t_o)$ is a nondegenerate proper subcontinuum of Z.

As a consequence of (1.80.2),

(4) $\mu^{-1}(t_o)$ is a subcontinuum of $C(Z)$ [this is true for any continuum--see (14.2)].

Note the following three facts about composants of indecomposable continua: By [186, (i), p.213]

(5) Z has uncountably many composants,

by [186, Theorem 5, p.212]

(6) the composants of Z are mutually disjoint,

and, by (6),

(7) Z is irreducible between any two points of different composants of Z.

Now, suppose Z is an S_4 space. Then, by (1) and (2), there is a selection $f: \mu^{-1}(t_o) \to Z$. By (4) and the continuity of f, we have that

(8) $f[\mu^{-1}(t_o)]$ is a subcontinuum of Z.

By using (1), (3), and (5), it follows easily that there exist A, B ε $\mu^{-1}(t_o)$ such that A and B are contained in different composants of Z. Thus, since $f(A) \varepsilon A$ and $f(B) \varepsilon B$, it follows using (7) and (8) that

(9) $f[\mu^{-1}(t_o)] = Z$.

Now, let $A_o \varepsilon \mu^{-1}(t_o)$. Let $a_o = f(A_o)$. Since $A_o \neq \{a_o\}$ [by (3)], there exists $a_1 \varepsilon A_o$ such that $a_1 \neq a_o$. Note that

(10) $f(A_o) \neq a_1 \varepsilon A_o$.

Since f is a selection, it follows easily using (2) and (10) that $a_1 \notin f[\mu^{-1}(t_o)]$. This contradicts (9). Therefore, Z is not an S_4 space and we have proved (5.14).

Except for results stated above, essentially nothing else is known about (5.13). In particular, it is not known:

(5.15) QUESTION [347, p.109]. If a (metric) continuum is an S_4 space, then must it be a dendrite? See below.

For locally connected continua, the answer to (5.15) is "yes" since no S_4 space can contain a circle [see the paragraph following (5.13)]. Hence, an affirmative answer to (5.15) would be obtained if it could be shown that a continuum which is an S_4 space must be locally connected. J. T. Rogers, Jr. obtained a partial result in this direction by showing that no continuum which is an

S_4 space can contain a sequence of mutually disjoint arcs which converges to a nondegenerate subcontinuum.

The answer to the following question is not known.

(5.16) QUESTION. Can a continuum which is an S_4 space contain an indecomposable continuum? In (5.14) we showed that a continuum which is an S_4 space can not contain an hereditarily indecomposable continuum.

Michael [225, 8.6] states that certain trees are S_4 spaces. However, the answer to the following question [the converse question to (5.15)] is not known.

(5.17) QUESTION. Is every dendrite an S_4 space?

Decompositions, whose members are compact, are natural classes of coverings to consider in relation to S_4 spaces. We make the following definition.

(5.18) DEFINITION. Call a Hausdorff space a $\begin{Bmatrix} u\text{-}space \\ c\text{-}space \end{Bmatrix}$ provided that it admits a selection for each $\begin{Bmatrix} \text{upper semi-continuous} \\ \text{continuous} \end{Bmatrix}$ decomposition of it into nonempty compact subsets (as usual, the selection's continuity is in terms of the Vietoris topology, not the decomposition topology).

As in the case of S_4 spaces, being a u-space is hereditary by closed subsets, no u-space can contain a circle, and (5.14) still holds with S_4 space replaced by u-space [the same proof works by using (1.80.1) and (1.77.2)]. An hereditarily indecomposable continuum can not be a c-space (by the same reasoning used to show it can not be a u-space), but it would seem that a c-space could contain an hereditarily indecomposable continuum. It is not known what spaces are u-spaces or c-spaces since these types of spaces have not been investigated. In particular, it is not known:

(5.19) QUESTIONS. What *continua* are u-spaces? What *continua* are c-spaces?

As can be seen from the proof of (5.14), Whitney maps play a role in studying selections. We define the following notion.

(5.20) DEFINITION. A continuum X is called a *Whitney map selection continuum* provided that, for every Whitney map μ for $C(X)$, there is a selection from $\mu^{-1}(t)$ into X for each $t \in [0, \mu(X)]$.

By the proof of (5.14), no Whitney map selection continuum can contain an hereditarily indecomposable continuum [hence, if finite-dimensional, such a continuum must be one-dimensional by (1.53)]. However, a Whitney map selection continuum can be indecomposable-- the solenoids are Whitney map selection continua, as is not difficult to prove. I do not know the answers to the following questions.

(5.21) QUESTIONS. What continua are Whitney map selection continua? What chainable continua and what circle-like continua are Whitney map selection continua? The reader is referred to Chapter XIV for background material which will be helpful in studying these questions.

The dual concept to a selection is given in the following definition, which was formulated by Krasinkiewicz and me in a conversation.

(5.22) DEFINITION. Let X be any continuum. A function $f: X \to 2^X$ is called a *coselection for* X provided that f is a mapping and $x \in f(x)$ for each $x \in X$. If f is a coselection for X and $f(x) \in \Lambda$ for each $x \in X$ [some $\Lambda \subset 2^X$], then f is called a Λ-*coselection for* X. If f is a Λ-coselection for X where $\Lambda \subset [2^X \setminus (F_1(X) \cup \{X\})]$, then f is called a *non-trivial coselection for* X. If f is a coselection for X, then the *mesh of* f, denoted mesh(f), is defined by [where diam means diameter of $f(x)$ as a subset of X]

   mesh(f) = l.u.b.{diam[f(x)]: $x \in X$}.

If X is a continuum such that given any $\varepsilon > 0$ there exists a non-trivial coselection [resp., Λ-coselection] for X of mesh less than ε, then X is called a *coselection space* [resp.,

Λ-*coselection space*].

It follows from the proof of (13.1) that, for any continuum X, there are non-trivial 2^X-coselections for X. By (1.80.2), (1.80.3), and the definition of η_t in (1.79), any hereditarily indecomposable continuum X is a C(X)-coselection space. There are very simple continua X which are not C(X)-coselection spaces. For example, such is the case for $X = X_1 \cup X_2 \cup X_3$ where

$X_1 = \{(x, \sin[1/x]) \in R^2 : 0 < x \leq 1\}$,
$X_2 = \{(x, \sin[1/x] - 2) \in R^2 : -1 \leq x < 0\}$,
$X_3 = \{(0, y) \in R^2 : -3 \leq y \leq +1\}$.

(5.23) QUESTIONS. When do coselections of the various types defined in (5.22) exist? What continua are coselection or C(X)-coselection spaces?

In [98], the authors investigate the question of the existence of (continuous) selections from the space of all nonempty closed subsets of a topological space Z. Among the results proved are: If Z is the reals [98, 5.1] or the rationals [98, 6.1], then there is no such selection; if Z is a zero-dimensional topologically complete metric space [for example, the irrationals], then there is such a selection. Young asks the following question:

(5.24) QUESTION [347, p.109]. What compact totally-disconnected Hausdorff spaces admit a (continuous) selection from their

(5.24) SELECTIONS 267

space of all nonempty closed subsets?

I have not even attempted a discussion of the very important and exhaustive treatment of selections in [217] through [220]. The reader is referred to these and related papers (for example, [222] through [224] and [423] through [425]); a discussion would of necessity be too lengthy for this book. Let us also mention that selection theory has proved to be a useful tool in other areas of mathematics. Of the hundreds of papers that could be cited, I have referenced [6], [143], [262], [314], and [432] through [435] for the interested reader.

VI.
RETRACTIONS

In (4.15) we posed the general question: When do there exist mappings of various types between two of the spaces X, 2^X, and C(X)? The previous chapter may be interpreted as giving some answers to this question for the case of selections. As pointed out after the proof of (5.10), selections can be thought of as being special types of retractions. So, the next most natural class of mappings to look at with respect to (4.15) are the retractions and, more generally, the r-maps. Recall [40]: Let X and Y be continua. A mapping r: X → Y is called an *r-map* provided there is a mapping g: Y → X such that r∘g is the identity mapping of Y onto Y. For an r-map as above it is easy to see that r sends X onto Y and g is one-to-one, hence a homeomorphism into X; Y is called an *r-image* of X. If Y ⊂ X and r(y) = y for each y ∈ Y, then Y is called a *retract* of X and r is said to be a *retraction*. It is easy to see that a continuum Y is an r-image of a continuum X if and only if there is an embedding Z of Y into X such that Z is a retract of X.

In this chapter we will be concerned with the following general question:

(6.1) QUESTION. Given any two of the spaces X, 2^X, and $C(X)$, when is there an r-map between them? Also, see (6.28) and (6.29).

The first questions in the literature about this topic appear in [243] and [250, 3.7]; they are special cases of (6.1):

(6.2) QUESTION [243, p.413]. What are necessary and sufficient conditions in order that the space of singletons $F_1(X)$ of a continuum X be a retract of 2^X or of $C(X)$?

(6.3) QUESTION [250, 3.7]. For what continua X is $C_1(X)$ a retract of 2^X? Recall that $C_1(X)$ means $C(X)$ as it is naturally embedded in 2^X by inclusion. Some important material related to this question is in (6.10). In particular, the example in (6.10.2) may be of interest to the reader at this point.

We will now survey what has been known about (6.1) through (6.3), and we will give a few new results [for example, (6.12)].

To begin with, Wojdyslawski [see (1.96)] proved that 2^X and $C(X)$ are absolute retracts when X is a locally connected continuum. Hence:

(6.4) THEOREM (comp., [243, p.413]). *For any locally con-*

nected continuum X, *the following five statements are equivalent:*

(6.4.1) X *is an r-image of* 2^X;

(6.4.2) X *is an r-image of* $C(X)$;

(6.4.3) $F_1(X)$ *is a retract of* 2^X;

(6.4.4) $F_1(X)$ *is a retract of* $C(X)$;

(6.4.5) X *is an absolute retract.*

Proof. See paragraph above.

Thus, (6.2) is answered for the case of locally connected continua, as is part of (6.1). A question related to (6.4) is in (6.28). Let us also note the following result:

(6.5) THEOREM. *If* X *is any locally connected continuum, then* $C_1(X)$ *is a retract of* 2^X *[see (6.12) for a specific retraction with special properties].*

Proof. Use (1.96).

Therefore, (6.3) is solved for locally connected continua. Also, we have the following result for r-images.

(6.6) THEOREM. *If* X *is any locally connected continuum, then* $C(X)$ *is a monotone open r-image of* 2^X.

Proof. Let X be any locally connected continuum and let I_∞ be the Hilbert cube. Curtis and Schori have shown [(1.97) and (1.98)]

$$2^X \cong I_\infty \cong C(X) \times I_\infty.$$

Let h denote any homeomorphism of 2^X onto $C(X) \times I_\infty$. Let $p \in I_\infty$ and let $\pi_p : C(X) \times I_\infty \xrightarrow{onto} C(X) \times \{p\}$ denote projection. Then it is easy to see $\pi_p \circ h$ is the desired monotone open r-map, since the composition of two monotone open r-maps is a monotone open r-map.

(6.7) QUESTION. Is $C_1(X)$ a monotone open retract of 2^X when X is any locally connected continuum? By (6.6), $C(X)$ is always such an r-image when X is locally connected. Also, by (6.12), $C_1(X)$ is always a monotone retract of 2^X when X is locally connected. However, the retract defined in the proof of (6.12) is not necessarily open--see the paragraph following the proof of (6.12).

(6.8) QUESTION. Same as (6.1) except require the r-map to be monotone, *or* open, *or* confluent, etc.

The next result is a slight generalization and strengthening of [243, Corollary 3].

(6.9) THEOREM (see sentence above). *Let* X *and* Y *be continua. If* Y *is an r-image of* $C(X)$ *or of* 2^X, *then* Y *is acyclic in all dimensions; also,* Y *is arcwise decomposable and weakly chainable.*

Proof. The first part of the theorem is a consequence of (1.172) through (1.174) and the fact that an r-image of an acyclic

(6.10) RETRACTIONS 273

space is acyclic [40, p.42]. The second part is a consequence of (4.7).

(6.10) SOME EXAMPLES AND QUESTIONS RELATED TO (6.3). When [250] was written, there were examples of continua X for which it was known that $C(X)$ does not have the fixed point property (see Chapter VII). However, there were no examples for which it was known that 2^X does not have the fixed point property. If $C(X)$ does not have the fixed point property and if there is a retraction as in (6.3), then clearly 2^X does not have the fixed point property. Thus, the question in (6.3) was partially motivated in [250] by the following question [250, 3.8]: For what continua X does 2^X have the fixed point property [see (7.9)]? For more discussion about this, see Chapter VII. As specifically regards (6.3), suffice it to say that until recently I did not know a single non-locally connected continuum X for which I could prove or disprove that $C_1(X)$ is a retract of 2^X. I suggested to Jim Lawson that it would be of interest to know if there is a retraction from 2^Z onto $C_1(Z)$, where Z is the continuum $(SP)_1$ defined in (8.22). Subsequently, Lawson [191] gave some conditions which would imply the existence of such a retraction. However, as Jack Goodykoontz has pointed out to me, there is an error in the proof of Proposition 4 of [191]. This means that the proofs in [191] that there are retractions from 2^X onto $C_1(X)$, where $X = (SP)_1$ and $X =$ the cone over $\{0, 1, 1/2, 1/3, \ldots, 1/n, \ldots\}$, are not correct. Thus, at this time, the answer to the following question is not known:

(6.10.1) QUESTION. Is there a non-locally connected continuum X such that there is a retraction from 2^X onto $C_1(X)$? In particular, what about $X = (SP)_1$ as defined in (8.22)? Some other special cases are asked about in (6.10.3).

There are general results in the literature which would seem to support the belief that there is always a retraction from 2^X onto $C_1(X)$. For example: 2^X and $C(X)$ are both arcwise connected [by (1.13)] and acyclic [by (1.172) through (1.174)], 2^X is contractible if and only if $C(X)$ is contractible [by (16.7)], if 2^X is connected im kleinen at $A \in C(X)$ then so is $C(X)$ ([124, Theorem 3]; for a stronger result, see Goodykoontz's result in (1.133) in this book), etc. However, Jack Goodykoontz [372] has recently obtained the following important and interesting example.

(6.10.2) EXAMPLE [372]. This is an example of a continuum X such that there is no retraction from 2^X onto $C_1(X)$. It is the first such example to be known. For two points $x, y \in R^3$, let $\widehat{xy}$ denote the convex arc in R^3 with end points x and y. Define the following points in R^3 [see Figure (6.10.2) below]: $a = (1, 0, 0)$, $b = (0, 1, 0)$, $c = (-1, 0, 0)$, $o = (0, 0, 0)$, $p = (0, 1/2, 0)$, and, for each $n = 1, 2, \ldots$, let $a_n = (1, 0, 1/n)$, $d_n = (1/n, 0, 1/n)$, $p_n = (0, 1/2, 1/n)$, $e_n = (-1/n, 0, 1/n)$, and $c_n = (-1, 0, 1/n)$. Let $X_o = \widehat{ac} \cup \widehat{op}$ and, for each $n = 1, 2, \ldots$, let

$$X_n = \widehat{a_n d_n} \cup \widehat{d_n p_n} \cup \widehat{p_n e_n} \cup \widehat{e_n c_n}.$$

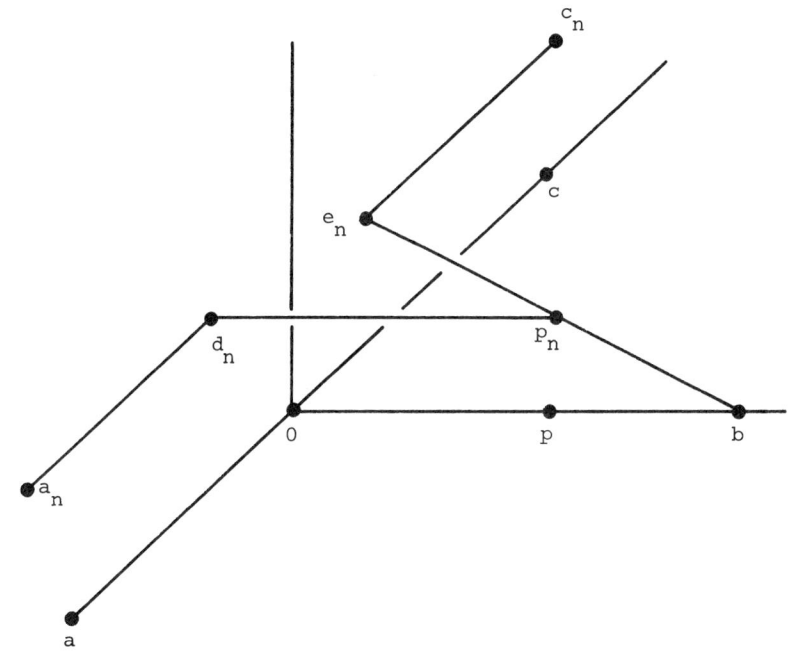

Figure (6.10.2)

Let $X = X_0 \cup [\cup_{n=1}^{\infty} X_n] \cup [\cup_{n=1}^{\infty} \widehat{bp_n}]$. Note that X is a continuum--in fact, X is a smooth dendroid [see (15.20)]. The following two facts will be used later:

(1) $X_n \in C_1(X)$ for each $n = 0,1,2,\ldots$;

(2) if $A \in C_1(X)$ such that $A \cap X_0 \neq \emptyset$ and $A \cap X_n \neq \emptyset$ for some $n \geq 1$, then $b \in A$.

Now, suppose there is a retraction r from 2^X onto $C_1(X)$. Let Λ be such that:

(3) Λ is an open subset of 2^X such that $X_0 \in \Lambda$

and

(4) $b \notin [\cup \Lambda]$.

By (1), $r(X_0) = X_0$. Thus, since r is continuous at X_0, we have by (3) that there exists Γ such that

(5) Γ is an open subset of 2^X,

(6) $X_o \in \Gamma$,

and

(7) $r[\Gamma] \subset [\Lambda \cap C_1(X)]$.

By (5), (6), and a simple sequence argument, there exists δ, $0 < \delta < 1/2$, such that for $q = (0, (1/2) - \delta, 0)$ and $Y_o = \widehat{ac} \cup \widehat{Oq}$,

(8) if $Z \in 2^X$ such that $Y_o \subset Z \subset X_o$, then $Z \in \Gamma$.

Note that $Y_o \in C_1(X)$ and, thus,

(9) $r(Y_o) = Y_o$.

For each $n = 1, 2, \ldots$, let v_n [resp., w_n] be the point on $\widehat{d_n p_n}$ [resp., on $\widehat{p_n e_n}$] whose second coordinate is $1/2 - \delta$ and let

$$Y_n = \widehat{a_n d_n} \cup \widehat{d_n v_n} \cup \widehat{w_n e_n} \cup \widehat{e_n c_n}.$$

Note that

(10) $Y_n \subset X_n$ for each $n = 1, 2, \ldots$

and

(11) $Y_n \to Y_o$ as $n \to \infty$.

By (9), (11), and the continuity of r,

(12) $r(Y_n) \to Y_o$ as $n \to \infty$.

Since Y_o is a subcontinuum of X_o such that $a, c \in Y_o$ and $p \notin Y_o$, it follows easily from the geometry that if a subcontinuum K of X is sufficiently close to Y_o then $K \subset X_o$. Hence, by (12), there exists a natural number N_1 such that

(13) $r(Y_n) \subset X_o$ for each $n \geq N_1$.

Since $Y_n \to Y_o$ and $X_n \to X_o$ as $n \to \infty$, it follows using (5), (8), and a simple sequence argument that for some natural number N_2

(14) if $Z_n \in 2^X$ and $Y_n \subset Z_n \subset X_n$ for any $n \geq N_2$, then $Z_n \in \Gamma$.

Let $j = \max\{N_1, N_2\}$. By (1), (10), and (1.8) there is an order arc α in 2^X from Y_j to X_j. Since $j \geq N_2$ and α is an order arc from Y_j to X_j, it follows easily from (14) that $\alpha \subset \Gamma$. Hence, by (7), $r[\alpha] \subset \Lambda$. Thus, by (4),

(15) $b \notin (\cup r[\alpha])$.

Since $r[\alpha]$ is a subcontinuum of $C_1(X)$, we have by (1.49) that

(16) $(\cup r[\alpha]) \in C_1(X)$.

Since $Y_j, X_j \in \alpha$ and $r(X_j) = X_j$ [by (1)],

(17) $[r(Y_j) \cup X_j] \subset (\cup r[\alpha])$.

Since $j \geq N_1$, $r(Y_j) \subset X_o$ by (13). Hence, by (17),

(18) $(\cup r[\alpha]) \cap X_o \neq \emptyset$ and $(\cup r[\alpha]) \cap X_j \neq \emptyset$.

By (16) and (18) we may apply (2) to obtain

(19) $b \in [\cup r[\alpha]]$.

Since (15) and (19) are incompatible, we have a contradiction. Therefore, there is no retraction from 2^X onto $C_1(X)$. This completes (6.10.2).

There are many simple continua X for which it is not known if there is a retraction from 2^X onto $C_1(X)$:

(6.10.3) QUESTIONS. Let X be the familiar $\sin[1/x]$-continuum defined in (1) of (8.22). Is $C_1(X)$ a retract of 2^X? In fact, except for the arc, it is not known whether or not $C_1(Y)$ is a retract of 2^Y for any given chainable continuum Y. Is $C_1(Y)$ a retract of 2^Y when Y is the pseudo-arc? Analogous questions are open when "retract" is replaced by "r-image"; in particular, [comp., (6.29)], it is not known whether or not $C(X)$ is an r-image of 2^X if X is the continuum in (6.10.2). No information is available for circle-like continua which are not the circle. Also, see (6.10.1).

The following result indicates that, when X is not locally connected, a retraction from 2^X onto $C_1(X)$ must be rather complicated. The result is original with the book, but has subsequently been written up for journal publication [395].

(6.11) THEOREM. *Let* X *be a continuum.*

(6.11.0) *If* $r: 2^X \xrightarrow{\text{onto}} \Omega$ *is a retraction onto some subcontinuum* Ω *of* $C_1(X)$ *such that* $r(A) \subset A$ *or* $r(A) \supset A$ *for each* $A \in 2^X$, *then*

(i) $r(A) \subset A$ *for each* $A \in 2^X$ *or*

(ii) $r(A) \supset A$ *for each* $A \in 2^X$.

(6.11.1) *There exists a mapping* m *from* 2^X *into* $C_1(X)$ *such that* $m(A) \subset A$ *for each* $A \in 2^X$ *if and only if* X *is an arc.*

(6.11.2) *If there is a retraction* $r: 2^X \xrightarrow{\text{onto}} C_1(X)$ *such that* $r(A) \supset A$ *for each* $A \in 2^X$, *then* X *is locally connected [the converse is also true-- see (6.12)].*

(6.11)   RETRACTIONS   279

Proof. To prove (6.11.0), let $\Lambda_1 = \{A \in 2^X : r(A) \subset A\}$ and let $\Lambda_2 = \{A \in 2^X : r(A) \supset A\}$. It is easy to see that Λ_1 and Λ_2 are closed subsets of 2^X and $\Lambda_1 \cup \Lambda_2 = 2^X$ [by the hypothesis of (6.11.0)]. Also, since $r(A) = A$ if and only if $A \in \Omega$, $\Lambda_1 \cap \Lambda_2 = \Omega$. Thus:

(1) $2^X \setminus \Omega = [2^X \setminus \Lambda_1] \cup [2^X \setminus \Lambda_2]$;

(2) $2^X \setminus \Lambda_1$ and $2^X \setminus \Lambda_2$ are open subsets of 2^X;

(3) $[2^X \setminus \Lambda_1] \cap [2^X \setminus \Lambda_2] = \emptyset$.

By (1.204.3),

(4) $2^X \setminus \Lambda$ is connected.

By (1) through (4), $2^X \setminus \Lambda_i = \emptyset$ for some $i = 1$ or 2. Hence, $\Lambda_i = 2^X$ for $i = 1$ or 2. This proves (6.11.0). To prove (6.11.1), assume m satisfies the hypotheses in (6.11.1). Let g denote the restriction of m to $F_2(X)$. Observe that, since $g(A) \subset A$ and $g(A) \in C_1(X)$ for each $A \in F_2(X)$, $g(A) \in F_1(X)$ for each $A \in F_2(X)$. Hence, letting $i: X \to F_1(X)$ be given by $i(x) = \{x\}$ for each $x \in X$, we see that $f = i^{-1} \circ g$ is a mapping defined on all of $F_2(X)$. Now, observe that f is a selection for $F_2(X)$. Thus, by (5.3), X is an arc. This proves half of (6.11.1). The other half is easy. Next, we prove (6.11.2). Assume that r satisfies the hypothesis of (6.11.2). Suppose X is not locally connected. Then, by (1.19), there is a point $p \in X$ such that X is not connected im kleinen at p. Let ε, K_n (n = 1,2,...), and K be as in the conclusion of (20.7). Then there is a sequence $\{p_n\}_{n=1}^{\infty}$ converging to p such that $p_n \in K_n$ for each $n = 1,2,...$. For each $n = 1,2,...$, let $A_n = \{p, p_n\}$. Note that

$A_n \cap K \neq \emptyset \neq A_n \cap K_n$ for each $n = 1, 2, \ldots$. Hence, since $r(A_n) \supset A_n$ [by the hypothesis in (6.11.2)],

(5) $r(A_n) \cap K \neq \emptyset \neq r(A_n) \cap K_n$ for each $n = 1, 2, \ldots$.

Since $r(A_n) \in C_1(X)$ for each $n = 1, 2, \ldots$, it follows from (5) and properties in (20.7) that

(6) $\text{diam}[r(A_n)] \geq \varepsilon$ for each $n = 1, 2, \ldots$.

Clearly, since $A_n \to \{p\}$ as $n \to \infty$ and since $r(\{p\}) = \{p\}$ [because r is a retraction],

(7) $r(A_n) \to \{p\}$ as $n \to \infty$.

However, (6) and (7) are incompatible [see (1.211.1)]. Therefore, X is locally connected and we have proved (6.11.2). This completes the proof of (6.11).

Let X denote any hereditarily unicoherent continuum. Define $\psi: 2^X \to C_1(X)$ by: For each $A \in 2^X$, $\psi(A)$ is the unique smallest subcontinuum of X containing A. In his doctoral thesis, Goodykoontz ([125, Theorem 3.2]; also [127, Theorem 1]) characterized dendrites among all hereditarily unicoherent continua X by showing X is a dendrite if and only if ψ is continuous. Let us observe that for a dendrite, ψ is a retraction of the type in (6.11.2). We have the following result (subsequent to obtaining the result here, it has been written up for journal publication [395]).

(6.12) THEOREM. *Let* X *be a continuum. Then: There is a retraction* $r: 2^X \xrightarrow{\text{onto}} C_1(X)$ *of the type in (6.11.2) if and only if* X *is locally connected. Furthermore, if* X *is locally con-*

(6.12) RETRACTIONS

nected, there is a retraction $r: 2^X \xrightarrow{onto} C_1(X)$ of the type in (6.11.2) such that r is monotone and, moreover, $r^{-1}(B)$ is contractible for each $B \in C_1(X)$.

Proof. To prove the parts which are not in (6.11.2), assume X is a locally connected continuum. By (0.65.4), X admits a convex metric ρ. Let K_ρ be defined as in (0.65.1). Define $\alpha_\rho: 2^X \to [0, +\infty)$ by

$\alpha_\rho(A) = \inf\{t \geq 0: K_\rho(t, A)$ is connected$\}$

for each $A \in 2^X$. We will prove the following three facts, the first and last of which will be used later in the proof of the theorem.

(i) $K_\rho(\alpha_\rho(A), A)$ is a subcontinuum of X.

(ii) If $A, B \in 2^X$, $t \geq 0$, $K_\rho(t, A)$ is connected, and $\eta \geq H_\rho(A, B)$, then $K_\rho(t + \eta, B)$ is connected.

(iii) α_ρ is continuous.

Proof of (i). By definition of $\alpha_\rho(A)$, there exists a sequence $\{t_n\}_{n=1}^\infty$ converging to $\alpha_\rho(A)$ such that, for each $n = 1, 2, \ldots$, $t_n \geq t_{n+1} \geq \alpha_\rho(A)$ and $K_\rho(t_n, A)$ is connected. It follows easily using the compactness of A that

$$K_\rho(\alpha_\rho(A), A) = \bigcap_{n=1}^\infty K_\rho(t_n, A).$$

The equation above is a representation of $K_\rho(\alpha_\rho(A), A)$ as a nested intersection of subcontinua of X. Hence, $K_\rho(\alpha_\rho(A), A)$ is a subcontinuum of X.

Proof of (ii). It follows easily from the hypotheses of (ii) that

(1) $A \subset K_\rho(t, A) \subset K_\rho(t + \eta, B)$ and

$K_\rho(t, A)$ is connected.

Now, let $x \in K_\rho(t + \eta, B)$. Then, there exists $b \in B$ such that $\rho(x, b) \leq t + \eta$. Since $b \in B$ and $H_\rho(A, B) \leq \eta$, there exists $a \in A$ such that $\rho(b, a) \leq \eta$. By using the result in (a) of (0.65.3) twice, there exist subsets γ_1 and γ_2 of X such that:

$x, b \in \gamma_1$ isometric to $[0, \rho(x, b)]$;

$b, a \in \gamma_2$ isometric to $[0, \rho(b, a)]$.

Let $G_x = \gamma_1 \cup \gamma_2$. Clearly G_x is connected, $x \in G_x$, and $A \cap G_x \neq \emptyset$. Also, using the isometric properties of γ_1 and γ_2, it follows easily that $G_x \subset K_\rho(t + \eta, B)$. Since $x \in K_\rho(t + \eta, B)$ was arbitrary, it follows using the properties of G_x and (1) that $K_\rho(t + \eta, B)$ is connected.

Proof of (iii). Let $\eta > 0$. Let $L_1, L_2 \in 2^X$ such that $H_\rho(L_1, L_2) \leq \eta$. By (i), $K_\rho(\alpha_\rho(L_1), L_1)$ is connected. Hence, by (ii), $K_\rho(\alpha_\rho(L_1) + \eta, L_2)$ is connected. Hence, by definition of $\alpha_\rho(L_2)$,

(2) $\alpha_\rho(L_2) \leq \alpha_\rho(L_1) + \eta$.

Interchanging the roles of L_1 and L_2 in the above argument, we obtain

(3) $\alpha_\rho(L_1) \leq \alpha_\rho(L_2) + \eta$.

Combining (2) and (3) we obtain $|\alpha_\rho(L_1) - \alpha_\rho(L_2)| \leq \eta$. Hence, we have shown that α_ρ is continuous, which proves (iii).

Thus, we have proved (i), (ii), and (iii). To complete the proof of (6.12), let

(4) $r(A) = K_\rho(\alpha_\rho(A), A)$ for each $A \in 2^X$.

By (i), the formula in (4) defines a function r from 2^X into $C_1(X)$. Since $K_\rho(0, A) = A$ for each $A \in 2^X$, $K_\rho(0, A)$ is connected for each $A \in C_1(X)$. Hence, by definition of $\alpha_\rho(A)$, we have that $\alpha_\rho(A) = 0$ for each $A \in C_1(X)$. Thus, by the formula in (4),

(5) $r(A) = K_\rho(0, A) = A$ for each $A \in C_1(X)$.

By the result in (f) of (0.65.3), $K_\rho: [0, +\infty) \times 2^X \to 2^X$ is continuous and, by (iii), α_ρ is continuous. Hence, it follows easily that $r: 2^X \to C_1(X)$ is continuous. Therefore, by (5), r is a retraction from 2^X onto $C_1(X)$. Also, by the formula in (4), $r(A) \supset A$ for each $A \in 2^X$. Hence, $r: 2^X \xrightarrow{\text{onto}} C_1(X)$ is a retraction of the type in (6.11.2). It remains to show that $r^{-1}(B)$ is contractible for each $B \in C_1(X)$. First, let us note the following fact which can be proved easily using (c) of (0.65.3) and the definition of α_ρ:

(6) If $A \in 2^X$ and $0 \leq s \leq 1$, then
$$\alpha_\rho(K(s \cdot \alpha_\rho(A), A)) = [1 - s] \cdot \alpha_\rho(A).$$

Now, let $B \in C_1(X)$ be fixed. It follows easily using the definition of r, (c) of (0.65.3), and (6) that

(7) If $A \in r^{-1}(B)$ and $0 \leq s \leq 1$, then
$$K_\rho(s \cdot \alpha_\rho(A), A) \in r^{-1}(B).$$

For each $(A, s) \in r^{-1}(B) \times [0, 1]$, let
$$h(A, s) = K_\rho(s \cdot \alpha_\rho(A), A).$$

By (7), the formula above defines a function h from $r^{-1}(B) \times [0, 1]$ into $r^{-1}(B)$. Since K_ρ and α_ρ are continuous, h is continuous. Clearly, $h(A, 0) = A$ and $h(A, 1) = r(A) = B$

for each $A \in r^{-1}(B)$. Therefore, $r^{-1}(B)$ is contractible. This completes the proof of (6.12).

In relation to (6.7), let us note that the retraction r defined in the proof of (6.12) is not necessarily an open mapping of 2^X onto $C_1(X)$. For example: Let $X = [0, 1]$ with the usual absolute-value metric ρ and let r be defined as in the proof of (6.12). Let $W = <U_1, U_2>$ where $U_1 = [0, 1/4)$, $U_2 = (3/4, 1]$, and $<>$ is as defined in (0.10). Then, W is an open subset of 2^X but, by the formula for r in the proof of (6.12), $r(A) = [0, 1]$ for each $A \in W$.

The following result is a corollary to (6.12) and to the proof of (6.11.2).

(6.13) COROLLARY. *Let X be a continuum. There is a retraction r from $F_2(X) \cup \Gamma$ onto some Γ, $F_1(X) \subset \Gamma \subset C_1(X)$, such that $r(A) \supset A$ for each $A \in F_2(X)$ if and only if X is locally connected.*

Proof. See sentence above.

(6.14) QUESTION. For what continua X is there a retraction from $F_2(X) \cup \Gamma$ onto some Γ, $F_1(X) \subset \Gamma \subset C_1(X)$?

The same question as (6.14) may be asked with $F_2(X) \cup \Gamma$ replaced by 2^X. Two other special cases of (6.14) are of interest:

(6.15) QUESTION [see (6.17.1)]. Take $\Gamma = F_1(X)$ in (6.14).

(6.16) QUESTION. Take $\Gamma = C_1(X)$ in (6.14).

A *topological groupoid* is a Hausdorff space G together with a continuous function $g: G \times G \to G$. If g is commutative $[g(x, y) = g(y, x)]$ and idempotent $[g(x, x) = x$, each $x]$, then G is called a *mean*; we will also call g a mean. Means were investigated in [10], [11], [12], [13], [94], [95], [226], [297], [298], and [299]. Let us note the following simple fact whose proof was asked for in (0.71.1).

(6.17) THEOREM. *A continuum* X *admits a mean if and only if there is a retraction from* $F_2(X)$ *onto* $F_1(X)$.

Hence, (6.15) is equivalent to the following question:

(6.17.1) QUESTION [see the papers on means cited above]. What continua X admit a mean?

A discussion of the general theory of means will not be given here. However, two of many results which have some bearing on (6.14) and (6.20) are stated below. They will be used in the discussion following (6.20).

(6.18) THEOREM [12, (2.4)]. *If a compact subset of the plane admits a mean, then it does not separate the plane [see (0.71.1)].*

(6.19) THEOREM [11]. *The familiar sin[1/x]-continuum* X, *defined in (1) of (8.22), admits no mean.*

(6.20) QUESTIONS. For what continua X is there a mapping from $F_2(X)$ into $C_1(X)$ which is the identity on $F_1(X)$? Does such a mapping exist for the familiar sin[1/x]-continuum in (1) of (8.22)? We call such a mapping a *pseudo-mean*.

The discussion in the next two paragraphs hopefully helps give some insight into (6.20) and its relationship to means.

Let S^1 denote the unit circle in the plane and let f be *any* mapping from $F_2(S^1)$ into $C_1(S^1)$ which is the identity on $F_1(S^1)$. Such an f exists, for example, by (6.12). As we will see, it follows from (6.18) that $f[F_2(S^1)] = C_1(S^1)$. To show this, first note that $C_1(S^1)$ is a 2-cell whose manifold boundary is $F_1(X)$ [see (0.55)]. Now, suppose $f[F_2(S^1)] \neq C_1(S^1)$. Then, there is a retraction k from $f[F_2(S^1)]$ onto $F_1(S^1)$. Thus, k∘f is a retraction from $F_2(S^1)$ onto $F_1(S^1)$. Hence, by (6.17), we have a contradiction to (6.18). Therefore, $f[F_2(S^1)] = C_1(S^1)$.

Let us take another example. Let X be the familiar sin[1/x]-continuum of (6.19) and let J denote the limit segment [i.e., $J = \{(0, y) \in R^2 : -1 \leq y \leq +1\}$]. I do not know if there is a mapping from $F_2(X)$ into $C_1(X)$ which is the identity of $F_1(X)$. But, assuming there is such a mapping f, it follows that

$J \in f[F_2(X)]$. To see this, suppose $J \notin f[F_2(X)]$. Rogers [280, p.284] has noted that the cone over X is homeomorphic to $C_1(X)$ -- see Chapter VIII for more discussion. Let $h: \text{Cone}(X) \xrightarrow{\text{onto}} C_1(X)$ be a homeomorphism [see (0.38) for definitions of symbols]. Note that the vertex v arcwise disconnects $\text{Cone}(X)$, so $h(v)$ arcwise disconnects $C_1(X)$. It is easy to see that J is the only member of $C_1(X)$ which arcwise disconnects $C_1(X)$ [see (11.5) and (11.6) for general results]. Hence, $h(v) = J$ [comp., (8.26)]. It follows from the proof in [240, 4.11] that h may be chosen so that $h[B(X)] = F_1(X)$, where $B(X)$ denotes the base of $\text{Cone}(X)$. Assume h is so chosen, i.e., $h[B(X)] = F_1(X)$ and, as we have noted must be true for any homeomorphism, $h(v) = J$. Let π denote the projection of the cone without v onto $B(X)$. Then, $h \circ \pi \circ h^{-1} \circ f$ is a retraction of $F_2(X)$ onto $F_1(X)$. Hence, by (6.17), we have a contradiction to (6.19).

This completes the discussion concerning (6.20).

Next let us note that, for any continuum X, $F_1(2^X)$ is a retract of 2^{2^X} (compose the union function of (1.48) and the inclusion map of 2^X into 2^{2^X}; this is specifically stated in [165, 1.1(b)]). Hence, in particular, 2^X admits a mean for any continuum X.

(6.21) QUESTION. For what continua X does $C(X)$ admit a mean? I do not know of any continuum X for which $C(X)$ does not admit a mean. I do not know if $C(S_o)$ admits a mean, where S_o is as in (1) of (8.22) [note: $C(S_o) \cong \text{Cone}(S_o)$--see Chapter VIII].

It would be interesting to know if $C(X)$ admits a mean when X is the continuum in (6.10.2). It is easy to see that the proof in (6.10.2) shows there is no retraction from $C_2(X)$ [see (0.48)] onto $C_1(X)$. However, this does not seem to imply there is no retraction from $F_2(C(X))$ onto $F_1(C(X))$ [comp., (0.71.2)].

(6.22) QUESTION. For what continua X is $F_1(C(X))$ a retract of $2^{C(X)}$? Clearly, by using (1.48) and (1.49), $F_1(C(X))$ is a retract of $C(C(X))$ for any continuum X.

We finish this chapter with the following questions.

(6.23) QUESTION. For what continua X is $C(A)$ an r-image of $C(X)$ for all subcontinua A of X?

(6.24) QUESTION. For what continua X is 2^A an r-image of 2^X for all subcontinua A of X?

Let us note that a homeomorphism is an r-map. Hence, the questions in (6.25) through (6.27) are special cases of (6.23) and (6.24). However, they are stated separately because they were the motivation for (6.23) and (6.24).

(6.25) QUESTION. For what continua X is $C(A) \cong C(X)$ for each nondegenerate subcontinuum A of X? Obviously [by (0.52)], any continuum X, all of whose nondegenerate subcontinua are homeomorphic to X, is such an X (for example, the arc and the pseudo-arc). Also [by (0.55)], a circle is such an X. Henderson [140] has proved that a decomposable continuum which is homeomorphic to each of its nondegenerate subcontinua must be an arc. This leads to the following question.

(6.26) QUESTION. For what *decomposable* continua X is $C(A) \cong C(X)$ for each nondegenerate subcontinuum A of X? If such an X contains an arc, then X must be an arc or a circle [as is easily seen using (1.92) and (1.109)]. Can there be such an X which contains no arc? It would be interesting if the answer to this last question were "no." The answer would then be a generalization of Henderson's theorem mentioned in (6.25). Also, if this generalization could be proved without using Henderson's theorem, a new proof for his theorem would be obtained. There is some hope for obtaining a hyperspace proof of Henderson's theorem since the arc structure for C(X) is a strong feature which is not available when working just in X as Henderson did. In particular, (14.8.1) might be helpful; also, see (11.18). Henderson's proof is difficult and a new proof of it would be of interest. To obtain his result, it may of course be assumed that X is hereditarily decomposable (instead of only decomposable as assumed above). Thus, there are many arcs in C(X) "parallel to and near $F_1(X)$" by (14.8.1).

(6.27) QUESTION. Same as (6.25) and (6.26) except replace

$C(X)$ and $C(A)$ by 2^X and 2^A respectively.

In relation to (6.1), (6.2), and (6.3) the following two questions are open.

(6.28) QUESTION. Let X be a continuum. Which of the following four statements are equivalent to each other?

(6.28.2) $F_1(X)$ is a retract of 2^X.

(6.28.2) $F_1(X)$ is a retract of $C(X)$.

(6.28.3) X is an r-image of 2^X.

(6.28.4) X is an r-image of $C(X)$.

By (6.4), they are all equivalent when X is a locally connected continuum.

(6.29) QUESTION. Let X be a continuum. Is $C_1(X)$ being a retract of 2^X equivalent to $C(X)$ being an r-image of 2^X? A continuum worth investigating in this connection is the continuum X in (6.10.2) [see (6.10.3)].

VII.

THE FIXED POINT PROPERTY

In 1952, B. Knaster posed the following question which appears as Problem 186 of The New Scottish Book [101]: If X is a continuum with the fixed point property, then does C(X) have the fixed point property? Recall that a space has the *fixed point property* provided any mapping of the space into itself leaves some point invariant.

The first partial answer to Knaster's question came ten years later when Segal [293] proved the following result:

(7.1) THEOREM [293, Theorem 3]. *If* X *is a chainable continuum, then* C(X) *has the fixed point property*.

In fact, Segal proves even more [293, Theorem 2]--he shows that when X is chainable, C(X) admits the structure of a quasi-complex. Then, since an acyclic quasi-complex has the fixed point property (as is well-known; see, for example, [293, p.237]), (7.1) follows from Segal's previous result that C(X) is acyclic [see (1.173)]. Some easier proofs of (7.1) are in [177] and [285].

After the publication of Segal's paper [293] there was a lot of interest in Knaster's question. However, no further results were obtained until 1972. Then, a major break-through occurred when Rogers proved [280, pp.282-284]:

(7.2) THEOREM [280, Theorem 8]. *Let* Z *be the circle-with-a-spiral given by* $Z = S^1 \cup \Lambda$, *where* S^1 *is the unit circle* [*see* (0.19)] *and* $\Lambda = \{[1 + (1/t)] \cdot e^{it} : t \geq +1\}$. *Then*, $C(Z)$ *is homeomorphic to* $Cone(Z)$ [*see Chapter VIII, especially the paragraph preceding (8.30);* $Z = (SP)_1$]. *Hence, since* $Cone(Z)$ *does not have the fixed point property* [170], $C(Z)$ *does not have the fixed point property* [280, p.282].

Thus, Rogers [280; (7.2) above] gave the first example of a continuum X such that $C(X)$ does not have the fixed point property. Since his continuum also does not have the fixed point property, it does not answer Knaster's question. However, by adding the inside of S^1 to Rogers' example, the following answer to Knaster's question is obtained.

(7.3) THEOREM [258]. *Let* $Y = Z \cup D$ *where* Z *is the continuum in (7.2) and* $D = \{(x, y) \in R^2 : x^2 + y^2 \leq 1\}$. *Then*, $C(Y)$ *does not have the fixed point property (even though* Y *does, a fact which appears in* [27, *Theorem 12*]).

Proof. We will define a retraction r from $C(Y)$ onto $C(Z)$ by combining a retraction f and a mapping g defined below. Then, (7.3) follows from (7.2).

Step 1: The retraction f. It is easy to see that for any

(7.3) FIXED POINT PROPERTY

$A \in \text{cl}[C(Y)\setminus C(D)]$, $A \cap Z$ is a subcontinuum of Z. Hence, by letting $f(A) = A \cap Z$ for each $A \in \text{cl}[C(Y)\setminus C(D)]$, we see that f is a function from $\text{cl}[C(Y)\setminus C(D)]$ onto $C(Z)$ and that $f(A) = A$ whenever $A \in C(Z)$. To see that f is continuous, and hence a retraction, first observe two (easy-to-verify) facts about f [Λ is as in (7.2)]:

(1) For each $A \in \text{cl}[C(Y)\setminus C(D)]$, $f(A) \subset A$;

(2) For each $A \in \text{cl}[C(Y)\setminus C(D)]$, $f(A) = A$ or
$f(A) = [A \cap \Lambda] \cup S^1$.

Recall that we are considering Y as contained in R^2 --the metric for Y is the one obtained from the metric defined in (0.19) for R^2. Denote this metric for Y by d, and let H_d denote the Hausdorff metric induced by d [as defined in (0.1)]. The following property of d is geometrically clear:

(3) For any point in Λ, the point nearest
it in D is on S^1.

By using (1) through (3), and by taking the cases indicated by (2), it can be verified that

(4) $H_d[f(A_1), f(A_2)] \leq H_d[A_1, A_2]$ for
any $A_1, A_2 \in \text{cl}[C(Y)\setminus C(D)]$.

By (4), f is continuous and, thus, we have shown that f is a retraction of $\text{cl}[C(Y)\setminus C(D)]$ onto $C(Z)$.

Step 2: The mapping g. Let
$$\mathcal{F} = \{A \in C(D) : S^1 \subset A\}$$
and, noting that
$$[C(S^1) \cup \mathcal{F}] \subset \text{cl}[C(Y)\setminus C(D)],$$

let $k: C(S^1) \cup \mathcal{F} \to C(S^1)$ be the restriction of f to $C(S^1) \cup \mathcal{F}$, i.e.,

(5) $\quad k(A) = f(A) \quad$ for each $A \in [C(S^1) \cup \mathcal{F}]$.

Since $C(S^1)$ is a 2-cell [by (0.55)] and since k is continuous, k can be extended to a continuous function $g: C(D) \to C(S^1)$ [186, p.339]. Since g is an extension of k, we have using (5) that

(6) $\quad g(A) = f(A) \quad$ for each $A \in [C(S^1) \cup \mathcal{F}]$.

Though it will not be important, note that g is a retraction from $C(D)$ onto $C(S^1)$.

Step 3: The retraction r. Note that

$$cl[C(Y)\setminus C(D)] = [C(Y)\setminus C(D)] \cup C(S^1) \cup \mathcal{F}$$

from which it follows easily that

(7) $\quad C(D) \cap cl[C(Y)\setminus C(D)] = C(S^1) \cup \mathcal{F}$.

By (6) and (7), the formula

$$r(A) = \begin{cases} f(A), & \text{if } A \in cl[C(Y)\setminus C(D)] \\ g(A), & \text{if } A \in C(D) \end{cases}$$

defines a function r on $C(Y)$. Thus, since f is a retraction from $cl[C(Y)\setminus C(D)]$ onto $C(Z)$ and since g is continuous on $C(D)$ such that $g[C(D)] = C(S^1) \subset C(Z)$, we have that r is a retraction from $C(Y)$ onto $C(Z)$. This completes the proof of (7.3) [use (7.2)].

Let us note that if $C(X)$ is a quasi-complex, then $C(X)$ has the fixed point property [see the paragraph following (7.1)]. Thus, as a consequence of (7.2) and (7.3), neither $C(Z)$ nor $C(Y)$ is a quasi-complex (see the Remark in [258, p.257]). Moreover, Y is acyclic and yet C(Y) is not a quasi-complex. Until [258], no example was known of a continuum X such that C(X) is not a quasi-complex. Segal [293] has asked the following question:

(7.4) QUESTION [293, p.248]. For what continua X is C(X) a quasi-complex? The only examples for which it has been proved that C(X) is not a quasi-complex are the two above.

Now let us return to (7.1). Krasinkiewicz [177] and Rogers [285] independently obtained easier proofs of it than the one in [293], and they proved the following result:

(7.5) THEOREM ([177, 4.2] and [285, Theorem 4]). *If* X *is a circle-like continuum, then* C(X) *has the fixed point property.*

Observe that if X is locally connected, then C(X) and 2^X are each absolute retracts [by (1.96)] and, therefore, have the fixed point property. On the other hand, assume X is an hereditarily indecomposable continuum. By (1.75), C(X) is contractible. Also, by (1.61), C(X) is uniquely arcwise connected. Young [348, Theorem 5] has proved that a contractible uniquely arcwise connected continuum has the fixed point property. This gives Rogers' proof [280, pp.284-285] of the following result:

(7.6) THEOREM [280, Remark on p.284]. *If* X *is any hereditarily indecomposable continuum, then* C(X) *has the fixed point property.*

Proof. See paragraph above.

Let us note that any arcwise connected subcontinuum of C(X) has the fixed point property when X is an hereditarily indecomposable continuum. This is easily proved by replacing (1.75) with (1.63) in the proof given above for (7.6). Let us also mention that Krasinkiewicz [176, p.180] gave another proof of (7.6)--see (19.25).

The result in (7.6) provides an interesting contrast. At one end of the spectrum are the locally connected continua, and at the other end are the hereditarily indecomposable continua--for all continua X at either end of the spectrum, C(X) has the fixed point property. For the continua in between, some do and some don't.

(7.7) QUESTION [280, p.283]. When does C(X), X a continuum, have the fixed point property?

Except for the results cited above, nothing is known about (7.7). In [285, p.234], Rogers poses the following question:

(7.8) QUESTION [285, p.234]. If X is a tree-like continuum, does C(X) have the fixed point property?

The following special case [see (0.28)] of (7.8) would seem to be of interest.

(7.8.1) QUESTION. Does $C(D)$ have the fixed point property for every dendroid D [see (7.8.3)]? Partial answers are given in (7.8.2). It is known [39, Theorem 2] that dendroids have the fixed point property. Hence, a negative answer to the question would be a "one-dimensional answer" to Knaster's question [comp., (7.3)]. In this connection, it would be of interest to know whether or not $C(X)$ has the fixed point property for every one-dimensional continuum X with the fixed point property [comp., (7.2) and the paragraph following it].

The following theorem answers (7.8.1) and the last question in (7.10) for some special dendroids.

(7.8.2) THEOREM. *If X is a fan or a smooth dendroid, then both $C(X)$ and 2^X have the fixed point property.*

Proof. Let X be a fan or a smooth dendroid and let d denote the metric for X. Let $\varepsilon > 0$. Then, by [441, Theorem 1] if X is a fan or by [442, Theorem 2] if X is a smooth dendroid, there is a finite graph $X_\varepsilon \subset X$ and a retraction r_ε from X onto X_ε such that $d(r_\varepsilon(x), x) < \varepsilon$ for each $x \in X$. Let $\hat{r}_\varepsilon : C(X) \to C(X_\varepsilon)$ be the induced map as defined in (0.49). Note that r_ε is a retraction of $C(X)$ onto $C(X_\varepsilon)$ and that $H(\hat{r}_\varepsilon(A), A) < \varepsilon$ for each $A \in C(X)$. Also note that $C(X_\varepsilon)$ has the fixed point property by (1.96). Therefore, since $\varepsilon > 0$ was arbitrary, it now follows that $C(X)$ has the fixed point property. The proof that 2^X has the fixed point property is similar--use

$r_\varepsilon^*: 2^X \to 2^{X_\varepsilon}$ given by $r_\varepsilon^*(A) = \{r_\varepsilon(a): a \in A\}$ for each $A \in 2^X$. This completes the proof of (7.8.2).

(7.8.3) QUESTION [442, p.261]. If X is a dendroid with metric d, then is there, for each $\varepsilon > 0$, a finite graph $X_\varepsilon \subset X$ and a retraction r_ε from X onto X_ε such that $d(r_\varepsilon(x), x) < \varepsilon$ for each $x \in X$? By using the proof of (7.8.2), an affirmative answer to this question would give an affirmative answer to the first question in (7.8.1). It would also give an affirmative answer to the last question in (7.10).

Now we turn our attention to 2^X. The following question was raised in [250]:

(7.9) QUESTION [250, 3.8]. When does 2^X, X a continuum, have the fixed point property? This question was in part the motivation for (6.3)--see (6.10).

When (7.9) was posed in [250], I did not know any continuum X for which I could show 2^X does not have the fixed point property. However, I strongly suspected that 2^Z does not have the fixed point property for Z as in (7.2). I discussed the problem of finding a retraction from 2^Z onto $C_1(Z)$ with Jim Lawson, which led to [191]. However [see (6.10)], it is not known if there is a retraction from 2^Z onto $C_1(Z)$. Moreover, at this time, I do not know of any 2^X for which it has been shown that 2^X does not have the fixed point property. Also, the following special cases of (7.9) are open.

(7.10) QUESTION. Does 2^X have the fixed point property for all hereditarily indecomposable continua X [comp., (7.6) above]? What about chainable [comp., (7.1)] or circle-like [comp., (7.5)] continua? What about *all* dendroids [see (7.8.2) and (7.8.3)]?

(7.11) QUESTION. Does 2^Y have the fixed point property, where Y is as in (7.3)? I conjecture that the answer is "no" and, in fact, that retractions can be defined from 2^Y onto 2^Z and from 2^Z onto $C_1(Z)$ where $Z \subset Y$ is as in (7.2); then, of course, (7.2) can be applied.

If the answer to (7.11) is "no," then Y is a continuum with the fixed point property such that 2^Y does not have the fixed point property.

The following is unsolved:

(7.12) QUESTION. Are the following two statements equivalent for any continuum X:

(7.12.1) 2^X has the fixed point property;

(7.12.2) C(X) has the fixed point property?

After (7.6) we mentioned that if X is an hereditarily indecomposable continuum, then any arcwise connected subcontinuum of

$C(X)$ has the fixed point property. Thus, the fixed point property holds for special subcontinua of certain hyperspaces $C(X)$. For any continuum X and any Whitney map μ for $C(X)$, it is known [see (14.21)] that $\mu^{-1}(t)$ is a subcontinuum of $C(X)$ for any $t \in [0, \mu(X)]$. These subcontinua are of interest, and will be discussed in depth in Chapter XIV. For now, let us note two facts: It follows immediately from (14.4) that $\mu^{-1}(t)$ has the fixed point property whenever X is a chainable continuum; however, as shown in (14.30), there is a circle-like continuum X such that X [and $C(X)$, by (7.5)] has the fixed point property but $\mu^{-1}(t)$ does not have the fixed point property for some t.

(7.13) QUESTION. For what continua X does $\mu^{-1}(t)$ have the fixed point property for every Whitney map μ for $C(X)$ and for each $t \in [0, \mu(X)]$? For some examples showing that X can have the fixed point property and $\mu^{-1}(t)$ may not, see (14.30) and (14.43.6). Let us also note that when X is a chainable continuum, $\mu^{-1}(t)$ has the fixed point property for each Whitney map μ for $C(X)$ and each t [as follows using (14.4)].

VIII.
RECOGNIZING HYPERSPACES
AS CONES

As can be seen from the previous chapter, knowledge that a hyperspace is (topologically) a cone can be an important ingredient in investigating the structure of hyperspaces. In this chapter we will discuss what is known about when a hyperspace is a cone, and we will raise some open questions. This is the first of three successive chapters on recognizing hyperspaces as other constructs.

I refer the reader to (0.38) for some notational conventions. Before beginning our discussion, let us note the following well-known fact about cones which will be used many times.

(8.0) LEMMA [well-known]. *If* X *is a continuum such that* $\dim[X] < \infty$, *then* $\dim[\text{Cone}(X)] = \dim[X] + 1$.

Proof. Let v denote the vertex of $\text{Cone}(X)$. Clearly

(1) $\quad \text{Cone}(X) \setminus \{v\} = \bigcup_{n=1}^{\infty} B_n$

where

(2) $\quad B_n$ is homeomorphic to $X \times [0, 1]$

for each $n = 1, 2, \ldots$.

Since X is compact and $[0, 1]$ is one-dimensional, we have by

(2) and [150, lines 13-14, p.34] that

(3) $\dim[B_n] = \dim[X] + 1$ for each $n = 1, 2, \ldots$.

Since B_n is a closed [in fact, by (2), compact] subset of $\text{Cone}(X) \setminus \{v\}$ for each $n = 1, 2, \ldots$, it follows using (1), (3), and the Sum Theorem [150, Theorem III2, p.30] that

(4) $\dim[\text{Cone}(X) \setminus \{v\}] = \dim[X] + 1$.

By (4) and [150, Corollary 2, p.32], $\dim[\text{Cone}(X)] = \dim[X] + 1$. This proves (8.0).

The first paper in which a result appeared about a hyperspace being a cone was [284]. The paper was mostly devoted to proving the result stated here in (3.3). However, to indicate the necessity that X in (3.3) be planar [comp., (3.6)], the following result was also proved [comp., (14.20)].

(8.1) THEOREM ([284, Theorem 2]; also, see [92]). *If* X *is a solenoid, then* $C(X) \cong \text{Cone}(X)$.

This was an important result; it led to the investigations in [280] which, in turn, led to the solution [258] of Knaster's question (see, here, Chapter VII) and to the discovery of many facts about the structure of hyperspaces.

The first paper devoted to a formal study of when a hyperspace is a cone was [280]. In section 1 of [280], Rogers studied continua X for which there is a special type of homeomorphism between $\text{Cone}(X)$ and $C(X)$; namely, he required that the homeomorphism

make the vertex v of the cone correspond with $X \in C(X)$, and the
base B(X) of the cone correspond with $F_1(X)$. We will call such a
homeomorphism a *Rogers homeomorphism*. Thus, a Rogers homeomorphism
may be thought of as a homeomorphism between the triples
(Cone(X), B(X), $\{v\}$) and (C(X), $F_1(X)$, $\{X\}$) [280, p.280]. If
there is a Rogers homeomorphism, then X is said to have the *cone =
hyperspace property* [280, p.280]. We emphasize that there are
continua X whose cone and hyperspace are homeomorphic but which do
not have the cone = hyperspace property. Such is the case, for example, for the continuum Z in (7.2). This is easy to see because
Cone(Z)\$\{v\}$ is not arcwise connected, whereas C(Z)\$\{Z\}$ is [the
last fact is easy to prove for Z; see (1.51) for a general theorem which applies].

The simplest continua which have the cone = hyperspace property
are an arc and a circle [see the homeomorphism g of (0.54) and the
homeomorphism h of (0.55)]. Rogers' proof [284, pp. 167-168]
of (8.1) shows that any solenoid has the cone = hyperspace property
as does the so-called "Buckethandle" continuum [defined in
(1.209.3)]. Rogers notes these facts in [280, p.280].

Now we give some results about the cone = hyperspace property.

(8.2) THEOREM [280, Theorem 1]. *If a continuum* X *has the
cone = hyperspace property, then each composant of* X *is arcwise
connected.*

Proof. Assume X is a continuum which has the cone = hyper-
space property. Let h: Cone(X) $\xrightarrow{\text{onto}}$ C(X) be a Rogers homeo-

morphism. Let $p \in X$ and let κ_p denote the composant in X of p. Let $q \in \kappa_p$ such that $q \neq p$. Then [see (0.32)], there is a proper subcontinuum Y of X such that $p, q \in Y$. Let

$$Y_o = \{(y, 0) \in B(X) : y \in Y\}.$$

Since h is a homeomorphism such that $h[B(X)] = F_1(X)$ and since Y_o is a proper subcontinuum of $B(X)$, $Z = \{x \in X: \{x\} \in h[Y_o]\}$ is a proper subcontinuum of X. Hence, using (1.13), $C(Z)$ is an arcwise connected subcontinuum of $C(X)$ and $X \notin C(Z)$. Thus, since $h(v) = X$, $h^{-1}[C(Z)]$ is an arcwise connected subcontinuum of Cone(X) such that $v \notin h^{-1}[C(Z)]$. Thus, $\pi(h^{-1}[C(Z)])$ is an arcwise connected subcontinuum of $B(X)$ where π is as defined in (0.38). Note that

$$(p, 0), (q, 0) \in \pi(h^{-1}[C(Z)]).$$

Hence, there is an arc $\alpha \subset \pi(h^{-1}[C(Z)])$ such that α has end points $(p, 0)$ and $(q, 0)$. If $\alpha = B(X)$, then it follows that $q \notin \kappa_p$. Therefore, $\alpha \neq B(X)$. Hence, letting

$$\beta = \{x \in X: (x, 0) \in \alpha\},$$

we see that β is an arc and $p, q \in \beta \subset \kappa_p$. This completes the proof of (8.2).

(8.3) COROLLARY [280, Corollary 1]. *If X is a decomposable continuum such that X has the cone = hyperspace property, then X is arcwise connected.*

Proof. Since X is a decomposable continuum, there are proper subcontinua A and B of X such that $X = A \cup B$. Let $p \in [A \cap B]$. Clearly the composant in X of p is X itself.

The corollary now follows from (8.2).

In the next result it is only assumed that Cone(X) and C(X) are homeomorphic, not (necessarily) that X has the cone = hyperspace property. As we will see [(8.17) and (8.18)], dim[X] < ∞ implies dim[X] = 1.

(8.4) THEOREM [280, Theorem 5]. *Let a continuum* X *satisfy* dim[X] < ∞, *each composant of* X *is arcwise connected, and* Cone(X) $\cong$ C(X). *Then* X *is an arc, arcwise connected circle-like continuum, or an indecomposable continuum such that each nondegenerate proper subcontinuum of* X *is an arc* [*see (8.7) for a stronger result when* X *is indecomposable*].

Proof. Let X be a continuum which satisfies the hypotheses of (8.4), i.e.,

(1) dim[X] < ∞;

(2) each composant of X is arcwise connected;

(3) Cone(X) $\cong$ C(X).

By (1) and (8.0),

(4) dim[Cone(X)] = dim[X] + 1 < ∞.

Suppose dim[X] ≥ 2. Then, by (1), we may apply (1.53) to see that X contains a proper nondegenerate hereditarily indecomposable subcontinuum. Hence, by using (2), it follows that X contains an n-odd for each n ≥ 2. Thus, by (1.100), C(X) contains an n-cell for each n ≥ 2. Therefore, dim[C(X)] = ∞ which, by (3), says dim[Cone(X)] = ∞. This last statement contradicts (4). We have proved [as is true more generally--see (8.18)]

(5) dim[X] = 1.

By (3) through (5),

(6) dim[C(X)] = 2.

The rest of the proof is technical, and the details will not be done here. However, for completeness, we mention the following result of Rogers [280, Theorem 3] which he used to finish the proof: If Z is a continuum whose composants are arcwise connected and such that dim[C(Z)] = 2, then Z must be one of the types of continua in the conclusion of (8.4).

(8.5) THEOREM [280, Theorem 7]. *Let* X *be a continuum such that* dim[X] $< \infty$ *and such that* X *has the cone = hyperspace property. Then, each nondegenerate proper subcontinuum of* X *is an arc* [*see (8.6) for a stronger formulation*].

Proof. Assume X is as in the hypotheses of (8.5). Then, by (8.2), X satisfies the hypotheses of (8.4). Hence, by (8.4), each proper subcontinuum of X is an arc or X is an arcwise connected circle-like continuum. Thus, it suffices to prove that

(*) the only arcwise connected circle-like continuum
     which can have the cone = hyperspace property is
     a circle.

To prove (*), let Z be an arcwise connected circle-like continuum which has the cone = hyperspace property. Let h: Cone(Z) $\xrightarrow{\text{onto}}$ C(Z) be a Rogers homeomorphism. Suppose Z is not a circle. We will use the structure theorem for arcwise connected circle-like continua

(8.5) HYPERSPACES AS CONES 307

given in [245, Theorem 6]. For readers not familiar with [245, Theorem 6], it roughly says that an arcwise connected circle-like continuum which is not a circle resembles (but is not necessarily homeomorphic to) the Warsaw circle [denoted S_2^1 in (8.22)]. Thus, if the reader follows this proof for the special case of S_2^1, he will understand the ideas involved. By [245, Theorem 6], it follows that

(1) there are points p of Cone(Z) arbitrarily close to the vertex v of Cone(Z) such that Cone(Z) is not connected im kleinen at p;

(2) there exists $\varepsilon > 0$ such that if $A \in [C(Z)\setminus\{Z\}]$ and $H(A, Z) < \varepsilon$, then A has a 2-cell neighborhood in $C(Z)$.

Actually, a stronger statement than (2) is true: Z itself has a 2-cell neighborhood in $C(Z)$--this can be proved using (14.31), but (2) is easier to verify and (2) is all that is needed. By (1) and (2), $h(v) \neq Z$ which is a contradiction. This proves (*) and completes the proof of (8.5).

Rogers did not state the next result, but it follows from the proof of (8.5). Also, it follows from the statement of (8.5) by the following simple argument. Assume X satisfies the hypotheses of (8.5) and that X is decomposable. Then, by (8.5), X is the union of two arcs and, hence, X is locally connected. Also, by (8.5), X contains no simple triod. Therefore, X must be an arc or a circle. Thus, we have the following result:

(8.6) THEOREM [280, proof of Theorem 7]. *Let* X *be a continuum such that* dim[X] $< \infty$ *and such that* X *has the cone = hyperspace property. Then* X *is an arc, a circle, or an indecomposable continuum such that each nondegenerate proper subcontinuum of* X *is an arc.*

Proof. See paragraph above.

If X is a continuum such the cone over X is homeomorphic to C(X), then we will say that X is a *C-H continuum*.

As is mentioned above, some finite-dimensional indecomposable continua are known to be C-H continua (the solenoids, for example). The following result shows that all such continua have a strong form of the cone = hyperspace property.

(8.7) THEOREM [239, 5.7]. *If* X *is a finite-dimensional indecomposable C-H continuum, then every homeomorphism between* Cone(X) *and* C(X) *is a Rogers homeomorphism. Hence,* X *has the cone = hyperspace property.*

By (8.5), we see that (8.7) is a stronger result than (8.4) when X is indecomposable.

Without the condition of finite-dimensionality, we also have the following two results.

(8.8) THEOREM [239, 5.2]. *If* X *is an indecomposable C-H continuum and if* h: Cone(X) $\xrightarrow{\text{onto}}$ C(X) *is a homeomorphism, then*

$h(v) = X$.

Proof. Let $W = C(X) \setminus C(h(v))$. We will show $W = \emptyset$. By (11.2), W is arcwise connected; clearly, W is an open subset of $C(X)$. Hence, $U = h^{-1}[W]$ is an open arcwise connected subset of Cone$(X) \setminus \{v\}$. Thus, $\pi[U]$ is an open arcwise connected subset of $B(X)$ [π and $B(X)$ as defined in (0.38)]. Since $B(X)$ is indecomposable,

(1) the composants of $B(X)$ are mutually

disjoint [186, Theorem 5, p.212];

(2) each composant of $B(X)$ is of the first

category (in itself) [186, Theorem 6, p.212].

Since $\pi[U]$ is arcwise connected, we have using (1) that $\pi[U]$ is contained in a composant of $B(X)$. Therefore, since $\pi[U]$ is an open subset of $B(X)$, it follows using (2) that $\pi[U] = \emptyset$. Hence, $W = \emptyset$ so $h(v) = X$. This proves (8.8).

(8.9) THEOREM [239, 5.4]. *If* X *is an indecomposable C-H continuum, then the composants of* X *coincide exactly with the arc components of* X.

When X is finite-dimensional, more can be said:

(8.10) THEOREM [239, 5.5]. *If* X *is a finite-dimensional indecomposable C-H continuum, then any composant of* X *is a one-to-one continuous image of* $[0, +\infty)$ *or* R^1 *[comp., (8.21)].*

Now we are prepared to state and discuss the following two questions.

(8.11) QUESTION. What finite-dimensional continua have the cone = hyperspace property?

(8.12) QUESTION [comp., (8.33)]. What are the finite-dimensional C-H continua? See (8.30).

By (8.6), (8.11) is open only for indecomposable continua. Furthermore, by (8.7), (8.11) and (8.12) are equivalent questions for indecomposable continua, i.e.,

(8.13) THEOREM. *Let* x *be a finite-dimensional indecomposable continuum. Then:* x *is a C-H continuum if and only if* x *has the cone = hyperspace property.*

It is suggested by (8.6) and (8.10) that finite-dimensional continua with the cone = hyperspace property are close to being chainable or circle-like. Thus, we have the following three questions.

(8.14) QUESTION. Must a finite-dimensional indecomposable C-H continuum be chainable or circle-like?

(8.15) QUESTION. What are the indecomposable chainable C-H

continua [also see (8.31)]? The "Buckethandle" continuum [defined in (1.209.3)] is an example of one, as was noted by Rogers [280, p.280].

(8.16) QUESTION. What are the indecomposable circle-like C-H continua [also see (8.32)]? The "Buckethandle" and the solenoids [by (8.1)] are examples of such continua.

For a large portion of this chapter, we have been focusing our attention on finite-dimensional C-H continua. It follows from (2.3) that such continua can be at most two-dimensional. In fact, they must be one-dimensional:

(8.17) THEOREM ([240] and [244, Lemma 3]). *If* X *is a finite-dimensional C-H continuum, then* dim[X] = 1. *In fact [244, Lemma 3],* X *can not contain a nondegenerate hereditarily indecomposable subcontinuum.*

The second part of (8.17) was not stated in [240], but it was used to prove the first part [see (1.53)]. Rogers [281, Theorem 8] has independently obtained the following much stronger result. His proof, which we give below, is quite elegant and yields considerable insight into the geometry of hyperspaces.

(8.18) THEOREM [281, Theorem 8]. *If* X *is a finite-dimensional C-H continuum, then* X *contains at most one nondegenerate indecomposable subcontinuum [hence, by (1.53), dim[X] = 1].*

Proof. The proof is in six steps. Let Y be a subcontinuum of X and let h be a homeomorphism from Cone(X) onto C(X).

(i) No arc component of $C(X)\setminus\{Y\}$ is homeomorphic to $[0, 1)$. To prove (i), let Γ be an arc component of $C(X)\setminus\{Y\}$. It is easy to see, using (1.8) and (1.11), that there exists $Z \in \Gamma$ such that $Z \notin F_1(X)$ [i.e., Z is nondegenerate]. If $Y \not\subset Z$, then $C(Z) \subset \Gamma$ [by (1.12)] and, since Z is nondegenerate, $\dim[C(Z)] \geq 2$ [by the first part of (2.1)]. If $Y \subset Z$, then (1.8) and (1.11) can be used to see that Γ contains a simple triod. In either case, Γ is not homeomorphic to $[0, 1)$.

(ii) No arc component of X is degenerate. If $\{x_o\}$ were an arc component of X, then

$$\Lambda = \{(x_o, t) \in \text{Cone}(X): 0 \leq t < 1\}$$

would be an arc component of $\text{Cone}(X)\setminus\{v\}$. Hence $h[\Lambda]$ would be an arc component of $C(X)\setminus\{h(v)\}$, a contradiction to (i) since $h[\Lambda] \cong [0, 1)$.

(iii) No point of Cone(X), except possibly v, arcwise disconnects Cone(X). This is clear from (ii) and the geometry of cones.

(iv) No member of C(X), except possibly h(v), arcwise disconnects C(X). This follows immediately from (iii) using h.

(v) If Y is a nondegenerate indecomposable subcontinuum of X, then $C(X)\setminus\{Y\}$ has uncountably many arc components. To prove (v), we first prove that:

(*) at most a finite number of composants $K_1, K_2, \ldots, K_j$

of Y have the property that some subcontinuum of X contains a point of X\Y and a point of the composant but does not contain Y.

Suppose (*) is false. Let n be a fixed natural number. Then there are n composants and n subcontinua of X with the property in (*). From these n subcontinua it is not difficult to produce n (possibly) new subcontinua each of which has the property in (*) and which are mutually disjoint. Denote these subcontinua by X_1, X_2,..., X_n. Then, $Y \cup [\cup_{i=1}^{n} X_i]$ forms an n-odd. Hence, by (1.100), C(X) contains an n-cell. Since n was arbitrary, we can now conclude that $\dim[C(X)] = \infty$. But, since $\dim[X] < \infty$ and Cone(X) $\cong$ C(X), we have using (8.0) that

$\dim[C(X)] = \dim[Cone(X)] < \infty$.

This is a contradiction, so we have proved (*). Now, let $\lambda_1 \neq \lambda_2$ be composants of Y different from any of the finitely many that may satisfy (*). Suppose there is an arc α in $C(X)\setminus\{Y\}$ given by a homeomorphism f: [0, 1] $\xrightarrow{onto}$ α such that f(0) is a subcontinuum of λ_1 and f(1) is a subcontinuum of λ_2. Suppose that $\alpha \subset C(Y)$. Then, $\cup\alpha$ is a subcontinuum [by (1.49)] of Y which intersects different composants of Y. Thus, since Y is indecomposable, $\cup\alpha = Y$ by [186, Theorem 5, p.212]. Therefore, by (1.50), $Y \varepsilon \alpha$. This contradicts the fact that $\alpha \subset [C(X)\setminus\{Y\}]$. Hence, $\alpha \not\subset C(Y)$. Now let

s_0 = l.u.b.{$s \varepsilon$ [0, 1]: f([0, s]) $\subset$ C(Y)}.

Note that $s_0 < 1$ and that f([0, s_0]) is an arc in C(Y). Thus, by using (1.49) and (1.50) as above, it follows that $\cup f([0, s_0])$

is a subcontinuum of λ_1. Hence, by continuity of union [see (1.48)], it follows that there exists $t_0 > s_0$ such that $[\cup f([s_0, t_0])] \not\supset Y$. It now follows that $\cup f([s_0, t_0])$ is a subcontinuum of X which contains a point of X\Y and a point of λ_1 but which does not contain Y. This proves λ_1 satisfies (*) and, therefore, is a contradiction. Thus, there is no arc α in $C(X)\setminus\{Y\}$ from a subcontinuum of λ_1 to a subcontinuum of λ_2. This completes the proof of (v).

(vi) **X can contain at most one nondegenerate indecomposable continuum.** This is a direct consequence of (iv) and (v).

This completes the proof of (8.18).

Let us observe that in the proof of (v) above the assumption that X was a C-H continuum was only used to deduce that $\dim[C(X)] < \infty$. Thus, the following result was proved.

(8.19) THEOREM [281, Proposition 5]. *Let X be a continuum such that* $\dim[C(X)] < \infty$. *If Y is a nondegenerate indecomposable subcontinuum of X, then $C(X)\setminus\{Y\}$ has uncountably many arc components.*

Proof. See paragraph above.

(8.20) SOME COMMENTS AND QUESTIONS. Let X be a finite-dimensional C-H continuum. As we will see in (8.23), it is completely known what X must be if X is hereditarily decomposable. So, for our purposes here, assume X contains a nondegenerate indecomposable subcontinuum Y. Note that, by (8.18), X contains only one such subcontinuum. In general, a nondegenerate subcontinuum of a C-H continuum need not be a C-H continuum. This is easily seen by examining various subcontinua of the continua listed in (8.23) (one such example is given in [281, Example 1]). However, I do not know the answer to the following question.

(8.20.1) QUESTION. Must Y be a C-H continuum?

Note that if the answer to (8.20.1) is "yes," then [by (8.7)] Y has the cone = hyperspace property and, thus [by (8.6)], each nondegenerate proper subcontinuum of Y is an arc. In attempting to answer (8.20.1), one is led to trying to answer the following question.

(8.20.2) QUESTION. If $h: \text{Cone}(X) \xrightarrow{\text{onto}} C(X)$ is a homeomorphism, then must $h[\text{Cone}(Y)] = C(Y)$?

For the rest of (8.20), I will discuss what I know about (8.20.2). The main information is summarized in (8.20.7). In (8.20.8), an affirmative answer to (8.20.2) is obtained with extra conditions on X. In (8.20.9), an affirmative answer to (8.20.2) is obtained under conditions which may not be restrictive [see the three sentences following (8.20.9)].

To begin our discussion of (8.20.2), let $h: \text{Cone}(X) \xrightarrow{\text{onto}} C(X)$ be a homeomorphism. By combining (iv) and (v) in the proof of

(8.18), we see that

(8.20.3) $h(v) = Y$.

By (11.2), $C(X)\setminus C(Y)$ is arcwise connected; clearly, $C(X)\setminus C(Y)$ is an open subset of $C(X)$. Hence, by (8.20.3), $h^{-1}[C(X)\setminus C(Y)]$ is an open arcwise connected subset of $Cone(X)\setminus\{v\}$. Thus, letting

$$Z = \pi(h^{-1}[C(X)\setminus C(Y)])$$

where π and $B(X)$ are as defined in (0.38), we see that

(1) Z is an open arcwise connected subset of $B(X)$.

We will show that

(*) $Z \cap B(Y) = \emptyset$.

To prove (*), suppose $Z \cap B(Y) \neq \emptyset$. Then, by (1), $Z \cap B(Y)$ is a nonempty open subset of $B(Y)$. Hence, since each composant of $B(Y)$ is dense in $B(Y)$ [186, p.209],

(2) Z intersects each composant of $B(Y)$.

Since Z is arcwise connected [by (1)], it is easy to see using (2) that X contains an n-odd for each $n \geq 2$. Hence, by (1.100), it follows that $\dim[C(X)] = \infty$. However, since $\dim[X] < \infty$, $\dim[Cone(X)] < \infty$ [by (8.0)] and therefore, since $Cone(X) \cong C(X)$, we have that $\dim[C(X)] < \infty$. Thus, we have a contradiction. Therefore, (*) holds. By (*), the definition of Z, and (8.20.3), it follows easily that $h^{-1}[C(X)\setminus C(Y)] \cap Cone(Y) = \emptyset$. Hence, $[C(X)\setminus C(Y)] \cap h[Cone(Y)] = \emptyset$. Therefore,

(8.20.4) $h[Cone(Y)] \subset C(Y)$.

Next we prove

(**) there exist at most two arc components Q_1 and Q_2 of $Cone(X)\setminus\{v\}$ such that $Q_i \cap Cone(Y) \neq \emptyset \neq$

(8.20.4) HYPERSPACES AS CONES 317

$Q_i \cap [\text{Cone}(X)\setminus\text{Cone}(Y)]$ for each $i = 1$ and 2.

To prove (**), suppose there exists three arc components Q_1, Q_2, and Q_3 of $\text{Cone}(X)\setminus\{v\}$ such that

$Q_i \cap \text{Cone}(Y) \neq \emptyset \neq Q_i \cap [\text{Cone}(X)\setminus\text{Cone}(Y)]$

for each $i \in \{1,2,3\}$. Then, for each $i \in \{1,2,3\}$, there exists an arc $A_i \subset Q_i$ such that

(3) $A_i \cap \text{Cone}(Y) \neq \emptyset \neq A_i \cap [\text{Cone}(X)\setminus\text{Cone}(Y)]$.

Since $A_i \subset Q_i$ for each $i \in \{1,2,3\}$ and since Q_1, Q_2, and Q_3 are distinct arc components of $\text{Cone}(X)\setminus\{v\}$, it is easy to see that if $j, k \in \{1,2,3\}$ such that $j \neq k$, then

(4) $\pi[A_j] \cap \pi[A_k] = \emptyset$.

It follows easily using (3) and (4) that X contains a triod. Hence, by (1.100), $C(X)$ contains a 3-cell. However, since $\dim[X] = 1$ [by (8.18)] and since $C(X) \cong \text{Cone}(X)$, we have by (8.0) that

(5) $\dim[C(X)] = \dim[\text{Cone}(X)] = 2$.

Hence, $C(X)$ does not contain a 3-cell (see [150, Corollary, p.42]). Thus, we have a contradiction. Therefore, (**) holds. Now we prove

(***) if Q is an arc component of $\text{Cone}(X)\setminus\{v\}$, then

$\pi[Q]$ can intersect at most two composants of $B(Y)$.

To prove (***), suppose Q is an arc component of $\text{Cone}(X)\setminus\{v\}$ such that $\pi[Q]$ intersects three (or more) composants of $B(Y)$. Then, since the composants of $B(Y)$ are mutually disjoint [186, Theorem 5, p.212], it follows that X contains a triod and hence, by (1.100), $C(X)$ contains a 3-cell. By (5) and [150, Corollary, p.42], $C(X)$ can not contain a 3-cell. Thus, we have a contradiction. Therefore, (***) holds. Since $B(Y)$ has uncountably many composants [186, (i), p.213], it follows using (**) and (***) that

(6) $\text{Cone}(Y)\setminus\{v\}$ contains uncountably many
arc components of $\text{Cone}(X)\setminus\{v\}$.

Now, let κ be a composant of Y and let

$$C(\kappa) = \{A \in C(Y) : A \subset \kappa\}.$$

Let $A_1, A_2 \in C(\kappa)$. It follows easily using the indecomposability of Y that there exists a proper subcontinuum L of Y such that $A_i \subset L$ for each $i = 1$ and 2. By (1.12), there is an arc in $C(L)$ from A_1 to A_2. Thus, since

$$C(L) \subset C(\kappa) \subset [C(X)\setminus\{Y\}],$$

we have proved that

(7) if κ is a composant of Y, then $C(\kappa)$ is
contained in an arc component of $C(X)\setminus\{Y\}$.

By (8.20.4), (6), and (7),

(8.20.5) $h[\text{Cone}(Y)] \supset C(\kappa)$ for uncountably many
composants κ of Y.

Since each composant of Y is dense in Y [186, p.209], we have from (8.20.5) that

(8.20.6) $h[\text{Cone}(Y)] \supset F_1(Y)$.

Summing up what we have shown in (8.20.3) through (8.20.6), we see that

(8.20.7) $F_1(Y) \subset h[\text{Cone}(Y)] \subset C(Y)$, $Y = h(v)$, and
$h[\text{Cone}(Y)] \supset C(\kappa)$ for uncountably many composants κ of Y.

With extra conditions on X, affirmative answers to (8.20.2) can be obtained. For example, by using (7), (8.20.4), and (8.20.6), it follows that:

(8.20.8) If each arc component of X is contained either in Y or in X\Y, then h[Cone(Y)] = C(Y).

Even though [by (**)] at most two arc components of X can fail to satisfy the hypothesis of (8.20.8), the hypothesis is still very strong. Note that, by (8.20.4) and (8.20.5), we have the following result:

(8.20.9) If C(κ) is dense in C(Y) for each composant κ of Y, then h[Cone(Y)] = C(Y).

In general, indecomposable continua do not have to satisfy the hypothesis of (8.20.9). However, for the situation we are dealing with, it may be that the hypothesis in (8.20.9) is automatically satisfied by Y. If this could be proved, then we would have an affirmative answer to (8.20.1) and (8.20.2).

The result in (8.21) below gives information about the structure of the arc components of any finite-dimensional C-H continuum [comp., (8.9) and (8.10)]. Let us note, as follows from (8.21.2) and (8.23), that there are exactly three finite-dimensional arcwise connected C-H continua.

(8.21) THEOREM [239, 2.5]. *If* X *is a finite-dimensional C-H continuum, then*

(8.21.1) *any arc component of* X *is a one-to-one continuous image of* [0, 1], [0, +∞) *or* R^1;

(8.21.2) *a compact arc component of* X *is an arc or an arcwise connected circle-like continuum.*

Proof. Assume X is a finite-dimensional C-H continuum. By (8.17) or (8.18), $\dim[X] = 1$. Hence, by (8.0), $\dim[\text{Cone}(X)] = 2$. Thus, since $\text{Cone}(X) \cong C(X)$, we have that $\dim[C(X)] = 2$. Hence, it follows using (1.100) that

(1) X is a-triodic.

Now, let L denote an arc component of X. Then, L is an arcwise connected metric space which, by (1), contains no simple triod. Hence, by [235, 3.2],

(2) L is a one-to-one continuous image of a connected subset G of R^1.

By (ii) of the proof of (8.18), L is nondegenerate. Hence, G in (2) must be homeomorphic to [0, 1], [0, +∞), or R^1. This proves (8.21.1). Now, to prove (8.21.2), assume

(3) L is a compact arc component of X

and assume

(4) L is not an arc.

It follows easily from the Structure Theorem for Real Curves [253, p.9] that any compact one-to-one continuous image of R^1 must contain a triod. Hence, by (1) and compactness of L [see (3)], L is not a one-to-one continuous image of R^1. Thus, by (3), (4), and (8.21.1), we see that L must be a compact one-to-one continuous image of [0, +∞). Hence, it follows easily using (1) and the Structure Theorem in [238, p.128] that L is an arcwise connected circle-like continuum. This proves (8.21.2) and completes the proof of (8.21).

For most of the rest of this chapter, we will focus our attention on hereditarily decomposable C-H continua. In the light of (8.18), this is not too severe a restriction within the class of finite-dimensional C-H continua.

The first results about hereditarily decomposable C-H continua were announced in [240]. Since then, these continua have been completely determined--there are *exactly eight* of them! The precise result appears in [239, 1.1] and will be stated below as (8.23). First, we define the following seven special continua.

(8.22) DEFINITION. Let

(1) $S_o = cl\{(x, \sin[1/x]): 0 < x \leq 1\}$;

(2) $S_1^1 = S^1 = \{(x, y) \in R^2: x^2 + y^2 = 1\}$;

(3) $S_2^1 = S_o$ with the points $(0, -1)$ and $(1, \sin[1])$ identified;

(4) $S_3^1 = S_o \cup \{(x, \sin[1/x]): -1 \leq x < 0\}$ with the points $(1, \sin[1])$ and $(-1, \sin[-1])$ identified;

(5) $(SP)_1 = S^1 \cup \{[1 + (1/t)] \cdot e^{it}: t \geq +1\}$;

(6) $(SP)_2 = (SP)_1 \cup \{[1 - (1/t)] \cdot e^{it}: t \geq +1\}$ with the points $2e^{i \cdot 1}$ and $(0, 0)$ identified;

(7) $(SP)_3 = (SP)_1 \cup \{[1 + (1/t)] \cdot e^{it}: t \leq -1\}$ with the points $2e^{i \cdot 1}$ and $(0, 0)$ identified.

In descriptive terms, S_2^1 is the usual $\sin[1/x]$-circle commonly called the Warsaw circle, S_3^1 is a particular compactification of

R^1 with an arc as the remainder, $(SP)_1$ is a circle with a counterclockwise-moving half-line spiraling down on it, $(SP)_2$ is a circle with a counterclockwise-moving line spiraling down on it, and $(SP)_3$ is a compactification of R^1 with half of R^1 spiraling down on S^1 in a counterclockwise direction and the other half of R^1 spiraling down on S^1 in a clockwise direction. We note that all the continua except $(SP)_3$ are embeddable in the plane.

We are now ready to state the following result:

(8.23) THEOREM [239, 1.1]. *If X is an hereditarily decomposable C-H continuum, then X is homeomorphic to one of the following eight continua: the closed unit interval* [0, 1], S_o, S_i^1 *for* i = 1, 2, *or* 3, *or* $(SP)_i$ *for* i = 1, 2, *or* 3. *Furthermore, each of these eight continua are C-H continua.*

Instead of giving the details in [239] of the proof of (8.23), we will indicate and discuss the ideas involved. The discussion will give some information not contained in the statement of (8.23) [for example, some facts about how a cone-to-hyperspace homeomorphism must behave—see (8.25) and (8.26)]. This discussion lasts until (8.30).

The first step in the proof in [239] of (8.23) is represented by the following less-specific result:

(8.24) LEMMA ([239, 2.7] and [281, Theorem 4]; see (8.24.1)). *If X is an hereditarily decomposable C-H continuum, then X is either*

   (1) *an arc;*

   (2) *an arcwise connected circle-like continuum;*

   (3) *a compactification of* [0, +∞) *with an arc or a circle as the remainder;*

   (4) *a compactification of* R^1 *with an arc or a circle as the remainder such that each end of the compactification is equal to the remainder* [*see below for definition of* end].

Let $Z = g[R^1] \cup Q$, g a homeomorphism, be a compactification of $g[R^1]$ with remainder Q. By an *end of the compactification* Z we mean

$$\bigcap_{n=1}^{\infty} cl[g([n, +\infty))] \quad \text{or} \quad \bigcap_{n=1}^{\infty} cl[g((-\infty, -n])];$$

thus, Z has two ends and the ends are independent of the homeomorphism g (except for possibly being switched, one for the other). The fact that the ends must be equal in (4) of (8.24) can be seen easily by examining the continuum

$$X = S_o \cup cl[\{(x, \sin[1/x] - 1): -1 \leq x < 0\}]$$

where S_o is as in (1) of (8.22)--no subcontinuum A of X [i.e., no A ∈ C(X)] arcwise disconnects C(X), so there is no place for the vertex to go under a supposed homeomorphism from Cone(X) onto C(X).

(8.24.1) REMARKS ABOUT (8.24). The result in (8.24) was obtained independently by Rogers and me in the summer of 1972--see [239, p.322] for more information. There appears to be a problem with Rogers' proof [281, p.285] of (8.24)--for a detailed discussion, see [239, pp.322-323]. Because of this presumed problem with Rogers' proof, I included my original proof of (8.24) in [239]. In [239], the theory concerning the structure of locally compact one-to-one continuous images of half-lines and lines is used. This theory is contained in [238] and [253].

The second step towards a proof of (8.23) is to determine some specific information about the behavior of any assumed homeomorphism h: Cone(X) $\xrightarrow{\text{onto}}$ C(X), where X is any of the continua in (2) through (4) of the conclusion of (8.24). This information is summarized below in (8.25) and (8.26).

(8.25) THEOREM [239, 2.6]. *If* X *is an arcwise connected circle-like C-H continuum and if* h: Cone(X) $\xrightarrow{\text{onto}}$ C(X) *is any homeomorphism, then* h[B(X)] = F_1(X). *Also* [239, proof of 3.11], *if* X *is not a circle*, h(v) = A *and* h[B(A)] = F_1(A) *where* A *is the arc of points of non-local connectivity of* X (*see* [245, *Theorem 6*]).

(8.26) THEOREM [239, 3.6, proof of 3.11, and 4.2]. *Assume* X *is a C-H continuum of the type in (3) or (4) of (8.24) and let* Q *denote the remainder. If* h: Cone(X) $\xrightarrow{\text{onto}}$ C(X) *is any homeomorphism, then:* (1) h(v) = Q; (2) h[Cone(Q)] = C(Q); (3) h[B(Q)] = F_1(Q).

(8.27)   HYPERSPACES AS CONES   325

Now recall that, for any continuum X, there is a *canonical contraction* λ of the cone over X to its vertex: Modulo identifications, the formula for λ is

$$\lambda[((x, t), s)] = (x, [1 - s] \cdot t + s)$$

for each $(x, t) \in \text{Cone}(X)$ and $s \in [0, 1]$. The third step in proving (8.23) is to use λ in conjunction with step two to deduce the existence of the following special contraction ψ of $F_1(X)$:

(8.27) LEMMA [239, 3.7, proof of 3.11, and 4.3]. *Assume* X *is a C-H continuum of the type in (2), (3), or (4) of (8.24) and assume* X *is not a circle. Let* N *denote*

A *in (8.25), if* X *is of type (2)*

or

Q *in (8.26), if* X *is of type (3) or (4).*

Then: There is a contraction $\psi: F_1(X) \times [0, 1] \to C(X)$ *of* $F_1(X)$ *to* N *in* C(X) *such that* ψ *is one-to-one on* $F_1(N) \times [0, 1)$.

Proof. Let $h: \text{Cone}(X) \xrightarrow{\text{onto}} C(X)$ be a homeomorphism. Define $\psi: F_1(X) \times [0, 1] \to C(X)$ by

$$\psi[(\{x\}, s)] = h \circ \lambda[(h^{-1}(\{x\}), s)]$$

for each $\{x\} \in F_1(X)$ and $s \in [0, 1]$, where λ is as defined above. Since by (8.25) and (8.26) $h(v) = N$, it follows using the formula for λ that ψ contracts $F_1(X)$ to N. Now let $(\{x_i\}, s_i) \in F_1(N) \times [0, 1)$ for $i = 1$ and 2. By (8.25) and (8.26), $h^{-1}(\{x_i\}) \in B(N)$ so let $(y_i, 0) = h^{-1}(\{x_i\})$ for $i = 1$ and 2. Assume $\psi[(\{x_1\}, s_1)] = \psi[(\{x_2\}, s_2)]$. Then, since h is one-to-one, $\lambda[((y_1, 0), s_1)] = \lambda[((y_2, 0), s_2)]$.

Thus, by using the formula for λ, we obtain $(y_1, s_1) = (y_2, s_2)$. Hence, since each $s_i < 1$, $y_1 = y_2$. So, $\{x_1\} = \{x_2\}$.

The fourth step in the proof of (8.23) is to find an intrinsic characterization of certain continua from among others, a characterization to which the invariant in (8.27) can be applied in order to distinguish the eight continua in (8.23) from all others of (8.24). Such a characterization is done in [239]:

(I) S_0 is characterized from among all compactifications of $[0, +\infty)$ with an arc as the remainder [239, 3.5].

(II) $(SP)_1$ is characterized from among all compactifications of $[0, +\infty)$ with a circle as the remainder [239, 4.5].

The characterizations (I) and (II) are technical and difficult to state in that they depend on special "chains" and on special embeddings of the compactifications in the plane ([239, 3.1 and 4.4]; also, see [238, pp.131-132], [253, 5.1 and 5.6], and [254, Lemma A]). However, we will do an example [see (8.29)] in the next step which should make things clearer. Parenthetically, let me mention that there has recently been an application of (I) outside of hyperspace theory--namely, (I) has been used to determine all the confluent images of S_0 [see (16.30)].

The fifth step in proving (8.23) is to use (I) of the fourth step to show that S_0 is the only possible compactification X of $[0, +\infty)$ with an arc as the remainder which can admit a contraction $\psi: F_1(X) \times [0, 1] \to C(X)$ as in (8.27) [239, 3.9]. The proof uses the topology of the plane! Namely, the following "coincidence"

(8.28)   HYPERSPACES AS CONES   327

lemma is used.

(8.28) LEMMA [239, 3.8]. *Let* $a \leq b$ *be real numbers and let* $\alpha: [0, w_1] \to C([a, b])$ *and* $\beta: [0, w_2] \to C([a, b])$ *be continuous functions such that* $\alpha(0) = \{a\}$, $b \in \alpha(w_1)$, $\beta(0) = \{b\}$, *and* $a \in \beta(w_2)$. *Then, there exist* $s \in [0, w_1]$ *and* $t \in [0, w_2]$ *such that* $\alpha(s) = \beta(t)$.

Proof. If $a < b$, $C([a, b])$ is a 2-cell [see (0.54)]. The following facts about $C([a, b])$ are easy to verify:

(i) $C_a([a, b]) = \{K \in C([a, b]): a \in K\}$ is an arc lying in the boundary of $C([a, b])$ with end points $\{a\}$ and $[a, b]$.

(ii) $C_b([a, b]) = \{K \in C([a, b]): b \in K\}$ is an arc lying in the boundary of $C([a, b])$ with end points $\{b\}$ and $[a, b]$.

(iii) $C_a([a, b]) \cap C_b([a, b]) = \{[a, b]\}$.

(iv) The simple closed curve S which is the boundary of $C([a, b])$ is given by
$$S = C_a([a, b]) \cup C_b([a, b]) \cup F_1([a, b]).$$

Now, let $\Gamma = \alpha([0, w_1])$ and let $\Lambda = \beta([0, w_2])$. Since $\alpha(0) = \{a\}$ and $\beta(0) = \{b\}$, we may assume for the purpose of proof that $\{b\} \notin \Gamma$ and $\{a\} \notin \Lambda$. Note that Γ is a subcontinuum of $C([a, b])$ intersecting S at $\{a\}$ and $\alpha(w_1)$ and Λ is a subcontinuum of $C([a, b])$ intersecting S at $\{b\}$ and $\beta(w_2)$. Furthermore, by (i) through (iv), $\{\{a\}, \alpha(w_1)\}$ disconnects S between $\{b\}$ and $\beta(w_2)$. Hence, by facts about the topology of the plane, $\Gamma \cap \Lambda \neq \emptyset$ (if $\Gamma \cap \Lambda = \emptyset$, then one can obtain two disjoint arcs γ and λ in $C([a, b])$ such that $\gamma \cap S = \{\{a\}, \alpha(w_1)\}$ and $\lambda \cap S = \{\{b\}, \beta(w_2)\}$; then, the θ-curve theorem [186, p.511] is contradicted). This proves (8.28).

A complete proof, showing how the fifth step is done using (I) and (8.28), can not be given since we have not given the characterization in (I). However, the following example is illustrative of the general idea (in [239, proof of 3.9]).

(8.29) EXAMPLE. Let $X = Q \cup G$ where

$$Q = \{(0, y) \in R^2 : 0 \leq y \leq +3/2\}$$

and

$$G = \{(x, y) \in R^2 : 0 < x \leq +1 \text{ and } y = \left|\frac{1}{2} + \sin[1/x]\right|\}.$$

Assume there is a contraction ψ as in (8.27). For convenience I will write points of Q as though they are real numbers [i.e., I will write y instead of $(0, y)$]. With this in mind, let $[a, b]$ be the closed subinterval of Q with end points $a = 0$ and $b = 1/2$. For each $i = 1, 2, \ldots,$ let $r_1^i = (6/(7\pi + 12i\pi), 0)$ and $r_2^i = (1/(2i\pi), 1/2)$. So, $\{r_1^i\}_{i=1}^{\infty}$ and $\{r_2^i\}_{i=1}^{\infty}$ are sequences in G converging to a and b respectively; also, because of the geometry, they have special properties which we will use. First note that, since ψ contracts to Q, the following numbers w_1^i and w_2^i exist: For each $i = 1, 2, \ldots,$ define w_1^i and w_2^i by

$$w_1^i = \text{g.l.b.}\{w \in [0, 1] : \text{some point of } \psi[(\{r_1^i\}, w)] \text{ has second coordinate equal to } 1/2\}$$

$$w_2^i = \text{g.l.b.}\{w \in [0, 1]: \text{ some point of } \psi[(\{r_2^i\}, w)] \text{ has second coordinate equal to } 0\}.$$

From the geometry it is clear that

(*) whenever $0 \le w \le w_j^i$ ($j = 1$ and 2), the second coordinate of each point of $\psi[(\{r_j^i\}, w)]$ lies between 0 and 1/2.

Assume without loss of generality that the sequences $\{w_j^i\}_{i=1}^{\infty}$ ($j = 1$ and 2) converge, $w_1^i \to w_1$ and $w_2^i \to w_2$. It follows, using the fact that ψ contracts to Q and using (*), that $w_1 < 1$ and $w_2 < 1$. Define α and β by:

$$\alpha(u) = \psi[(\{a\}, u)], \text{ each } u \in [0, w_1];$$

$$\beta(v) = \psi[(\{b\}, v)], \text{ each } v \in [0, w_2].$$

It follows, using (*) and convergence properties, that α sends $[0, w_1]$ into $C([a, b])$ and β sends $[0, w_2]$ into $C([a, b])$. Clearly $\alpha(0) = \{a\}$ and, since $w_1^i \to w_1$, $b = 1/2 \in \alpha(w_1)$. Also $\beta(0) = \{b\}$ and, since $w_2^i \to w_2$, $a = 0 \in \beta(w_2)$. Hence, by (8.28), there exist $s \in [0, w_1]$ and $t \in [0, w_2]$ such that $\alpha(s) = \beta(t)$, i.e.,

$$\psi[(\{a\}, s)] = \psi[(\{b\}, t)].$$

But, since $s \le w_1 < 1$ and $t \le w_2 < 1$, this contradicts the one-to-oneness of ψ on $F_1(Q) \times [0, 1)$. Therefore, there is no contraction ψ as in (8.27), so X is not a C-H continuum. This completes (8.29).

At this point we have essentially covered the major ideas behind the proof in [239] of (8.23). We now indicate how these ideas are used to show the eight continua listed in (8.23) are the only possible hereditarily decomposable C-H continua. To show S_1^1 and S_2^1 are the only possible continua of type (2) in (8.24) which are C-H continua, the characterization of arcwise connected circle-like continua in [245, Theorem 6] is used; this characterization says an arcwise connected circle-like continuum is a circle or a compactification of a half-line, with an arc as remainder, followed by identifying the "first point" of the half-line with an end point of the remainder. Such a continuum can be embedded in the plane in a special way (see [239, proof of 3.11] or [253, p.43]). The characterization and embedding just mentioned allow the use of (I) of step four and (8.27) to do a proof [239, proof of 3.11] similar to the one indicated by the fifth step above. The proof that S_3^1 is the only possible C-H compactification of R^1 with an arc as the remainder proceeds along the following lines: It is first observed [239, proof of 3.11] that any compactification of R^1 with an arc as the remainder can be put in the plane in a special way by using [253, Lemma 5.1]. Such an embedding in the plane again allows the use of (I) of step four and (8.27). As regards $(SP)_1$: (II) of step four and (8.28) are used to show that $(SP)_1$ is the only possible compactification of $[0, +\infty)$ with a circle as the remainder which can admit a contraction ψ as in (8.27) [239, 4.6]. The result for $(SP)_1$, and its proof, can then be applied [239, 4.8] to show that any C-H compactification of R^1 with a circle as the remainder must by $(SP)_2$ or $(SP)_3$.

We have now completed our discussion of why the eight continua listed in (8.23) are the only possible hereditarily decomposable C-H continua. In [280, pp.283-284] Rogers describes a "wrapping process" to indicate why $\text{Cone}[(SP)_1]$ and $C[(SP)_1]$ are homeomorphic. Rogers also says [280, p.284] that this "wrapping process" can be modified to give a proof that $\text{Cone}(S_o)$ and $C(S_o)$ are homeomorphic. Later [281, p.287] Rogers states (without proof) that certain of the other eight continua in (8.23) are C-H continua. On the basis of what is done in [280, pp.283-284], it might appear that a modification of the "wrapping process" would yield a similar result for certain other continua, such as the one in (8.29). Hence, I felt that more detailed proofs to show each of the eight continua is a C-H continua would be helpful. Such a proof for S_o was done in [239, 4.11], which can be extended and/or modified to give precise proofs for the others. The proof uses a different procedure than Rogers' "wrapping process."

This completes the discussion of the ideas involved in the proof in [239] of (8.23).

(8.30) REMARK. By combining (8.6), (8.7), (8.18), (8.21), and (8.23) we see how close we are to determining *all* finite-dimensional C-H continua [i.e., to answering (8.12)]. If an affirmative answer to (8.20.1) could be obtained, then the structure of all such continua could essentially be determined.

It is now an appropriate time to ask some questions which are more general than some previously asked questions.

(8.31) QUESTION [244, Problem 2]. What are the chainable C-H continua [comp., (8.14) and (8.15)]?

(8.32) QUESTION. What are the circle-like C-H continua [comp., (8.14) and (8.16)]?

(8.33) QUESTION [244, Problem 2]. What are the C-H continua?

From (8.7) and (8.25) we see that, for certain C-H continua X, any homeomorphism h: Cone(X) $\xrightarrow{onto}$ C(X) must take the base of Cone(X) onto the singletons in C(X). It can be seen from the proof in [239, 4.11] that, for any of the eight hereditarily decomposable C-H continua X of (8.23), there is a homeomorphism h: Cone(X) $\xrightarrow{onto}$ C(X) such that $h[B(X)] = F_1(X)$. I do not know an example of a C-H continuum X for which there is no such homeomorphism:

(8.34) QUESTION. If X is a C-H continuum, must there exist a homeomorphism h: Cone(X) $\xrightarrow{onto}$ C(X) such that $h[B(X)] = F_1(X)$?

So far we have been discussing C-H continua, i.e., continua X such that $C(X) \cong Cone(X)$. The following questions concern more general situations.

(8.35) QUESTIONS. For what continua X does there exist a continuum Y such that $C(X) \cong Cone(Y)$? If X is any given such continuum, what can be said about the class of continua Y such that $C(X) \cong Cone(Y)$? Answers to these questions would even be of interest under the additional assumption that $\dim[C(X)] < \infty$. Under this assumption I can show that if X is a locally connected continuum and Y is a continuum such that $C(X) \cong Cone(Y)$, then X and Y must be arcs or circles. The proof uses local cut points and ideas and results in [85]. For examples related to this and the two questions above, we refer the reader to the proof of (10.2) and to (10.16).

(8.36) QUESTION. For what continua X is $2^X \cong Cone(X)$? Two comments are appropriate. First: If X is not arcwise connected and $h: Cone(X) \xrightarrow{onto} 2^X$ is any homeomorphism, then $h(v) \varepsilon C(X)$; this follows from (11.3). Second: If X is a locally connected continuum and $2^X \cong Cone(X)$, then $Cone(X) \cong I_\infty$ by (1.97).

(8.37) QUESTION. Same as (8.35) with $C(X)$ replaced by 2^X. Recall, $\dim[2^X] = \infty$ by (1.95).

IX.
RECOGNIZING HYPERSPACES
AS SUSPENSIONS

In the previous chapter, we presented a number of results concerning when C(X) is homeomorphic to the cone over X for a continuum X. As we saw, much is known about this when $\dim[X] < \infty$ [see (8.30)]. Thus, a natural question to ask is: For what finite-dimensional continua X is C(X) homeomorphic to the suspension Sus(X) of X? This is completely answered in the following result:

(9.1) THEOREM [252, 1.1]. *If* X *is a finite-dimensional continuum such that* C(X) $\cong$ Sus(X), *then* X *is an arc (and conversely).*

In (9.1.1) below, we will give a proof of (9.1). First, we mention two useful general properties of suspensions. Then, we state two hyperspace results [(9.2) and (9.3)]. The result in (9.2) is not specifically about hyperspaces which are suspensions. It is used to prove (9.3), which is a general result about continua whose hyperspaces and suspensions are homeomorphic.

The following two properties of the suspension over any continuum Y make up the essential ingredients for the proof of (9.1):

(1) Let v_1 and v_2 denote the vertices of Sus(Y) [see (0.39) for precise definition]. Then, Sus(X)\\{v_i} is arcwise connected for each i = 1 and 2.

(2) Let H^n[Y, Z] denote Čech cohomology of Y with integer coefficients Z. Then, H^2[Sus(Y), Z] is isomorphic to H^1[Y, Z].

The following general fact is proved in [252]. The proof is difficult in a technical sense, and will not be done here. It uses (14.27) and (11.7). A question related to (9.2) is in (11.17).

(9.2) THEOREM [252, 2.4]. *Let Y be any continuum. If A_1, A_2 ∈ C(Y) such that C(Y)\\{A_i} is arcwise connected for each i = 1 and 2, then C(Y)\\{A_1, A_2} is arcwise connected.*

This theorem, together with (1) above, gives the following result. It should be compared with the situation for C-H continua in Chapter VIII.

(9.3) THEOREM [252, 2.5]. *If Y is any continuum such that C(Y) $\cong$ Sus(Y), then Y must be arcwise connected.*

Proof. See paragraph above.

Now we are ready to prove (9.1).

(9.1.1) PROOF OF (9.1) [252]. Let X be a finite-dimensional

(9.5)   HYPERSPACES AS SUSPENSIONS   337

continuum such that $C(X) \cong \text{Sus}(X)$. Then, $\dim[C(X)] < \infty$ and, by (9.3), X is arcwise connected. Therefore, by (1) of (2.5), $\dim[X] = 1$. Thus, $\dim[\text{Sus}(X)] = 2$ and, hence, $\dim[C(X)] = 2$. So, by (1.100), X is a-triodic. We have now shown that X is an arcwise connected a-triodic continuum. Therefore, by [254, Theorem 2], X must be an arc or an arcwise connected circle-like continuum. Since $C(X) \cong \text{Sus}(X)$, we have from (2) above that $H^2[C(X), Z]$ is isomorphic to $H^1[X, Z]$. By (1.174), $H^2[C(X), Z]$ is zero. Therefore, $H^1[X, Z]$ is zero. However, if X is an arcwise connected circle-like continuum, then [245, Theorem 6] can be used to show that $H^1[X, Z]$ is isomorphic to Z (see the Remark in [245, p.233]). Hence, X must be an arc. This completes the proof of (9.1).

Let us note that, since (9.3) is applicable to infinite-dimensional continua Y, (9.3) is of interest apart from (9.1).

(9.4) QUESTION [252]. For what infinite-dimensional continua X is $C(X) \cong \text{Sus}(X)$? For example note that, since $C(I_\infty) \cong I_\infty$ [by the second part of (1.98)], $C(I_\infty) \cong \text{Sus}(I_\infty)$. In (10.16) we will give another example.

(9.5) QUESTIONS. For what continua X does there exist a continuum Y such that $C(X) \cong \text{Sus}(Y)$ [see (10.16)]? If X is any given such continuum, what can be said about the class of continua Y such that $C(X) \cong \text{Sus}(Y)$? If X and Y are continua such that $C(X) \cong Sus(Y)$ and $\dim[C(X)] < \infty$, then must X be an arc or a circle, equivalently--must Y be an arc?

(9.6) QUESTION. For what continua X is $2^X \cong Sus(X)$? It appears that (9.2) can be modified by replacing $C(Y)$ with 2^Y. If so, then (9.3) with $C(Y)$ replaced by 2^Y would be valid, and we would know that the continua X such that $2^X \cong Sus(X)$ must be arcwise connected.

(9.7) QUESTION. Same as the first two questions in (9.5) with $C(X)$ replaced by 2^X.

X.
RECOGNIZING HYPERSPACES
AS CARTESIAN PRODUCTS

In the previous two chapters we have discussed when a hyperspace is a cone or a suspension. In this chapter we will be concerned with the general question: When is a hyperspace homeomorphic to a cartesian product of nondegenerate continua?

Clearly, any space is homeomorphic to the cartesian product of itself and a one-point space. So, whenever we say $C(X)$ [resp., 2^X] *is a product*, or write $C(X) \cong Y \times Z$ [resp., $2^X \cong Y \times Z$], we will mean that $C(X)$ [resp., 2^X] is homeomorphic to the cartesian product of *nondegenerate* continua Y and Z.

The first result about when $C(X)$ is a product appears in Duda's paper [85]. The main emphasis of [85] is an analysis of the polyhedral structure of $C(X)$ when X is any finite graph. However, as a by-product of this analysis, Duda obtained a result [85, 9.7] which, using (1.109), may be stated as follows:

(10.1) THEOREM [see paragraph above]. *Let* X *be a locally connected continuum. If* $C(X)$ *is a finite-dimensional product, then* X *is an arc or a circle (and conversely).*

We will not give a proof of (10.1), but the following simple example illustrates ideas in Duda's proof [85, p.285].

(10.2) EXAMPLE [see (0.57)]. Let conv denote closed convex hull [see (0.63.5)]. Let X be a simple triod; for convenience, we let X be the simple triod in R^3 given by

$$X = [\text{conv}\{(2,0,0), (0,2,0)\}] \cup [\text{conv}\{(1,1,0), (2,2,0)\}].$$

To obtain a picture of C(X), first define the following three sets:

$K = \{(x_1, x_2, x_3) \in R^3 : x_i \in \{0, 1\} \text{ for each } i = 1,2,3\}$,
$L = \{(2,0,0), (0,2,0), (0,0,0)\}$,
$M = \{(1,1,0), (2,2,0), (1,1,1)\}$.

Then, it is easy to verify that

$$C(X) \cong [\text{conv}(K)] \cup [\text{conv}(L)] \cup [\text{conv}(M)].$$

More specifically: Conv(K) represents all the subcontinua of X with the branch point (1, 1, 0) in them, conv(L) represents all the subcontinua of the arc conv$\{(2,0,0), (0,2,0)\}$, and conv(M) represents all the subcontinua of the arc conv$\{(1,1,0), (2,2,0)\}$. Now, suppose C(X) is a product. Then, there are nondegenerate continua Y and Z and a homeomorphism h: C(X) $\xrightarrow{\text{onto}}$ Y × Z. Note that since C(X) is locally connected, so are Y and Z. Let

$$T = \{A \in C(X) : \dim_A [C(X)] = 2\} = C(X) \setminus \text{conv}(K).$$

Let G = h[T]. Let π_Y and π_Z denote the projection maps of Y × Z onto Y and Z respectively. Clearly,

$$G \subset (\pi_Y[G] \times \pi_Z[G]).$$

Note that, since T is precisely those members of C(X) which

have a 2-cell neighborhood in $C(X)$,

(1) G consists of precisely those points of $Y \times Z$ which have a 2-cell neighborhood in $Y \times Z$.

Note the following general fact: If the product of two nondegenerate continua is embeddable in a 2-cell, then the two continua are arcs or one is an arc and the other is a circle (a complete proof is in [244, Lemma 2]). From this general fact it follows using (1) and the local connectivity of Y and Z that

(2) each point in $\pi_Y[G]$ and each point in $\pi_Z[G]$ has a neighborhood (in Y and Z, respectively) which is an arc.

It follows easily using (1) and (2) that

$$(\pi_Y[G] \times \pi_Z[G]) \subset G.$$

Hence, we have shown

(3) $G = \pi_Y[G] \times \pi_Z[G]$.

Next, let $D = \text{cl}[G]$ and look at $h^{-1}[D] = \text{cl}[T]$ (comp., [85, p.279 and p.285]). Clearly, $h^{-1}[D]$ is the union of three (convex) 2-cells Q_1, Q_2, and Q_3 such that

$$Q_i \cap Q_j = \{(1,1,0)\} = \bigcap_{k=1}^{3} Q_k$$

whenever $i \neq j$. Hence, D is a locally connected continuum and $h((1,1,0))$ is a local cut point of D. Also, since $D = \text{cl}[G]$, simple computations using (3) and the fact that π_Y and π_Z are closed maps give $D = \pi_Y[D] \times \pi_Z[D]$. This is a contradiction, for it is well-known and easy to prove that the product of any two nondegenerate locally connected continua can have no local cut point.

Therefore, C(X) is not a product. This completes (10.2).

As far as I know, Duda's result in (10.1) above is the only theorem *now* in the literature concerning when a hyperspace is a product. In a recently completed paper [241], this subject has been investigated. We now give some of the results in this paper.

First, recall that cones and suspensions are formed from products with an arc. Thus, the following result is related, in a sense, to the topics covered in the previous two chapters.

(10.3) THEOREM [241, 2.8]. *If* X *is a finite-dimensional continuum such that* $C(X) \cong X \times [0, 1]$, *then* X *is an arc.*

The result in (10.3) is a corollary to (10.5) below. Before proving (10.5), let us note the following lemma.

(10.4) LEMMA [241, 2.1]. *If* $C(X) \cong Y \times Z$, *then* Y *and* Z *are each arcwise connected and acyclic [see (10.23)].*

Proof. Since C(X) is arcwise connected [by (1.12)] and acyclic [by (1.174)], $Y \times Z$ is also. Thus, by using the projection maps of $Y \times Z$ onto Y and Z, the lemma follows.

(10.5) THEOREM [241, 2.7]. *If* $\dim[C(X)] < \infty$ *and if* $C(X) \cong X \times Z$, *then* X *must be an arc.*

Proof. By (10.4), X is arcwise connected. Therefore, since $\dim[C(X)] < \infty$, we have by part (1) of (2.5) that $\dim[X] = 1$.

Hence, X is a one-dimensional, arcwise connected, and acyclic [by (10.4)] continuum. So, by [39, p.17], X is a dendroid. Let us note the following lemma, a proof of which is in [241].

(10.5.1) LEMMA [241, 2.6]. *Let* D *be a dendroid. Then* C(D) *is finite-dimensional if and only if* D *is a finite graph.*

Since X is a dendroid and $\dim[C(X)] < \infty$, we have by (10.5.1) that X is a finite graph. Thus, X is a locally connected continuum such that $C(X) \cong X \times Z$ is finite-dimensional. It now follows, using (10.1) and (10.4), that X is an arc. This proves (10.5).

In (10.3) and (10.5) we assumed X was one of the factors. Now we give some results in which the factors are arbitrary.

(10.6) LEMMA [241, 2.3]. *If* C(X) *is a product, then no member of* C(X) *can arcwise disconnect* C(X) *[see (10.24)].*

Proof. Let $C(X) \cong Y \times Z$. By (10.4), Y and Z are each arcwise connected. As is easy to prove, no product of nondegenerate arcwise connected spaces can become non-arcwise connected by removing a point.

(10.7) THEOREM [241, 2.4]. *If* C(X) *is a product, then* X *is decomposable [see (10.25)].*

Proof. The theorem is an easy consequence of (10.6) and (1.51).

The reader should compare (10.7), the next result, and the situation for C-H continua in Chapter VIII.

(10.8) THEOREM [241, 2.5]. *If* C(X) *is a finite-dimensional product, then* X *is an hereditarily decomposable continuum and, hence [see (1.53)], one-dimensional.*

Proof. The theorem follows immediately from (10.6) and (8.19).

I do not know the answer to the following question:

(10.9) QUESTION [241, 2.0]. If C(X) is a finite-dimensional product, then must X be an arc or a circle? We have seen that the answer is "yes" when X is locally connected. We will show the answer is "yes" under other conditions, the most general one being when X is a-triodic [see (10.12)].

The proof of (10.12) begins with the following special case, which shows the answer to (10.9) is "yes" for chainable continua and for circle-like continua.

(10.10) LEMMA [241, 2.2]. *If* C(X) *is a product where* X *is chainable or circle-like, then* X *is an arc or a circle (respectively).*

Proof. Let C(X) = Y × Z. By (2.1) dim[C(X)] = 2. It follows that dim[Y] = 1 = dim[Z]. Hence, by (10.4) and [39, p.17], Y and Z are each dendroids. We will show that Y and Z are both arcs. Since X is an inverse limit of arcs or circles, $C(X)$ is an inverse limit of 2-cells [by (0.54), (0.55), and (1.169)]. Thus, by (1.162), we see that any subcontinuum of $Y \times Z$ can be ε-mapped into R^2 for each $\varepsilon > 0$. Now suppose Y is not an arc. Then, since Y is a dendroid, Y must contain a simple triod T (this follows from [39, Lemma 3]; also, see [245, Lemma 2] for a complete proof). Let A be an arc in Z. Then, $T \times A$ contains a continuum M homeomorphic to

$$\{(x,y,0) \in R^3 : x^2 + y^2 \leq 1\} \cup \{(0,0,z) \in R^3 : 0 \leq z \leq 1\}.$$

But, Bennett [23, Theorem 3] has shown that, for small enough $\varepsilon > 0$, there is no ε-map of M into R^2. Thus, we have a contradiction. This proves Y must be an arc. By a similar argument, Z must be an arc. Therefore, $C(X)$ is a 2-cell. Thus, it follows [see (1.208.4)] that X must be an arc or a circle. This proves (10.10).

The proof in [241] of the following lemma is tedious and will be omitted here. A related result with additional information is in (11.16).

(10.11) **LEMMA** [241, 2.10]. *Let* X *be an a-triodic continuum such that given any nondegenerate proper subcontinuum* E *of* X, *there exists a subcontinuum* Y_o *of* X *such that*

$$Y_o \cap E \neq \emptyset \neq Y_o \cap [X \backslash E] \text{ and } Y_o \not\supseteq E.$$

Then, each proper subcontinuum of X *is unicoherent.*

(10.12) THEOREM [241, 2.13]. *If C(X) is a finite-dimensional product and X is a-triodic, then X is an arc or a circle [see (11.16) for a more general result].*

Proof. Let E be any given nondegenerate proper subcontinuum of X. By (10.6), $C(X) \setminus \{E\}$ is arcwise connected. Hence, by (11.5), there exists a subcontinuum Y_o of X such that

$Y_o \cap E \neq \emptyset \neq Y_o \cap [X \setminus E]$ and $Y_o \not\supset E$.

Hence, the hypotheses of (10.11) are all satisfied; so, by (10.11) we have that

(1) each proper subcontinuum of X is unicoherent.

By (10.8), X is hereditarily decomposable. Bing [32, Theorem 11] showed that an hereditarily decomposable continuum is chainable if and only if it is hereditarily unicoherent and a-triodic. Clearly: Each proper subcontinuum of X is hereditarily decomposable [because X is], hereditarily unicoherent [by (1)], and a-triodic [because X is]. Hence, by Bing's theorem [32, Theorem 11],

(2) each proper subcontinuum of X is chainable.

Now, Ingram [151, Theorem 4] showed that a decomposable continuum X satisfying (2) must be chainable or circle-like. Hence, since X is a decomposable continuum satisfying (2), X is chainable or circle-like. Thus, (10.10) now applies, and we have proved (10.12).

(10.13) COROLLARY [241, 2.14]. *If C(X) is a product and $\dim[C(X)] = 2$, then X is an arc or a circle.*

Proof. Since $\dim[C(X)] = 2$, we have by (1.100) that X does not contain a 3-odd, i.e., X is a-triodic by (0.21). Hence,

(10.13) follows from (10.12).

For completeness we mention the following result which is easy to prove using some ideas from the proof of (10.5).

(10.14) THEOREM [241, 2.9]. *If* $C(X)$ *is a finite-dimensional product where* X *is arcwise connected and acyclic, then* X *is an arc.*

This completes the discussion of when $C(X)$ is a finite-dimensional product. Except for results above, no information is known concerning (10.9).

We turn our attention to the case when X is a locally connected continuum and $C(X)$ is an infinite-dimensional product. Observe that if X is any locally connected continuum which contains no free arc [see (0.18) for definition], then $C(X)$ is a product because $C(X) \cong I_\infty$ [by (1.98)]. Hence, we will be interested in locally connected continua X which contain free arcs, and in conditions imposed on the free arcs by the assumption that $C(X)$ is a product. We will give such conditions in (10.20)--they may be "good enough" to be reversible [see (10.21)]. First, we give an example of a locally connected continuum X which *contains* a free arc and whose hyperspace $C(X)$ *is* an infinite-dimensional product. We go into some detail in the verifications; the techniques used shed some light on the problems involved in answering (10.21).

For use in the example, we adopt the following notation. The definition of I_∞ is in (0.40). Let $\tau_j: I_\infty \to [0, 2^{-j}]$ denote projection, let $\theta = (0, 0,\ldots, 0,\ldots) \in \ell_2$, and let $J_\infty = \Pi_{j=1}^{\infty}[a_j, b_j] \subset \ell_2$ where $[a_1, b_1] = [-2^{-1}, 0]$ and $[a_j, b_j] = [0, 2^{-j}]$ for each $j > 1$. Also: A function f is said to be a *homeomorphism of* $(A_1, A_2,\ldots, A_n)$ *onto* $(B_1, B_2,\ldots, B_n)$, written

$$f: (A_1, A_2,\ldots, A_n) \xrightarrow{\text{onto}} (B_1, B_2,\ldots, B_n)$$

provided that f maps A_i onto B_i homeomorphically for each $i = 1, 2,\ldots, n$. I refer the reader to (1.214.1), [1], and [354] for the definition of and some facts about property Z.

(10.15) EXAMPLE [241, 3.1]. Let $X = B^2 \cup L$ where B^2 is as in (0.19) and $L = \{(x, 0) \in R^2: 1 \leq x \leq 2\}$. Let

$\Omega = \{A \in C(L): (1, 0) \in A\}$,

$\Gamma = C(B^2)$,

$C(X; A) = \{K \in C(X): K \cap L = A\}$ for each $A \in \Omega$,

$\Lambda = \cup \{C(X; A): A \in \Omega\}$.

Note that

$\Gamma \cap \Lambda = C(X; \{(1, 0)\})$.

We first prove (i) through (iii) below.

(i) $\Gamma \cap \Lambda$ has property Z in Γ.

Proof of (i). Let U be a homotopically trivial, nonempty open subset of Γ, note that $U \setminus [\Gamma \cap \Lambda] \neq \emptyset$ (because $\Gamma \cap \Lambda$ is nowhere dense in Γ), and let $f: S^k \to U \setminus [\Gamma \cap \Lambda]$ be continuous. Since U is homotopically trivial, there is an extension $F: B^{k+1} \to U$ of f. For each ε, $0 < \varepsilon < 1$, let

$$\hat{r}_\varepsilon: B^2 \xrightarrow{onto} B^2 \setminus \{(x, y) \in B^2: [(x-1)^2 + y^2]^{\frac{1}{2}} < \varepsilon\}$$

be a retraction; for each r_ε, let $\hat{r}_\varepsilon: \Gamma \to \Gamma$ be the induced map given by [as in (0.49)]

$$\hat{r}_\varepsilon(M) = \{r_\varepsilon(m): m \in M\} \quad (= r_\varepsilon[M]) \quad \text{for each } M \in \Gamma.$$

Now, since $F(B^{k+1})$ is a compact subset of U and $f(S^k)$ is a compact subset of $U \setminus [\Gamma \cap \Lambda]$, it is easy to see that there exists ε' such that $\hat{r}_{\varepsilon'}[F(B^{k+1})] \subset U$ and $\hat{r}_{\varepsilon'}(M) = M$ for each $M \in f(S^k)$. Hence, $r_{\varepsilon'} \circ F$ is an extension of f and $r_{\varepsilon'} \circ F$ maps B^{k+1} into $U \setminus [\Gamma \cap \Lambda]$. This proves (i).

(ii) There is a homeomorphism

$$h_1: (\Gamma, \Gamma \cap \Lambda, \{(1, 0)\}) \xrightarrow{onto} (I_\infty, \tau_1^{-1}(0), \theta).$$

Proof of (ii). By (1.98), there is a homeomorphism $k_1: \Gamma \xrightarrow{onto} I_\infty$. By (i), $k_1[\Gamma \cap \Lambda]$ has property Z in I_∞. Therefore, since $k_1[\Gamma \cap \Lambda] \cong \tau_1^{-1}(0)$ [by (1.99.3)] and $\tau_1^{-1}(0)$ has property Z in I_∞ [1, Theorem 8.2 with $K = \emptyset$], there is a homeomorphism

$$k_2: (I_\infty, k_1[\Gamma \cap \Lambda]) \xrightarrow{onto} (I_\infty, \tau_1^{-1}(0))$$

by [1, Corollary 10.3]. Now, by the homogeneity of the cube $\tau_1^{-1}(0)$, there is a homeomorphism

$$k_3: (\tau_1^{-1}(0), k_2 \circ k_1(\{(1, 0)\})) \xrightarrow{onto} (\tau_1^{-1}(0), \theta).$$

Let $\bar{k}_3: I_\infty \xrightarrow{onto} I_\infty$ be the extension of k_3 given by:

$$\bar{k}_3((x_1, x_2, \ldots, x_n, \ldots)) = (x_1, y_2, y_3, \ldots)$$

where

$$(0, y_2, y_3, \ldots, y_n, \ldots) = k_3((0, x_2, x_3, \ldots, x_n, \ldots)).$$

Let $h_1 = \bar{k}_3 \circ k_2 \circ k_1$.

(iii) Let $h = h_1 |[\Gamma \cap \Lambda]$ (the restriction of h_1 to $\Gamma \cap \Lambda$) and let $\alpha = \{(t, 0, 0, \ldots) \in J_\infty : -2^{-1} \le t \le 0\}$. Then, h can be extended to a homeomorphism

$$h_2 : (\Lambda, \Omega) \xrightarrow{\text{onto}} (J_\infty, \alpha).$$

Proof of (iii). It is easy to see that Ω is an arc and that $\{(1, 0)\}$ is an end point of Ω; hence, there is a homeomorphism $g_1 : (\Omega, \{(1, 0)\}) \xrightarrow{\text{onto}} (\alpha, \theta)$. Now note that for each $K \in \Lambda$, $[K \cap B^2] \in [\Gamma \cap \Lambda]$ and the function $g_2 : \Lambda \xrightarrow{\text{onto}} \Gamma \cap \Lambda$, given by $g_2(K) = K \cap B^2$ for each $K \in \Lambda$, is a continuous retraction. Also, it is easy to see that the function $g_3 : \Lambda \xrightarrow{\text{onto}} \alpha$, given by $g_3(K) = g_1(K \cap L)$ for each $K \in \Lambda$, is continuous. Define $h_2 : \Lambda \to J_\infty$ by

$$h_2(K) = g_3(K) + h \circ g_2(K) \quad \text{for each } K \in \Lambda,$$

where $+$ means vector addition in ℓ_2. Since g_2 is a retraction of Λ onto $\Gamma \cap \Lambda$ and $g_3(K) = \theta$ whenever $K \in [\Gamma \cap \Lambda]$, h_2 is an extension of h. It is also easy to verify that h_2 is a homeomorphism of (Λ, Ω) onto (J_∞, α). This proves (iii).

Now, let

$$G_1(K) = \begin{cases} h_1(K), & K \in \Gamma \\ h_2(K), & K \in \Lambda. \end{cases}$$

Then, we have from (ii) and (iii) that

$$G_1 : (\Gamma \cup \Lambda, \Omega) \xrightarrow{\text{onto}} (I_\infty \cup J_\infty, \alpha) \text{ is a homeomorphism. Let}$$

$\beta = \{(t, 0, 0, \ldots) \in [I_\infty \cup J_\infty] : -2^{-1} \le t \le 2^{-1}\}$.

Since α and β each have property Z in the cube $I_\infty \cup J_\infty$ (use [1, Theorem 8.2 with $K = \emptyset$]), there is [1, Corollary 10.3] a homeomorphism $G_2 : (I_\infty \cup J_\infty, \alpha) \xrightarrow{\text{onto}} (I_\infty \cup J_\infty, \beta)$. Let $G = G_2 \circ G_1$.

Now note that $C(X) = \Gamma \cup \Lambda \cup C(L)$, $C(L)$ is a 2-cell such that $C(L) \cap [\Gamma \cup \Lambda] = \Omega$, and Ω is an arc in the boundary of $C(L)$. Let

$$\Delta = \{(t_1, t_2, 0, 0, \ldots) \in \ell_2: -2^{-1} \le t_1 \le 2^{-1} \text{ and } -1 \le t_2 \le 0\}.$$

Let g denote the restriction of G to Ω. We note the following simple consequence of the Schönflies theorem [186, p.535]: If A is a 2-cell in R^2, B is any 2-cell, λ is an arc in the (manifold) boundary of B, and f is a homeomorphism of λ onto an arc in the boundary of A, then f can be extended to a homeomorphism of all of B onto A. Hence, since Δ may be considered as in R^2, g can be extended to a homeomorphism $\bar{G}: C(L) \xrightarrow{\text{onto}} \Delta$. Now, define $e: C(X) \xrightarrow{\text{onto}} [I_\infty \cup J_\infty \cup \Delta]$ by

$$e(K) = \begin{cases} G(K), & K \in [\Gamma \cup \Lambda] \\ \bar{G}(K), & K \in C(L). \end{cases}$$

It is easy to verify that e is a homeomorphism of $C(X)$ onto $I_\infty \cup J_\infty \cup \Delta$. Therefore, letting $\gamma = \{(2^{-1}, t_2, 0, 0, \ldots) \in \Delta\}$, we see that $C(X) \cong [\tau_1^{-1}(2^{-1}) \cup \gamma] \times \text{arc}$.

(10.16) REMARK. Let X be as in (10.15). Let us note that we showed $C(X) \cong [I_\infty \cup \rho] \times \text{arc}$ where

$$\rho = \{(2^{-1}, t, 0, 0, \ldots) \in \ell_2: -1 \le t \le 0\}.$$

In relation to (8.35) and (9.5) note that

$$\text{Cone}(I_\infty \cup \rho) \cong C(X) \cong \text{Sus}(I_\infty \cup \rho).$$

The same techniques as those used for the verifications in (10.15) show that [see (8.33) and (9.4)]

Cone$(I_\infty \cup \rho) \cong C(I_\infty \cup \rho) \cong$ Sus$(I_\infty \cup \rho)$.

Slight changes in the continuum in (10.15) can result in locally connected continua Y such that C(Y) is infinite-dimensional but is not a product. We give two such examples now. Though their properties can be deduced directly from (10.20), they can also be verified by using the techniques in (10.2). The examples will provide some basic intuition, and will illustrate some of the subtleties of (10.20.3).

(10.17) EXAMPLE [241, 3.2 and 3.16]. Let X denote the continuum in (10.15) and let Y = X ∪ K where
$$K = \{(x, x - 1) \in R^2 : 1 \le x \le 2\}.$$
Then, C(Y) is not product. Let us note that, for
$$Y' = X \cup \{(0, y) \in R^2 : 1 \le y \le 2\}$$
where X is as in (10.15), C(Y') is a product as can be verified using techniques in (10.15).

(10.18) EXAMPLE [241, 3.3 and 3.17]. Let Y = $B^2 \cup S$ where B^2 is as in (0.19) and
$$S = \{(x, y) \in R^2 : (x - 2)^2 + y^2 = 1\}.$$
Then, C(Y) is not a product. However, for
$$Y'' = Y' \cup \{(x, -x + 2) \in R^2 : 0 \le x \le 2\}$$
where Y' is as in (10.17), C(Y'') is a product.

Having presented the examples in (10.15) through (10.18) above, we state the main theorem. First, we give the following notation.

(10.19) NOTATION. Let X be a continuum. If J is an arc in X with end points p and q, let $\tilde{J} = J\setminus\{p, q\}$. Let $\psi[X] = \cup\{\tilde{J}: J$ is a free arc in $X\}$.

(10.20) THEOREM [241, 3.15]. *If X is a locally connected continuum such that C(X) is a product, then one of the following must hold:*

(10.20.1) X *is a circle;*

(10.20.2) X *contains no free arc;*

(10.20.3) *the closure of any component of $\psi[X]$ is a free arc in X which is disjoint from any free arc (in X) not contained in it.*

(10.21) QUESTION [241, 3.19]. If X is any locally connected continuum satisfying (10.20.3), then must C(X) be a product? It is clear from what was done in (10.15) that knowledge about the union of Hilbert cubes which intersect in a Hilbert cube is intimately connected with this question. Hence, more information about unions of Hilbert cubes than is currently available would seem to be necessary in order to answer the question (Dick Sher [296] has recently shown the conjecture in [2, p.213] is false; however, his example sheds no light on the question above). One further comment: The question above is stated in the context of locally connected continua. It may in fact be true that if C(X) is a product, then X *is* locally connected. If this is true and if the answer to the above question is "yes," then a complete characterization of those continua X such that C(X) is a product would be obtained.

We have now completed the discussion of continua X such that C(X) is a product. Let us briefly consider the case of 2^X.

(10.22) QUESTIONS. For what continua X is 2^X a product? Note that if X is any locally connected continuum, then 2^X is a product because $2^X \cong I_\infty$ [by (1.97)]. If X is a continuum such that 2^X is a product, then must X be locally connected [comp., comment near end of (10.21)]?

We note the following facts in relation to (10.22). First, since 2^X is arcwise connected [by (1.9)] and acyclic [by (1.174)], we have the following result which is analogous to (10.4).

(10.23) THEOREM. *If $2^X \cong Y \times Z$, then Y and Z are each arcwise connected and acyclic.*

Proof. See paragraph above.

The following result is analogous to (10.6) and can be proved the same way [using (10.23) in place of (10.4)].

(10.24) THEOREM. *If 2^X is a product, then no member of 2^X can arcwise disconnect 2^X.*

Proof. See paragraph above.

It is known ([234, 4.3]; (11.4) here) that a continuum X is decomposable if and only if $2^X\setminus\{X\}$ is arcwise connected. Using this fact and (10.24), we obtain the following result analogous to (10.7).

(10.25) THEOREM. *If 2^X is a product, then X is decomposable.*

Proof. See paragraph above.

In [191] Lawson asks the following questions: "Is $C(X)$ a factor of 2^X? Indeed, is it the case that $C(X) \times I_\infty \cong 2^X$? The questions are motivated in Lawson's paper [191] by the results stated here in (1.97) and (1.98). These two results imply that $C(X) \times I_\infty \cong 2^X$ if X is a locally connected continuum. As can be seen from (10.25), the answer to both of Lawson's questions is "no"; for if X is any indecomposable continuum, 2^X is not a product of any (nondegenerate) continua. Also, there are many "simple" non-locally connected continua X for which there is not even a continuous function from $C(X) \times I_\infty$ onto 2^X. By (1.38), such is the case for any non-locally connected continuum X such that $C(X)$ does not contain an open set with uncountably many components. Two specific examples are the familiar $\sin[1/x]$-continuum [see (1.42)] and the cone over $\{0, 1, 1/2, 1/3,\ldots, 1/n,\ldots\}$. I do not know a single example of a non-locally connected continuum X such that $C(X) \times I_\infty$ is homeomorphic to 2^X. Thus, the following question is unanswered.

(10.26) QUESTION. If X is a continuum such that $C(X) \times I_\infty \cong 2^X$, then must X be locally connected?

Also, I do not know the answer to the following question.

(10.27) QUESTION. If X is a continuum such that $C(X) \cong 2^X$, then must X be locally connected? If the answer is "yes", then by (1.97) and (1.98) we would know all such continua X.

XI.
ARCWISE DISCONNECTING
HYPERSPACES

As can be seen from Chapters VIII, IX, and X, knowledge about points of a hyperspace which arcwise disconnect the hyperspace can be important, especially in determining structural information about the hyperspace or, about the continuum X itself. The first person to prove a result about this was Kelley [165, section 8]:

(11.1) THEOREM [165, 8.2]. *Let* X *be a continuum. Then* X *is indecomposable if and only if* $C(X)\setminus\{x\}$ *is not arcwise connected.*

Proof. See (1.51).

The next person to make significant use of points which arcwise disconnect a hyperspace was Rogers [281]--see (8.18) and its proof and (8.19).

The first systematic study of points which arcwise disconnect hyperspaces was done in [234, section 4]. A number of seemingly fundamental results were proved, including characterizations of such points [here, see (11.5) and (11.7)]. Also, the subject was studied for 2^X instead of just $C(X)$. We will now present most

357

of the results in [234, section 4]. For some results and applications related to material in this chapter, see Chapters VIII through X and Chapter XII.

Observe the following simple lemma which will be used many times.

(11.2) **LEMMA** ([234, 4.1] and, for the case of $C(X)\setminus C(E)$, [281, Proposition 1]). *If* E *is a subcontinuum of a continuum* X, *then* $2^X\setminus 2^E$ *and* $C(X)\setminus C(E)$ *are each arcwise connected.*

Proof. The lemma for $2^X\setminus 2^E$ follows easily by using (1.25) and, for $C(X)\setminus C(E)$, by using (1.25) and (1.26).

Recall that a point p of an arcwise connected continuum M is said to *arcwise disconnect* M provided $M\setminus\{p\}$ is not arcwise connected. Thus, a *point* which arcwise disconnects a hyperspace, say 2^X, is a nonempty compact subset A of X such that $2^X\setminus\{A\}$ is not arcwise connected. The following result shows that such a point A must actually be a subcontinuum of X.

(11.3) **THEOREM** [234, 4.2]. *Let* $A \in 2^X$. *If* $2^X\setminus\{A\}$ *is not arcwise connected, then* $A \in C(X)$.

Proof. Assume $A \notin C(X)$. Let K and L be disjoint nonempty compact sets such that $A = K \cup L$. Let $B \in [2^X\setminus\{A\}]$, $B \neq X$. It suffices to show there exists an arc from B to X missing A. If $B \not\subset A$, then a segment from B to X, as guaranteed by (1.25), would be such an arc. So, for the purpose of proof, assume $B \subset A$.

We also assume, without loss of generality, that $B \cap K \neq \emptyset$ and $B \cap L \neq L$ [indeed, $B \cap L$ may be empty]. Now, by (1.25), there is a segment $\sigma: [0, 1] \to 2^X$ from $B \cap K$ to X. Let

$$t_0 = \text{l.u.b.} \{t \in [0, 1]: \sigma(t) \subset K\}.$$

Using the properties of K and L and the definition of a segment [see (1.15), specifically (1.15.1) and (1.15.4)], it follows easily that

(i) $\sigma(t) \cap [X \setminus A] \neq \emptyset$ for each $t > t_0$,

(ii) $B \cup \sigma(t) \neq A$ for any $t \leq t_0$.

Define $f: [0, 1] \to 2^X$ by $f(t) = \sigma(t) \cup B$ for each $t \in [0, 1]$. By (i) and (ii), $f(t) \neq A$ for any $t \in [0, 1]$. Also f is continuous, $f(0) = B$, and $f(1) = X$. It now follows that $2^X \setminus \{A\}$ is arcwise connected. This completes the proof of (11.3).

The next result is analogous to (11.1). An application of it was given at the end of Chapter X, where it was used to prove (10.25) and, thus, answer Lawson's questions.

(11.4) THEOREM [234, 4.3]. *A continuum* X *is indecomposable if and only if* $2^X \setminus \{x\}$ *is not arcwise connected.*

Proof. Assume X is indecomposable. Let x and y be points of disjoint composants of X [186, p.212]. Let $f: [0, 1] \to 2^X$ be any continuous function such that $f(0) = \{x\}$ and $f(1) = \{y\}$. We will show $f(t_0) = X$ for some $t_0 \in [0, 1]$. First note that, since f is continuous and $f(0) \in C(X)$, it follows from (1.49) that $[\cup f([0, 1])] \in C(X)$ for each $t \in [0, 1]$.

Thus, letting

$$g(t) = \cup f([0, 1]) \quad \text{for each} \quad t \in [0, 1],$$

it follows from the continuity of f and of union [see (1.48)] that g is a continuous function from $[0, 1]$ into $C(X)$. Since $x, y \in g(1) \in C(X)$, we see that $g(1)$ is a subcontinuum of X such that $g(1)$ intersects two different composants of X. Thus, since the composants of X are mutually disjoint [186, p.212], $g(1) = X$. Hence, the number t_o given by

$$t_o = \text{g.l.b.}\{t \in [0, 1]: g(t) = X\}$$

exists. Note that $g(t_o) = X$ and $t_o > 0$; also, $g(t)$ is nowhere dense in X for each $t < t_o$ (because each such $g(t)$ is a proper subcontinuum of the indecomposable continuum X; see [186, p.207]). Now note that, for each $t < t_o$,

$$X = g(t_o) = g(t) \cup [\cup f([t, t_o])]$$

and

$\cup f([t, t_o])$ is compact.

These last two facts, together with the nowhere denseness of $g(t)$ in X for $t < t_o$, give:

$$\cup f([t, t_o]) = X \quad \text{for each} \quad t < t_o.$$

Therefore, by a simple continuity argument, $f(t_o) = X$. This proves $2^X \setminus \{X\}$ is not arcwise connected, and completes the proof of half of (11.4). The proof of the other half involves some tedious case-taking and will not be done here (a complete proof is in [234]).

The following theorem is the first of two characterization

results [the other one, (11.7), is more general]. It gives necessary and sufficient conditions in order that a proper decomposable subcontinuum of X be an arcwise-disconnecting point of 2^X or (equivalently) of C(X).

(11.5) THEOREM [234, 4.4]. *Let* E *be a nondegenerate proper subcontinuum of a continuum* X. *Consider the following three statements:*

(11.5.1) *If* Y *is any subcontinuum of* X *such that* Y ∩ E ≠ ∅ ≠ Y ∩ [X\E], *then* Y ⊃ E.

(11.5.2) $2^X \setminus \{E\}$ *is not arcwise connected.*

(11.5.3) C(X)\\{E} *is not arcwise connected.*

Then, (11.5.1) implies (11.5.2) and (11.5.1) implies (11.5.3). Furthermore, if E *is decomposable, all three statements are equivalent.*

Proof. Assume (11.5.1) holds. We show simultaneously that (11.5.2) and (11.5.3) each holds. Let
$$\begin{cases} f\colon [0, 1] \to 2^X \\ f\colon [0, 1] \to C(X) \end{cases}$$
be continuous such that $f(0) \in [C(E)\setminus\{E\}]$ and $f(1) \not\subset E$. Let

t_0 = l.u.b.{t ∈ [0, 1]: f(t) ⊂ E}.

Note that, since f is continuous and $f(1) \not\subset E$, $t_0 < 1$. Choose t_1 such that $t_0 < t_1 \leq 1$. Note that:

(i) $f(t_0) \subset E$,

(ii) $[\cup f([t_0, y_1])] \not\subset E$.

Now, let $\alpha = \{\cup f([t_0, t]): t_0 \leq t \leq t_1\}$. Using the continuity of f and of union [see (1.48)], we see that α is a continuum.

Hence, by (1.4), α is an order arc. Therefore, by (1.19), there is a segment $\sigma \colon [0, 1] \xrightarrow{\text{onto}} \alpha$. Note that [by (1.15.4)] $\sigma(0) = f(t_o)$ and $\sigma(1) = \cup f([t_o, t_1])$. So, by (1.25), each component of $\sigma(1)$ intersects $\sigma(0)$. By (ii) above, there is a component K of $\sigma(1)$ such that $K \not\subset E$. Therefore, since $\sigma(0) = f(t_o) \subset E$ [by (i) above], we have:

$$K \cap E \neq \emptyset \neq K \cap [X \backslash E].$$

Hence, by (11.5.1) $K \supset E$. Certainly, then,

$$[\cup f([t_o, t_1])] \supset E.$$

Since t_1 was arbitrarily chosen in $(t_o, 1]$, a simple continuity argument now shows that $f(t_o) \supset E$. Therefore, by (i), $f(t_o) = E$. It now follows that (11.5.2) and (11.5.3) each holds. To prove the second part of (11.5), assume E is decomposable. Assume (11.5.1) does not hold. We show simultaneously that neither (11.5.2) nor (11.5.3) holds. Since (11.5.1) does not hold, there exists a subcontinuum Y_o of X such that

$$Y_o \cap E \neq \emptyset \neq Y_o \cap [X \backslash E] \quad \text{and} \quad Y_o \not\subset E.$$

Now, let $A \in \begin{Bmatrix} 2^E \backslash \{E\} \\ C(E) \backslash \{E\} \end{Bmatrix}$. By $\begin{Bmatrix} (11.4) \\ (11.1) \end{Bmatrix}$, there is an arc γ in $\begin{Bmatrix} 2^E \backslash \{E\} \\ C(E) \backslash \{E\} \end{Bmatrix}$ with end points A and a component L of $Y_o \cap E$. By (1.25) and (1.26), there is a segment $\sigma_1 \colon [0, 1] \to C(Y_o)$ from L to Y_o. Note that since $Y_o \not\subset E$ and $\sigma_1(t) \subset Y_o$ for all $t \in [0, 1]$, $\sigma_1(t) \neq E$ for any $t \in [0, 1]$. Hence,

$$[\gamma \cup \sigma_1([0, 1])] \subset \begin{Bmatrix} 2^X \backslash \{E\} \\ C(X) \backslash \{E\} \end{Bmatrix}.$$

Hence, there is an arc in $\begin{Bmatrix} 2^X \setminus \{E\} \\ C(X) \setminus \{E\} \end{Bmatrix}$ with end points A and Y_o. Recall that $Y_o \not\subset E$. Also note that $2^X \setminus 2^E$ and $C(X) \setminus C(E)$ are each arcwise connected by (11.2). Now we are done.

The fact that (11.5.3) implies (11.5.1) when E is decomposable was also obtained by Rogers [281, Proposition 3].

Let us note the following corollary to (11.5) [see (11.9)].

(11.6) COROLLARY [234, 4.9]. *If E is a decomposable subcontinuum of a continuum X such that $2^X \setminus \{E\}$ [equivalently, $C(X) \setminus \{E\}$] is not arcwise connected, then E is nowhere dense in any subcontinuum of X which properly contains E.*

Proof. Assume there is a subcontinuum Z of X such that Z properly contains E and E is not nowhere dense in Z. Let Y_o be a component of $cl[Z \setminus E]$. By (20.1), $Y_o \cap E \neq \emptyset$. Since E is not nowhere dense in Z, $Y_o \not\subset E$. Thus, (11.5.1) is violated. Therefore, since E is decomposable, $2^X \setminus \{E\}$ [equivalently, $C(X) \setminus \{E\}$] is arcwise connected by the second part of (11.5).

The next theorem is the main characterization result. A proof may be found in [234]. The proof begins by showing, as is not difficult, that (11.7.1) is equivalent to (11.5.1) when E is decomposable. Then, the bulk of the proof deals with the case when E is indecomposable. The details are not included here.

(11.7) THEOREM [234, 4.6]. *Let* E *be a nondegenerate proper subcontinuum of a continuum* X. *Then, the following three statements are equivalent:*

(11.7.1) *There is a dense subset* D *of* E *such that if* Y *is a subcontinuum of* X *satisfying* Y ∩ D ≠ ∅ ≠ Y ∩ [X\E], *then* Y ⊃ E.

(11.7.2) $2^X \setminus \{E\}$ *is not arcwise connected.*

(11.7.3) $C(X) \setminus \{E\}$ *is not arcwise connected.*

In (11.3) we showed that if a point of 2^X arcwise disconnects 2^X, then it is a point of C(X). The following result shows even more.

(11.8) COROLLARY (see the comment immediately preceding 4.6 of [234]). *For any* $A \in 2^X$, *the following two statements are equivalent:*

(1) $2^X \setminus \{A\}$ *is not arcwise connected.*

(2) $C(X) \setminus \{A\}$ *is not arcwise connected.*

Proof. Assume (1) or (2) holds. Then, by (11.3), $A \in C(X)$. Also, as is easy to see from (11.2), $A \notin F_1(X)$. Hence, if $A \neq X$, the equivalence is due to (11.7). If $A = X$, then the equivalence follows from (11.1) and (11.4).

The following example illustrates some of the aspects of (11.5) through (11.7) and (11.10).

(11.9) EXAMPLE [234, 4.5]. Let $X = E \cup Y$ where E is an indecomposable continuum, Y is any continuum, and $E \cap Y$ consists of only a single point p. It is clear that (11.5.1) is not satisfied by E and Y. However, (11.7.1) does hold as can be seen by taking D to be any composant of E not having p in it. Hence, (11.7.2) and (11.7.3) hold. Note that the conclusion of (11.6) does not hold for E, which explains why it was assumed in (11.6) that E was decomposable.

The next result shows that if (11.7.1) holds, there is a maximal dense set for which it holds.

(11.10) THEOREM. *Let* E *be a nondegenerate proper subcontinuum of a continuum* X *such that* E *arcwise disconnects* 2^X *[equivalently,* $C(X)$*]. Let*

$D(E) = \{p \in E:$ *if* Y *is a subcontinuum of* X *such that* $p \in Y$ *and* $Y \cap [X \backslash E] \neq \emptyset,$ *then* $Y \supset E\}.$

Then: $D(E)$ *is the union of some, perhaps all, composants in* E.

Proof. By (11.7.1), $D(E) \neq \emptyset$. Assume E is decomposable. Then, by the second part of (11.5), (11.5.1) holds so $D(E) = E$ and E is a composant of E. Next, assume E is indecomposable. Then, (11.7.1) holds. Let D be as in (11.7.1). Let $q \in D$ and let $\kappa(q)$ denote the composant in E of q. Let $p \in \kappa(q)$. Let Y be a subcontinuum of X such that $p \in Y$ and $Y \cap [X \backslash E] \neq \emptyset$. Let M be a proper subcontinuum of E such that $p, q \in M$. Then, M is nowhere dense in E [186, p.207]. Since $q \in [D \cap (Y \cup M)]$, we have by (11.7.1) that $[Y \cup M] \supset E$. Therefore, since M is nowhere dense in E, it follows that $Y \supset E$. Thus, $p \in D(E)$ and we have proved $\kappa(q) \subset D(E)$. The result now follows.

The following easy consequence of the proof of (11.10) shows that to prove a given indecomposable subcontinuum of X arcwise disconnects 2^X or $C(X)$, only one point "needs to be found" instead of the dense subset in (11.7.1).

(11.11) COROLLARY [234, 4.7]. *Let* E *be a nondegenerate proper subcontinuum of a continuum* X. *Consider the following statement:*

(11.11.1) *There exists* $p \in E$ *such that if* Y *is a subcontinuum of* X *such that* $p \in Y$ *and* $Y \cap [X \setminus E] \neq \emptyset$, *then* $Y \supset E$.

Then: (11.7.1) implies (11.11.1) and if E *is indecomposable, (11.7.1) and (11.11.1) are equivalent.*

(11.12) REMARK. Easy-to-find examples show that (11.11.1) does not in general imply (11.7.1). Simply take $X = [0, 1]$, $E = [0, 1/2]$, and $p = 0$.

(11.13) THEOREM [234, 4.8]. *Let* E, A, *and* B *be subcontinua of a continuum* X. *Then, the following two statements are equivalent:*

(1) *If α is an arc in* $C(X)$ *such that*
 $A, B \in \alpha$, *then* $E \in \alpha$.

(2) *If α is an arc in* 2^X *such that*
 $A, B \in \alpha$, *then* $E \in \alpha$.

Proof. Clearly (2) implies (1). Now, assume (1) holds. Throughout the proof assume $A \neq E \neq B$, since otherwise (2) would trivially hold. Also, if $A, B \in [C(X)\setminus C(E)]$, then (1) would not hold by (11.2). Hence, $A \subset E$ or $B \subset E$ (or both). Without loss of generality assume $A \subset E$. We take two cases.

Case I: E is decomposable. Then, by (11.4), $2^E\setminus\{E\}$ is arcwise connected. By (11.2), $2^X\setminus 2^E$ is arcwise connected. Since (1) holds and $A \neq E \neq B$, $C(X)\setminus\{E\}$ is not arcwise connected. Thus, by (11.8), $2^X\setminus\{E\}$ is not arcwise connected. Note that
$$2^X\setminus\{E\} = [2^E\setminus\{E\}] \cup [2^X\setminus 2^E].$$
Since (1) holds and $A \in [2^E\setminus\{E\}]$ which is arcwise connected, $B \notin [2^E\setminus\{E\}]$. Hence, since $B \neq E$, $B \in [2^X\setminus 2^E]$. It now follows that (2) holds.

Case II: E is indecomposable. Let $f: [0, 1] \to 2^X$ be a continuous function such that $f(0) = A$ and $f(1) = B$. Since $f(0), f(1) \in C(X)$, it follows from (1.49) that $\cup f([0, t])$ and $\cup f([t, 1])$ are each subcontinua of X for any $t \in [0, 1]$. Thus, letting

$$g(t) = \begin{cases} \cup f([0, t]) &, 0 \leq t \leq 1 \\ \cup f([t-1, 1]) &, 1 \leq t \leq 2 \end{cases}$$

we see that g is a continuous function from $[0, 2]$ into $C(X)$ such that $g(0) = A$ and $g(2) = B$. Hence, since (1) holds, there exists $t \in [0, 2]$ such that $g(t) = E$. Let

$$t_o = \text{g.l.b.}\{t \in [0, 2]: g(t) = E\}.$$

First assume $t_o \leq 1$. Then, note that $g(t) \in C(E)$ for each $t \in [0, t_o]$. Hence, we can proceed as we did with the function g in the proof of (11.4), to show that $f(t_o) = E$. Next, assume $t_o > 1$. Then, let

$$s_o = \text{l.u.b.}\{t \in [1, 2]: g(t) = E\}.$$

Then, except for minor notational changes, proceed as we just did for t_o and we obtain $f(s_o) = E$. This completes the proof that (2) holds.

The theorem just proved gives relationships between the arc components of $2^X\setminus\{E\}$ and those of $C(X)\setminus\{E\}$. For example:

(11.14) COROLLARY (see the comment immediately preceding 4.8 of [234]). *Let* E *be a subcontinuum of a continuum* X. *If* Γ *is an arc component of* $2^X\setminus\{E\}$ *and if* Γ ∩ C(X) ≠ ∅, *then* Γ ∩ C(X) *is an arc component of* C(X)\{E}.

It is possible to have every nondegenerate subcontinuum of a continuum X be an arcwise-disconnecting point of 2^X. In fact, the next result shows that the class of continua X with this property is precisely the class of hereditarily indecomposable continua.

(11.15) THEOREM [234, 4.11]. *For a continuum* X, *the fol-*

lowing three statements are equivalent:

(11.15.1) *X is hereditarily indecomposable.*

(11.15.2) *For any nondegenerate subcontinuum E of X, $2^X \setminus \{E\}$ is not arcwise connected.*

(11.15.3) *For any nondegenerate subcontinuum E of X, $C(X) \setminus \{E\}$ is not arcwise connected.*

Proof. By (11.8), it suffices to prove (11.15.1) is equivalent to (11.15.3). Assume (11.15.1) holds. Observe the following fact which is easy to prove using (1.25) and (1.26): If A is any nondegenerate subcontinuum of a continuum Z, then A is a non-end point of some arc in C(Z). From this fact and (1.61) it is easy to see that (11.15.1) implies (11.15.3). So, assume (11.15.1) does not hold.[1] Let M be a decomposable subcontinuum of X. Let A and Y be proper subcontinua of M such that $M = A \cup Y$. Let E be a proper subcontinuum of M such that $E \supset A \neq E$; such a continuum E exists as a consequence of (20.1). Since $A \setminus Y$ is a nonempty open subset of M and $[A \setminus Y] \subset E \subset M$, it follows that A is a proper subcontinuum of E with interior in E. Hence, E is decomposable [186, p.207]. Note that $A \subset E$, $Y \cap A \neq \emptyset$, $M = A \cup Y$, $Y \not\subset A$, and $M \neq E \subset M$. It follows from these five facts that

$Y \cap E \neq \emptyset \neq Y \cap [X \setminus E]$ and $Y \not\subset E$.

Thus, (11.15.1) is violated. Therefore, since E is decomposable, we have by the second part of (11.5) that $C(X) \setminus \{E\}$ is arcwise connected, i.e., (11.15.3) does not hold. This completes the proof of (11.15).

For some related information about (11.14) and (11.15), see (12.29) and (12.30).

We have given essentially all facts that are known about points which arcwise disconnect a hyperspace. We will give some relationships between this and arcwise accessibility in the next chapter. For the present, let us note the following application of the theory to the structure of continua.

(11.16) THEOREM [241, 2.12]. *Let* X *be an a-triodic continuum such that* $C(X)\setminus\{E\}$ *is arcwise connected for each proper subcontinuum* E *of* X. *Then:*

(1) *Each proper subcontinuum of* X *is unicoherent.*

(2) *If each proper subcontinuum of* X *is decomposable, then each proper subcontinuum of* X *is chainable and, hence,* X *is either chainable, circle-like, or indecomposable.*

(3) *If* X *is hereditarily decomposable, then* X *is chainable or circle-like.*

The proof of (11.16) can be deduced from the proof of (10.12).

Now let us return to (9.2). As was mentioned, its proof in [252] uses the result stated here in (11.7).

(11.17) QUESTIONS. How can (9.2) be generalized? For example: Is it still true with the two sets A_1 and A_2 replaced by n sets, n finite? What about countably many? What about a collection $\{A_\lambda : \lambda \in \Lambda\}$ which is a compact zero-dimensional subset of the hyperspace? The last of these questions is somewhat motivated by (2.15).

(11.18) QUESTION. Assume X is any chainable continuum which is not an arc; then, must X contain a subcontinuum Y such that $C(Y)\setminus\{E\}$ is not arcwise connected for some subcontinuum E of Y? Of course [see (11.1)], the question is only open for the class of hereditarily decomposable chainable continua.[2] This may seem like an off-beat question, but an affirmative answer may lead to a hyperspace proof of Henderson's result [140]. Hopefully, a "simpler" proof than the one Henderson gave would evolve. Some material which may lead to generalizations of Henderson's result is in (6.23) through (6.27).

Finally, we remark that it would be interesting to have more information about the arc components of $C(X)\setminus\{E\}$ and $2^X\setminus\{E\}$ when E arcwise disconnects. Some facts about this may be found in (1.52) and (8.19) here (also, see [234], [239], and [281]).

NOTES

1. It might appear to the reader that (1.61) directly gives the fact that (11.15.3) implies (11.15.1). However, there exist arcwise connected continua Y such that each point of Y arcwise disconnects Y but Y is not uniquely arcwise connected. For example: Let W denote the continuum in (3) of (8.22) and let $p = (0, +1) \in W$, let M denote the Cantor middle-thirds set, let f be any mapping from $\{(m, p) \in M \times W : m \in M\}$ onto a circle S [f exists by (0.69.4)], and let Y be the continuum obtained by attaching $M \times W$ to S by f, i.e., $Y = (M \times W) \cup_f S$. The

example was mentioned to me by E. D. Tymchatyn.

2. Recently J. Grispolakis and E. D. Tymchatyn have shown that the answer to (11.18) is "yes". Considering the case when X is hereditarily decomposable, they use (11.5) and ideas in [186, pp.199-204]. Their paper, to be entitled "Irreducible continua with degenerate end-tranches and arcwise accessibility," is currently under preparation. Other results in it are mentioned in Footnote 1 on p.393.

XII.
ARCWISE AND SEGMENTWISE ACCESSIBILITY

Let $\Sigma \subset 2^X$. A member A of $C_1(X)$ is said to be $\begin{Bmatrix} arcwise \\ segmentwise \end{Bmatrix}$ *accessible from* $\Sigma \backslash C_1(X)$ [234] provided there is a $\begin{Bmatrix} \text{homeomorphism} \\ \text{segment} \end{Bmatrix}$ $\sigma: [0, 1] \to \Sigma$ such that $\sigma(t) \in [\Sigma \backslash C_1(X)]$ for all $t < 1$ and $\sigma(1) = A$ [the notion of segmentwise accessibility is due to Joe Stiles--see (12.33)]. We will say that A is accessible *beginning with* K [234] provided there is a σ as above such that $\sigma(0) = K$. In this chapter, Σ will be 2^X or $C_2(X)$. We will give some answers to the following question:

(12.1) QUESTION [234, 1.3]. Which members of $C_1(X)$ are arcwise accessible from $2^X \backslash C_1(X)$?

This question was raised and investigated in [234]. As we will see [for example, (12.14); comp., (12.15)] it is connected with the topic covered in Chapter XI. In fact, the results in Chapter XI grew out of the investigation in [234] of accessibility.

It might appear to the reader that: "Each member of $C_1(X)$ is arcwise accessible from $2^X \setminus C_1(X)$." Though the statement in quotes is false [even for rational continua; see (12.16)], it is "almost true" as the following result shows. This result and (12.4) were a consequence of considering a question raised by Joe Stiles [see (12.33)].

(12.2) THEOREM [234, 2.2]. *Every nondegenerate member of* $C_1(X)$ *is segmentwise accessible from* $C_2(X) \setminus C_1(X)$ *beginning with a two-point set.*

Proof [see (12.3)]. Let $M \in C_1(X)$ such that M is nondegenerate. It follows from a theorem of Bing [33, Theorem 5] that there exists a point $p \in M$ and a sequence $\{M_n\}_{n=1}^{\infty}$ of nondegenerate subcontinua M_n of M such that (for details, see [234, proof of 2.2]):

(i) $M_n \subset M_{n+1} \neq M_n$ for each $n = 1, 2, \ldots$;

(ii) $p \in [M \setminus \bigcup_{n=1}^{\infty} M_n]$;

(iii) $\{M_n\}_{n=1}^{\infty}$ converges to M.

Let $q \in M_1$. By (1.25) and (1.26), there is a segment $\sigma_1: [0, 1] \to C(M_1)$ from $\{q\}$ to M_1. Also, since $M_n \subset M_{n+1}$ are continua for each $n = 1, 2, \ldots$, we have using (1.25) and (1.26) that there are segments $\sigma_{n+1}: [0, 1] \to C(M_{n+1})$ from M_n to M_{n+1} for each $n = 1, 2, \ldots$. For each $n = 1, 2, \ldots$, let

$$I_n = \left[\frac{2^{n-1} - 1}{2^{n-1}}, \frac{2^n - 1}{2^n} \right]$$

and let $f_n: I_n \to 2^M$ be given by

$$f_n(t) = \sigma_n(2^n \cdot t + 2 - 2^n) \cup \{p\}.$$

Geometrically, f_n is σ_n "speeded up" and with the point p adjoined. Define $f: [0, 1] \to 2^M$ by

$$f(t) = \begin{cases} f_n(t), & t \in I_n \\ M, & t = 1. \end{cases}$$

It follows from (i), (iii), and the fact that M_1 is nondegenerate [so, f_1 is one-to-one] that f is a homeomorphism. Furthermore, by (i), $f(s) \subset f(t)$ whenever $0 \le s \le t$. Hence, the range of f is an order arc [by (1.2)]. Also, by (ii) and the definition of f and f_n for each $n = 1, 2, \ldots$, it follows that $f(0) = \{q, p\}$, $f(1) = M$, and $f(t) \in [C_2(M) \setminus C_1(M)]$ for each $t < 1$. The theorem now follows from the fact that an order arc is the range of a segment [see (1.19)].

(12.3) REMARK [234, 2.3]. In (12.2) we showed segmentwise accessibility beginning with a *two-point* set. The following argument, which is somewhat simpler than the one given for (12.2), shows segmentwise accessibility beginning with a *countable* set: Let $M \in C_1(X)$ such that M is nondegenerate, let $x \in M$, let $\sigma: [0, 1] \to C(M)$ be a segment from $\{x\}$ to M [use (1.25) and (1.26)], let $x_n \in [M \setminus \sigma([n-1]/n)]$ for each $n = 1, 2, \ldots$, and let $\{x_{n(i)}\}_{i=1}^{\infty}$ be a convergent subsequence of $\{x_n\}_{n=1}^{\infty}$, $x_{n(i)} \to x_o$. Define $f: [0, 1] \to 2^M$ by

$$f(t) = \sigma(t) \cup \{x_{n(i)} : i = 1, 2, \ldots\} \cup \{x_o\}$$

and use (1.19).

Let us note the following consequence of (12.2).

(12.4) COROLLARY [234, 2.1 and 2.2]. *A member* M *of* $C_1(X)$ *is segmentwise accessible from* $2^X \setminus C_1(X)$ *if and only if* M *is nondegenerate.*

Proof. To prove the half not implied by (12.2) let $M \in F_1(X)$, $M = \{x\}$. Let $\sigma: [0, 1] \to 2^X$ be any segment such that $\sigma(1) = M$. By (1.15.4), $\sigma(t) \subset \sigma(1)$ for each $t < 1$. Therefore, since $\sigma(1) = \{x\}$, $\sigma(t) = \sigma(1)$ for each $t \in [0, 1]$. The result follows.

Thus, (12.4) shows exactly which members of $C_1(X)$ are segmentwise accessible from $2^X \setminus C_1(X)$. In particular, since a segment is a special type of arc, (12.4) reduces (12.1) to the following question:

(12.5) QUESTION [234, 1.2]. *When is a member of* $F_1(X)$ *arcwise accessible from* $2^X \setminus C_1(X)$?

Beginning in (12.8), we give some answers to (12.5) and pose some related questions. First, note the following two lemmas.

(12.6) LEMMA. *Let* Y *be an indecomposable continuum and let* $\Lambda \subset 2^Y$ *be an arcwise connected continuum. If* $\cup \Lambda = Y$ *and if* $\Lambda \cap C(Y) \neq \emptyset$, *then* $Y \in \Lambda$.

Proof. The lemma was proved in (1.50).

The next lemma is a generalization of [234, 3.2].

(12.7) LEMMA. *Let* X *be any continuum and let* $\Lambda \subset 2^X$ *be an arcwise connected continuum such that* $\Lambda \cap C(X) \neq \emptyset$. *Then,* $\cup \Lambda$ *is a subcontinuum of* X *such that* $[\cup \Lambda] \in \Lambda$ *or* $\cup \Lambda$ *is decomposable.*

Proof. Let $Y = \cup \Lambda$. Since $\Lambda \cap C(X) \neq \emptyset$, we have by (1.49) that Y is a subcontinuum of X. Assume Y is indecomposable. Then, (12.6) can be applied to see that $Y \in \Lambda$. This proves (12.7).

Now we give our first result related to (12.5).

(12.8) THEOREM [234, 3.2]. *Let* $x_o \in X$. *If* $\{x_o\}$ *is arcwise accessible from* $2^X \setminus C_1(X)$, *then* x_o *belongs to arbitrarily small decomposable subcontinua of* X. *In fact: If* $f: [0, 1] \to 2^X$ *is continuous such that* $f(t) \in [2^X \setminus C_1(X)]$ *for all* $t < 1$ *and* $f(1) = \{x_o\}$, *then* $\cup f([t, 1])$ *is a decomposable subcontinuum of* X, *with* x_o *in it, for each* $t < 1$.

Proof. Let $f: [0, 1] \to 2^X$ be continuous such that $f(t) \in [2^X \setminus C_1(X)]$ for all $t < 1$ and $f(1) = \{x_o\}$. Let $t_o < 1$. By taking $\Lambda = f([t_o, 1])$ we see that Λ satisfies the hypotheses in (12.7). So, by (12.7), $\cup \Lambda$ is a subcontinuum of X such that $[\cup \Lambda] \in \Lambda$ or $\cup \Lambda$ is decomposable. Suppose $[\cup \Lambda] \in \Lambda$. Then, since $[\cup \Lambda] \in C_1(X)$, $\cup \Lambda = f(1) = \{x_o\}$. This implies $f(t_o) = \{x_o\}$ which can not happen since $t_o < 1$. Therefore, $\cup \Lambda$ is decomposable.

(12.9) COROLLARY [234, 3.4]. *If X is an hereditarily indecomposable continuum, then no member of $F_1(X)$ is arcwise accessible from $2^X \setminus C_1(X)$.*

(12.10) EXAMPLE [234, 3.5]. The converse of (12.9) is false. For example: Let $X = K \cup M$, where K and M are each hereditarily indecomposable continua and $K \cap M$ is a *nondegenerate* proper subcontinuum of each. It is not difficult to see that any decomposable subcontinuum of X must contain $K \cap M$ (a complete proof of this detail is in [234, 3.5]). It follows from this and (12.8) that no member of $F_1(X)$ is arcwise accessible from $2^X \setminus C_1(X)$. Note that X is decomposable.

In (12.10) it was crucial that $K \cap M$ not just consist of one point, as the next result shows [see (12.27) and (12.28) for more general results].

(12.11) THEOREM [234, 3.9]. *If p is a cut point of a continuum X, then $\{p\}$ is arcwise accessible from $C_2(X) \setminus C_1(X)$ beginning with a two-point set.*

Proof. We indicate the idea of the proof; the details are in [234]. Since p is a cut point of X, there are disjoint nonempty open subsets U and V of X such that $X \setminus \{p\} = U \cup V$. Let $A = U \cup \{p\}$ and let $B = V \cup \{p\}$. As is well-known and easy to prove, A and B are each continua. Let $a_1 \in [A \setminus \{p\}]$ and let $b_1 \in [B \setminus \{p\}]$. Using (1.25), take segments σ_{11}, σ_{12}, and

σ_{13} defined on [0, 1] into C(X) as follows: Let σ_{11} be a segment from $\{a_1\}$ to A, let σ_{12} be a segment from A_1 to A where A_1 is a nondegenerate subcontinuum of A such that $p \in A_1$ and diam$[A_1] < 2^{-1}$, and let σ_{13} be a segment from $\{a_2\}$ to A_1 where $a_2 \in [A_1 \setminus \{p\}]$. Now define $f_1: [0, 3] \to C_2(X) \setminus C_1(X)$ by

$$f_1(t) = \begin{cases} \sigma_{11}(t) \cup \{b_1\} , & 0 \le t \le 1 \\ \sigma_{12}(2 - t) \cup \{b_1\} , & 1 \le t \le 2 \\ \sigma_{13}(3 - t) \cup \{b_1\} , & 2 \le t \le 3 . \end{cases}$$

Next, using (1.25), take segments σ_{21}, σ_{22}, and σ_{23} from [0, 1] into C(X) as follows: Let σ_{21} be a segment from $\{b_1\}$ to B, let σ_{22} be a segment from B_1 to B where B_1 is a nondegenerate subcontinuum of B such that $p \in B_1$ and diam$[B_1] < 2^{-1}$, and let σ_{23} be a segment from $\{b_2\}$ to B_1 where $b_2 \in [B_1 \setminus \{p\}]$. Now define $g_1: [3, 6] \to C_2(X) \setminus C_1(X)$ by

$$g_1(t) = \begin{cases} \sigma_{21}(t - 3) \cup \{a_2\} , & 3 \le t \le 4 \\ \sigma_{22}(5 - t) \cup \{a_2\} , & 4 \le t \le 5 \\ \sigma_{23}(6 - t) \cup \{a_2\} , & 5 \le t \le 6 . \end{cases}$$

Analogously, obtain f_2 inside $A_1 \cup \{b_2\}$, g_2 inside $B_1 \cup \{a_3\}$, etc. making sure that A_n and B_n each have diameter $< 2^{-n}$. Having done this for each $n = 1, 2, \ldots$, we obtain a mapping k from $[0, \infty)$ into $C_2(X) \setminus C_1(X)$ which can be extended to ∞ by defining $k(\infty) = \{p\}$. Careful use of these ideas to define k gives a precise proof of (12.11).

By (12.11), we know that there is a singleton which is arcwise accessible from $2^X \setminus C_1(X)$ if X contains a subcontinuum with a cut point. This, together with the following lemma, leads us to study rational continua in connection with (12.5). Recall [338, p.82-83] that a continuum is said to be *rational* provided each point belongs to arbitrarily small open sets whose boundaries are (at most) countable.

(12.12) LEMMA [234, 3.10]. *Every rational continuum contains a subcontinuum with a cut point.*

Proof. Let Q be a rational continuum. Let U be a nonempty open subset of Q such that $cl[U] \neq Q$ and $Bd[U] = cl[U] \setminus U$ is countable. Let $V = Q \setminus cl[U]$ and let $W = Q \setminus cl[V]$. It is easy to see that

(1) $Bd[V] \subset Bd[U]$.

Hence, since $Bd[U]$ is countable,

(2) $Bd[V]$ is countable.

It is not difficult to verify that

(3) $Bd[V] = Bd[W]$.

Therefore, since every compact countable (metric) space has an isolated point, it follows using (2) and (3) that

(4) there is an isolated point p of $Bd[V] = Bd[W]$.

Now, (4) allows us to apply Theorem 8 of [186, p.185] twice--once to obtain a continuum $A \subset [V \cup \{p\}]$ such that $p \in A \neq \{p\}$, and again to obtain a continuum $B \subset [W \cup \{p\}]$ such that $p \in B \neq \{p\}$. Since $A \cap B = \{p\}$, $A \cup B$ is a subcontinuum of Q with a cut point, namely p.

(12.13) THEOREM [234, 3.11]. *If Q is a rational continuum, then there is a dense subset D of Q such that, for each $p \in D$, $\{p\}$ is arcwise accessible from $C_2(Q)\setminus C_1(Q)$ beginning with a two-point set.*[1] *In fact, such a subset D exists so that $Q\setminus D$ is punctiform [= contains no (nondegenerate) continuum, or equivalently, D is continuumwise dense].*

Proof. Let $D = \{p \in Q\colon p$ is a cut point of some subcontinuum of $Q\}$. By (12.11), the accessibility part of (12.13) holds for each $\{p\}$, $p \in D$. Now: The property of being a rational continuum is hereditary. Hence, by (12.12), D intersects every nondegenerate subcontinuum of Q, i.e., $Q\setminus D$ is punctiform [therefore, D is a dense subset of Q by (20.1)].

Thus, (12.13) shows that many members of $F_1(Q)$ are arcwise accessible from $2^Q\setminus C_1(Q)$ when Q is any rational continuum. So, we are led to ask the following question: Is there a rational continuum Q_o such that $\{q_o\}$ is not arcwise accessible from $2^{Q_o}\setminus C_1(Q_o)$ for some $q_o \in Q_o$? Such a continuum is constructed in [234, 3.12]; in fact, it is a rational chainable continuum Q_o which has a *dense* set of points q such that $\{q\}$ is not arcwise accessible from $2^{Q_o}\setminus C_1(Q_o)$. In the paragraph below, (12.14), (12.16), and (12.17), we indicate the ideas behind the construction of Q_o and the verification of its properties. The reader is referred to [234] for more details.

The construction of Q_o in (12.16) is done using inverse limits. The theorem in (12.14) will be useful [in (12.17)] for verifying that certain singletons are not arcwise accessible from $2^{Q_o}\backslash C_1(Q_o)$. Observe [as will be illustrated in (12.17)] that (11.5.1) is exactly the type of condition which can be used, in conjunction with (1.160.1), to show that a subcontinuum of the inverse limit space arcwise disconnects the hyperspace of the inverse limit space.

(12.14) THEOREM [234, 4.13]. *Let* X *be an hereditarily decomposable continuum and let* $x_o \in X$. *If* x_o *belongs to arbitrarily small subcontinua of* X *each of which arcwise disconnects* 2^X *[equivalently, by (11.8), C(X)], then* $\{x_o\}$ *is not arcwise accessible from* $2^X\backslash C_1(X)$.

Proof. Assume $\{x_o\}$ is arcwise accessible from $2^X\backslash C_1(X)$. Let $f: [0, 1] \to 2^X$ be continuous such that $f(t) \in [2^X\backslash C_1(X)]$ for all $t < 1$ and $f(1) = \{x_o\}$. Let

$\eta = \text{diam}[\cup f([0, 1])]$.

Let K be a nondegenerate subcontinuum of X such that $x_o \in K$ and $\text{diam}[K] < \eta$ [K exists by (20.1)]. Then, clearly, $[\cup f([0, 1])] \not\subset K$. Hence, there exists $s_o \in [0, 1]$ such that $f(s_o) \not\subset K$. Thus,

(1) $f([0, 1]) \cap [2^X\backslash 2^K] \neq \emptyset$.

Also, since $f(1) = \{x_o\}$ and $x_o \in K$,

(2) $f(1) \in [2^K\backslash\{K\}]$.

Finally note that since $f(t) \notin C_1(X)$ for any $t < 1$ and $f(1) = \{x_o\} \neq K \in C_1(X)$,

(3) $f([0, 1]) \subset [2^X\backslash\{K\}]$.

(12.16)   ARCWISE, SEGMENTWISE ACCESSIBILITY   383

Now, note that, by using (1.25), $2^X \backslash 2^K$ is arcwise connected and, by (11.4), $2^K \backslash \{K\}$ is arcwise connected. By these facts and (1) through (3) above, we have that $2^X \backslash \{K\}$ is arcwise connected. Since K was any continuum of diameter less than η, (12.14) is proved.

(12.15) EXAMPLE [234, see comment just before 4.13]. Let $X = Z \cup Y$ where Z is an hereditarily indecomposable continuum, Y is any continuum, and $Z \cap Y$ consists of just one point x_o. Then, by considering arbitrarily small nondegenerate subcontinua of Z containing x_o and using (11.7), it can be seen that x_o (but not X) satisfies all the hypotheses of (12.14). However, by (12.11), $\{x_o\}$ is arcwise accessible from $2^X \backslash C_1(X)$. Thus, we have a host of examples which explain why hereditary decomposability was assumed in (12.14).

(12.16) *Indication of the construction in* [234, 3.12] *of a rational continuum* Q_o *which contains a dense* [*in fact, continuumwise-dense*] *set of points* z_o *such that* $\{z_o\}$ *is not arcwise accessible from* $2^{Q_o} \backslash C_1(Q_o)$. Let[2]
$$X_1 = \text{cl}\{(x, \sin[1/x]) \in R^2 : 0 < |x| \leq 1\}.$$
Let $D_1 = \{x_1^1, x_2^1, \ldots, x_i^1, \ldots\}$ be a countable continuumwise-dense subset of X_1, i.e., D_1 intersects every nondegenerate subcontinuum of X_1. Also, assume that none of the (four) end points of the arc components of X_1 are in D_1 [this assumption will play a role in describing the bonding maps]. Now, let X_2 be

the continuum obtained by "inserting a copy X_1' of X_1 in X_1 at x_1^1." Let J denote the arc in X_1' which corresponds to the arc $\{(0, y): |y| \leq 1\}$ in X_1. Let $f_1: X_2 \xrightarrow{\text{onto}} X_1$ be the natural map which "shrinks J to the point x_1^1" and is a homeomorphism "near the identity" on $X_2 \setminus J$. Next, let

$$D_2 = \{x_1^2, x_2^2, \ldots, x_i^2, \ldots\}$$

be a countable continuumwise-dense subset of X_2 such that none of the (six) end points of the arc components of X_2 are in D_2. Now, let X_3 be the continuum obtained by "inserting copies of X_1 in X_2 at $f_1^{-1}(x_2^1)$ and at x_1^2" [unless $f_1(x_1^2) = x_2^1$, in which case only one copy is "inserted"]. Let $f_2: X_3 \xrightarrow{\text{onto}} X_2$ be defined in a manner similar to the way f_1 was defined. Next, let D_3 be a countable continuumwise-dense subset of X_3 such that none of the end points of the arc components of X_3 are in D_3. We obtain X_4 and $f_3: X_4 \xrightarrow{\text{onto}} X_3$ by the same type of procedure used to obtain the previous continua and maps, this time making sure that copies of X_1 are "inserted in X_3 at $(f_1 \circ f_2)^{-1}(x_3^1)$, $f_2^{-1}(x_2^2)$, and at the first enumerated point of D_3." Continuing this process, we obtain an inverse sequence $\{X_n, f_n\}_{n=1}^{\infty}$. The continuum Q_0 is the inverse limit of $\{X_n, f_n\}_{n=1}^{\infty}$. For some more details and some pictures of the construction of Q_0, see [234, 3.12].

(12.17) VERIFICATION OF THE PROPERTIES OF Q_0 [234]. Note that for any given n and any $x \in [X_n \setminus D_n]$, $(f_n \circ f_{n+1} \circ \cdots \circ f_{n+k})^{-1}(x)$ consists of only one point for each $k = 0, 1, 2, \ldots$. Using this, and the fact that each D_n is countable, the rationality of Q_o can be verified from the following lemma [in particular, by considering X_n as being in R^2 and using the fact that D_n is countable, the base B_n in the lemma can be produced].

(12.17.1) LEMMA [234, 3.15]. *Let* Y *be the inverse limit of* $\{Y_n, g_n\}_{n=1}^{\infty}$ *where* Y_n *is a continuum and* g_n *continuously maps* Y_{n+1} *onto* Y_n *for each* $n = 1, 2, \ldots$. *Also assume that, for each* $n = 1, 2, \ldots$, Y_n *has a base* $B_n = \{U_i^n : i = 1, 2, \ldots\}$ *of open sets such that, for each* $i = 1, 2, \ldots$: *(1)* $Bd[U_i^n]$ *is countable;* *(2)* $(g_n \circ g_{n+1} \circ \cdots \circ g_{n+k})^{-1}(y)$ *consists of only one point for each* $y \in Bd[U_i^n]$ *and each* $k = 0, 1, 2, \ldots$. *Then,* Y *is a rational continuum.*

Next, let $Z_o = \{(y_1, y_2, \ldots) \in Q_o : \text{for some } n, y_n \in D_n\}$. Since D_n is continuumwise-dense in X_n for each $n = 1, 2, \ldots$, it follows easily [use (1.160.1)] that Z_o is continuumwise-dense in Q_o. It remains to show that $\{z\}$ is not arcwise accessible from $2^{Q_o} \setminus C_1(Q_o)$ for any $z \in Z_o$. Let $z_o = (z_1, z_2, \ldots) \in Z_o$. Since Q_o is a rational continuum [therefore, hereditarily decomposable], it suffices by (12.14) to show that z_o belongs to arbitrarily small subcontinua of Q_o each of which arcwise disconnects 2^{Q_o}. Hence, by (11.5), it suffices to show that z_o belongs to arbitrarily small subcontinua of Q_o each of which satisfies (11.5.1). This follows by considering, for larger and larger i, the subcontinua E^i of Q_o given by: E^i is the inverse limit of

$\{E_n^i, f_n|E_{n+1}^i\}_{n=1}^{\infty}$ where

$$E_n^i = \begin{cases} \{z_n\} & , 1 \leq n \leq i \\ (f_i \circ f_{i+1} \circ \cdots \circ f_{n-1})^{-1}(z_i) & , n > i \end{cases}$$

and the "vertical line" denotes the restriction of f_n to E_{n+1}^i. Since $E^1 \supset E^2 \supset \cdots \supset E^i \supset \cdots$ and since $\cap_{i=1}^{\infty} E^i = \{z_o\}$, it follows easily that, for large enough i, the continua E^i are "as small as we wish." The geometry of the construction of Q_o, together with (1.160.1), can be used to show quite easily that each E^i ($i = 1,2,\ldots$) satisfies (11.5.1) [with E in (11.5.1) replaced by any given E^i]--the idea is as follows. Fix i. If n is large enough so that E_n^i is nondegenerate, then E_n^i is "buried" so as to satisfy (11.5.1). Thus, applying (1.160.1) to a given subcontinuum Y of Q_o satisfying $Y \cap E^i \neq \emptyset \neq Y \cap [Q_o \setminus E^i]$, it follows that $Y \supset E^i$. This completes (12.17).

(12.18) REMARK. Note that the continuum Q_o of (12.16) is chainable because it is an inverse limit of chainable continua. Also, since Q_o has a continuumwise-dense subset Z_o of points z_o such that $\{z_o\}$ is not arcwise accessible from $2^{Q_o} \setminus C_1(Q_o)$, it is clear that Q_o contains no arc. Hence, Q_o is a rational chainable continuum which contains no arc.[2] The construction in (12.16) is similar to constructions done in [3] to obtain other types of continua. Using the procedures in [3], Andrews [5] constructed a rational chainable continuum no two of whose nondegenerate subcontinua are homeomorphic.

Recall that a rational continuum is hereditarily decomposable. Thus, in view of (12.13), the following question is natural to ask.

(12.19) QUESTION [234, 6.1]. For any hereditarily decomposable continuum X, must there be a point $x \in X$ such that $\{x\}$ is arcwise accessible from $2^X \backslash C_1(X)$?[1]

I do not know the answer to (12.19) but, if the following question has a negative answer, then the answer to (12.19) is "yes" [by (12.11)].

(12.20) QUESTION [234, 6.2]. Is there an hereditarily decomposable continuum such that no subcontinuum of it has a cut point?[1] At first, one might suspect Whyburn's example [339] to give an affirmative answer to this question. However, Whyburn's continuum is rational [168].

If the answer to (12.20) is "yes" then, by (12.12), the answer to the following question is also "yes."

(12.21) QUESTION [234, 6.3]. Is there an hereditarily decomposable continuum which contains no rational continuum?[1]

We have given some necessary conditions and some sufficient conditions in order that a singleton be arcwise accessible from $2^X \backslash C_1(X)$.

(12.22) QUESTIONS [234, 6.4]. What are conditions, which are at the same time both necessary and sufficient, in order that $\{x_o\}$ be arcwise accessible from $2^X \backslash C_1(X)$ for a given point x_o of a continuum X? What about such conditions when X is rational or when X is hereditarily decomposable? For hereditarily decomposable continua, is the converse of (12.14) true?

For a given continuum X, let

$AA[X] = \{x \in X: \{x\}$ is arcwise accessible from $2^X \backslash C_1(X)\}$.

As we see from (12.13), AA[X] is always continuumwise dense in X when X is rational. On the other hand, by (12.16), there exist rational continua X such that X\AA[X] is also continuumwise dense in X.

(12.23) QUESTION [234, 6.5]. For what rational continua X is $AA[X] = X$?[1]

(12.24) QUESTION. What is the Borel type [185, p.47] of AA[X] when X is a rational continuum or, more generally, when X is any continuum? In fact, is AA[X] always a Borel set?

(12.25) QUESTION [234, 6.6]. What hereditarily decomposable continua X have the property that each [or some or no] singleton is arcwise accessible from $2^X \backslash C_1(X)$?[1] It may be that each hereditarily decomposable continuum has some arcwise-accessible singleton; see (12.19).

In (12.2) we showed that every nondegenerate member of $C_1(X)$ is arcwise accessible from $C_2(X) \setminus C_1(X)$. In other results, whenever we show that a given singleton $\{x\}$ is arcwise accessible from $2^X \setminus C_1(X)$, we actually show $\{x\}$ is arcwise accessible from $C_2(X) \setminus C_1(X)$. I do not know the answer to the following question.

(12.26) QUESTION [234, 6.7]. Is there a continuum X such that, for some $x_o \in X$, $\{x_o\}$ is arcwise accessible from $2^X \setminus C_1(X)$ but not from $C_2(X) \setminus C_1(X)$?

Let us return to (12.11). In [234, 3.7 and 3.8], two results are proved which are more general than (12.11). We now state them without proof.

(12.27) THEOREM [234, 3.7]. *Let p be a point of a continuum X such that there exist subcontinua K_n and L_n of X, $n = 1, 2, \ldots$, satisfying*

(1) $\bigcap_{n=1}^{\infty} K_n = \{p\} \bigcap_{n=1}^{\infty} L_n$

and

(2) $K_i \not\subset [\bigcup_{n=1}^{\infty} L_n]$ *and* $L_i \not\subset [\bigcup_{n=1}^{\infty} K_n]$ *for any* $i = 1, 2, \ldots$.

Then, $\{p\}$ is arcwise accessible from $C_2(X) \setminus C_1(X)$ beginning with a two-point set.

(12.28) COROLLARY [234, 3.8]. *Let p be a point of a continuum X such that there exist nondegenerate subcontinua A and B of X such that $\{p\}$ is a component of $A \cap B$. Then $\{p\}$ is* *arcwise accessible from* $C_2(X) \setminus C_1(X)$ *beginning with a two-point set.*

Now, let us return to (11.14). It says that an arc component of $2^X \setminus \{E\}$ which intersects $C(X)$ does so in an arc component of $C(X) \setminus \{E\}$. The following stronger result holds when X is hereditarily indecomposable [also see (12.31)].

(12.29) THEOREM [234, 5.2]. *Let X be any hereditarily indecomposable continuum. If Γ is any arcwise connected subset of 2^X, then $\Gamma \cap C(X)$ is arcwise connected [perhaps empty].*

The theorem has some geometric interpretations. Let X be any hereditarily indecomposable continuum. If an arc begins in $2^X \setminus C_1(X)$, "enters" $C_1(X)$, and then "leaves" $C_1(X)$, it can never "return" to $C_1(X)$. Furthermore, the arc must "enter" $C_1(X)$ from below," i.e., if t_0 is the first time the arc h "touches" $C_1(X)$, then $h(t) \subset h(t_0)$ for each $t \leq t_0$. So, if an arcwise connected subset Λ of 2^X just touches $C_1(X)$ once, say at A, then Λ is "always below" A, i.e., $\Lambda \subset 2^A$ [see the notion of touching set defined in the next chapter]. These statements can be verified using (12.29) and (1.25). They represent generalizations of (12.9) and should be compared with (13.1), which shows that analogous statements for non-arcwise connected subcontinua of 2^X are false. Also, the theorem in (12.30) gives more-specific information.

In (11.15), we characterized hereditary indecomposability in terms of arcwise disconnecting. The following result gives characterizations in terms of arcwise accessibility. It also gives information about where arcs, which begin in $2^X \setminus C_1(X)$ and end in $C_1(X)$, must begin and end [see (12.32)].

(12.30) THEOREM [234, 5.4]. *For a continuum* X, *the following three statements are equivalent:*

(12.30.1) X *is hereditarily indecomposable.*

(12.30.2) *Given* $K \in [2^X \setminus C_1(X)]$, *there exists one and only one* $A \in C(X)$ *which is arcwise accessible from* $2^X \setminus C_1(X)$ *beginning with* K.

(12.30.3) *If* $A \in C_1(X)$ *and* $K \in [2^X \setminus C_1(X)]$, *then* A *is arcwise accessible from* $2^X \setminus C_1(X)$ *beginning with* K *if and only if* A *is irreducible about* K.

The next result, part of which is (12.29), also characterizes hereditary indecomposability.

(12.31) THEOREM [234, 5.5]. *For a continuum* X, *the following three statements are equivalent:*

(12.31.1) X *is hereditarily indecomposable.*

(12.31.2) *If* Γ *is any arcwise connected subset of* 2^X, *then* Γ ∩ C(X) *is arcwise connected.*

(12.31.3) *If Γ is any arcwise connected subset of 2^X, then $\Gamma \cap C(X)$ is connected.*

Proof. Assume X contains a decomposable subcontinuum Y. Let A and B be proper subcontinua of Y such that $Y = A \cup B$. Let $p \in [A \backslash B]$. By (1.25), there are three segments σ_1, σ_2, and σ_3 defined on [0, 1] satisfying: $\sigma_1(0) = A \cap B$ and $\sigma_1(1) = A$, $\sigma_2(0) = A \cap B$ and $\sigma_2(1) = B$, $\sigma_3(0) = B$ and $\sigma_3(1) = Y$. Define $f: [0, 3] \to 2^Y$ by

$$f(t) = \begin{cases} \sigma_1(1-t) \cup \{p\}, & 0 \le t \le 1 \\ \sigma_2(t-1) \cup \{p\}, & 1 \le t \le 2 \\ \sigma_3(t-2) \cup \{p\}, & 2 \le t \le 3 \end{cases}$$

By letting $\Gamma = f([0, 3])$, it is easy to see that $\Gamma \cap C(X)$ is not connected. This proves (12.31.3) implies (12.31.1). Hence, (12.31.2) implies (12.31.1). By (12.29), the other implications hold.

The following questions are partially motivated by (12.30).

(12.32) QUESTIONS. Given a continuum X and an $A \in C_1(X)$, what can be said about a $K \in [2^X \backslash C_1(X)]$ such that A is arcwise accessible from $2^X \backslash C_1(X)$ beginning with K? What can be said about the set of all such K? On the other hand, given a $K \in [2^X \backslash C_1(X)]$, what can be said about an $A \in C_1(X)$ such that A is arcwise accessible from $2^X \backslash C_1(X)$ beginning with K? What can be said about the set of all such A? One such A, for example, is a continuum which is irreducible about K.

We have now given most of the information which is known about arcwise accessibility in hyperspaces. However, there are a few results which have not been discussed (notably [234, 4.14.1 and 4.14.2]). The reader is referred to [234] for any omissions.

(12.33) CONCLUDING COMMENTS. The topic covered in this chapter was begun when Joe Stiles, then a graduate student at Tulane University, asked me the following question [234, 1.1]: Is each member of $C_1(X)$ segmentwise accessible from $2^X \setminus C_1(X)$? Formally, the question is easy to answer (in the negative)--we did so in the *proof* of (12.4). However, slight variations on the question open many avenues for investigation. When I learned from Stiles that he was not going to pursue his original question, I decided to look at it. I first obtained the simple fact that no singleton can be segmentwise accessible. Then I proved (12.2) and began investigating arcwise accessibility. I hope that other people will get interested in this, for there is much left to do. A related topic will be discussed in the next chapter.

NOTES

1. J. Grispolakis and E. D. Tymchatyn have generalized (12.13) to Suslinian continua. They have also shown that the answer to (12.19) is yes for chainable continua and for Suslinian continua. They have answered (12.20), hence (12.21), affirmatively. In relation to (12.23) and (12.25), they have shown that $AA[X] = X$ for rational continua of finite rim-type. Their paper, which will be entitled "Irreducible continua with degenerate end-tranches and arcwise accessibility," is currently under preparation. See Footnote 2 on p.372.

2. The construction in (12.16) is a modification of one used by Janiszewski in "Über die Begriffe Linie und Fläche," Proc. Cambridge Internat. Congr. Math., 2(1912), pp.126-128. See (12.18) for other historical comments. With respect to the construction in (12.16), it is important to use half-lines closing down on both sides of an arc [see X_1 , etc.]. If just one half-line were used, then it may be that each singleton is arcwise accessible.

XIII.

CONTINUUMWISE ACCESSIBILITY

In (12.2) we saw that every nondegenerate member of $C_1(X)$ is arcwise accessible from $2^X \backslash C_1(X)$. We also saw that, for some continua X, no singleton is arcwise accessible from $2^X \backslash C_1(X)$ [see (12.9) and (12.10)]. Thus, the following question arises: Is every singleton continuumwise accessible from $2^X \backslash C_1(X)$ for any continuum X? In (13.1) we will show that the answer is "yes;" in fact, we will show even more. First, we give the following definition [234, 6.8]: Let $\Sigma \subset 2^X$. A member A of $C_1(X)$ is said to be *continuumwise accessible from* $\Sigma \backslash C_1(X)$ provided there is a nondegenerate subcontinuum Γ of Σ such that $\Gamma \cap C_1(X) = \{A\}$.

(13.1) THEOREM [234, 6.8.1]. *Let* X *be a continuum. Then any member of* $C_1(X)$ *is continuumwise accessible from* $C_2(X) \backslash C_1(X)$. *Furthermore, given any* $A \in [C_1(X) \backslash \{X\}]$, *there exists a nondegenerate subcontinuum* Γ *of* $C_2(X)$ *such that the following two statements hold:*

(13.1.1) $\Gamma \cap C_1(X) = \{A\}$;

(13.1.2) Γ *is a monotone continuous image of* X. *Moreover, if* $A \in F_1(X)$, Γ *may be chosen as above and so as to be homeomorphic to* X.

Proof. Assume $A \in [C_1(X)\setminus\{X\}]$. Let $\Gamma \subset 2^X$ be given by $\Gamma = \{A \cup \{x\}: x \in X\}$. Define $f: X \xrightarrow{\text{onto}} \Gamma$ by $f(x) = A \cup \{x\}$ for each $x \in X$. Then, letting d denote the metric for X and H_d the Hausdorff metric for 2^X induced by d [specifically as defined in (0.1)], it follows easily that

$$H_d[f(x), f(y)] \leq d(x, y)$$

for all $x, y \in X$. Hence, f is continuous. Also, since for $G = [A \cup \{x\}] \in \Gamma$

$$f^{-1}(G) = \begin{cases} A, & \text{if } x \in A \\ \{x\}, & \text{if } x \notin A, \end{cases}$$

f is monotone. Note that Γ is nondegenerate since $A \neq X$. Thus, we have proved that (13.1.1) and (13.1.2) each hold. Moreover, if $A \in F_1(X)$, f is one-to-one and, therefore, a homeomorphism. This proves all the theorem for the case when $A \in [C_1(X)\setminus\{X\}]$. But, when $A = X$, A is arcwise accessible from $C_2(X)\setminus C_1(X)$ by (12.2). So, all of (13.1) is proved.

The proof of (13.1) gives rise to some other facts and some questions.

Assume a continuum X and an $A \in [C_1(X)\setminus\{X\}]$ are given. Let Γ_A denote the Γ in the proof of (13.1), i.e.,

$$\Gamma_A = \{A \cup \{x\}: x \in X\} \subset 2^X.$$

On the other hand, let X/A denote the upper semi-continuous decomposition of X whose only nondegenerate member is A. Define $g: X/A \xrightarrow{\text{onto}} A$ by

$$g(D) = A \cup D \text{ for each } D \in [X/A].$$

From the definition of the decomposition topology for $X/_A$ and the definition of the topology for 2^X, it is clear that g is a homeomorphism. Thus, though the decomposition topology is in general smaller than the Hausdorff metric topology relativized to the elements of the decomposition as they sit by inclusion in 2^X [see (1.77)], we have a specific model in 2^X of a particular upper semi-continuous decomposition of X. In general, the way an upper semi-continuous decomposition of X is shown to be metrizable is by appealing to the Urysohn metrization theorem (see, for example, the proof in [338, pp.123-125]). It would be interesting if one could find a way to product a *canonical* picture in 2^X of any upper semi-continuous decomposition of X, and thus have another (perhaps, clearer) way of visualizing the metric for the decomposition space. Of course, since 2^X contains a Hilbert cube [by (1.95)], any such decomposition space is embeddable in 2^X.

Let us say that a nondegenerate subcontinuum Γ of 2^X is a *touching set* provided $\Gamma \cap C_1(X) = \{A\}$ for some $A \in C_1(X)$, in which case Γ is said to be a *touching set at* A [comp., the paragraph following (12.29)]. Thus, when A is nondegenerate, there is always an *arc* which is a touching set at A [by (12.2)]; however, when $A \in F_1(X)$, this may not be the case [for example, see (12.9)]. Nonetheless, there are always touching sets [by (13.1)]. In the proof of (13.1), we produced some touching sets which, even when A is nondegenerate, are different from arcs in most cases. The following question is motivated by the proof of (13.1):

(13.2) QUESTION [234, 6.10]. What monotone upper semi-continuous decompositions of X can be touching sets?

Of course, there is the following general question:

(13.3) QUESTION [234, 6.9]. What subcontinua of 2^X can be touching sets? It might be true that if A is nondegenerate, any nondegenerate continuum is a touching set at A. This would certainly be true if one could produce a Hilbert cube which is a touching set at A.

Of particular interest with respect to (13.2) and (13.3) is the case when X is hereditarily indecomposable. Some observations about this were made immediately following (12.29).

Another special case of interest with respect to (13.2) and (13.3) is when the touching takes place at a member of $F_1(X)$.

Except for results in [234], most of which have been presented here in Chapters XII and XIII, no other work has been done on touching sets--the subject is virtually untouched (sorry about that!).

XIV.
WHITNEY PROPERTIES

Let P be a topological property. Then P is said to be a *Whitney property* [179] provided that whenever a continuum X has property P, so does $\mu^{-1}(t)$ for each Whitney map μ for $C(X)$ and each t, $0 \le t < \mu(X)$. The reader is referred to (0.50) for the definition of Whitney map and the symbol μ. In particular, when the symbol μ is used with no further comment, μ denotes *any* given Whitney map for $C(X)$.

A.
Basic Whitney Properties

The notion of Whitney property was not formalized until recently. However, some earlier results on what are now called Whitney properties led to the isolation of the concept. We begin this section by giving these earlier results (using the language introduced above). There are essentially nine of them, given below as (14.1) through (14.9). The first of them is contained in some proofs in [165, section 8] (it was explicitly stated in [93, p.1032]):

(14.1) THEOREM [165]. *The property of being an hereditarily indecomposable continuum is a Whitney property [for a "converse" result, see (14.54)].*

Proof. Let X be an hereditarily indecomposable continuum and let $0 \leq t < \mu(X)$. By (1.80.2), $\mu^{-1}(t)$ is a monotone continuous image of X and, by part (a) of (1.213.1), $\mu^{-1}(t)$ is nondegenerate. Hence, $\mu^{-1}(t)$ is a continuum [this is also true in general but by a different proof--see (14.2)]. Therefore, since $\mu^{-1}(t)$ is a monotone continuous image of X, (14.1) follows by using (1.207.1).

After Kelley's paper [165] there were no papers with results about the sets $\mu^{-1}(t)$ until [93]. In [93], two properties were shown to be Whitney properties (see the next two theorems).

(14.2) THEOREM [93, p.1032]. *The property of being a continuum is a Whitney property [comp., (14.61) where it is shown that Whitney maps for 2^X need not be monotone].*

Before giving the proof of (14.2), we make some comments. In the proof of (14.1), we showed $\mu^{-1}(t)$ is a continuum by showing that it is a continuous image of X. It is not always true that there is a mapping from X onto $\mu^{-1}(t)$ --see (14.60). Thus, the proof of (14.2) must be done with a different method than that used in the proof of (14.1).

Proof of (14.2). Let X be a continuum and let $0 \leq t < \mu(X)$. If $t = 0$, then $\mu^{-1}(t) = F_1(X) \cong X$. So, for the purpose of proof, assume $t \neq 0$. Then, it follows easily from (1.8) and (1.11) that $\mu^{-1}([0, t])$ is the union of all order arcs in $C(X)$ beginning with a member of $F_1(X)$ and ending with a member of $\mu^{-1}(t)$. Hence, $\mu^{-1}([0, t])$ is connected. Thus, since $\mu^{-1}([0, t])$ is compact [by the continuity of μ and (0.8)],

(1) $\mu^{-1}([0, t])$ is a subcontinuum of $C(X)$.

Also, by (1.8) and (1.11), it follows easily that $\mu^{-1}([t, \mu(X)])$ is the union of all order arcs in $C(X)$ beginning with a member of $\mu^{-1}(t)$ and ending with X. Hence, $\mu^{-1}([t, \mu(X)])$ is arcwise connected. Thus, since $\mu^{-1}([t, \mu(X)])$ is compact [by the continuity of μ and (0.8)],

(2) $\mu^{-1}([t, \mu(X)])$ is an (arcwise connected) subcontinuum of $C(X)$.

Clearly,

(3) $\mu^{-1}([0, t]) \cup \mu^{-1}([t, \mu(X)]) = C(X)$

and

(4) $\mu^{-1}([0, t]) \cap \mu^{-1}([t, \mu(X)]) = \mu^{-1}(t)$.

By (1) through (4) and (1.176), $\mu^{-1}(t)$ is connected. Clearly, $\mu^{-1}(t)$ is compact and, since $t < \mu(X)$, $\mu^{-1}(t)$ is nondegenerate by part (a) of (1.213.1). Therefore, $\mu^{-1}(t)$ is a continuum. This proves (14.2).

Note that (14.2) says that when one studies the sets $\mu^{-1}(t)$, one is studying continua. Thus, in a sense, (14.2) is basic to much of the material in this chapter.

(14.3) THEOREM [93, p.1032]. *The property of being a pseudo-arc is a Whitney property [for a "converse" result, see (14.54)].*

Proof. Let X be a pseudo-arc and let $0 \leq t < \mu(X)$. By (1.80.2), $\mu^{-1}(t)$ is a monotone continuous image of X and, by part (a) of (1.213.1), $\mu^{-1}(t)$ is nondegenerate. Bing [26, Theorem 4] proved that any nondegenerate monotone continuous image of a pseudo-arc is again a pseudo-arc. Hence, $\mu^{-1}(t)$ is a pseudo-arc.

The next results to appear about Whitney properties are the following four results in [177].

(14.4) THEOREM [177, 6.2(a)]. *The property of being a chainable continuum is a Whitney property.*

(14.5) THEOREM [177, 6.2(b)]. *The property of being a proper circle-like continuum [see (0.24)] is a Whitney property.*

(14.6) THEOREM [177, 6.4(a)]. *The property of being an arc is a Whitney property.*

Proof. See below.

For a more general result than (14.6), see (14.8). For a "converse" to (14.6), see (14.50).

(14.7) THEOREM [177, 6.4(b)]. *The property of being a circle is a Whitney property.*
Proof. See below.

For a "converse" to (14.7), see (14.51). Also, the reader should compare (14.6) and (14.7) with (14.41).

Krasinkiewicz's proofs in [177, pp.162-164] of (14.6) and (14.7) seem to be more complicated than is necessary. Simpler proofs were given in [249, section 2], which we now present here. Also, we mention that another proof of (14.6) is indicated in (14.72.2).

Proof of (14.6) and (14.7). [249, pp.397-398]. We prove (14.6) and (14.7) simultaneously. We will first prove the results for the unit interval [0, 1] and the unit circle S^1. Let μ denote a Whitney map for $C(X)$ where $X = [0, 1]$ or S^1. Let $0 < t < \mu(X)$. Note that each member of $\mu^{-1}(t)$ is an arc, since it must be a nondegenerate proper subcontinuum of X. Define $m: \mu^{-1}(t) \to X$ by

$m(A) = $ the midpoint of A

for each $A \in \mu^{-1}(t)$. It is easy to see that m is continuous. Also, as we now show, m is one-to-one. Assume $A_1, A_2 \in \mu^{-1}(t)$ such that $m(A_1) = m(A_2)$. Then, A_1 and A_2 are arcs in X with the same midpoint. Hence, as is geometrically obvious, one of them must be contained in the other. Therefore, since $\mu(A_1) = \mu(A_2)$, we have $A_1 = A_2$ by (0.50.a). So, m is one-to-one and, thus, a homeomorphism of $\mu^{-1}(t)$ into X. Since $\mu^{-1}(t)$ is a continuum [by (14.2)], we are done in the case when $X = [0, 1]$ (note that m does *not map onto* $[0, 1]$). Now, assume $X = S^1$. We will be done in this case if we show m maps *onto* S^1. Let $p \in S^1$. Let

$$\Lambda(p) = \{\{p\}, S^1\} \cup A(p)$$

where

$$A(p) = \{A \in C(S^1) : A \text{ is an arc with midpoint } p\}.$$

It is easy to see that $\Lambda(p)$ is an arc with end points $\{p\}$ and S^1. Therefore, since $\mu(\{p\}) = 0$ [by (0.50.b)] and $\mu(S^1) = \mu(X)$, it follows from the continuity of μ that there exists $A_o \in A(p)$ such that $\mu(A_o) = t$. Clearly, $m(A_o) = p$. This proves that m maps $\mu^{-1}(t)$ onto S^1, and we are done in the case when $X = S^1$. To complete the proof of (14.6) and (14.7), it suffices to prove the following general fact:

(#) *If* Y *and* Z *are homeomorphic continua and if* $\mu_Y^{-1}(t)$ *has a certain topological property* P *for all Whitney maps* μ_Y *for* $C(Y)$ *and each* $t \in [0, \mu_Y(Y))$, *then* $\mu_Z^{-1}(t)$ *has property* P *for all Whitney maps* μ_Z *for* $C(Z)$ *and each* $t \in [0, \mu_Z(Z))$.

To prove (#), let h: $Y \xrightarrow{\text{onto}} Z$ be a homeomorphism and let μ_Z be any given Whitney map for $C(Z)$. Let $\hat{h}: C(Y) \to C(Z)$ be the induced map [as defined in (0.49)]. Since h is a homeomorphism, it is easy to see that μ_Y, given by $\mu_Y = \mu_Z \circ \hat{h}$, is a Whitney map for $C(Y)$ [for a more general result, see (0.51)]. Since, as is easy to see, $\hat{h}$ is a homeomorphism of $C(Y)$ onto $C(Z)$, it follows easily that

$$\mu_Z^{-1}(t) = \hat{h}[\mu_Y^{-1}(t)] \text{ and, thus, } \mu_Z^{-1}(t) \cong \mu_Y^{-1}(t)$$

for each $t \in [0, \mu_Z(Z)] = [0, \mu_Y(Y)]$. The result in (#) now follows. This completes the proof of (14.6) and (14.7).

Rogers [206, pp.574-575] also gave proofs [of (14.6) and (14.7)] which are different than the ones in [177, pp.162-164]. Rogers' proofs of these results are interesting in that they use some facts about the topology of the plane. However, the "simplest" proofs seem to be the ones just given; also, see (14.72.2).

After [177], the next paper to appear about Whitney properties was [249]. We now state and give proofs for the two main results in [249].

(14.8) THEOREM [see (14.10)]. *The property of being an arcwise connected continuum is a Whitney property [see (14.75)].*

Proof [249, pp.398-400]. First, we prove the following lemma [see (14.32) for a more general result]:

(14.8.1) LEMMA [see (14.10)]. *Let X be any continuum and let $0 < t_o < \mu(X)$. If $A, B \in \mu^{-1}(t_o)$ such that $A \cap B \neq \emptyset$ and $A \neq B$, then there is an arc*

$$\alpha \subset [\mu^{-1}(t_o) \cap C(A \cup B)]$$

such that α has end points A and B. Moreover [249, proof of Lemma 1] if K is any given component of $A \cap B$, then α may be chosen as above so that $K \subset L$ for each $L \in \alpha$.

Proof of (14.8.1) [249]. Assume, for convenience, that $\mu(X) = 1$; as the reader can easily see from (1.16), there is no loss of generality in doing this. Now, let K denote a component of A ∩ B. By (1.25) and (1.26), there exist segments $\sigma_1 \colon [0, 1] \to C(A)$ from K to A and $\sigma_2 \colon [0, 1] \to C(B)$ from K to B. Fix $t \in [0, 1]$. Since

$$\mu[\sigma_1(t) \cup \sigma_2(0)] = \mu[\sigma_1(t)] \le \mu(A) = t_o$$

and

$$\mu[\sigma_1(t) \cup \sigma_2(1)] = \mu[\sigma_1(t) \cup B] \ge \mu(B) = t_o,$$

it follows that there exists an $s \in [0, 1]$ such that

(#) $\quad \mu[\sigma_1(t) \cup \sigma_2(s)] = t_o.$

For any given $t \in [0, 1]$ there can be many choices for an s satisfying (#)--let s_t denote one such choice for each t. Then, given $t \in [0, 1]$ we have

$$\mu[\sigma_1(t) \cup \sigma_2(s_t)] = t_o.$$

Define $g \colon [0, 1] \to \mu^{-1}(t_o)$ by

$$g(t) = \sigma_1(t) \cup \sigma_2(s_t)$$

for each $t \in [0, 1]$. We will prove that g is continuous. To do this fix $t \in [0, 1]$ and let $\{t(n)\}_{n=1}^\infty$ be a sequence in $[0, 1]$ such that $t(n) \to t$ as $n \to \infty$. Since the corresponding sequence $\{s_{t(n)}\}_{n=1}^\infty$ has a convergent subsequence, assume without loss of generality that $\{s_{t(n)}\}_{n=1}^\infty$ itself converges and denote its limit by r. By the continuity of σ_1 and σ_2 [see (1.15.1)] and the definition of g we have

(i) $\quad g(t(n)) = \sigma_1(t(n)) \cup \sigma_2(s_{t(n)}) \to \sigma_1(t) \cup \sigma_2(r)$

as $n \to \infty$.

Thus, since $\mu^{-1}(t_o)$ is compact and $\mu[g(t(n))] = t_o$ for each $n = 1, 2, \ldots,$

(ii) $\mu[\sigma_1(t) \cup \sigma_2(r)] = t_o$.

Observe that, by definition of g,

(iii) $[\sigma_1(t) \cup \sigma_2(r)] \subset g(t)$ if $r \leq s_t$,

(iv) $g(t) \subset [\sigma_1(t) \cup \sigma_2(r)]$ if $s_t \leq r$.

Hence, by (ii) through (iv), we see that $\sigma_1(t) \cup \sigma_2(r)$ and $g(t)$ are subcontinua of X with the same μ-value such that at least one of them is contained in the other. Hence [by (0.50.a)], they are equal, i.e., $g(t) = \sigma_1(t) \cup \sigma_2(r)$. Therefore, by (i),

(v) $g(t(n)) \to g(t)$ as $n \to \infty$.

This completes the proof that g is continuous. It is easy to verify that $g(0) = B$ and $g(1) = A$. Thus, (14.8.1) now follows since there is an arc $\alpha \subset g([0, 1]) \subset \mu^{-1}(t_o)$ with end points A and B.

Proof of (14.8) [249]. Assume X is an arcwise connected continuum. Let μ be a Whitney map for $C(X)$ and let $0 < t_o < \mu(X)$. By (14.2), $\mu^{-1}(t_o)$ is a continuum. Hence, it suffices to show $\mu^{-1}(t_o)$ is arcwise connected. Let $A, B \in \mu^{-1}(t_o)$. If $A \cap B \neq \emptyset$ and $A \neq B$, then there is an arc in $\mu^{-1}(t_o)$ with end points A and B by (14.8.1). Hence, for the purpose of proof, assume $A \cap B = \emptyset$. Since X is arcwise connected, there is an arc γ in X from a point $a \in A$ to a point $b \in B$. We consider two cases.

Case I: $\mu(\gamma) \leq t_o$. Then, using (1.8) and (1.11) together with the continuity of μ, it follows easily that there is a subcontinuum B_o of B such that $b \in B_o$ and $[\gamma \cup B_o] \in \mu^{-1}(t_o)$. Now (14.8.1) can be applied twice, once to A and $\gamma \cup B_o$ and once to $\gamma \cup B_o$ and B to see that there is an arc in $\mu^{-1}(t_o)$ with end points A and B.

Case II: $\mu(\gamma) > t_o$. Then, there are subarcs γ_a and γ_b of γ such that $a \in \gamma_a$, $b \in \gamma_b$, and γ_a, $\gamma_b \in \mu^{-1}(t_o)$. Since the restriction of μ to $C(\gamma)$ is a Whitney map for $C(\gamma)$, we have by (14.6) that there is an arc in $\mu^{-1}(t_o) \cap C(\gamma)$ with end points γ_a and γ_b. Assuming $A \neq \gamma_a$, (14.8.1) can be applied to give an arc in $\mu^{-1}(t_o)$ with end points A and γ_a. Assuming $\gamma_b \neq B$, (14.8.1) can be applied to give an arc in $\mu^{-1}(t_o)$ with end points γ_b and B. It follows that there is an arc in $\mu^{-1}(t_o)$ with end points A and B. This completes the proof of Case II and of (14.8).

For a "converse" to the following theorem, see (14.47).

(14.9) THEOREM [249, Theorem 3]. *The property of being a locally connected continuum is a Whitney property* [*see (14.75)*].

Proof. Assume X is a locally connected continuum. Let $0 < t_o < \mu(X)$. By (14.2), $\mu^{-1}(t_o)$ is a continuum. Hence, it suffices to prove that $\mu^{-1}(t_o)$ is locally connected. Let $A_o \in \mu^{-1}(t_o)$ and let $\{A_n\}_{n=1}^{\infty}$ be any sequence of members A_n of $\mu^{-1}(t_o) \setminus \{A_o\}$ such that $A_n \to A_o$ as $n \to \infty$. Then, by using the local connectivity of X and going to a subsequence of $\{A_n\}_{n=1}^{\infty}$ if necessary, there exist arcs γ_n in X such that, for each $n = 1, 2, \ldots,$

(i) $\gamma_n \cap A_n \neq \emptyset \neq \gamma_n \cap A_o$

and the diameter of γ_n is small enough so that

(ii) $\mu(\gamma_n) < [2^{-n}] \cdot t_o$.

For each $n = 1,2,\ldots$, let $X_n = A_n \cup \gamma_n \cup A_o$. Note that, by (ii), $\mu(\gamma_n) < t_o$ for each $n = 1,2,\ldots$. Thus, by using (14.8.1) at most twice, it follows that there are arcs

$$\Gamma_n \subset [\mu^{-1}(t_o) \cap C(X_n)]$$

with end points A_n and A_o for each $n = 1,2,\ldots$. Since $A_n \to A_o$ and [by (ii)] $\text{diam}[\gamma_n] \to 0$ as $n \to \infty$, it follows easily that $X_n \to A_o$. Now we show that

(*) $C(X_n) \cap \mu^{-1}(t_o) \to \{A_o\}$ as $n \to \infty$.

To prove (*), first note that $A_o \in [C(X_n) \cap \mu^{-1}(t_o)]$ for each $n = 1,2,\ldots$. Hence,

(a) $A_o \in \liminf[C(X_n) \cap \mu^{-1}(t_o)]$.

Now, let $B \in \limsup[C(X_n) \cap \mu^{-1}(t_o)]$. Then there is a sequence $\{B_i\}_{i=1}^{\infty}$ converging to B, where

$$B_i \in [C(X_{n(i)}) \cap \mu^{-1}(t_o)]$$

for each $i = 1,2,\ldots$. Since $B_i \in \mu^{-1}(t_o)$ for each $i = 1,2,\ldots$, $B \in \mu^{-1}(t_o)$. Since $X_{n(i)} \to A_o$ as $i \to \infty$ and since $B_i \subset X_{n(i)}$ for each $i = 1,2,\ldots$, $B \subset A_o$. So, $\mu(B) = t_o = \mu(A_o)$ and $B \subset A_o$. Thus, by (0.50.a), $B = A_o$. This proves

(b) $\limsup[C(X_n) \cap \mu^{-1}(t_o)] \subset \{A_o\}$,

and (*) now follows from (a) and (b). It follows easily from (*) that $\text{diam}[\Gamma_n] \to 0$ as $n \to \infty$. Therefore, since A_o was an arbitrary member of $\mu^{-1}(t_o)$ and since $\{A_n\}_{n=1}^{\infty}$ was an arbitrary sequence in $\mu^{-1}(t_o) \setminus \{A_o\}$ converging to A_o, it follows that $\mu^{-1}(t_o)$ is locally connected.

(14.10) REMARK. The results in (14.8) and (14.8.1) appeared in [249] and in [286]. The proofs in [286] of these results seem to contain a gap (compare lines 5-7 in the proof of Theorem 3.6 of [286] with the italicized statement following (#) in the proof of (14.8.1) above). It may be that this presumed gap in proofs in [286] can be filled in by a careful use of the Moore Theorem [186, p.533]. One would have to show that the Moore Theorem is applicable and that, after going to the decomposition space, the rest of the proof in [286]--including the last sentence in the proof of Theorem 3.7 of [286]--can be carried out.

We have now presented the results that led to formalizing the concept "Whitney property" in [179]. As we see, there are two categories:

G-general properties

| (14.1) | (14.4) | (14.8) |
| (14.2) | (14.5) | (14.9) |

T-invariance of topological type

| (14.3) | (14.6) | (14.7) |

Next we give some more Whitney properties in category G, and then some more in category T [with respect to category T, see (14.39.1)].

As we saw in (14.1), the property of being hereditarily indecomposable is a Whitney property. As is easy to see, the property of being hereditarily decomposable is not a Whitney property. However:

(14.11) **THEOREM** [179, 3.4]. *The property of being a decomposable continuum is a Whitney property.*

Proof. First we prove the following lemma.

(14.11.1) **LEMMA** [179, 3.2]. *Let* X *be any continuum, let* $A \in C(X)$, *and let* $0 \le t_o \le \mu(X)$. *Let* [*as in (0.68.5)*]
$$X(A, \mu, t_o) = \{M \in \mu^{-1}(t_o) : M \cap A \ne \emptyset\}.$$
Then, $X(A, \mu, t_o)$ *is a subcontinuum of* $\mu^{-1}(t_o)$. *Furthermore:*

(1) *If* $\mu(A) \le t_o$, *then* $X(A, \mu, t_o)$ *is an arcwise connected subcontinuum (perhaps degenerate) of* $\mu^{-1}(t_o)$;

(2) *If* $\mu(A) > t_o$, *then* $\Lambda = \mu^{-1}(t_o) \cap C(A)$ *is a subcontinuum[1] of* $X(A, \mu, t_o)$ *and each member (if any) of* $X(A, \mu, t_o) \setminus \Lambda$ *can be joined to a member of* Λ *by an arc* $\alpha \subset X(A, \mu, t_o)$.

Proof of (14.11.1). By (0.68.5), $X(A, \mu, t_o)$ is compact. Hence, the fact that $X(A, \mu, t_o)$ is a subcontinuum of $\mu^{-1}(t_o)$ follows once we prove (1) and (2). To prove (1), assume $\mu(A) \le t_o$. By using (1.8) and (1.11) [if necessary, i.e., if $\mu(A) < t_o$], we see that there exists $M_o \supset A$ such that $M_o \in \mu^{-1}(t_o)$. Note that $M_o \in X(A, \mu, t_o)$. Now, let $M \in X(A, \mu, t_o)$ such that $M \ne M_o$. Let $p \in [M \cap A]$. Then, since $M_o \supset A$, $p \in [M \cap M_o]$. Hence, since $M, M_o \in \mu^{-1}(t_o)$ and $M \ne M_o$, it follows easily from (14.8.1) that there is an arc γ in $\mu^{-1}(t_o)$ with end points M and M_o such that $p \in L$ for each $L \in \gamma$. Since $p \in A$, $\gamma \subset X(A, \mu, t_o)$. Since M was an arbitrary member of $X(A, \mu, t_o)$ different from M_o, we have proved (1). To prove (2), assume $\mu(A) > t_o$, and let $\Lambda = \mu^{-1}(t_o) \cap C(A)$. Let μ' denote the restriction of μ to $C(A)$. Note that

(*) μ' is a Whitney map for $C(A)$ and $\Lambda = (\mu')^{-1}(t_o)$.

It follows easily using (*) and (14.2) that Λ is a subcontinuum of $X(A, \mu, t_o)$. Now, let $M_1 \in [X(A, \mu, t_o) \setminus \Lambda]$. Since $M_1 \in X(A, \mu, t_o)$, $M_1 \cap A \neq \emptyset$. It follows from (*) that $\cup \Lambda = A$. Hence, there exists $M_2 \in \Lambda$ such that $M_1 \cap M_2 \neq \emptyset$. Let $p \in [M_1 \cap M_2]$. By using (14.8.1) [comp., the proof of part (1) above], it follows that there is an arc in $X(A, \mu, t_o)$ from M_1 to M_2. Therefore, since $M_2 \in \Lambda$, this proves (2) and completes the proof of (14.11.1).

Proof of (14.11). Let X be a decomposable continuum and let $0 \leq t_o < \mu(X)$. Since $\mu^{-1}(t_o)$ is a continuum [by (14.2)], it suffices to show $\mu^{-1}(t_o)$ is decomposable. Let A and B be proper subcontinua of X such that $X = A \cup B$. Let

(a) $\begin{cases} X(A, \mu, t_o) = \{K \in \mu^{-1}(t_o): K \cap A \neq \emptyset\} \\ X(B, \mu, t_o) = \{K \in \mu^{-1}(t_o): K \cap B \neq \emptyset\}. \end{cases}$

Since $X = A \cup B$, it is clear that

$\mu^{-1}(t_o) = X(A, \mu, t_o) \cup X(B, \mu, t_o)$.

Also, by (14.11.1), $X(A, \mu, t_o)$ and $X(B, \mu, t_o)$ are subcontinua of $\mu^{-1}(t_o)$. Hence, if

$X(A, \mu, t_o) \neq \mu^{-1}(t_o) \neq X(B, \mu, t_o)$,

we are done. However, it can be that one of the sets in (a) is all of $\mu^{-1}(t_o)$ [see (1.213.4)]. So, for the purpose of proof, assume (without loss of generality) that

(#) $X(A, \mu, t_o) = \mu^{-1}(t_o)$.

Consider the case when $\mu(A) \leq t_o$. Then, by part (1) of (14.11.1), $X(A, \mu, t_o)$ is arcwise connected. Hence, by (#), $\mu^{-1}(t_o)$ is an arcwise connected continuum and, therefore [186, p.212], decomposable. Next, consider the case when $\mu(A) > t_o$. Let Λ be as in part (2) of (14.11.1). Then, since $A \neq X$, $\Lambda \neq \mu^{-1}(t_o)$. Hence, by (#) and part (2) of (14.11.1), it follows from [186, p.212] that $\mu^{-1}(t_o)$ is decomposable. This completes the proof of (14.11).

Though both decomposability and hereditary indecomposability are Whitney properties [(14.11) and (14.1), respectively], indecomposability is not. Some theorems illustrating that indecomposability is not a Whitney property for circle-like continua are given in [179] and in [286, 5.3 and 5.4]. By using a general result [stated here in (14.34)], the authors of [179] also obtained many non-circle-like examples. They [179] also determined very specific information about certain indecomposable circle-like continua. Let us take a particular circle-like example (comp., [286, 5.3 and 5.4]):

(14.12) EXAMPLE [179, 5.4]. Let X be a pseudo-arc and let $Y = X/_{\{a,b\}}$ be the continuum obtained by identifying two points a and b in different composants of X. Note that Y is an indecomposable circle-like continuum. Let μ_o be any Whitney map for $C(Y)$. Let $0 < t < \mu_o(Y)$. Let $\nu: X \xrightarrow{\text{onto}} Y$ be the natural map. Let $\hat{\nu}: C(X) \to C(Y)$ be the induced map as defined in (0.49). Since
$$\nu^{-1}[\nu(x)] = \{x\} \text{ for all } x \in [X\setminus\{a, b\}],$$
it follows [see (0.51)] that $\mu_1 = \mu_o \circ \hat{\nu}$ is a Whitney map for $C(X)$.

Hence, by (14.3),

(i) $\mu_1^{-1}(t)$ is a pseudo-arc.

By (i) above and (4) of (14.34),

(ii) $\mathrm{cl}[\mu_o^{-1}(t)\setminus\{M \in \mu_o^{-1}(t): \{a, b\} \in M\}]$ is a pseudo-arc.

Furthermore, by (6) of (14.34),

(iii) $\{M \in \mu_o^{-1}(t): \{a, b\} \in M\}$ is an arc with end points $\hat{\nu}(A_t)$ and $\hat{\nu}(B_t)$, where A_t and B_t are the unique members of $\mu_1^{-1}(t)$ such that $a \in A_t$ and $b \in B_t$; also,

$$\{\hat{\nu}(A_t), \hat{\nu}(B_t)\} = \mathrm{Bd}_{\mu_o^{-1}(t)} [\{M \in \mu_o^{-1}(t): \{a, b\} \in M\}]$$

where $\mathrm{Bd}_{\mu_o^{-1}(t)}$ denotes boundary relative to $\mu_o^{-1}(t)$.

It follows easily from (ii) and (iii) that $\mu_o^{-1}(t)$ is homeomorphic to the union of a pseudo-arc P and an arc A such that A ∩ P consists of exactly two points which are end points of A and which lie in distinct composants of P. Thus, we have determined the topological type of each of the levels $\mu_o^{-1}(t)$ where μ_o is *any* Whitney map for C(Y) [see (14.25)]. Some applications of this are in the paragraph following (14.45). For now we note that, since Y is indecomposable and circle-like, unicoherence and indecomposability are not Whitney properties even for circle-like continua.

Though indecomposability is not a Whitney property for circle-like continua [see (14.12)], it *is* a Whitney property *for chainable continua.* This last fact, which we prove here in (14.13), was proved in [179]--it answered a question in [286, Question 2, p.584].

(14.13) THEOREM [179, 4.3]. *The property of being an indecomposable chainable continuum is a Whitney property.*

Proof. First, we prove the lemma in (14.13.1). This lemma shows that for chainable continua X, each $\mu^{-1}(t)$ is an "irreducible cover" of X by subcontinua of $\mu^{-1}(t)$. This "covering property" will be discussed extensively in section J of this chapter.

(14.13.1) LEMMA [179, 4.2]. *Let* X *be any chainable continuum, let* μ *be a Whitney map for* C(X), *and let* $t \in [0, \mu(X)]$. *If* Λ *is a subcontinuum of* $\mu^{-1}(t)$ *such that* $\cup \Lambda = X$, *then* $\Lambda = \mu^{-1}(t)$. *In other words:* $X \in CP$, *as defined in (14.14).*

Proof of (14.13.1). Let Λ be a subcontinuum of $\mu^{-1}(t)$ such that $\cup \Lambda = X$. Let $A \in \mu^{-1}(t)$. We will show $A \in \Lambda$. To do this, it suffices to prove

(*) given any $\varepsilon > 0$, there exists $L_\varepsilon \in \Lambda$ such that $H(A, L_\varepsilon) < \varepsilon$.

To prove (*), let $\varepsilon > 0$. A simple sequence argument shows that

(1) there exists $\eta = \eta(\varepsilon) > 0$ such that if $K, L \in \mu^{-1}(t)$ and $L \subset N(\eta, K)$, then $H(K, L) < \varepsilon$ [177, 2.3].

Let $\{U_1, U_2, \ldots, U_n\}$ be a chain [see (0.22)] of open subsets of X covering X such that

(2) $\text{diam}[U_i] < \eta$ for each $i = 1, 2, \ldots, n$.

Let $m = \min\{i: U_i \cap A \neq \emptyset\}$, $s = \max\{i: U_i \cap A \neq \emptyset\}$, and let

$\Lambda_1 = \{L \in \Lambda: L \cap U_i \neq \emptyset$ for some $i \leq m\}$,

$\Lambda_2 = \{L \in \Lambda: L \cap U_i \neq \emptyset$ for some $i \geq s\}$.

We consider two cases.

Case I: $\Lambda_1 \cup \Lambda_2 \neq \Lambda$. Then there exists $L_o \in \Lambda$ such that

(3) $L_o \subset [\bigcup_{i=m+1}^{s-1} U_i]$.

Since $\{U_1, U_2, \ldots, U_n\}$ is a chain and since A is a subcontinuum of X such that, $A \cap U_m \neq \emptyset \neq A \cap U_s$, it follows easily that

(4) $A \cap U_i \neq \emptyset$ for each $i = m+1, m+2, \ldots, s-1$.

It follows easily from (2), (3), and (4) that

(5) $L_o \subset N(\eta, A)$.

Since $L_o, A \in \mu^{-1}(t)$ and since (5) holds, we have by (1) that

(6) $H(A, L_o) < \varepsilon$.

Since $L_o \in \Lambda$ and (6) holds, $L_\varepsilon = L_o$ satisfies (*).

Case II: $\Lambda_1 \cup \Lambda_2 = \Lambda$. It follows easily from the definitions of Λ_1 and Λ_2 that Λ_1 and Λ_2 are open subsets of Λ and that, since $\cup \Lambda = X$, $\Lambda_1 \neq \emptyset \neq \Lambda_2$. Thus, since $\Lambda = \Lambda_1 \cup \Lambda_2$ is a continuum, we have that $\Lambda_1 \cap \Lambda_2 \neq \emptyset$. Let $K_o \in [\Lambda_1 \cap \Lambda_2]$. Then, there exists $i' \leq m$ and $j' \geq s$ such that $K_o \cap U_{i'} \neq \emptyset \neq K_o \cap U_{j'}$. Hence, since $\{U_1, U_2, \ldots, U_n\}$ is a chain covering X and since K_o is a subcontinuum of X,

(7) $K_o \cap U_i \neq \emptyset$ for each $i = i', i'+1, \ldots, j'$.

Therefore, since $A \subset [\bigcup_{i=m}^{s} U_i]$ and since $i' \leq m$ and $j' \geq s$, it follows easily using (2) and (7) that

(8) $A \subset N(\eta, K_o)$.

Since K_o, $A \in \mu^{-1}(t)$ and since (8) holds, we have by (1) that

(9) $H(A, K_o) < \varepsilon$.

Since $K_o \in \Lambda$ and (9) holds, $L_\varepsilon = K_o$ satisfies (*).

From what has been shown in Case I and Case II, we have proved (*). This completes the proof of (14.13.1). Now we use (14.13.1) to prove (14.13).

Proof of (14.13). Assume that X is an indecomposable chainable continuum. Let $0 < t < \mu(X)$. Let Γ_1 and Γ_2 be subcontinua of $\mu^{-1}(t)$ such that $\mu^{-1}(t) = \Gamma_1 \cup \Gamma_2$. Let $X_1 = \cup \Gamma_1$ and let $X_2 = \cup \Gamma_2$. By (1.49), X_1 and X_2 are subcontinua of X. Also

$$X_1 \cup X_2 = [\cup \Gamma_1] \cup [\cup \Gamma_2] = \cup \mu^{-1}(t) = X.$$

Thus, since X is indecomposable, $X_1 = X$ or $X_2 = X$. Assume without loss of generality that $X_1 = X$. Then, since X is chainable, we have by (14.13.1) that $\Gamma_1 = \mu^{-1}(t)$. It now follows easily from what we have shown, that $\mu^{-1}(t)$ is indecomposable. Since X is a chainable continuum, we have by (14.4) that $\mu^{-1}(t)$ is a chainable continuum. Therefore, we have proved (14.13).

Another proof of (14.13) is indicated in (14.72.4).

(14.14) REMARK. In (14.13.1) we proved that every chainable continuum has the following property (which was first investigated in [179]): A continuum X is said to have the *covering property*, written $X \in CP$, provided that no proper subcontinuum of $\mu^{-1}(t)$ covers X for any Whitney map μ for $C(X)$ and any $t \in [0, \mu(X)]$. We will return to the covering property in section J of this chapter. For now, let us note the following theorem; the first part is a simple consequence of (1.78) and the second part is not difficult to prove.

(14.14.1) THEOREM [179, section 6]. *If* X *is any hereditarily indecomposable continuum, then* $X \in CP$. *If* $X \in CP$, *then* X *is unicoherent.*

Let us also note the following fact, which follows easily from the last part of the proof of (14.13).

(14.14.2) THEOREM [267, Proposition 34]. *If* $X \in CP$ *and if* X *is an indecomposable continuum, then* $\mu^{-1}(t)$ *is an indecomposable continuum for each* t, $0 \le t < \mu(X)$. *Hence, indecomposability is a Whitney property for continua with the covering property.*

We have seen in (14.4) that chainability is a Whitney property. It is easy to see, using a simple triod, that being tree-like is not a Whitney property [286, 5.5]. However, Krasinkiewicz proved the following result:

(14.15) THEOREM [176, 4.2]. *The property of being an hereditarily indecomposable tree-like continuum is a Whitney property*

(14.17.1)   WHITNEY PROPERTIES   419

[see (14.17)].

Proof. Let X be an hereditarily indecomposable tree-like continuum and let $0 < t < \mu(X)$. By (14.1), $\mu^{-1}(t)$ is an hereditarily indecomposable continuum. Thus, we need to know $\mu^{-1}(t)$ is tree-like. By (1.80.2), $\mu^{-1}(t)$ is a monotone continuous image of X. In [287, Theorem 1], Rosen proved that any monotone continuous image of any tree-like continuum is tree-like. Hence, $\mu^{-1}(t)$ is tree-like and we have proved (14.15).

(14.16) REMARK. In the proof of (14.15), we used [287, Theorem 1]. Let us mention the following stronger result: Any *confluent* image of a tree-like continuum is tree-like [202]. We also mention that Krasinkiewicz gave a proof of [287, Theorem 1] in [176, pp.183-184].

(14.17) REMARK. Krasinkiewicz used (14.15) to give the following partial answer to a question in [93, Question 3]--(2.7) here:

(14.17.1) THEOREM [176, 4.1]. *If* X *is an hereditarily indecomposable tree-like continuum, then* $\dim[C(X)] = 2$.

Proof. By (14.15),
$$\dim[\mu^{-1}(t)] = \begin{cases} 1, & 0 \le t < \mu(X) \\ 0, & t = \mu(X) \end{cases}$$
where μ is a given Whitney map for $C(X)$. Hence, by (#) on p.215, we have that

$\dim[C(X)] \le 1 + 1 = 2$.

Therefore, by (2.1) [or (2.2)], dim[C(X)] = 2. This proves (14.17.1).

It is easy to see, (1.213.7), that planarity and nonplanarity are not Whitney properties. It will be seen in (14.43.6) that having a certain shape and having a certain Čech cohomology are not Whitney properties. As a trivial consequence of (14.4), all the above-mentioned properties are Whitney properties for chainable continua. The following result says they are also Whitney properties for circle-like continua.

(14.18) THEOREM [see each part]. *If* X *is a circle-like continuum, then:*

(14.18.1) *The shape of* X *is equal to the shape of* $\mu^{-1}(t)$ *whenever* $0 \leq t < \mu(X)$ *[178, 3.3(1)]*;

(14.18.2) *The Čech cohomology groups (over the integers) of* X *are the same as those of* $\mu^{-1}(t)$ *whenever* $0 \leq t < \mu(X)$ *[286, 4.7]*.

Hence, for the class of circle-like continua, the following two properties are Whitney properties:

(14.18.3) *planarity ([178, 3.3(2)] and [286, 4.10])*

(14.18.4) *nonplanarity ([178, 3.3] and [286, 4.8])*.

We mention that (14.18.3) answers a question raised in [177, p.164]. Also, the reader should compare (14.18.1) and (14.18.2) with (14.29) and (14.43.6).

(14.20)   WHITNEY PROPERTIES   421

The following result is a combination of some results in Ann Petrus' dissertation.

(14.19) THEOREM [267, Section 1]. *The following are Whitney properties: aposyndesis, finite aposyndesis, mutual aposyndesis, and semi-aposyndesis* [*see (14.74)*].

It is clear that $\mu^{-1}(t)$ does not have to be aposyndetic, etc. However, $C(X)$ is always aposyndetic whether X is or not [see (1.138)].

In the paragraph following (14.10), Whitney properties were divided into two categories. Since then, we have given more results about Whitney properties in category G. Now, we give two more Whitney properties in category T.

Rogers proved that if X is a solenoid, then $C(X) \cong \text{Cone}(X)$ [see (8.1)]. His homeomorphism, in the proof in [284, pp.167-168] makes the base and the vertex of the cone correspond, respectively, to the singletons $F_1(X) = \mu^{-1}(0)$ and to $X \in C(X)$ [as they must, by the first part of (8.7)]. However, his homeomorphism does not make all the other "levels," especially those near the "top," match up. In [179, 4.5], it was shown that *all* "levels" can be preserved with the *same* homeomorphism [see (14.20) below]. As a consequence, it follows that the property of being a particular solenoid is a Whitney property [see (14.21)].

(14.20) THEOREM [179, 4.5]. *Let Σ be any given solenoid* and let μ be a given Whitney map for $C(\Sigma)$ such that (without loss of generality) $\mu(\Sigma) = 1$. Let p: $\text{Cone}(\Sigma) \xrightarrow{\text{onto}} [0, 1]$ denote the standard projection mapping. Then, there exists a homeomorphism h: $\text{Cone}(\Sigma) \xrightarrow{\text{onto}} C(\Sigma)$ such that the diagram

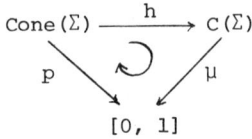

commutes. Furthermore, for each $x \in \Sigma$, $h((x, t_1)) \subset h((x, t_2))$ whenever $0 \leq t_1 \leq t_2 \leq 1$.

(14.21) COROLLARY [179, 4.5]. *The property of being any particular solenoid is a Whitney property.*

Proof. Use (14.20).

The proof, in [179], for (14.20) above is a delicate inverse limit procedure, which does not seem to be easily modified in order to obtain an affirmative answer to the following question:

(14.22) QUESTION. Let X be the "Buckethandle" continuum [defined in (1.209.3)]. Let μ be a Whitney map for $C(X)$ such that $\mu(X) = 1$. Is there a homeomorphism h: $\text{Cone}(X) \xrightarrow{\text{onto}} C(X)$ such that the diagram [where p is the standard projection mapping]

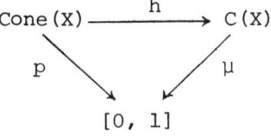

commutes? It is known that Cone(X) and C(X) are homeomorphic

[280, p.280], and that any cone-to-hyperspace homeomorphism must take the base and the vertex of the cone onto $\mu^{-1}(0)$ and $X \in C(X)$ respectively [by (8.7)].

A *pseudo-solenoid* (see [16, p.791] or [286, p.580]) is any hereditarily indecomposable circle-like continuum which is not chainable. A *pseudo-circle* is a planar pseudo-solenoid. It is known that there is only one pseudo-circle [104].

(14.23) THEOREM ([286, 4.11]; also, see (14.24) below). *The property of being any particular pseudo-solenoid is a Whitney property.*

Proof. Let X be a given pseudo-solenoid and let $0 \le t < \mu(X)$. By (14.1) and (14.5), $\mu^{-1}(t)$ is a pseudo-solenoid and, by (14.18.2), $\mu^{-1}(t)$ is cohomologically equivalent to X. Fearnley [103] has shown that cohomology determines the pseudo-solenoid. Hence, $\mu^{-1}(t) \cong X$.

For a "converse" to (14.23), see (14.54).

(14.24) REMARK. In particular, by (14.23), the property of being a *pseudo-circle* is a Whitney property. This fact is explicitly stated in [178, 3.4] and in the second part of [286, 4.11]. The more general fact in (14.23) is in [286, 4.11]--it is also deducible from [178, 3.3(1)], which is (14.18.1) above.

(14.25) REMARK. If X is any particular pseudo-solenoid, then results analogous to the one in (14.12) about the levels $\mu^{-1}(t)$ can be obtained by using (14.23).

B.
Some Non-Whitney Properties, Structure Results, and Questions

We have seen that a number of properties are Whitney properties. Now, we give some general theorems and examples which show that certain topological properties are not Whitney properties. We have already seen, in (14.12), that unicoherence and indecomposability are not Whitney properties.

In (14.5) we saw that the property of being proper circle-like is a Whitney property. However, the property of being circle-like is not. In [286, 5.1], Rogers proved the following result.

(14.26) THEOREM [286, 5.1]. *If X is a decomposable continuum which is both chainable and circle-like, then there exists* $t_o < \mu(X)$ *such that* $\mu^{-1}(t)$ *is not circle-like for any* $t > t_o$.

In the proof in [286] of (14.26), Rogers showed that $\mu^{-1}(t)$ is chainable (and not circle-like) when $t > t_o$. For his choice of t_o, $\mu^{-1}(t)$ can be a variety of chainable continua (see the proof in [286]). Let us note the following result [also, see (14.28) and (14.31)]:

(14.27) **THEOREM** [179, 3.5]. *If X is a decomposable continuum, then there exists $t_a < \mu(X)$ such that $\mu^{-1}(t)$ is arcwise connected for each $t \geq t_a$.*

Proof. Let A and B be proper subcontinua of X such that $X = A \cup B$. Let $t_a = \max\{\mu(A), \mu(B)\}$. Assume for the purpose of proof that $t_a = \mu(A)$. Let $t \geq t_a$. Let [as in (0.68.5)]

$$X(A, \mu, t) = \{K \in \mu^{-1}(t) : K \cap A \neq \emptyset\}.$$

By part (1) of (14.11.1), $X(A, \mu, t)$ is arcwise connected. Thus, we will have proved (14.27) once we show

(*) $X(A, \mu, t) = \mu^{-1}(t)$.

To prove (*), let $K \in \mu^{-1}(t)$. Suppose $K \cap A = \emptyset$. Then, since $X = A \cup B$ and $A \cap B \neq \emptyset$, $K \subset B$ and $K \neq B$. Hence, by (0.50.a), $\mu(K) < \mu(B)$. But,

$$\mu(K) = t \geq t_a \geq \mu(B).$$

Thus, we have a contradiction. Therefore, $K \cap A \neq \emptyset$. Hence, since $K \in \mu^{-1}(t)$, $K \in X(A, \mu, t)$. We have proved $\mu^{-1}(t) \subset X(A, \mu, t)$. Clearly, then, (*) holds. This completes the proof of (14.27).

As a consequence of (14.27), we obtain more definitive information than is in (14.26): Assume X is a decomposable chainable continuum and let t_a be as in (14.27). Let $t \geq t_a$ such that $t < \mu(X)$. Then, by (14.4), (14.27), and (0.23.1), $\mu^{-1}(t)$ is an arc [see (14.31)].

(14.28) **THEOREM** [179, 5.3]. *Let X be a continuum such that*

$X = X_1 \cup X_2$ *where* X_1 *and* X_2 *are proper subcontinua of* X *satisfying:*

(1) $X_1 \cap X_2$ *is a subcontinuum of* X;

(2) $X_1 \cap X_2$ *is terminal in both* X_1 *and* X_2;

(3) *each subcontinuum of* X *intersecting both* $X_1 \setminus [X_1 \cap X_2]$ *and* $X_2 \setminus [X_1 \cap X_2]$ *contains* $X_1 \cap X_2$.

Then, letting $t_o = \max\{\mu(X_1), \mu(X_2)\}$, $\mu^{-1}(t)$ *is an arc for each* $t \in (t_o, \mu(X))$.

We will not prove (14.28) here--it follows from the observation that each member of $\mu^{-1}(t)$ intersects $X_1 \cap X_2$ when $t > t_o$.

A specific example of a circle-like continuum X such that $\mu^{-1}(t)$ is not circle-like is easily obtained by letting X be the union of two copies of the "Buckethandle" [defined in (1.209.3)] such that the copies intersect only in the point $(0,0,\ldots,0,\ldots)$. Then, (14.27) or (14.28) may be applied to see that $\mu^{-1}(t)$ is an arc for large enough t [(14.27) may be applied by using the argument following the proof of (14.27)]. Thus we see that being circle-like is not a Whitney property, even though being proper circle-like is.

In (14.12), we saw that unicoherence is not a Whitney property. In [286, 5.6], Rogers gives another example of this. It is also true that non-unicoherence is not a Whitney property, as the following example shows.

(14.29) EXAMPLE [179, 5.5]. Let

$Y = \text{cl}\{(x, \sin[1/x]): 0 < x \leq 1\}$,

$Z = \text{cl}\{(x - 1, \sin[1/x]): -1 \leq x < 0\}$,

and let X be the continuum obtained from $Y \cup Z$ by identifying $(-1, -1)$ with $(0, -1)$ and $(-1, 1)$ with $(0, 1)$. Let μ be a Whitney map for $C(X)$. By taking X_1 and X_2 to be the subcontinua of X whose union is X and whose intersection is precisely the circle in X, we see that all the hypotheses of (14.28) are satisfied. Hence, letting t_o be as in (14.28), we see that $\mu^{-1}(t)$ is an arc for each $t \in (t_o, \mu(X))$. Thus, X is a non-unicoherent continuum such that $\mu^{-1}(t)$ is unicoherent for some values of t. Let us also note that X is a cyclic continuum with nontrivial shape such that eventually $\mu^{-1}(t)$ is an arc. The reader should compare this with (14.18.1) and (14.18.2).

Next, we give an example which shows that the fixed point property is not a Whitney property. This is the first such example. Recently, an example [see (14.43.6)] has been obtained which shows that the fixed point property is not a Whitney property for the class of absolute retracts.

(14.30) EXAMPLE [179, 5.6]. Let X be the Warsaw circle, i.e., X is the continuum in (8.22) denoted by S_2^1. It is well-known that X has the fixed point property [27, Theorem 13]. Let μ be a Whitney map for $C(X)$. It is not difficult to see that for large enough $t < \mu(X)$, $\mu^{-1}(t)$ is a circle [see (14.31)] and, hence, does not have the fixed point property. Let us note that not only does X have the fixed point property, but so does C(X) by (7.5); also, C(X) $\cong$ Cone(X) [see Chapter VIII, especially (8.23) and the paragraph preceding (8.30)]. For a general question about the fixed point property for $\mu^{-1}(t)$, see (7.13).

In (14.30) we observed that $\mu^{-1}(t)$ is a circle for large enough $t < \mu(X)$, where X was a particular decomposable proper circle-like continuum. Note the following general theorem.

(14.31) THEOREM [179, 4.4]. *If* x *is a decomposable chainable [resp., proper circle-like] continuum, then there exists* $t_o < \mu(X)$ *such that* $\mu^{-1}(t)$ *is an arc [resp., a circle] whenever* $t_o \le t < \mu(X)$.

The "chainable part" of (14.31) was proved in the paragraph preceding (14.28). The reader is referred to [179] for a proof of the "circle-like part" of (14.31). However, let us note that it can be seen (intuitively) why the "circle-like part" is true by using (14.27) and the structure of all arcwise connected circle-like continua given in [245, Theorem 6]. The proof in [179] is along somewhat different lines.

We take this opportunity to state without proof two results in [179] about the general structure of hyperspaces. These results were used in [179] to prove some of the results above.

We will use the following notation. Let θ denote $(0, 0, \ldots, 0, \ldots) \in \ell_2$ and, for each $n = 1, 2, \ldots$, let e_n denote the point of ℓ_2 whose $j\underline{\text{th}}$ coordinate is zero if $j \neq n$ and one if $j = n$. If $S \subset \ell_2$, then let conv S denote the closed convex hull of S. For each $n = 1, 2, \ldots$, let

$$\Sigma^n = \text{conv}\{\theta, e_1, \ldots, e_n\}$$

and, for each $s \in [0, 1]$, let

$$\Sigma^n_s = \text{conv}\{s \cdot e_1, s \cdot e_2, \ldots, s \cdot e_n\}.$$

Note that $\Sigma^1 \subset \Sigma^2 \subset \cdots \subset \Sigma^n \subset \cdots$.

The following theorem is a generalization of (14.8.1). It is also related to (1.145).

(14.32) THEOREM [179, 3.1]. *Fix* $t_0 \in [0, \mu(X)]$. *Assume there exist* $A_1, A_2, \ldots, A_n \in \mu^{-1}(t_0)$ *such that* $\cap_{i=1}^n A_i \neq \emptyset$. *Let* K *be a subcontinuum of* $\cap_{i=1}^n A_i$. *Then there is a continuous function* $f_n: \Sigma^n \to C(X)$ *such that*

(1) $K \subset f_n(p) \subset [\cup_{i=1}^n A_i]$ *for each* $p \in \Sigma^n$;

(2) $f_n(e_i) = A_i$ *for each* $i = 1, 2, \ldots, n$;

(3) $f_n(\theta) = K$;

(4) $f_n[\Sigma^n_s] \subset \mu^{-1}[(1 - s) \cdot \mu(K) + s \cdot t_0]$ *for each* $s \in [0, 1]$;

(5) *if* $A_i \neq K = A_i \cap A_j$ *for all* $i, j = 1, 2, \ldots, n$ *with* $i \neq j$, *then* f_n *is a homeomorphism*.

When all the hypotheses of (14.32), including those in (5), are satisfied, we will call a function $f_n: \Sigma^n \to C(X)$ satisfying (1) through (5) of (14.32) an *n-segplex with vertices* K, $A_1, A_2, \ldots, A_n$.

Before stating the second general result of [179], we give the following application of (14.32).

(14.33) THEOREM [179, 3.3]. *If* X *contains an n-odd, then there exists* $t_o \in (0, \mu(X))$ *such that* $\mu^{-1}(t_o)$ *contains an (n-1)-cell* [*comp., (1.213.9)*].

Proof. Let $M \subset X$ be an n-odd. Then, by (0.21), there is a subcontinuum K of M such that $M\backslash K$ is the union of n nonempty mutually separated sets $S_1, S_2, \ldots, S_n$. For each $i = 1, 2, \ldots, n$, let $B_i = S_i \cup K$. Note that $B_i \in C(X)$, each $i = 1, 2, \ldots, n$. Let $t_o = \min\{\mu(B_i): i = 1, 2, \ldots, n\}$. For each $i = 1, 2, \ldots, n$, let A_i be a subcontinuum of B_i such that $\mu(A_i) = t_o$ [such subcontinua A_i can be seen to exist by using (1.8), (1.11), and the continuity of μ]. By (14.32), there exists an n-segplex $f_n: \Sigma^n \to C(X)$ with vertices K, $A_1, A_2, \ldots, A_n$. Since (4) of (14.32) guarantees that $f_n[\Sigma_1^n] \subset \mu^{-1}(t_o)$ and since Σ_1^n is an (n-1)-cell, we have proved (14.33).

For the next result, we adopt the following notation. For a continuum Z and a point $z \in Z$, let

$Z(z) = \{K \in C(Z): z \in K\}$,

and, if μ is any Whitney map for $C(Z)$, let

$Z(z, \mu, t) = Z(z) \cap \mu^{-1}(t) = Z(\{z\}, \mu, t)$,

the last equality being valid by virtue of the definition in (0.68.5).

The following theorem can be applied to obtain some of the results in [286, especially 5.3 and 5.4] and some more facts related to them.

(14.34) THEOREM [179, 5.1]. *Let X be a continuum which is irreducible between points a and b and assume a and b are each terminal in X. Let $Y = X/\{a,b\}$ be the continuum obtained from X by identifying a and b and let p be the corresponding identification point in Y, i.e., $p = \{a, b\} \in Y$. Let $\nu: X \to Y$ be the quotient map and let μ be a Whitney map for $C(Y)$. Then (for notation, see paragraph above):*

(1) $\mu_1 = \mu \circ \hat{\nu}$ is a Whitney map for $C(X)$;

(2) $\hat{\nu}$ embeds $C(X)\setminus\{\{a\}, \{b\}\}$ into $C(Y)$;

(3) $S = \hat{\nu}[X(a)] \cup \hat{\nu}[X(b)]$ is a circle;

(4) for each $0 < t < \mu(X)$, $\hat{\nu}$ maps $\mu_1^{-1}(t)$ homeomorphically onto $cl[\mu^{-1}(t)\setminus Y(p, \mu, t)]$;

(5) $Y(p)$ is a 2-cell and S is its manifold boundary;

(6) for each $0 < t < \mu(X)$, $Y(p, \mu, t)$ is an arc with end points $\hat{\nu}(A_t)$ and $\hat{\nu}(B_t)$, where $\{A_t\} = X(a, \mu_1, t)$ and $\{B_t\} = X(b, \mu_1, t)$. Moreover, $\{\hat{\nu}(A_t), \hat{\nu}(B_t)\} = Bd_{\mu^{-1}(t)}[Y(p, \mu, t)]$, where $Bd_{\mu^{-1}(t)}$ denotes boundary relative to $\mu^{-1}(t)$;

(7) $C(Y) = \hat{\nu}[C(X)] \cup Y(p)$ and $\hat{\nu}[C(X)] \cap Y(p) = S$.

Now, we state and discuss some questions about Whitney properties.

(14.35) QUESTION [286, Question 3]. How are other properties of continua reflected in the continua $\mu^{-1}(t)$?

(14.36) QUESTION [176, section 6]. For a given topological property, determine whether or not it is a Whitney property. In particular, is homogeneity (resp., λ connectedness, weak chainability) a Whitney property? In [179, section 6], it was asked if some other specific properties are Whitney properties. However, it has recently been shown that they are not--the examples are given here in (14.42.1), (14.43.6), and (14.43.7).

(14.37) REMARK. Let $H^n[Y]$ denote the reduced n^{th} Alexander-Čech cohomology group of Y. Rogers [279] has recently shown: (1) For any continuum X and any Whitney map μ for C(X), there is an induced injection $\gamma^*: H^1[\mu^{-1}(t)] \to H^1[X]$ for any $t \in [0, \mu(X)]$; (2) Being acyclic is a Whitney property for the class of one-dimensional continua [279, Corollary 7]. This gives a partial affirmative answer to the question: Is acyclicness a Whitney property [179, section 6]? As we will see in (14.43.6), it is not a Whitney property (for all continua). For results related to [279], see [426].

(14.38) QUESTION. Let X denote any compact convex subset of a Banach space such that $\dim[X] > 1$. Let d be the metric for X obtained from the norm on the Banach space. Let μ_d denote the restriction to C(X) of the Whitney map constructed (using d) as

(14.39) WHITNEY PROPERTIES 433

in (0.50.1). Is $\mu_d^{-1}(t)$ a Hilbert cube whenever $0 < t < \mu(X)$? By the second part of (1.98), $C(X)$ is a Hilbert cube and, by (14.9), $\mu_d^{-1}(t)$ is a locally connected continuum. Also, by using (14.43.8), it can be seen that $\mu_d^{-1}(t)$ is contractible. Let us note that, by (14.43.6), the answer to the above question can be "no" when μ_d is replaced by a Whitney map constructed as in (0.50.1) using a metric other than the one obtained from the norm. Some possible applications of an affirmative answer to the above question are mentioned in (14.42.2) and (14.42.3). Some related questions are in (14.43.11) and (14.43.12).

Some other non-Whitney properties are given in the next two sections of this chapter--see especially (14.42.1), (14.43.6), and (14.43.7).

C.
Whitney Stable Continua

Several years ago, Joe Quinn asked the following question: For what continua X is $\mu^{-1}(t)$ homeomorphic to X for all $t \in [0, \mu(X)]$? This question was stated in [249, p.407] and in [286, p.584]. In (14.39) below, we give a more precise version of Quinn's question. This leads to the notions of Whitney stable and weak Whitney stable continua.

(14.39) QUESTION [see paragraph above]. For what continua X

is it true that for $\begin{Bmatrix} \text{some Whitney map} \\ \text{all Whitney maps} \end{Bmatrix}$ μ for $C(X)$, $\mu^{-1}(t) \cong X$

for all $t \in [0, \mu(X))$?

In this section, we discuss what is known about (14.39) [some related material is in (14.58) through (14.60) and in (14.73.34)]. In order to formulate statements concisely, we give the following definition:

(14.39.1) DEFINITION. A continuum X is said to be $\begin{Bmatrix} \textit{weak Whitney stable} \\ \textit{Whitney stable} \end{Bmatrix}$ provided that for $\begin{Bmatrix} \text{some Whitney map} \\ \text{all Whitney maps} \end{Bmatrix}$ μ for $C(X)$, $\mu^{-1}(t)$ is homeomorphic to X for all $t \in [0, \mu(X))$. Note that X is Whitney stable if and only if the property "homeomorphic to X" is a Whitney property.

In [249, p.408], the following partial answer was given to (14.39):

(14.40) THEOREM [249, Theorem 5]. *If X is a finite-dimensional arcwise connected continuum such that X is weak Whitney stable, then X must be an arc or a circle (and, conversely, an arc and a circle are Whitney stable).*

Proof. Let μ be a Whitney map for $C(X)$ such that $\mu^{-1}(t) \cong X$ for each $t \in [0, \mu(X))$. Then, $\dim[\mu^{-1}(t)] = \dim[X]$ if $0 \le t < \mu(X)$. Also, since $\mu^{-1}[\mu(X)] = \{X\}$, $\dim[\mu^{-1}(t)] = 0$

if $t = \mu(X)$. Hence, by (#) on p.215,

(a) $\dim[C(X)] \leq \dim[X] + 1 < \infty$.

Since X is arcwise connected, we have by (a) and by part (1) of (2.5) that $\dim[X] = 1$. Hence, by (a) and (2.1), $\dim[C(X)] = 2$. Therefore, by (1.100) (or, see [283, Corollary 1]), X is a-triodic. This proves X is an arcwise connected a-triodic continuum. Hence, by [254, Theorem 2], X must be an arc or an arcwise connected circle-like continuum. It is easy to see using [245, Theorem 6] that if X is an arcwise connected circle-like continuum, then $\mu^{-1}(t)$ is a circle for t different from, but sufficiently close to, $\mu(X)$ [this also follows directly from (14.31)]. This completes the proof of (14.40), except to mention that the converse is due to (14.6) and (14.7).

The proof in [249, pp.408-409] of (14.40) is somewhat more complicated than the one just given.

In [179], (14.40) was extended to finite-dimensional *decomposable* continua:

(14.41) THEOREM [179, 3.6]. *If X is a finite-dimensional decomposable continuum such that X is weak Whitney stable, then X must be an arc or a circle (and, conversely, an arc and a circle are Whitney stable).*

Proof. By (14.27), $\mu^{-1}(t)$ is arcwise connected for some $t \in [0, \mu(X))$. Hence, since $\mu^{-1}(t) \cong X$, X is arcwise connected and the result follows from (14.40).

We note that (14.41) completely answers (14.39) for the class of finite-dimensional decomposable continua.

(14.42) REMARK. We have seen examples of Whitney stable continua: The arc (14.6), the circle (14.7), the pseudo-arc (14.3), any particular solenoid (14.21), and any particular pseudo-solenoid (14.23). In [179, section 6], it was asked if being a Hilbert cube is a Whitney property, i.e.: Is the Hilbert cube Whitney stable? We now show that the answer is "no":

(14.42.1) THEOREM. *There is a Hilbert cube Q and a Whitney map γ for $C(Q)$ such that $\gamma^{-1}(t_o)$ can be retracted onto a 2-sphere for some t_o. Hence, being a Hilbert cube is not a Whitney property--equivalently, the Hilbert cube is not Whitney stable.*

Proof. First, we prove the following special extension result in (*):

(*) LEMMA. *Let X and Y be continua, let $p \in Y$, consider $X \times \{p\}$ in the natural way as a subspace of $X \times Y$, and let $\pi_p: X \times Y \to X \times \{p\}$ denote projection, i.e., $\pi_p(x, y) = (x, p)$ for each $(x, y) \in X \times Y$. Then: If μ_p is a Whitney map for $C(X \times \{p\})$, μ_p can be extended to a Whitney map γ for $C(X \times Y)$. Furthermore, γ may be chosen so that $\gamma(\pi_p[A]) \leq \gamma(A)$ for each $A \in C(X \times Y)$.*

Proof of (*). Define $s: C(X \times Y) \to [0, +\infty)$ by [D denotes any metric for $X \times Y$ giving the product topology]

(14.42.1) WHITNEY PROPERTIES 437

$$s(A) = \sup\{D((x, y), (x, p)): (x, y) \in A\}$$

for each $A \in C(X \times Y)$. Clearly:

(1) s is continuous;

(2) $s(A) = 0$ if and only if $A \in C(X \times \{p\})$;

(3) if $A, B \in C(X \times Y)$ such that $A \subset B$, then $s(A) \leq s(B)$.

Let μ_1 denote any Whitney map for $C(X \times Y)$. Define $\gamma: C(X \times Y) \to [0, +\infty)$ by

$$\gamma(A) = s(A) \cdot \mu_1(A) + \mu_p(\pi_p[A])$$

for each $A \in C(X \times Y)$. We show γ is as required in (*). It follows easily using (1) that

(4) γ is continuous.

Assume $A, B \in C(X \times Y)$ such that $A \subset B$ and $A \neq B$. By (3),

(5) $s(A) \leq s(B)$.

Since μ_1 is a Whitney map for $C(X \times Y)$,

(6) $\mu_1(A) < \mu_1(B)$.

Since μ_p is a Whitney map for $C(X \times \{p\})$ and since $\pi_p[A] \subset \pi_p[B]$,

(7) $\mu_p(\pi_p[A]) \leq \mu_p(\pi_p[B])$.

By (5) through (7) and the formula for γ,

(8) $\gamma(A) < \gamma(B)$ for any $A, B \in C(X \times Y)$ such that $A \subset B$ and $A \neq B$.

Clearly, from the formula for γ,

(9) $\gamma(\{(x, y)\}) = 0$ for each $(x, y) \in X \times Y$.

By (4), (8), and (9),

(10) γ is a Whitney map for $C(X \times Y)$.

It follows easily from (2) and the formula for γ that

(11) $\gamma(A) = \mu_p(A)$ for each $A \in C(X \times \{p\})$

and

(12) $\gamma(\pi_p[A]) \leq \gamma(A)$ for each $A \in C(X \times Y)$.

By (10) through (12), we have proved (*).

Proof of (14.42.1). We will make use of the following recent result, due to Ann Petrus [see (14.43.6), especially (8) of the proof]:

(#) For $X = \{(x, y, z) \in S^2 : -1 \leq z \leq +\frac{1}{2}\}$, there is a Whitney map μ for $C(X)$ and a t_o such that there is a retraction from $\mu^{-1}(t_o)$ onto a 2-sphere.

Let X, μ, and t_o be as in (#), let d denote the metric for X obtained from the usual Euclidean metric for R^3 [defined in (0.19)], let $Y = I_\infty$, let $p = (0, 0, \ldots, 0, \ldots) \in Y$, let $Q = X \times Y$ [note that Q is a Hilbert cube since X is a 2-cell], let $X \times \{p\}$ and π_p be as in (*), and let D be any metric for Q giving the product topology and such that $j_p : X \times \{p\} \to X$, defined by

$j_p(x, p) = x$ for each $(x, p) \in X \times \{p\}$

is an isometry [thus, D may be taken to be the usual "Euclidean" product metric]. Define $\mu_p : C(X \times \{p\}) \to [0, +\infty)$ by $\mu_p = \mu \circ \hat{j}_p$, where $\hat{j}_p$ is the induced map as defined in (0.49). Clearly, since j_p is a homeomorphism, μ_p is a Whitney map for $C(X \times \{p\})$. Let γ be as guaranteed by (*). Note that $\mu_p^{-1}(t_o)$ is homeomorphic to $\mu^{-1}(t_o)$ [by the restriction of $\hat{j}_p$ to $\mu_p^{-1}(t_o)$]. Hence, by (#),

(14.42.1)   WHITNEY PROPERTIES

it suffices to show that

(##) there is a retraction r from $\gamma^{-1}(t_o)$ onto $\mu_p^{-1}(t_o)$.

To prove (##), let δ denote the metric for $X \times \{p\}$ obtained from D and let K_δ be as defined in (0.65.1). The following two properties [(i) and (ii) below] of K_δ will be used in the proof. The one in (i) follows from (0.65.6), since (X, d) and $(X \times \{p\}, \delta)$ are isometric and K_d, as defined in (0.65.1), is easily seen to be continuous [from the geometry of (X, d), it is clear that (X, d) satisfies (2) of (0.65.1)]. The one in (ii) is a simple consequence of the fact that $K_d(t, A) \in C(X)$ for each $(t, A) \in [0, +\infty) \times C(X)$ and the fact that (X, d) is isometric to $(X \times \{p\}, \delta)$.

(i) K_δ is continuous;

(ii) $K_\delta(t, A) \in C(X \times \{p\})$ for each

$(t, A) \in [0, +\infty) \times C(X \times \{p\})$.

We will make use of the auxiliary function

$\tau: \mu_p^{-1}([0, t_o]) \to [0, +\infty)$

defined by: For each $A \in \mu_p^{-1}([0, t_o])$, there exists one [by (i), (ii), and the continuity of μ_p] and only one [since $t_o < \mu_p(X \times \{p\})$] number $\tau(A)$ such that $K_\delta(\tau(A), A) \in \mu_p^{-1}(t_o)$. Noting that the values of τ are bounded, specifically

$0 \leq \tau(A) < \text{diam}[X \times \{p\}]$ for each $A \in \mu_p^{-1}([0, t_o])$,

an easy sequence argument using (i) and the compactness of $\mu_p^{-1}(t_o)$ shows that τ is continuous. Now, we define r. Let $A \in \gamma^{-1}(t_o)$. By the last part of (*), $\gamma(\pi_p[A]) \leq t_o$. Hence, since γ is an extension of μ_p [by (*)],

$\mu_p(\pi_p[A]) = \gamma(\pi_p[A]) \leq t_o$.

Thus, $\pi_p[A]$ is in the domain of τ. Let

$r(A) = K_\delta(\tau(\pi_p[A]), \pi_p[A])$.

Since K_δ, τ, and π_p are continuous, r is continuous. By definition of τ, $r(A) \in \mu_p^{-1}(t_o)$ for each $A \in \gamma^{-1}(t_o)$. Finally, let $A \in \mu_p^{-1}(t_o)$. Then, since $K_\delta(0, A) = A$, $K_\delta(0, A) \in \mu_p^{-1}(t_o)$. Hence, by definition of τ, we have that $\tau(A) = 0$. Thus, since $\pi_p[A] = A$, we have that $r(A) = A$. Therefore, we have proved (##). This completes the proof of (14.42.1).

(14.42.2) QUESTION. Is there a continuum which is weak Whitney stable but not Whitney stable? By (14.42.1), the Hilbert cube is not Whitney stable; it may be weak Whitney stable--see (14.38). Is there a *finite-dimensional* continuum which is weak Whitney stable but not Whitney stable? The problem is that a continuum can admit essentially different Whitney maps--see the next section of this chapter.

(14.42.3) QUESTION (comp., [249, P.4, p.407]). If a continuum X is weak Whitney stable [resp., Whitney stable], then is every (nondegenerate) subcontinuum of X weak Whitney stable [resp., Whitney stable]? In other words: Are these properties hereditary? It might be that the Hilbert cube is weak Whitney stable [see (14.42.2)]; if so, then the property weak Whitney stable is not hereditary. Is the property weak Whitney stable [resp., Whitney stable] hereditary for finite-dimensional continua? Note that the

(14.42.3)

continua mentioned at the beginning of (14.42) are hereditarily Whitney stable.

D.
Essentially Different
Whitney Maps for $C(X)$

In this section, we will be concerned with the following question [249, p.407]: Do the topological types of Whitney map point inverses in $C(X)$ depend on the Whitney map, i.e., if μ and μ' are Whitney maps for $C(X)$, then must each member of either of the collections

$\{\mu^{-1}(t): 0 \leq t \leq \mu(X)\}$

or

$\{(\mu')^{-1}(t): 0 \leq t \leq \mu'(X)\}$

be homeomorphic to some member of the other collection? In [249, Example 1 and Remark on p.396], an example was given which showed that the analogous question with $C(X)$ replaced by 2^X has a negative answer--the continuum X was an arc. This example will be given later in this chapter [see (14.61) and (14.62)]. Note that, by (14.6), an arc can not be used to provide a negative answer to the question for $C(X)$. This question was answered negatively by Petrus [267] with several examples, some of which Petrus also used to provide answers to some questions raised in [179, section 6]. In (14.43.1) through (14.43.7), we give many of Petrus' examples [see (14.74)]. We use the following terminology.

(14.43) DEFINITION [267, p.31]. We say that a continuum X *admits essentially different Whitney maps* provided that there exist Whitney maps μ and μ' for $C(X)$ such that there is some member Λ of one of the collections

$$\{\mu^{-1}(t): 0 \leq t \leq \mu(X)\}$$

or

$$\{(\mu')^{-1}(t): 0 \leq t \leq \mu'(X)\}$$

such that Λ is not homeomorphic to any member of the other collection.

The first example shows that a simple triod admits essentially different Whitney maps [for a general fact, see (14.43.5)].

(14.43.1) EXAMPLE [267, Example 35(a)]. Let X be the simple triod given by $X = A_1 \cup A_2 \cup A_3$, where

$A_1 = \{(x, 0) \in R^2: -1 \leq x \leq 0\}$,
$A_2 = \{(x, 0) \in R^2: 0 \leq x \leq +1\}$,
$A_3 = \{(0, y) \in R^2: 0 \leq y \leq +1\}$.

Let Y be the simple triod given by [see (14.43.3)], $Y = X \cup B$ where

$B = \{(x, 0) \in R^2: +1 \leq x \leq +2\}$.

Let d denote the metric, for both X and Y, obtained from the usual Euclidean metric for R^2 [see (0.19)]. Let μ_X [resp., μ_Y] denote the restriction to $C(X)$ [resp., $C(Y)$] of the Whitney map constructed using d as in (0.50.1). For use here, we note the following fact about μ_X and μ_Y [see (0.68.1)]:

(14.43.2)   WHITNEY PROPERTIES   443

  (i) any two isometric subcontinua of X [resp., Y]
      have the same value under μ_X [resp., μ_Y].

Let B^2 be as in (0.19) and let α_1 [resp., α_2, α_3] denote the convex arc in R^2 with end points $(-1, 0)$ and $(-2, 0)$ [resp., $(1, 0)$ and $(2, 0)$, $(0, 1)$, and $(0, 2)$]. Now, let $t_o = \mu_X(A_1)$. By using (i) and the formulas for μ_X and μ_Y, it is easy to see that (1) and (2) below hold:

  (1) $\mu_X^{-1}(t)$ is homeomorphic to X [when $t = 0$],
      $B^2 \cup \alpha_1 \cup \alpha_2 \cup \alpha_3$ [when $0 < t < t_o$], B^2 [when $t_o \leq t < \mu_X(X)$], a point [when $t = \mu_X(X)$];

  (2) $\mu_Y^{-1}(t_o)$ is homeomorphic to $B^2 \cup \alpha_1$ [actually, $\mu_Y^{-1}(t_o) = \mu_X^{-1}(t_o) \cup \beta$ where β is the arc in $\mu_Y^{-1}(t_o)$ with end points B and A_2].

Hence, $\mu_Y^{-1}(t_o)$ is not homeomorphic to any of the contnnua in (1). Therefore, a simple triod admits essentially different Whitney maps [as is precisely verified by letting $\mu'_X = \mu_Y \circ \hat{h}$, where h is a homeomorphism of X onto Y].

By (14.3) and (14.6), neither a pseudo-arc nor an arc can admit essentially different Whitney maps. Also, the familiar sin[1/x]-continuum S_o defined in (8.22) does not admit essentially different Whitney maps [249, p.404]. However, as the next example shows, there are chainable continua which admit essentially different Whitney maps.

(14.43.2) EXAMPLE [267, Example 35(e)]. Let X be the

chainable continuum given by $X = X_1 \cup X_2$ where

$X_1 = \text{cl}\{(x, \sin[1/x]) \in R^2 : 0 < x \le 1\}$,

$X_2 = \text{cl}\{(x, 2 + \sin[1/x]) \in R^2 : -1 \le x < 0\}$.

Let d denote the metric for X obtained from the usual Euclidean metric for R^2 [see (0.19)]. Let μ denote the restriction to $C(X)$ of the Whitney map constructed using d as in (0.50.1). It can be shown (and was essentially noted in [286, Example 4.1]) that:

(1) $\mu^{-1}(t)$ is homeomorphic to X [when $t = 0$], an arc [for $t < \mu(X)$ and "near" $\mu(X)$], a point [when $t = \mu(X)$], or the continuum $X_1 \cup A \cup Z$ where

$A = \{(0, y) \in R^2 : +1 \le y \le +2\}$,

$Z = \text{cl}\{(x, 3 + \sin[1/x]) \in R^2 : -1 \le x < 0\}$.

We define another Whitney map μ' for $C(X)$ as follows: First, for each $i = 1$ and 2, define $f_i : C(X) \to [0, +\infty)$ by the formula

$$f_i(A) = \begin{cases} \mu(A \cap X_i) , & \text{if } A \cap X_i \ne \emptyset \\ 0 , & \text{if } A \cap X_i = \emptyset . \end{cases}$$

Note that since X is hereditarily unicoherent, the first part of the formula "makes sense." Also, it is easy to see that f_1 and f_2 are each continuous. Now, define [see (14.43.3)] $\mu' : C(X) \to [0, +\infty)$ by

$\mu'(A) = 2 \cdot f_1(A) + f_2(A)$ for each $A \in C(X)$.

It is easy to verify that μ' is a Whitney map for $C(X)$. Let $B_1 = \{(0, y) \in R^2 : -1 \le y \le 1\}$, $B_2 = \{(0, y) \in R^2 : 1 \le y \le 3\}$, and let $t_o = \mu'(B_2)$. We now show why $(\mu')^{-1}(t_o)$ is not homeomorphic to any of the continua in (1). Note that, since B_1 and B_2 are isometric, $\mu(B_1) = \mu(B_2)$ by (0.68.1). Using this fact, simple computations show that $\mu'(B_1) = 2t_o$. It follows from this that $\Lambda_1 = (\mu')^{-1}(t_o) \cap C(X_1) \cong X_1$. Since $\mu'(B_2) = t_o$, it follows easily that $\Lambda_2 = (\mu')^{-1}(t_o) \cap C(X_2)$ is an arc with B_2 as one end point. It follows from (14.6) that $\Gamma = (\mu')^{-1}(t_o) \cap C(B_1 \cup B_2)$ is an arc with B_2 as one end point. Finally, it is easy to prove that $\Lambda_1 \cup \Lambda_2 \cup \Gamma = (\mu')^{-1}(t_o)$. The information above, together with some easy-to-see details, can be used to show that $(\mu')^{-1}(t_o) \cong X_1 \cup \alpha$ where α is the convex arc in R^2 with end points $(0, 1)$ and $(-1, 1)$. Hence, $(\mu')^{-1}(t_o)$ is not homeomorphic to any of the continua in (1). Therefore, X admits essentially different Whitney maps.

(14.43.3) REMARK. Petrus [267, Example 35(a)] used the type of procedure in (14.43.2) for a simple triod. Instead, in (14.43.1) we "lengthened" one leg of a simple triod and used the Whitney map in (0.50.1) again. The effect is "the same"--in (14.43.2), we could have taken two half-lines closing down on arcs of different lengths and used the Whitney map in (0.50.1) again.

(14.43.4) EXAMPLE [267, Example 35(f)]. In (14.43.2), we saw that there are chainable continua which admit essentially different Whitney maps. So do some circle-like continua: Let Y be the circle-like continuum obtained from the continuum X in (14.43.2) by joining the points $p = (1, \sin[1])$ and $q = (-1, 2 + \sin[-1])$ with an arc A_o having end points p and q and otherwise being disjoint from X. Then, using the ideas in (14.43.2), it follows that $Y = X \cup A_o$ admits essentially different Whitney maps.

In (14.43.1), we showed that a simple triod admits essentially different Whitney maps. By (14.6) and (14.7), neither an arc nor a circle can admit essentially different Whitney maps. Petrus has proved the following result.

(14.43.5) THEOREM [267, Proposition 38]. *Let* X *be a one-dimensional compact connected polyhedron. If* X *is not an arc or a circle, then* X *admits essentially different Whitney maps [and conversely].*

In (14.43.6), we come to one of Petrus' most interesting examples. It answers many questions raised in [179, section 6]. Let us note that when B^2 has the metric in (0.19) and when μ is the restriction to $C(B^2)$ of the Whitney map constructed in (0.50.1), $\mu^{-1}(t)$ is contractible for each $t \in [0, \mu(B^2)]$. This can be seen by using (14.43.8). In the next example, a 2-cell X and a Whitney map μ for $C(X)$ are given such that, among other things, $\mu^{-1}(t_o)$ is not contractible for some t_o. I have recently found a way to use this example to answer a question in [179, section 6]—see (14.42.1).

(14.43.6) EXAMPLE [267, Example 46]. Let S^2 be the 2-sphere in R^3 [see (0.19)] and let
$$X = \{(x, y, z) \in S^2 : -1 \leq z \leq +\tfrac{1}{2}\} .$$

Thus, X is a 2-cell. Let μ denote the restriction to $C(X)$ of the Whitney map constructed as in (0.50.1)--the metric d for X is the usual Euclidean metric for R^3 [see (0.19)] restricted to $X \times X$. Let

$$t_o = \mu(\{(x, y, +\tfrac{1}{2}) \in X\}).$$

We will define a homeomorphism g from S^2 into $\mu^{-1}(t_o)$ such that $g[S^2]$ is a retract of $\mu^{-1}(t_o)$. For each $z_o \in [-1, +1]$, let

$$S^2(z_o) = \{(x, y, z_o) \in S^2\}.$$

Note that, from the formula for μ, (1) and (2) below hold:

(1) $\mu(S^2(-1/2)) = \mu(S^2(+1/2)) = t_o$,

(2) $\mu(S^2(z)) > t_o$ whenever $-\tfrac{1}{2} < z < +\tfrac{1}{2}$.

For each $(x, y, z) \in [S^2 \setminus \{(0, 0, -1), (0, 0, +1)\}]$, let

(#) $G(x, y, z)$ denote the longitudinal great circle in S^2 containing $\{(x, y, z), (0, 0, -1), (0, 0, +1)\}$

and let

(##) $\psi(x, y, z)$ denote that unique point in $G(x, y, z) \cap S^2(\tfrac{z}{2})$ which is nearest to (x, y, z).

Now we define g. First, let

(3) $g(0, 0, -1) = S^2(-\tfrac{1}{2})$ and $g(0, 0, +1) = S^2(+\tfrac{1}{2})$.

Next, let $(x, y, z) \in [S^2 \setminus \{(0, 0, -1), (0, 0, +1)\}]$. Then, by (2), $\mu(S^2(z/2)) > t_o$. Thus, since $S^2(z/2)$ is a circle, it follows [using (0.50.a)] that there is one and only one subarc $A(x, y, z)$ of $S^2(z/2)$ such that $\mu(A(x, y, z)) = t_o$ and such that $\psi(x, y, z)$ is the midpoint of $A(x, y, z)$. We define:

(4) If $(x, y, z) \in [S^2 \setminus \{(0, 0, -1), (0, 0, +1)\}]$, then

$g(x, y, z) = A(x, y, z)$ [as above].

The formulas in (3) and (4) define a function g from S^2 into $\mu^{-1}(t_o)$ [in particular, (1) implies the values of g in (3) are members of $\mu^{-1}(t_o)$]. It is easy to see that g is a homeomorphism, i.e., g is an embedding of S^2 into $\mu^{-1}(t_o)$. To show that $g[S^2]$ is a retract of $\mu^{-1}(t_o)$, Petrus defines a continuous function f from $\mu^{-1}(t_o)$ onto S^2 such that $f \circ g$ is the identity mapping on S^2 [i.e., S^2 is an r-image of $\mu^{-1}(t_o)$--see the beginning of Chapter VI]. The mapping f is defined with the help of the following fact, which is more general than (2):

(5) If A is a subcontinuum of $\cup \{S^2(z): -\frac{1}{2} \leq z \leq +\frac{1}{2}\}$ such that $A \neq S^2(-\frac{1}{2})$, $A \neq S^2(+\frac{1}{2})$, and A "projects along longitudinal great circles" onto $S^2(0)$, then $\mu(A) > t_o$.

Instead of giving an analytical definition of f, we will give a descriptive one (an analytical definition is in [267, p.53]). Let $A \in \mu^{-1}(t_o)$. If

(6) $A \cap [\cup\{S^2(z): -1 \leq z \leq -\frac{1}{2}\}] \neq \emptyset$,

then let $f(A) = (0, 0, -1)$. If

(7) $A = S^2(+\frac{1}{2})$,

then let $f(A) = (0, 0, +1)$. Now, assume A is not as in (6) or (7). Then, since $\mu(A) = t_o$, we have by (5) that A "projects along longitudinal great circles" onto a subarc A_o of $S^2(0)$; let $m(A)$ denote the midpoint of A_o. Letting

$p(A) = \text{g.l.b.}\{z: A \cap S^2(z) \neq \emptyset\}$,

we see that, since A does not satisfy (6), $p(A) > -1/2$. Also, note that $p(A) \leq +1/2$, as would be true for any subcontinuum of

X [actually, since A does not satisfy (7), it follows from (0.50.a) that $p(A) < +1/2$]. Hence,

$$-1 \leq 2p(A) \leq +1$$

and so $S^2(2p(A))$ "makes sense." Let $f(A)$ denote that unique point in $G(m(A)) \cap S^2(2p(A))$ which is nearest to $m(A)$. This completes the descriptive definition of f. It can be seen that f is a continuous function from $\mu^{-1}(t_o)$ onto S^2 and that $f \circ g: S^2 \xrightarrow{\text{onto}} S^2$ is the identity mapping. Therefore, $g[S^2]$ is a retract of $\mu^{-1}(t_o)$ [the mapping $g \circ f$ being the desired retraction]. Hence, since g is a homeomorphism, we have shown:

(8) There is a retraction of $\mu^{-1}(t_o)$ onto a 2-sphere.

There are a number of consequences of (8). Recalling that X is a 2-cell, (8) shows that the following properties are not Whitney properties (thus answering some questions in [179, section 6]): Acyclicness in the sense of Čech cohomology, being an absolute retract, contractibility, and having trivial shape. With respect to acyclicness, we call the reader's attention to (14.37) where it is mentioned that acyclicness is a Whitney property for the class of one-dimensional continua. In (14.30), there is an example which shows that the fixed point property is not a Whitney property. As we see from (8) above, the fixed point property is not a Whitney property *even* for the class of absolute retracts (this last fact is not explicitly mentioned in [267]). Finally, though perhaps anticlimactically, we note that a 2-cell admits essentially different Whitney maps [see paragraph preceding this example].

The next example answers a question raised in [179, section 6] by showing that the property of being an absolute neighborhood retract is not a Whitney property. It also shows that the property of being locally contractible is not a Whitney property.

(14.43.7) EXAMPLE [267, Example 50]. Let

$$A_o = \{(x, y, 0) \in R^3 : x^2 + y^2 = 1\}$$

and, for each $n = 1, 2, \ldots$, let

$$A_n = \{(x, y, 2^{1-n}) \in R^3 : x^2 + y^2 = 1\}.$$

For each $n = 1, 2, \ldots$, let X_n be the closed "bulging annulus" bounded by the circles A_n and A_{n+1} and contained in the 2-sphere in R^3 with center $(0, 0, 3[2^{-n-1}])$ and radius $[1 + 2^{-n}]^{\frac{1}{2}}$. Let

$$X = cl[\bigcup_{n=1}^{\infty} X_n] = [\bigcup_{n=1}^{\infty} X_n] \cup A_o.$$

Assume the metric d for X is the usual Euclidean metric for R^3 [see (0.19)] restricted to $X \times X$, and let μ denote the restriction to $C(X)$ of the Whitney map constructed as in (0.50.1). Let

$$t_o = \mu(A_o) \quad [= \mu(A_n) \text{ for each } n = 1, 2, \ldots].$$

Petrus has shown that $\mu^{-1}(t_o)$ is not locally contractible. The basic ideas used to do this are similar to those in (14.43.6), and the details will not be done here. Since $X \tilde{=} S^1 \times [0, 1]$, the example shows that being an absolute neighborhood retract is not a Whitney property (absolute neighborhood retracts are locally contractible [40, 3.3, p.87]). This answers a question raised in [179, section 6]. Furthermore, as Petrus has mentioned,

$$Y = X \cup \{(x, y, 1) \in R^3 : x^2 + y^2 \leq 1\}$$

is a 2-cell such that, for μ defined as in (0.50.1) and for t_o as above, $\mu^{-1}(t_o)$ is not locally contractible.

In (14.43.6) we saw that contractibility is not a Whitney property even for the class of absolute retracts. However, as we will see in (14.43.10), it is a Whitney property for the class of one-dimensional absolute retracts. First, let us note the following general result which gives sufficient conditions in order that a given "level" $\mu^{-1}(t_o)$ be contractible.

(14.43.8) THEOREM [267, Proposition 42]. *Let* X *be a contractible continuum, let* μ *be a given Whitney map for* C(X), *and let* $t_o \in [0, \mu(X)]$. *If there exists a homotopy* h: X × [0, 1] → X *contracting* X *to a point* p $\in$ X *such that*

$\mu(h[A \times \{s\}]) \le t_o$ *for each* $A \in \mu^{-1}(t_o)$

and each s $\in$ [0, 1],

then $\mu^{-1}(t_o)$ *is contractible.*

(14.43.9) COROLLARY [267, Corollary 43]. *Let* X *be a contractible continuum, let* μ *be a given Whitney map for* C(X), *and let* $t_o \in [0, \mu(X)]$. *If there exists a homotopy* h: X × [0, 1] → X *contracting* X *to a point* p $\in$ X *such that*

h[A × {s}] $\in$ [C(A) ∪ F_1(X)] *for each* $A \in \mu^{-1}(t_o)$

and each s $\in$ [0, 1]

then $\mu^{-1}(t_o)$ *is contractible.*

Proof. By (0.50.a), the hypotheses of (14.43.9) imply those

of (14.43.8).

(14.43.10) COROLLARY [267, Corollary 44]. *The property of being contractible is a Whitney property for the class of dendrites.*

Proof. For any dendrite X, the hypotheses of (14.43.9) can be shown to hold for any Whitney map μ for $C(X)$ and any $t_o \in [0, \mu(X)]$.

The following questions are motivated by the material in (14.43.1) through (14.43.10).

(14.43.11) QUESTIONS. What continua admit essentially different Whitney maps? Are the arc and the circle the only locally connected continua which do not admit essentially different Whitney maps? We note that the Hilbert cube admits essentially different Whitney maps [comp., (14.38) and (14.42.1)]. Are the arc and the circle the only finite-dimensional locally connected continua which do not admit essentially different Whitney maps? What chainable or circle-like continua do not admit essentially different Whitney maps?

(14.43.12) QUESTION. If X is an absolute retract such that $\dim[X] \geq 2$, then does X admit essentially different Whitney maps? In particular: Are there Whitney maps μ and μ' for $C(X)$ such that $\mu^{-1}(t)$ is contractible for each $t \in [0, \mu(X)]$ but $(\mu')^{-1}(t_o)$ is not contractible for some $t_o \in [0, \mu'(X)]$? This is true for the Hilbert cube [comp., (14.38) and (14.42.1)].

E.
Whitney-Reversible Properties

In [93, p.1032], it was shown that for any continuum X and any Whitney map μ for C(X), μ is both monotone [see (14.2)] and open [see part (d) of (1.213.1)]. By [186, p.68], this is equivalent to saying:

(14.44) THEOREM [93, p.1032]. *If* X *is any continuum and* μ *is any Whitney map for* C(X), *then* $\{\mu^{-1}(t): 0 \leq t \leq \mu(X)\}$ *is a monotone continuous decomposition of* C(X).

The reader should compare (14.44) with (14.61), where it is shown that Whitney maps for 2^X do not (necessarily) give such a decomposition of 2^X.

We have been discussing relationships between the elements of $\mu^{-1}(t)$ of the decomposition space of (14.44). In particular: *A Whitney property is a topological property which if it holds for the level* $\mu^{-1}(0)$ *then it holds for every level* $\mu^{-1}(t)$, *where* $0 < t < \mu(X)$ *and* μ *is any Whitney map for* C(X). This way of viewing Whitney properties motivates the following definition of a "converse" notion:

(14.45) DEFINITION. Let P be a topological property. Then P is said to be a $\begin{cases} \text{strong Whitney-reversible property} \\ \text{Whitney-reversible property} \end{cases}$ provided that whenever X is a continuum such that $\mu^{-1}(t)$ has property P for $\begin{Bmatrix} \text{some Whitney map} \\ \text{all Whitney maps} \end{Bmatrix}$ μ for $C(X)$ and all $0 < t < \mu(X)$, then X has property P.

Note first that there are topological properties which are strong Whitney-reversible properties but which fail dramatically to be Whitney properties. For example: By (14.46), the property of being an indecomposable continuum is a strong Whitney-reversible property--however, in (14.12) we gave an example of an indecomposable continuum Y such that $\mu_o^{-1}(t)$ is decomposable for *all* $0 < t < \mu_o(Y)$ and *any* Whitney map μ_o for $C(Y)$. Also note that, by (14.12), decomposability and non-unicoherence are not Whitney-reversible properties. However [see (14.46)], unicoherence is a strong Whitney-reversible property. Also, as can be seen from (14.14.2), decomposability is a strong Whitney-reversible property for $X \in CP$.

We give some results below about Whitney-reversible properties. Except for (14.47) and the example mentioned in (14.48), no work on this topic has appeared in the literature. Also, the terminology in (14.45) originates here.

(14.46) THEOREM. *The property of being an indecomposable continuum is a strong Whitney-reversible property, as is the property of being a unicoherent continuum. In fact:*

(1) *If there exists a Whitney map* μ *for* $C(X)$ *and some* $t < \mu(X)$ *such that* $\mu^{-1}(t)$ *is indecomposable, then* X *is indecomposable--more generally, if* P *is a Whitney property and* $\mu^{-1}(t)$ *has the property "not* P*" for some* $t < \mu(X)$ *and some* μ*, then* X *has the property "not* P*."*

(2) *If there is a sequence* $\{t_n\}_{n=1}^{\infty}$ *such that* $t_n \to 0$ *as* $n \to \infty$ *and* $\mu^{-1}(t_n)$ *is unicoherent for each* $n = 1, 2, \ldots$, *then* X *is unicoherent.*

Proof. Part (1) follows immediately from (14.11). To prove part (2), assume $\{t_n\}_{n=1}^{\infty}$ is as in (2). Assume X is not unicoherent. Then, there exist subcontinua A and B of X such that $X = A \cup B$ and $A \cap B$ is not connected. Let K and L be disjoint nonempty compact sets such that $A \cap B = K \cup L$. Then, since

$$0 < \text{g.l.b.}\{d(x, y): x \in K \text{ and } y \in L\},$$

it follows easily using (0.8) and the properties of μ in (0.50) that there exists a natural number j such that

(i) $\{M \in \mu^{-1}(t_j): M \cap K \neq \emptyset\} \cap \{M \in \mu^{-1}(t_j): M \cap L \neq \emptyset\} = \emptyset$.

Let $X(A, \mu, t_j)$ and $X(B, \mu, t_j)$ be as defined in (0.68.5). By (14.11.1),

(ii) $X(A, \mu, t_j)$ and $X(B, \mu, t_j)$ are subcontinua of $\mu^{-1}(t_j)$.

Since $X = A \cup B$,

(iii) $\mu^{-1}(t_j) = X(A, \mu, t_j) \cup X(B, \mu, t_j)$.

Since $X \setminus [A \cap B]$ is not connected and since $\mu^{-1}(t_j) \subset C(X)$, it follows easily that

(iv) $X(A, \mu, t_j) \cap X(B, \mu, t_j) =$
$\{M \in \mu^{-1}(t_j): M \cap [A \cap B] \neq \emptyset\}$.

Since $A \cap B = K \cup L$, it follows using (iv) that

(v) $X(A, \mu, t_j) \cap X(B, \mu, t_j) = \{M \in \mu^{-1}(t_j):$
$M \cap K \neq \emptyset\} \cup \{M \in \mu^{-1}(t_j): M \cap L \neq \emptyset\}$.

It follows easily, using the compactness of K and L, that the two sets on the right-hand side of the equation in (v) are compact. Also, they are disjoint [by (i)] and nonempty [use (1.8), (1.11), continuity of μ, and the fact that $K \neq \emptyset \neq L$]. Hence, by (v), $X(A, \mu, t_j) \cap X(B, \mu, t_j)$ is not connected. Thus, by (ii) and (iii), $\mu^{-1}(t_j)$ is not unicoherent. Therefore, we have proved part (2) of (14.46). This completes the proof of (14.46).

The next result is the first theorem in the literature on Whitney-reversible properties. Recall [see (14.9)] that the property of being a locally connected continuum is a Whitney property. It is also a strong Whitney-reversible property:

(14.47) THEOREM [249, Theorem 4]. *Let* X *be a continuum and let* μ *be a Whitney map for* C(X). *If there is a sequence* $\{t_n\}_{n=1}^{\infty}$ *such that* $t_n \to 0$ *as* $n \to \infty$ *and* $\mu^{-1}(t_n)$ *is locally connected for each* $n = 1, 2, \ldots$, *then* X *is locally connected. Hence, being a locally connected continuum is a strong Whitney-reversible property.*

Proof. It is well-known (see, for example [338, p.23]) that a continuum Z is locally connected if and only if given any

$\varepsilon > 0$, Z can be written as the union of finitely many subcontinua each of diameter less than ε. So, to prove X is locally connected, let $\varepsilon > 0$. It follows from the compactness of $C(X)$ [see (0.8)], (1.211.1), and the properties of μ in (0.50) that there exists a particular t_n, denoted by s_o, such that

(1) $\operatorname{diam}[A] < \varepsilon/2$ for each $A \in \mu^{-1}(s_o)$.

Since $\mu^{-1}(s_o)$ is a locally connected continuum [$\mu^{-1}(s_o)$ is a continuum by (14.2)], there exist finitely many subcontinua $\Gamma_1, \Gamma_2, \ldots, \Gamma_k$ of $\mu^{-1}(s_o)$ such that $\mu^{-1}(s_o) = \cup_{i=1}^{k} \Gamma_i$ and

(2) $\operatorname{diam}[\Gamma_i] < \varepsilon/2$ for each $i = 1, 2, \ldots, k$.

Now, for each $i = 1, 2, \ldots, k$, let $G_i = \cup \Gamma_i$. By (1.49), G_i is a subcontinuum of X for each $i = 1, 2, \ldots, k$. Also, since $\mu^{-1}(s_o) = \cup_{i=1}^{k} \Gamma_i$ and $\cup \mu^{-1}(s_o) = X$ [by part (b) of (1.213.1)], it follows that $X = \cup_{i=1}^{k} G_i$. Easy computations, using (1), (2), and the Hausdorff metric as defined in (0.1), give that $\operatorname{diam}[G_i] < \varepsilon$ for each $i = 1, 2, \ldots, k$. This completes the proof of (14.47).

(14.48) REMARK. By (14.9) and (14.47), we see that the property of being a locally connected continuum is both a Whitney property and a strong Whitney-reversible property. In (14.8), we showed that the property of being an arcwise connected continuum is a Whitney property; however, it is not a Whitney-reversible property. An example showing this, due to Joe Quinn, is in [249, Example 3].

(14.49) THEOREM. *Let* X *be a continuum and let* μ *be a Whitney map for* $C(X)$. *If there is a sequence* $\{t_n\}_{n=1}^{\infty}$ *such that* $t_n \to 0$ *as* $n \to \infty$ *and* $\mu^{-1}(t_n)$ *is a-triodic for each* $n = 1, 2, \ldots$, *then* X *is a-triodic. Hence, the property of being a-triodic is a strong Whitney-reversible property.*

Proof. We first prove the following stronger result, which will also be used in (14.55.1).

(14.49.1) THEOREM. *If* X *is a triod, then there exists* $\delta > 0$ *such that* $\mu^{-1}(t)$ *is a triod for each* $0 \leq t < \delta$. *Hence, the property of not being a triod is a strong Whitney-reversible property.*

Proof of (14.49.1). Assume X is a triod so that, by (0.21), there is a subcontinuum N of X such that $X \setminus N$ is the union of three nonempty mutually separated sets S_1, S_2, and S_3. For each $i = 1, 2,$ and 3, let $A_i = S_i \cup N$. Each A_i is a subcontinuum of X. Choose $\delta > 0$ such that

(1) $[\mu^{-1}(\delta) \cap C(A_i)] \setminus X(N, \mu, \delta) \neq \emptyset$ for each
     $i = 1, 2,$ and 3,

where $X(N, \mu, \delta)$ is as defined in (0.68.5). Let t_o be such that $0 \leq t_o < \delta$. We will prove that $\mu^{-1}(t_o)$ is a triod. By (14.11.1),

(2) $X(N, \mu, t_o)$ is a subcontinuum of $\mu^{-1}(t_o)$.

For each $i = 1, 2,$ and 3, let

$U_i = \{B \in \mu^{-1}(t_o): B \subset S_i\}$.

It follows easily from (1) that

(3) $U_i \neq \emptyset$ for each $i = 1, 2$, and 3.

Furthermore, since S_i is an open subset of X for each $i = 1$, 2, and 3, it follows easily that

(4) U_i is an open subset of $\mu^{-1}(t_o)$ for each $i = 1, 2$, and 3.

Also, since $X = [\cup_{i=1}^{3} S_i] \cup N$ and the sets S_1, S_2, and S_3 are mutually separated, it follows easily that

(5) $\mu^{-1}(t_o) \setminus X(N, \mu, t_o) = \bigcup_{i=1}^{3} U_i$

and

(6) U_1, U_2, and U_3 are mutually disjoint.

Since $\mu^{-1}(t_o)$ is a continuum [by (14.2)], we have by (2) through (6) that $\mu^{-1}(t_o)$ is a triod. This completes the proof of (14.49.1).

Proof of (14.49). The theorem in (14.49) follows easily from (14.49.1) and the fact that if X contains a triod M, then the restriction of μ to $C(M)$ is a Whitney map for $C(M)$. This completes the proof of (14.49).

The results in (14.50) and (14.51) below should be compared with the one in (1.213.8).

(14.50) COROLLARY. *Let* X *be a continuum and let* μ *be a Whitney map for* $C(X)$. *If there is a sequence* $\{t_n\}_{n=1}^{\infty}$ *such that* $t_n \to 0$ *as* $n \to \infty$ *and* $\mu^{-1}(t_n)$ *is an arc for each* $n = 1, 2, \ldots$, *then* X *is an arc. Hence, the property of being an arc is a strong*

Whitney-reversible property.

Proof. By (14.47), X is a locally connected continuum and, by (14.49), X is a-triodic. Hence, X is an arc or a circle. By (14.7) and the fact that $\mu^{-1}(t_n)$ is an arc, X is not a circle. Therefore, X is an arc. This proves (14.50).

(14.51) COROLLARY. *Let* X *be a continuum and let* μ *be a Whitney map for* C(X). *If there is a sequence* $\{t_n\}_{n=1}^{\infty}$ *such that* $t_n \to 0$ *as* $n \to \infty$ *and* $\mu^{-1}(t_n)$ *is a circle for each* $n = 1, 2, \ldots$, *then* X *is a circle. Hence, the property of being a circle is a strong Whitney-reversible property.*

Proof. The proof is analogous to the proof of (14.50); use (14.6) instead of (14.7).

(14.52) THEOREM. *Let* X *be a continuum and let* μ *be a Whitney map for* C(X). *If there is a sequence* $\{t_n\}_{n=1}^{\infty}$ *such that* $t_n \to 0$ *as* $n \to \infty$ *and* $\mu^{-1}(t_n)$ *contains no arc for each* $n = 1, 2, \ldots$, *then* X *contains no arc. Hence, the property of not containing an arc is a strong Whitney-reversible property* [see *(14.55.2)*].

Proof. Assume X is a continuum such that X contains an arc A. Let μ be any Whitney map for C(X). Then, since the restriction of μ to C(A) is a Whitney map for C(A), it follows using (14.6) that $\mu^{-1}(t_n)$ contains an arc for some t_n. This proves (14.52).

(14.53) THEOREM. *Let* X *be a continuum and let* μ *be a Whitney map for* C(X). *If there is a sequence* $\{t_n\}_{n=1}^{\infty}$ *such that* $t_n \to 0$ *as* $n \to \infty$ *and* $\mu^{-1}(t_n)$ *contains no circle for each* $n = 1, 2, \ldots$, *then* X *contains no circle. Hence, the property of not containing a circle is a strong Whitney-reversible property.*

Proof. The proof is analogous to the proof of (14.52); use (14.7) instead of (14.6).

(14.54) THEOREM. *Let* X *be a continuum and let* μ *be a Whitney map for* C(X). *If there is a sequence* $\{t_n\}_{n=1}^{\infty}$ *such that* $t_n \to 0$ *as* $n \to \infty$ *and* $\mu^{-1}(t_n)$ *is*

(1) hereditarily indecomposable,

(2) a pseudo-arc,

or

(3) any particular pseudo-solenoid

for each $n = 1, 2, \ldots$, *then* X *is as in (1), (2), or (3) respectively. Hence, the property of being as in (1) [resp., (2) and (3)] is a strong Whitney-reversible property.*

Proof. The proof of (1) is similar to the proof of (14.52)-- use (14.11) instead of (14.6). To prove (2), let μ be some Whitney map for C(X) such that $\mu^{-1}(t_n)$ is a pseudo-arc for each $n = 1, 2, \ldots$. By (1), X is an hereditarily indecomposable continuum. Hence, by (1.80.3) and the fact that $t_n \to 0$ as $n \to \infty$, there is an ε-mapping of X onto a pseudo-arc for each $\varepsilon > 0$. It now follows easily that X is chainable. Hence, since X is hereditarily indecomposable, X is a pseudo-arc [see (0.23)]. This proves (2). To prove (3), let μ be some Whitney map for $C(X)$ such that $\mu^{-1}(t_n)$ is a particular pseudo-solenoid for each $n = 1,2,\ldots$. Proceeding as above we see that X is an hereditarily indecomposable circle-like continuum. Hence, if X were chainable, X would be a pseudo-arc and thus, by (14.3), so would $\mu^{-1}(t_n)$ for each $n = 1,2,\ldots$. Hence, X is not chainable and, therefore, X is a pseudo-solenoid. Furthermore, by (14.23), X is the particular pseudo-solenoid that $\mu^{-1}(t_n)$ is for each $n = 1,2,\ldots$. This proves (3) and completes the proof of (14.54).

The next theorem gives three more strong Whitney-reversible properties. The proof uses some later results.

(14.55) THEOREM. *Each of the following three properties is a strong Whitney-reversible property: (i) being irreducible; (ii) being in Class[W]; (iii) having CP.*

Proof. Let μ be some Whitney map for $C(X)$. Assume $\mu^{-1}(t)$ is irreducible for all $0 < t < \mu(X)$. Then, since (1') of (14.73.5) implies (1) of (14.73.5), $\mu^{-1}(0)$ is irreducible. Hence, since $\mu^{-1}(0) \approx X$, X is irreducible. Next, assume $\mu^{-1}(t) \in$ Class[W] for all $0 < t < \mu(X)$. Then, by a statement in (14.73.25), $\mu^{-1}(t)$ is irreducible for all $0 < t < \mu(X)$. Thus, since (1') of (14.73.5) implies (3) of (14.73.5), $X \in$ CP. Hence, by (14.73.21), we have that $X \in$ Class[W]. Finally, assume $\mu^{-1}(t) \in$ CP for all $0 < t < \mu(X)$. Then, by the second part of (14.73.1) [with X replaced by $\mu^{-1}(t)$], $\mu^{-1}(t)$ is irreducible for all $0 < t < \mu(X)$. Thus, since (1') of (14.73.5) implies (3) of (14.73.5), $X \in CP$. Therefore, we have proved (14.55).

(14.55.1) REMARK AND QUESTION. Part (i) of (14.55) gives a partial answer to the question in (14.73.23). In connection with (14.73.23), also note the following: Assume there is a sequence $\{t_n\}_{n=1}^{\infty}$ such that $t_n \to 0$ as $n \to \infty$ and $\mu^{-1}(t_n)$ is irreducible for each $n = 1, 2, \ldots$. Then, $\mu^{-1}(t_n)$ is not a triod for any $n = 1, 2, \ldots$. Thus, by (14.49.1), X is not a triod. Hence, if X is unicoherent, X is irreducible by [306, Theorem 3.2]. Therefore, we have proved that, for unicoherent continua X, "sequences suffice" to force irreducibility of X. I do not know if "sequences suffice" to prove irreducibility of X for non-unicoherent continua X. In fact, I do not know the answer to the following general question: Is there a strong Whitney-reversible property P, a continuum X, and a sequence $\{t_n\}_{n=1}^{\infty}$ such that $t_n \to 0$ as $n \to \infty$, $\mu^{-1}(t_n)$ has property P for each $n = 1, 2, \ldots$, but X does not have property P? In particular, what about the properties in (14.55)?

(14.55.2) REMARK. Let X be a continuum and let μ be a Whitney map for $C(X)$. It follows easily using (14.1) and (14.8.1) that $\mu^{-1}(t)$ contains no arc for any $0 < t < \mu(X)$ if and only if X is hereditarily indecomposable (comp., [286, Theorem 4.4]). However, the hypotheses in the first part of (14.52) can hold and yet X can be decomposable. The types of continua X in (12.10) can be used to see this.

We have given some results which illustrate the concepts introduced in (14.45). Many questions remain to be answered:

(14.56) QUESTION. Is there a Whitney-reversible property which is not a strong Whitney-reversible property? The problem is, of course, that a continuum can admit essentially different Whitney maps [see (14.43)].

(14.57) QUESTION. For a given topological property determine whether or not it is a strong Whitney-reversible property [or, Whitney-reversible property]. Specifically, what about the following properties: Acyclic, absolute neighborhood retract, absolute retract, chainable, circle-like or proper circle-like, contractibility, hereditarily decomposable[2], λ connected, one-dimensional[2], being a particular solenoid, weakly chainable? Perhaps $X = \cup_{n=1}^{\infty} X_n$, where each X_n is the circle in R^2 with center $([n-1]/n, 0)$ and radius $1/n$, can be used to see that being an absolute neighborhood retract is not a Whitney-reversible property; for $t > 0$, all but finitely many of the "holes" in X seem to disappear in $\mu^{-1}(t)$.

(14.60)

F.
Mappings between the Levels $\mu^{-1}(t)$

In (14.39) through (14.42), we discussed the situation when the continua $\mu^{-1}(t)$, $0 \leq t < \mu(X)$, are all mutually homeomorphic. A more general situation of interest is the following:

(14.58) QUESTIONS [see (14.60)]. For what continua X are there mappings between all the levels $\mu^{-1}(t)$, $0 \leq t < \mu(X)$, for all [resp., some] Whitney maps μ for $C(X)$? For a given t and s, when is there a mapping from $\mu^{-1}(t)$ onto $\mu^{-1}(s)$?

A special case of (14.58) is the question of the existence of mappings between $\mu^{-1}(t)$ and $X \cong \mu^{-1}(0)$. Let us note the following result.

(14.59) THEOREM. *If X is a decomposable continuum such that, for some Whitney map μ for $C(X)$, there is a mapping from $\mu^{-1}(t)$ onto X for each $0 \leq t < \mu(X)$, then X is arcwise connected.*

Proof. The theorem follows immediately from (14.27).

Let us note that mappings between all the levels $\mu^{-1}(t)$, $0 \leq t < \mu(X)$, exist when X is a locally connected continuum by (14.9).

(14.60) REMARK. The first paper concerned with the existence of mappings as in (14.58) was [249, p.407], where the following question was asked: Is there always a mapping from X onto $\mu^{-1}(t)$ for any $t \in [0, \mu(X)]$? In [286, 4.1], Rogers gave a negative answer to this question--his continuum X is the one in (14.43.2). Let us note that there is not always a mapping from $\mu^{-1}(t)$ onto X --such is the case for $X = S_o$ [defined in (8.22)] and any t *near enough to* $\mu(X)$ [249, p.407]. In fact, there are continua X such that there is no mapping from $\mu^{-1}(t)$ onto X for *any* t > 0 --such must be the case for continua as in [249, Example 3]-- see (14.48).

G.
Properties of Whitney Maps for 2^X

In this section, we turn our attention to Whitney maps for 2^X. Recall that a fundamental fact about Whitney maps for C(X) is that they are *always* monotone, i.e., the property of being a continuum is a Whitney property [see (14.2)]. Moreover [see (14.44)], *any* Whitney map for C(X) is also open. However, Whitney maps for 2^X need not have either of these two pleasant properties. The following example shows this.

(14.61) EXAMPLE [249, Example 1]. Let Y be the polygonal arc in R^2 given by

(14.61) WHITNEY PROPERTIES 467

$Y = \{(x, y) \in R^2: \max\{|x|, |y|\} = 1$ and such that

if $x = 1$ then $y \leq 0$ or $y \geq 1/2\}$.

Let the metric d for Y be the usual Euclidean metric for R^2 [see (0.19)] restricted to $Y \times Y$. Let ω denote the Whitney map for 2^Y constructed using d as in (0.50.1). Let $A_0 = \{(1, 0), (1, 1/2)\}$ and let

$$W = \{K \in 2^Y: H(A_0, K) < 1/4\}.$$

Let $B \in W$ such that $B \neq A_0$. By the geometry, there exists a two-element subset B_0 of B such that $B_0 \neq A_0$ and $B_0 \in W$. Again by the geometry,

(1) $\quad \text{diam}[A_0] < \text{diam}[B_0]$.

From the way $\omega_2, \omega_3, \ldots, \omega_n, \ldots$ are defined in (0.50.1) and from the fact that A_0 and B_0 are each two-element sets, we have that

(2) $\quad \omega_2(A_0) = \text{diam}[A_0]$,

(3) $\quad \omega_2(B_0) = \text{diam}[B_0]$,

(4) $\quad \omega_n(A_0) = 0 = \omega_n(B_0)$ for each $n \geq 3$.

By (2) through (4) and by the formula for ω in (0.50.1),

(5) $\quad \omega(A_0) = \dfrac{\text{diam}[A_0]}{2}$ and $\omega(B_0) = \dfrac{\text{diam}[B_0]}{2}$.

By (1) and (5),

(6) $\quad \omega(A_0) < \omega(B_0)$.

Since $B_0 \subset B$, $\omega(B_0) \leq \omega(B)$. Hence, by (6),

(7) $\quad \omega(A_0) < \omega(B)$.

Let $t_0 = \omega(A_0)$. Since B was an arbitrary member of W different from A_0, it follows from (7) that

(8) $\omega[W] \subset [t_o, \omega(Y)]$ and $\omega^{-1}(t_o) \cap W = \{A_o\}$.

Since W is an open subset of 2^Y, it follows immediately from the second part of (8) that A_o is an isolated point of $\omega^{-1}(t_o)$. Hence, $\omega^{-1}(t_o)$ is not connected. Therefore, ω is not monotone. Since W is an open subset of 2^Y and $t_o \neq 0$, it follows immediately from (8) that $\omega[W]$ is not an open subset of $[0, \omega(Y)]$. Therefore, ω is not an open mapping. Thus we see that Whitney maps for 2^X (may) behave quite differently than those for $C(X)$.

(14.62) REMARK. In (14.61), we gave an example of an arc Y and a Whitney map ω for 2^Y such that $\omega^{-1}(t_o)$ is not connected for some t_o. On the other hand [249, Remark on p.396], let $Z = [0, 1]$ with the usual "absolute-value distance" d. Let ω' denote the Whitney map for 2^Z constructed using d as in (0.50.1). It can then be verified that ω' is both monotone and open [see (14.66) for a more general fact]. In particular, all point inverses under ω' are connected. Thus we see that, *even* when X is an arc, the topological types of Whitney map point inverses in 2^X depend on the Whitney map. Of course, when X is an arc [but not in general--see (14.43.1)], the topological types of Whitney map point inverses in $C(X)$ do not depend on the Whitney map for $C(X)$ [by (14.6)].

(14.63) QUESTION. Given any continuum X, is there a Whitney map for 2^X which is monotone? See (14.66).

(14.64) QUESTION. Given any continuum X, is there a Whitney map for 2^X which is open? See (14.65) - (14.67) and (14.69).

The following result shows where the trouble spots are for the openness of any Whitney map for 2^X.

(14.65) THEOREM. *Let* X *be a continuum and let* ω *denote any Whitney map for* 2^X. *Let* $A \in 2^X$ *such that* A *consists of just one point or* A *is infinite. Then: Given any* $\varepsilon > 0$, *there exists an open subset* U *of* $[0, \omega(X)]$ *such that* $\omega(A) \in U \subset \omega[B_H(\varepsilon, A)]$, *where* $B_H(\varepsilon, A) = \{K \in 2^X: H(A, K) < \varepsilon\}$.

Proof. It is easy to see, using the fact that the restriction of ω to $C(X)$ is an open mapping [see (14.44)], that (14.65) is true when $A \in C(X)$. Hence, for the rest of the proof, we may assume $A \notin C(X)$ --actually, the rest of the proof will be done under the assumption that $A \neq X$ and A is infinite. Let $\varepsilon > 0$. Let $a_o \in A$ and let M denote the component of $cl[N(\varepsilon/2, A)]$ such that $a_o \in M$. Since $A \neq X$, it follows using (20.1) that

$A \neq [A \cup M] \supset A$.

Hence, by (0.50.a), $\omega(A \cup M) > \omega(A)$. Thus, it follows easily that there is a finite set F such that

(1) $F \subset A$,

(2) $a_o \in F$,

(3) $A \subset N(\varepsilon, F)$,

(4) $\omega(F \cup M) > \omega(A)$.

Since (2) holds and since M is a continuum such that $a_o \in M$, it follows easily that $A_0 = F$ and $A_1 = F \cup M$ satisfy (1.25.2). Hence, by (1.25), there is a segment $\sigma: [0, 1] \to 2^X$ such that $\sigma(0) = F$ and $\sigma(1) = F \cup M$. By (1.15.4),

(5) $\quad F \subset \sigma(t) \subset [F \cup M]\quad$ for each $t \in [0, 1]$.

Since $M \subset cl[N(\varepsilon/2, A)] \subset N(\varepsilon, A)$ and $F \subset A$ [by (1)];

(6) $\quad [F \cup M] \subset N(\varepsilon, A)$.

By (5) and (6),

(7) $\quad \sigma(t) \subset N(\varepsilon, A) \quad$ for each $t \in [0, 1]$.

By (3) and (5),

(8) $\quad A \subset N[\varepsilon, \sigma(t)] \quad$ for each $t \in [0, 1]$.

By (7), (8), and the formula for H in (0.1),

(9) $\quad \sigma(t) \in B(\varepsilon, A) \quad$ for each $t \in [0, 1]$.

It follows easily using (1.15.3) that

(10) $\quad \omega[\{\sigma(t): 0 < t < 1\}]$ is the nonempty [by (4)]
    open interval $(\omega(F), \omega(F \cup M))$.

Now note that, since F is finite and A is infinite, $F \ne A$. Hence, by (1) and (0.50.a),

(11) $\quad \omega(F) < \omega(A)$.

By (4) and (11),

(12) $\quad \omega(A) \in (\omega(F), \omega(F \cup M))$.

By (9) and (10),

(13) $\quad (\omega(F), \omega(F \cup M)) \subset \omega[B(\varepsilon, A)]$.

Therefore, by taking $U = (\omega(F), \omega(F \cup M))$, we see from (12) and (13) that we have proved (14.65).

Thus, we see that the "trouble spots" referred to in the

sentence preceding (14.65) are the finite nondegenerate sets.

Let Z and ω' be as in (14.62), where we said that ω' is both monotone and open. Noting that Z is convex (in the classical sense), we have the following generalization:

(14.66) THEOREM. *Let* X *be a compact convex (in the classical sense) subset of a normed linear space* L *with norm* $||\ ||$. *Let* ω *denote the Whitney map for* 2^X *constructed using* $||\ ||$ *as in (0.50.1). Then,* ω *is both monotone and open.*

Idea of proof. Using the convexity of X, it can be shown that $\omega^{-1}([0, t])$ is connected for any $t \leq \omega(X)$. Thus, since 2^X is unicoherent [by (1.176)], the proof for (14.2) may be applied to show that ω is monotone. We will not prove the openness here, except to mention that the proof is not difficult if (14.65) is used.

(14.67) REMARK. In (14.66) we saw that continua X, which are convex in the classical sense, admit monotone open Whitney maps for 2^X. It is known [see (0.65.4)] that every locally connected continuum admits a convex metric [for definition, see (0.65.3)]. In view of (14.66), one might suspect that the Whitney map ω for 2^X, constructed as in (0.50.1) by using a convex metric for a given locally connected continuum X, is open. This is not necessarily the case. For example: Let $X = S^1$ [see (0.19)], let ρ denote the arc length metric for X, and let ω be as we have indicated above. The metric ρ is convex, but ω is not an open mapping. To see this, let $x_1 = (1, 0)$, $x_2 = (-1/2, \sqrt{3}/2)$, $x_3 = (-1/2, -\sqrt{3}/2)$, $A_o = \{x_1, x_2, x_3\}$, and $U = \langle U_1, U_2, U_3 \rangle$ where $\langle \rangle$ is as defined in (0.10) and, for each $i = 1, 2$, and 3,

$$U_i = \{z \in X: \rho(x_i, z) < \pi/4\}.$$

By using (0.50.a) and the formula for ω in (0.50.1), it can be shown that

(1) if $B \in 2^X$ such that $B \in U$, then $\omega(B) \geq \omega(A_o)$.

It follows easily from (1) that $\omega[U]$ is not an open subset of $[0, \omega(X)]$. Therefore, ω is not an open mapping. We also mention that the diameter mapping with respect to ρ [comp., (1.211.3)] is not an open mapping from 2^X onto $[0, \pi]$. In fact, I do not know the answers to the following questions.

(14.68) QUESTION. Is there a circle Y such that the diameter mapping on 2^Y is an open mapping? More generally, for what continua X is there (for some metric for X) an open diameter mapping for 2^X?

(14.69) QUESTION. Is there an open Whitney map for 2^Y where Y is a circle [see (14.64)]?

(14.70) QUESTION. When are there mappings of any of the types mentioned in (4.15) from 2^X onto $[0, 1]$? Of course, if X is a locally connected continuum, all these types of mappings exist because $2^X \cong I_\infty$ by (1.97). Hence, the question is for non-locally connected continua. Note that there is always a monotone open mapping from $C(X)$ onto $[0, 1]$ by (14.44).

H.
Spaces of Whitney Maps

We adopt the following notation:

(14.71) NOTATION. For a continuum X, let $W[C(X)]$ (resp., $W[2^X]$) denote the space of all Whitney maps for $C(X)$ (resp., 2^X) with the "sup metric."

The spaces in (14.71) have not been investigated before now. Note that, under pointwise multiplication, they each form a topological semigroup [see (0.68.4)]. We assume this semigroup structure from now on. Let Y and Z be homeomorphic continua and let $h: Y \xrightarrow{\text{onto}} Z$ be a homeomorphism. Let $\hat{h}: C(Y) \to C(Z)$ be the induced map as defined in (0.49). Since h is a homeomorphism of Y onto Z, $\hat{h}$ is a homeomorphism of $C(Y)$ onto $C(Z)$. For each $\mu \in W[C(Y)]$, let
$$j(\mu) = \mu \circ \hat{h}^{-1}.$$
Since h is a homeomorphism, it is easy to prove directly [or, use (0.51.2)] that the formula above defines a function
$$j: W[C(Y)] \to W[C(Z)].$$
Furthermore, it is easy to verify that j is a homeomorphism and a semigroup isomorphism onto $W[C(Z)]$. Thus, if Y and Z are homeomorphic, $W[C(Y)]$ and $W[C(Z)]$ are topologically and alge-

braically isomorphic; by a similar argument, so are $W[2^Y]$ and $W[2^Z]$. It would be interesting to have some more information about these semigroups and their relationships to one another. In particular, I do now know the answers to the questions in (14.71.1) and (14.71.2).

(14.71.1) QUESTION. Are the converses of the results in (14.71) true? In other words: If Y and Z are continua such that $W[C(Y)]$ (resp., $W[2^Y]$) is homeomorphic, or both homeomorphic and algebraically isomorphic, to $W[C(Z)]$ (resp., $W[2^Z]$), then must Y and Z be homeomorphic?

(14.71.2) QUESTION. For what continua X are $W[C(X)]$ and $W[2^X]$ homeomorphic and/or algebraically isomorphic?

Let us note the following easy-to-prove fact [see (0.68.4)]:

(14.71.3) THEOREM. *For any continuum* X, $W[C(X)]$ *(resp., $W[2^X]$) is a positive convex cone without vertex in the Banach space of all continuous real-valued functions on* $C(X)$ *(resp., 2^X) with the "sup norm." Hence,* $W[C(X)]$ *and* $W[2^X]$ *are each contractible and locally contractible.*

(14.71.4) QUESTION. What other topological properties, besides those in (14.71.3), do the spaces $W[C(X)]$ and $W[2^X]$ possess? For example, are these spaces topologically complete?

(14.71.5) QUESTION. Is the mapping $f: W[2^X] \to W[C(X)]$ onto $W[C(X)]$, where $f(\omega)$ denotes the restriction of ω to $C(X)$ for each $\omega \in W[2^X]$? In other words: Can every Whitney map for $C(X)$ be extended to a Whitney map for 2^X?

I.
Cut Points of $\mu^{-1}(t)$

Bruce Hughes has characterized those $A \in \mu^{-1}(t)$ such that A is a cut point of $\mu^{-1}(t)$. In this section, we state his result [see (14.72.1)] and mention some applications and some related results.

(14.72.0) REMARK. Let X be a continuum and let μ be a Whitney map for $C(X)$. Recall [from (14.2)] that $\mu^{-1}(t)$ is a continuum for each $t \in [0, \mu(X))$. It may be that $\mu^{-1}(t)$ has no cut points [for example, see (14.7)]. The result in (14.72.1) determines, in terms of a special type of separation of X by A, when an $A \in \mu^{-1}(t)$ is a cut point of $\mu^{-1}(t)$.

(14.72.1) THEOREM [148, Theorem 2.1]. *Let X be a continuum, let μ be a Whitney map for $C(X)$, and let $t \in [0, \mu(X)]$. Let $A \in \mu^{-1}(t)$. Then: $\mu^{-1}(t)\setminus\{A\}$ is not connected if and only if there exist disjoint nonempty open subsets X_1 and X_2 of X such that $X\setminus A = X_1 \cup X_2$ and such that, for any $B \in \mu^{-1}(t)$,*
$B \subset [X_1 \cup A]$ *or* $B \subset [X_2 \cup A]$.

(14.72.2) REMARK. Bruce Hughes [148, Corollary 2.2] has used (14.72.1) to give another proof of (14.6). His proof uses the characterization of the arc as the only continuum with exactly two noncut points.

By using (14.72.1) and (14.8.1), Hughes obtaines the following interesting result. Recall that a continuum Z is said to be *irreducible* provided that there exist two points p, q ε Z such that no proper subcontinuum of Z contains p, q. When this happens, we say Z is *irreducible about* p *and* q.

(14.72.3) THEOREM [148, Theorem 3.1]. *Let* X *be a continuum. If there is a Whitney map* μ *for* C(X) *such that, for some* $t_o \in [0, \mu(X))$, $\mu^{-1}(t_o)$ *is decomposable and irreducible [about two members of* $\mu^{-1}(t_o)$], *then* X *is decomposable.*

Proof. Let μ and t_o be as in the hypotheses. Since $\mu^{-1}(t_o)$ is decomposable, there exist proper subcontinua Λ and Γ of $\mu^{-1}(t_o)$ such that $\mu^{-1}(t_o) = \Lambda \cup \Gamma$. By (1.49), $L = \cup \Lambda$ and $G = \cup \Gamma$ are subcontinua of X. Also, since $\mu^{-1}(t_o) = \Lambda \cup \Gamma$,

$$L \cup G = [\cup \Lambda] \cup [\cup \Gamma] = \cup \mu^{-1}(t_o) = X.$$

Since $\mu^{-1}(t_o)$ is irreducible and since Λ and Γ are proper subcontinua of $\mu^{-1}(t_o)$, we have that $L \neq X \neq G$ by (14.73.2) [note: An argument using (14.8.1) and (14.72.1), and not using (14.73.2), has been given in [148] to prove $L \neq X \neq G$]. Thus we have that $X = L \cup G$ where L and G are proper subcontinua of X. Therefore, X is decomposable. This proves (14.72.3).

The continuum Y in (14.12) shows the necessity of the assumption in (14.72.3) that $\mu^{-1}(t_o)$ be irreducible.

(14.72.4) REMARK. Hughes [148, Corollary 3.2] gives another proof of (14.13) using (14.4), (14.72.3), and a result of Sorgenfrey [306, Theorem 3.2] which implies any chainable continuum is irreducible about some two points.

J.
The Covering Property
(Revisited)

The covering property has been defined here in (14.14). It originated in [179] for the purpose of proving that indecomposability is a Whitney property for chainable continua [here, (14.13)]. The following question about the covering property was asked in [179, section 6]:

(14.73) QUESTION [179, section 6]. What class of continua does the covering property characterize?

Some answers to (14.73), and many other facts, will be presented in this section. Bruce Hughes and Ann Petrus became interested in the covering property and, before writing up their results for publication, they have been kind enough to allow their results and questions to be included here.

The first two results below give basic and interesting facts about the covering property. They are summarized in (14.73.3), which shows that the covering property is equivalent to $\mu^{-1}(t)$ being irreducible for each μ and each $t \in [0, \mu(X))$. In (14.73.4), it is shown that the covering property is, in a sense, independent of μ. The results in (14.73.1) through (14.73.4) are then summarized, with some added information, in (14.73.5).

The proof of the following result uses (14.73.18). For the purpose of emphasizing the next two results, we have postponed the statement and proof of the result in (14.73.18). Note that (14.73.18) is a direct consequence of (14.73.1)--of course, (14.73.18) will be proved without using (14.73.1).

(14.73.1) THEOREM [Bruce Hughes]. *If* $X \in CP$, *then* $\mu^{-1}(t)$ *is irreducible for any Whitney map* μ *for* $C(X)$ *and each* $t \in [0, \mu(X))$. *Hence,* $X \in CP$ *implies* X *is irreducible.*

Proof. Assume $X \in CP$. By the second part of (14.14.1), X is unicoherent and, by the first part of (14.73.18), X is not a triod. Hence, by [306, Theorem 3.2], X is irreducible about, say, p and q. Now, let $t \in [0, \mu(X))$ be fixed and let $A, B \in \mu^{-1}(t)$ such that $p \in A$ and $q \in B$ [A and B exist by (1.25)]. Then, since X is irreducible about p and q, it is easy to see using the covering property that $\mu^{-1}(t)$ is irreducible about A and B.

For use below, we adopt the following terminology and notation.

Let X be a continuum and let μ be a given Whitney map for $C(X)$. We say that X *has the covering property relative to* μ, written $X \in CP(\mu)$, provided that no proper subcontinuum of $\mu^{-1}(t)$ covers X for any $t \in [0, \mu(X)]$. As we will see in (14.73.4), $CP(\mu)$ for some μ is equivalent to CP.

(14.73.2) THEOREM [Bruce Hughes]. *If* μ *is a Whitney map for* $C(X)$ *such that* $\mu^{-1}(t_o)$ *is irreducible for some* $t_o \in [0, \mu(X))$, *then no proper subcontinuum of* $\mu^{-1}(t_o)$ *covers* X [*written:* $X \in CP(\mu)$ *at* t_o].

Proof. Let μ and t_o be as in the hypothesis and let M_1, $M_2 \in \mu^{-1}(t_o)$ such that $\mu^{-1}(t_o)$ is irreducible about M_1 and M_2. Let Γ be a subcontinuum of $\mu^{-1}(t_o)$ such that $\cup \Gamma = X$. We will show that M_1, $M_2 \in \Gamma$, from which it follows that $\Gamma = \mu^{-1}(t_o)$. Let $i \in \{1, 2\}$. Choose α_i as follows. If $M_i \in \Gamma$, then let $\alpha_i = \{M_i\}$. Next, assume $M_i \notin \Gamma$. Since $\cup \Gamma = X$, there exists $G_i \in \Gamma$ such that $M_i \cap G_i \neq \emptyset$. Note that, since $M_i \notin \Gamma$, $M_i \neq G_i$. Hence, by (14.8.1), there exists an arc $\alpha \subset \mu^{-1}(t_o)$ with end points M_i and G_i. Assuming without loss of generality that $\alpha \cap \Gamma = \{G_i\}$, let $\alpha_i = \alpha$. We have thus chosen α_1 and α_2. Now, for the purpose of proof, suppose $M_1 \notin \Gamma$. Then, $M_1 \neq G_1$. Hence, since M_1, $G_1 \in \mu^{-1}(t_o)$, we have by (0.50.a) that $M_1 \notsubset G_1$. Thus, there exists $x_1 \in [M_1 \setminus G_1]$. Let $f: [0, 1] \xrightarrow{onto} \alpha_1$ be a homeomorphism such that $f(0) = M_1$ and $f(1) = G_1$. Since $S = \{s \in [0, 1]: x_1 \in f(s)\}$ is nonempty and compact, $s_1 = \text{l.u.b.}(S) \in S$ and therefore, since $x_1 \notin G_1 = f(1)$, $s_1 < 1$.

Since $\cup\Gamma = X$, there exists $X_1 \in \Gamma$ such that $x_1 \in X_1$. Note that, since $f(s_1) \notin \Gamma$ [by the way we chose α_1 and the fact that $s_1 < 1$], $X_1 \neq f(s_1)$. Hence, since $x_1 \in [f(s_1) \cap X_1]$, we have by (14.8.1) that there is an arc $\beta \subset \mu^{-1}(t_o)$ with end points $f(s_1)$ and X_1 such that $x_1 \in B$ for each $B \in \beta$. The fact that $x_1 \in B$ for each $B \in \beta$ and the definition of s_1 imply

(1) $\beta \cap \alpha_1 \subset f([0, s_1])$.

Since $s_1 < 1$ and $\alpha_1 \cap \Gamma = \{G_1\} = \{f(1)\}$, we have that

(2) $f([0, s_1]) \cap \Gamma = \emptyset$.

Since $f(s_1) \in [\beta \cap \alpha_1]$,

(3) $\beta \cap \alpha_1 \neq \emptyset$

and, since $X_1 \in [\beta \cap \Gamma]$,

(4) $\beta \cap \Gamma \neq \emptyset$.

Thus, since β is connected and since $f([0, s_1])$ and Γ are compact, it follows from (1) through (4) that

(5) $\beta \not\subset \alpha_1 \cup \Gamma$.

Note that, since $\alpha_1 \cup \Gamma \cup \alpha_2$ is a subcontinuum of $\mu^{-1}(t_o)$ such that $M_1, M_2 \in [\alpha_1 \cup \Gamma \cup \alpha_2]$,

(6) $\alpha_1 \cup \Gamma \cup \alpha_2 = \mu^{-1}(t_o)$.

Since $\beta \subset \mu^{-1}(t_o)$, we have by (5) and (6) that

(7) $\beta \cap \alpha_2 \neq \emptyset$.

It follows using (3) and (7), that $\alpha_1 \cup \beta \cup \alpha_2$ is a subcontinuum of $\mu^{-1}(t_o)$. Hence, since $M_1 \in \alpha_1$ and $M_2 \in \alpha_2$,

(8) $\alpha_1 \cup \beta \cup \alpha_2 = \mu^{-1}(t_o)$.

By (8), $\mu^{-1}(t_o)$ is an arcwise connected continuum (in fact, locally connected). Thus, since $\mu^{-1}(t_o)$ is irreducible about M_1 and

M_2,

(9) $\mu^{-1}(t_o)$ is an arc with end points M_1 and M_2.

Since $f(0) = M_1$ and $s_1 < 1$, it follows easily using (1), (3), and (9), that $\beta \subset f([0, s_1])$. Hence, by (4), $f([0, s_1]) \cap \Gamma \neq \emptyset$. This contradicts (2). Therefore, $M_1 \in \Gamma$. A similar argument shows that $M_2 \in \Gamma$. Therefore, $\Gamma = \mu^{-1}(t_o)$. We have proved (14.73.2).

An application of (14.73.2) is given in the proof of (14.73.13).

Bruce Hughes has asked me if the converse of (14.73.2) is true. I have the following example to show that it is not. Let $X = \cup_{i=1}^{4} X_i$ where X_1 is an hereditarily indecomposable continuum and, for each $i \in \{2, 3, 4\}$, X_i is an arc satisfying: $X_i \cap X_1 = \{x_i\}$ where x_i is an end point of X_i and, if $j \in \{2, 3, 4\}$ such that $j \neq i$, $X_i \cap X_j = \emptyset$ and x_i and x_j are points of different composants of X_1. Let μ be any Whitney map for $C(X)$. Let $t_o = \mu(X_1)$. Then, it is easy to see that $\mu^{-1}(t_o)$ is a simple triod (as was observed in [267, Example 14(b)]). Thus, $\mu^{-1}(t_o)$ is not irreducible. It is also easy to see that no proper subcontinuum of $\mu^{-1}(t_o)$ covers X --in fact, $\sigma_{t_o,\mu}$ is one-to-one on $C(\mu^{-1}(t_o))$ [comp., (14.73.13) and the paragraph following its proof].

In spite of the example in the paragraph above, note the following simple consequence of (14.73.1) and (14.73.2).

(14.73.3) THEOREM [Bruce Hughes]. *A continuum* $X \in CP$ *if and only if* $\mu^{-1}(t)$ *is irreducible for each Whitney map* μ *for* $C(X)$ *and for each* $t \in [0, \mu(X))$.

Petrus has shown that the covering property relative to μ is actually a property of the continuum, i.e., is independent of the choice of μ. This result, which is stated next, should be useful in determining when continua have the covering property.

(14.73.4) THEOREM [267, Proposition 18]. *Let* X *be a continuum and let* μ *be a Whitney map for* $C(X)$. *If* $X \in CP(\mu)$, *then* $X \in CP(\mu_1)$ *for any Whitney map* μ_1 *for* $C(X)$. *In other words: If* $X \in CP(\mu)$, *then* X *has the covering property.*

Proof. Assume $X \in CP(\mu)$ and let μ_1 be a Whitney map for $C(X)$. Let $t_1 \in [0, \mu_1(X)]$ and let Λ_1 be a proper subcontinuum of $\mu_1^{-1}(t_1)$. We will show that $\cup \Lambda_1 \neq X$. Let $A_0 \in [\mu_1^{-1}(t_1) \setminus \Lambda_1]$. Let $t_0 = \mu(A_0)$. Define Λ by

$$\Lambda = \{B \in \mu^{-1}(t_0): \text{ for some } K \in \Lambda_1, K \subset B \text{ or } K \supset B\}.$$

We will prove the facts in (*) and (**) below from which, since $X \in CP(\mu)$, it follows that $\cup \Lambda_1 \neq X$.

(*) $[\cup \Lambda_1] \subset [\cup \Lambda]$;

(**) Λ is a proper subcontinuum of $\mu^{-1}(t_0)$.

Proof of (*). Let $x \in [\cup \Lambda_1]$. Then there exists $K \in \Lambda_1$ such that $x \in K$. We consider two cases-- $\mu(K) \geq t_0$ and $\mu(K) < t_0$. First assume $\mu(K) \geq t_0$. Then, since $\{x\} \subset K \in C(X)$ and μ is continuous, it follows using (1.8) and (1.11) that there exists $B \in \mu^{-1}(t_o)$ such that $x \in B \subset K$. Hence, $B \in \Lambda$ and therefore, since $x \in B$, we have $x \in [\cup \Lambda]$. Next assume $\mu(K) < t_o$. Then, since $X \supset K \in C(X)$ and μ is continuous, it follows using (1.8) and (1.11) that there exists $B \in \mu^{-1}(t_o)$ such that $K \subset B$. Hence, $B \in \Lambda$ and therefore, since $x \in K \subset B$, we have $x \in [\cup \Lambda]$. We have now proved (*).

Proof of (**). By definition, $\Lambda \subset \mu^{-1}(t_o)$. To see that $\Lambda \neq \mu^{-1}(t_o)$, note that, by definition of t_o, $A_o \in \mu^{-1}(t_o)$. We show that $A_o \notin \Lambda$. Supoose $A_o \in \Lambda$. Then, by definition of Λ, there exists $K \in \Lambda_1$ such that $K \subset A_o$ or $K \supset A_o$. Hence, since $\Lambda_1 \subset \mu_1^{-1}(t_1)$ and $A_o \in \mu_1^{-1}(t_1)$, we have by (0.50.a) that $K = A_o$. Thus, since $K \in \Lambda_1$, we have that $A_o \in \Lambda_1$. This is a contradiction, since $A_o \notin \Lambda_1$. Therefore, $A_o \notin \Lambda$. This proves that $\Lambda \neq \mu^{-1}(t_o)$. An easy sequence argument, using the compactness of Λ_1, shows that Λ is compact; we leave the details to the reader. We complete the proof of (**) by showing that Λ is connected. Suppose Λ is not connected. Then $\Lambda = \Gamma_1 \cup \Gamma_2$ where Γ_1 and Γ_2 are disjoint nonempty compact subsets of Λ. For each $i = 1$ and 2, define Γ_i^1 by

$$\Gamma_i^1 = \{A \in \Lambda_1 : \text{for some } G \in \Gamma_i, G \subset A \text{ or } G \supset A\}.$$

We will show that $\Lambda_1 = \Gamma_1^1 \cup \Gamma_2^1$ and that Γ_1^1 and Γ_2^1 are nonempty compact disjoint sets. To show $\Lambda_1 = \Gamma_1^1 \cup \Gamma_2^1$, it suffices to show $\Lambda_1 \subset [\Gamma_1^1 \cup \Gamma_2^1]$. To do this, let $L_1 \in \Lambda_1$. By (1.8), (1.11), and the continuity of μ, there exists $M \in \mu^{-1}(t_o)$ such that $L_1 \subset M$ or $L_1 \supset M$. Hence, by definition of Λ, $M \in \Lambda$. Hence, since $\Lambda = \Gamma_1 \cup \Gamma_2$, $M \in [\Gamma_1 \cup \Gamma_2]$. But $M \in \Gamma_i$ implies, by definition of Γ_i^1, that $L_1 \in \Gamma_i^1$. We had proved $\Lambda_1 = \Gamma_1^1 \cup \Gamma_2^1$. Using the definition of Λ and the fact that Γ_1 is a nonempty subset of Λ, it follows easily that Γ_1^1 is nonempty. Similarly, Γ_2^1 is nonempty. An easy sequence argument, using the compactness of Γ_i, shows that Γ_i^1 is compact for each $i = 1$ and 2. Next we show that $\Gamma_1^1 \cap \Gamma_2^1 = \emptyset$. Suppose there exists an $A \in [\Gamma_1^1 \cap \Gamma_2^1]$. We consider two cases-- $\mu(A) \geq t_o$ and $\mu(A) < t_o$. First assume $\mu(A) \geq t_o$. Then, since $A \in \Gamma_i^1$ and $\Gamma_i \subset \Lambda \subset \mu^{-1}(t_o)$, it follows from (0.50.a) and the definition of Γ_i^1 that there exists $G_i \in \Gamma_i$ such that $G_i \subset A$ for each $i = 1$ and 2. Hence, $G_i \in [C(A) \cap \mu^{-1}(t_o)]$ for each $i = 1$ and 2. By part (2) of (14.11.1), $C(A) \cap \mu^{-1}(t_o)$ is a subcontinuum of $\mu^{-1}(t_o)$. Thus, since $A \in \Lambda_1$, it follows from the definition of Λ that $[C(A) \cap \mu^{-1}(t_o)] \subset \Lambda$. Therefore, we have: $C(A) \cap \mu^{-1}(t_o)$ is a subcontinuum of Λ and, since $G_i \in \Gamma_i$ and $G_i \in [C(A) \cap \mu^{-1}(t_o)]$ for each $i = 1$ and 2, $[C(A) \cap \mu^{-1}(t_o)] \cap \Gamma_i \neq \emptyset$ for each $i = 1$ and 2. This contradicts the choice of Γ_1 and Γ_2. Next assume $\mu(A) < t_o$. Then, since $A \in \Gamma_i^1$ and $\Gamma_i \subset \Lambda \subset \mu^{-1}(t_o)$, it follows from (0.50.a) and the definition of Γ_i^1 that there exists $G_i' \in \Gamma_i$ such that $G_i' \supset A$ for each $i = 1$ and 2. Thus, since G_1', $G_2' \in \mu^{-1}(t_o)$ and since A is a subcontinuum of $G_1' \cap G_2'$ [note also that, since $\Gamma_1 \cap \Gamma_2 = \emptyset$, $G_1' \neq G_2'$], we have by (14.8.1) that there is an arc $\alpha \subset \mu^{-1}(t_o)$ such that α has end points G_1' and G_2' and such that $A \subset L$ for each $L \in \alpha$. Hence, since $A \in \Lambda_1$, it follows from the definition of Λ that $\alpha \subset \Lambda$. Therefore, since $G_i' \in [\Gamma_i \cap \alpha]$ for each $i = 1$ and 2, we have a contradiction to the choice of Γ_1 and Γ_2. We have now proved that $\Gamma_1^1 \cap \Gamma_2^1 = \emptyset$. Thus, we have shown Λ_1 is not connected, a contradiction. Therefore, Λ is connected. This completes the proof of (**) and of (14.73.4).

As a consequence of (14.73.2) through (14.73.4), we have the following result. An application of it has been given in the proof of (14.55).

(14.73.5) COROLLARY. *For any continuum* X, *the following five statements are equivalent:*

(1) *For some Whitney map* μ *for* C(X), $\mu^{-1}(t)$
is irreducible for each $t \in [0, \mu(X))$.

(1') *For some Whitney map* μ *for* C(X), $\mu^{-1}(t)$
is irreducible for each $t \in (0, \mu(X))$.

(2) *For any Whitney map* μ *for* C(X), $\mu^{-1}(t)$
is irreducible for each $t \in [0, \mu(X))$.

(2') *For any Whitney map* μ *for* C(X), $\mu^{-1}(t)$
is irreducible for each $t \in (0, \mu(X))$.

(3) X $\in$ CP.

Proof. By (14.73.3), (2) and (3) are equivalent. By (14.73.2), (1) implies X $\in$ CP(μ) which implies (3) by (14.73.4). Note that no proper subcontinuum of $\mu^{-1}(0)$ covers X. Hence, by (14.73.2), (1') [resp., (2')] implies X $\in$ CP(μ) for some [resp., any] μ which implies, by (14.73.4) [resp., by the definition of CP], that (3) [resp., (3)] holds. Clearly, (2) implies (1), (1) implies (1'), and (2) implies (2'). Therefore, we have proved (14.73.5).

Petrus has studied continua which have the *covering property hereditarily*, i.e., each (nondegenerate) subcontinuum has the covering property. If a continuum X has the covering property hereditarily, then we will write $X \in CPH$. Let us note the following result, which should be useful in determining when continua have CPH.

(14.73.6) THEOREM [267, Corollary 19]. *If there exists a Whitney map μ for $C(X)$ such that, for any (nondegenerate) subcontinuum A of X,*

$$A \in CP(\mu|C(A))$$

where $\mu|C(A)$ denotes the restriction of μ to $C(A)$, then $X \in CPH$. Hence: If $X \in CPH(\mu)$, then $X \in CPH$.

Proof. The theorem follows easily from (14.73.4).

The mapping, defined below in (14.73.7), plays a significant role in investigating CPH.

(14.73.7) NOTATION [176, p.176]. Let X be a continuum and let μ be a Whitney map for $C(X)$. For each $t \in [0, \mu(X)]$ consider $C(\mu^{-1}(t))$ as a subset of $C(C(X)) \subset 2^{2^X}$ by inclusion. Recall from (1.48) that the union function

$$\cup: 2^{2^X} \to 2^X$$

is continuous. For each $t \in [0, \mu(X)]$, let $\sigma_{t,\mu}$ denote the restriction of $\cup$ to $C(\mu^{-1}(t))$. When there is no confusion, we will simply use the symbol σ_t instead of $\sigma_{t,\mu}$. Note that, for each $t \in [0, \mu(X)]$, σ_t is continuous and that, by (1.49) and (0.50.a),

$$\sigma_t: C(\mu^{-1}(t)) \to \mu^{-1}([t, \mu(X)]).$$

In fact, $\sigma_t[C(\mu^{-1}(t))] = \mu^{-1}([t, \mu(X)])$ as we now prove. Some other facts about σ_t are given in section L of this chapter.

(14.73.8) THEOREM [176, 1.2]. *For any continuum* X *and any Whitney map* μ *for* $C(X)$, $\sigma_{t,\mu}$ $[= \sigma_t]$ *is a mapping from* $C(\mu^{-1}(t))$ *onto* $\mu^{-1}([t, \mu(X)])$ *for each* $t \in [0, \mu(X)]$.

Proof. Let $t \in [0, \mu(X)]$. We have already observed in (14.73.7) that σ_t is a mapping of $C(\mu^{-1}(t))$ into $\mu^{-1}([t, \mu(X)])$. Hence, we need only show

(*) $\mu^{-1}([t, \mu(X)]) \subset \sigma_t[C(\mu^{-1}(t))]$.

To prove (*), let $A \in \mu^{-1}([t, \mu(X)])$. Let

$\Lambda = C(A) \cap \mu^{-1}(t)$.

By part (c) of (1.213.1), $\cup\Lambda = A$. By part (2) of (14.11.1), Λ is a subcontinuum of $\mu^{-1}(t)$, i.e., $\Lambda \in C(\mu^{-1}(t))$. This means Λ is in the domain of σ_t. Thus, by the definition of σ_t and the fact that $\cup\Lambda = A$, we have that $\sigma_t(\Lambda) = A$. Since $A \in \mu^{-1}([t, \mu(X)])$ was arbitrary, we have proved (*). This completes the proof of (14.73.8).

The following theorem restates the covering property in terms of the mappings σ_t.

(14.73.9) THEOREM [267, p.23]. *Let* X *be a continuum and let* μ *be a Whitney map for* C(X). *Then:* X *has the covering property if and only if* $\sigma_{t,\mu}^{-1}(X) = \{\mu^{-1}(t)\}$ *for each* $t \in [0, \mu(X)]$.

Proof. The theorem follows easily from (14.73.4), part (b) of (1.213.1), the definition of the covering property [in (14.14)], and the definition of $\sigma_{t,\mu}$ for each $t \in [0, \mu(X)]$.

The following interesting and useful theorem gives the relationship between CPH and the mappings σ_t. A more general version of it and some related facts are in (14.76.5). Some applications of it are in (14.73.11) and in (14.73.13) through (14.73.16). Also, see (14.76.3), (14.76.6), (14.76.8), and (14.76.11).

(14.73.10) THEOREM [267, p.23]. *Let* X *be a continuum and let* μ *be a Whitney map for* C(X). *Then:* X $\in$ CPH *if and only if* $\sigma_{t,\mu}$ *is one-to-one for each* $t \in [0, \mu(X)]$. *Hence,* X $\in$ CPH *if and only if* $\sigma_{t,\mu}$ *is a homeomorphism of* $C(\mu^{-1}(t))$ *onto* $\mu^{-1}([t, \mu(X)])$ *for each* $t \in [0, \mu(X)]$.

Proof. Assume X $\in$ CPH. Let $t \in [0, \mu(X)]$ and let $A \in \sigma_{t,\mu}[C(\mu^{-1}(t))]$. For the purpose of proving that $\sigma_{t,\mu}$ is one-to-one, assume A is nondegenerate [if $A \in F_1(X)$, then $\sigma_{t,\mu}^{-1}(A) = \{\{A\}\}$]. Then, since X $\in$ CPH, A has the covering property. Hence, since the restriction μ_0 of μ to C(A) is a Whitney map for C(A) and since $t \leq \mu_0(A)$, we have by (14.73.9) that $\sigma_{t,\mu_0}^{-1}(A) = \{\mu_0^{-1}(t)\}$. Hence, by the formulas for $\sigma_{t,\mu}$ and

(14.73.11)   WHITNEY PROPERTIES   489

σ_{t,μ_o}, $\sigma_{t,\mu}^{-1}(A) = \{\mu_o^{-1}(t)\}$. This proves $\sigma_{t,\mu}$ is one-to-one. Conversely, assume $\sigma_{t,\mu}$ is one-to-one for each $t \in [0, \mu(X)]$. Then it follows easily that, for any (nondegenerate) subcontinuum A of X,

A $\in$ CP($\mu|$C(A))

where $\mu|$C(A) denotes the restriction of μ to C(A). Hence, by (14.73.6), X $\in$ CPH. This completes the proof of the first part of (14.73.10). The second part is a simple consequence of the first part and (14.73.8).

The following result, which is less specific than (14.73.10), is stated for the purpose of motivating later material. In particular, it gives rise to the notion of a Whitney hyperspace [see (14.73.28) through (14.73.30)].

(14.73.11) COROLLARY [267, Corollary 21]. *If* X $\in$ CPH, *then* $\mu^{-1}([t, \mu(X)]) \cong C(\mu^{-1}(t))$ *for any Whitney map* μ *for* C(X) *and any* $t \in [0, \mu(X)]$.
   Proof. See (14.73.10).

The result in (14.73.11) gives a sufficient condition for "tops" of hyperspaces to again be hyperspaces. Furthermore, it shows what the "tops" are hyperspaces of. Related matters will be discussed in (14.73.28) through (14.73.34).

To aid in some applications, let us note the following reformulation of (14.13.1) and the first part of (14.14.1). For some other continua having CPH, see (14.73.17).

(14.73.12) THEOREM ([179, section 6], where it is stated for CP). *Any hereditarily indecomposable continuum has CPH and any chainable continuum has CPH.*

Proof. The first part of (14.73.12) is immediate from the first part of (14.14.1). The second part of (14.73.12) follows immediately from (14.13.1) and the well-known and easy-to-prove fact that any subcontinuum of a chainable continuum is chainable.

A continuum is said to be *hereditarily irreducible* provided that each of its nondegenerate subcontinua is irreducible.

(14.73.13) THEOREM. *Let* X *be a continuum and let* μ *be a Whitney map for* C(X). *If* $\mu^{-1}(t)$ *is hereditarily irreducible for some* $t \in [0, \mu(X))$, *then* $\sigma_{t,\mu}$ *is one-to-one and, hence,* $\sigma_{t,\mu}$ *is a homeomorphism of* $C(\mu^{-1}(t))$ *onto* $\mu^{-1}([t, \mu(X)])$. *Thus:*

(1) [148, Theorem 4.3]. *If* $\mu^{-1}(t)$ *is an arc for some* $t \in [0, \mu(X))$, *then* $\mu^{-1}([t, \mu(X)])$ *is a 2-cell* [comp., (1.213.2)];

(2) [148, Corollary 4.4]. *If* X *is a decomposable chainable continuum, then there exists* $t < \mu(X)$ *such that* $\mu^{-1}([t, \mu(X)])$ *is a 2-cell*.

Proof. Assume $t \in [0, \mu(X))$ such that $\mu^{-1}(t)$ is hereditarily irreducible. Let σ_o denote $\sigma_{t,\mu}$. By (14.73.8), σ_o is a mapping from $C(\mu^{-1}(t))$ onto $\mu^{-1}([t, \mu(X)])$. For the purpose of proving that σ_0 is one-to-one, let $A \in \mu^{-1}((t, \mu(X)])$ [if $\mu(A) = t$, then $\sigma_0^{-1}(A) = \{\{A\}\}$]. Let μ_1 denote the restriction of μ to $C(A)$. Then, since μ_1 is a Whitney map for $C(A)$ and $t < \mu(A)$, we have by (14.2) and part (a) of (1.213.1) that

$$\mu_1^{-1}(t) = \mu^{-1}(t) \cap C(A)$$

is a nondegenerate subcontinuum of $\mu^{-1}(t)$. Hence, since $\mu^{-1}(t)$ is hereditarily irreducible, $\mu_1^{-1}(t)$ is irreducible. Thus, (14.73.2) may be applied [with X replaced by A and μ replaced by μ_1] to see that no proper subcontinuum of $\mu_1^{-1}(t)$ covers A. Therefore, since $\sigma_0[\mu_1^{-1}(t)] = A$ [by part (c) of (1.213.1)], it follows easily that $\sigma_0^{-1}(A) = \{\mu_1^{-1}(t)\}$. This completes the proof that σ_0 is one-to-one. We have proved the part of (14.73.13) which precedes (1) and (2). Since an arc is hereditarily irreducible, part (1) now follows using (0.54). By (14.31), part (2) is a consequence of part (1). This completes the proof of (14.73.13).

Let us note that the converse to the main part of (14.73.13) is false--in fact, $\sigma_{t,\mu}$ can be one-to-one for some $t \in [0, \mu(X))$ and yet $\mu^{-1}(t)$ may not even be irreducible. An example showing this is in the second paragraph following the proof of (14.73.2).

Let us also note that part (2) of (14.73.13) can be proved without using part (1). First, by (14.31), there exists $t < \mu(X)$ such that $\mu^{-1}(t)$ is an arc. Then, since a chainable continuum has CPH [by (14.73.12)], we may apply (14.73.11) and (0.54) to see that $\mu^{-1}([t, \mu(X)])$ is a 2-cell.

The next three results are applications of (14.73.10).

(14.73.14) THEOREM [176, 1.3]. *If* X *is an hereditarily indecomposable continuum and if* μ *is a Whitney map for* $C(X)$, *then* $C(\mu^{-1}(t)) \cong \mu^{-1}([t, \mu(X)])$ *for each* $t \in [0, \mu(X)]$; *in fact,*

$$\sigma_{t,\mu} : C(\mu^{-1}(t)) \xrightarrow{onto} \mu^{-1}([t, \mu(X)])$$

is a homeomorphism for any $t \in [0, \mu(X)]$.

Proof. By the first part of (14.73.12), $X \in$ CPH. Hence, the result follows from (14.73.10).

The next two results give particular instances in which (14.73.14) can be applied to obtain the exact topological type of $\mu^{-1}([t, \mu(X)])$. They give rise to the notion of an invariant Whitney hyperspace [see (14.73.31) through (14.73.34)].

(14.73.15) COROLLARY (comp., [261, Lemma 2.1]). *If* X *is a pseudo-arc and if* μ *is a Whitney map for* $C(X)$, *then* $\mu^{-1}([t, \mu(X)]) \cong C(X)$ *for each* $t < \mu(X)$.

Proof. The result follows directly from (14.3) and (14.73.14).

(14.73.16) COROLLARY. *If* X *is any particular pseudo-solenoid and if* μ *is a Whitney map for* $C(X)$, *then* $\mu^{-1}([t, \mu(X)]) \cong C(X)$ *for each* $t < \mu(X)$.

Proof. The result follows directly from (14.23) and (14.73.14).

In (14.73.12) we have some classes of continua which have CPH. In the next result, another class of such continua is given.

(14.73.17) THEOREM [267, Proposition 24]. *If* X *is any non-planar circle-like continuum, then* X ε CPH.

Let X be a non-planar solenoid [i.e., a solenoid which is not a circle]. Then X is one of the continua covered by (14.73.17). Petrus [267, p.26] has given an especially simple proof that X ε CPH. Her proof uses (14.21), the fact that each nondegenerate proper subcontinuum of a solenoid is an arc [see (1.209.4)], and [165, 8.1] (a more general version of which is in (1.50)). Let us note that since X ε CPH, it follows from (14.21) and (14.73.11) that

$$\mu^{-1}([t, \mu(X)]) \cong C(X)$$

for any Whitney map μ for $C(X)$ and any $t < \mu(X)$. However, this last fact also follows directly from (14.20). Also note that this last fact is true when X is a circle [see (1.213.2)].

In (14.73.12) and (14.73.17) we have obtained some examples of continua which have CPH. The next three results show ways in which the class of continua having CPH is limited.

Bruce Hughes has an example of a continuum which has the covering property and which is not a-triodic. Note the following result; after its proof, more will be said about Hughes' example.

(14.73.18) THEOREM (first part due to Bruce Hughes; second

part in [267, Proposition 27]). *If* $X \in CP$, *then* X *is not a triod. If* $X \in CPH$, *then* X *is a-triodic.*

Proof [see paragraph preceding (14.73.1)]. Assume X is a triod [see (0.21)]. Let μ be a Whitney map for $C(X)$. By (0.21), there is a subcontnnuum N of X such that $X \setminus N$ is the union of three nonempty mutually separated sets S_1, S_2, and S_3. For each $i = 1, 2,$ and 3, let $B_i = S_i \cup N$. Note that B_i is a subcontinuum of X for each $i = 1, 2,$ and 3. Let t_o be any fixed number satisfying

$$\mu(N) < t_o \leq \min\{\mu(B_i): i = 1, 2, 3\}.$$

Note that $B_1 \cup B_2$, $B_2 \cup B_3$, and $B_1 \cup B_3$ are subcontinua of X. Thus, by (14.2) and the fact that the restriction of μ to $C(B_i \cup B_j)$, any $i \neq j$, is a Whitney map, we have that

(1) $\mu^{-1}(t_o) \cap C(B_i \cup B_j)$ is a subcontinuum of $\mu^{-1}(t_o)$, any $i \neq j$.

Let $\Gamma_1 = \mu^{-1}(t_o) \cap C(B_1 \cup B_2)$ and $\Gamma_2 = \mu^{-1}(t_o) \cap C(B_2 \cup B_3)$. From the choice of t_o, it follows easily using (1.8) and (1.11) that there is a subcontinuum K of B_2 such that $\mu(K) = t_o$. Hence, $K \in [\Gamma_1 \cap \Gamma_2]$ and, thus, $\Gamma_1 \cap \Gamma_2 \neq \emptyset$. Therefore, by (1),

(2) $\Gamma_1 \cup \Gamma_2 = \Gamma$ is a subcontinuum of $\mu^{-1}(t_o)$.

Since the restriction of μ to $C(B_1 \cup B_2)$ [resp., $C(B_2 \cup B_3)$] is a Whitney map for $C(B_1 \cup B_2)$ [resp., $C(B_2 \cup B_3)$], $\cup \Gamma_1 = B_1 \cup B_2$ and $\cup \Gamma_2 = B_2 \cup B_3$. Hence, since $\Gamma = \Gamma_1 \cup \Gamma_2$,

(3) $\cup \Gamma = [\cup \Gamma_1] \cup [\cup \Gamma_2] = B_1 \cup B_2 \cup B_3 = X$.

Since N is a proper subcontinuum of B_1 and since $\mu(N) < t_o$, it follows using (1.8) and (1.11) that there is a subcontinuum L_1 of B_1 such that $N \subset L_1 \neq N$ and $\mu(L_1) < t_o$. Again using (1.8) and (1.11), let α be an order arc in $C(B_3)$ from N to B_3. Note that, since $A \supset N$ for any $A \in \alpha$ and $L_1 \supset N$, $L_1 \cup A$ is a subcontinuum of X for each $A \in \alpha$. Since $\mu(L_1 \cup N) = \mu(L_1) < t_0$ and $\mu(L_1 \cup B_3) \geq \mu(B_3) \geq t_0$, there exists $A_0 \in \alpha$ such that $\mu(L_1 \cup A_0) = t_0$. Clearly, $N \subset A_0 \neq N$. Thus, since $A_0 \subset B_3$, we have that $A_0 \cap S_3 \neq \emptyset$. Hence,

(4) $[L_1 \cup A_0] \cap S_3 \neq \emptyset$.

Since $L_1 \subset B_1$ and $N \subset L_1 \neq N$, we have that $L_1 \cap S_1 \neq \emptyset$. Hence,

(5) $[L_1 \cup A_0] \cap S_1 \neq \emptyset$.

By (4), $[L_1 \cup A_0] \not\subset \Gamma_1$ and, by (5), $[L_1 \cup A_0] \not\subset \Gamma_2$. Thus, $[L_1 \cup A_0] \not\subset \Gamma$. Therefore, we have

(6) $[L_1 \cup A_0] \in \mu^{-1}(t_0)$ and $[L_1 \cup A_0] \not\subset \Gamma$.

By (3) and (6), $X \notin CP$. This proves the first part of (14.73.18). Since the second part is a trivial consequence of the first part, this completes the proof of (14.73.18).

In the paragraph preceding (14.73.18), we mentioned that Hughes has an example of a continuum which has the covering property and which is not a-triodic. His example is obtained as a compactification of a half-line G [i.e., $G \cong [0, +\infty)$] with a simple triod T as the remainder such that each subcontinuum of T is a limit of arcs in G. This is accomplished by choosing a countable dense subset Λ of $C(T)$ and then approximating the members of Λ, as subcontinua of T, while spiraling G in on T. The construction easily generalizes to prove the following fact: *Any continuum can be embedded in a continuum with the covering property.*

The next theorem completely determines those hereditarily decomposable continua which have CPH.

(14.73.19) THEOREM. *Assume* X *is an hereditarily decomposable continuum. Then:* X ε CPH *if and only if* X *is a chainable continuum.*

Proof. By the second part of (14.14.1), we have that:

(1) If a continuum has CPH, then it is hereditarily unicoherent.

Bing [32, Theorem 11] has proved that an hereditarily decomposable continuum is chainable if and only if it is a-triodic and hereditarily unicoherent. This fact, together with (14.73.18) and (1), proves half of (14.73.19). The other half is contained in (14.73.12).

Though a continuum having the covering property must be unicoherent [by (14.14.1)], it need not be hereditarily unicoherent. This can be seen, as was observed in [267, Example 28], by using the continuum $(SP)_1$ defined in (8.22).

The next result shows that the arc is the only arcwise connected continuum having CP.

(14.73.20) THEOREM. *Assume* X *is an arcwise connected continuum. Then:* X ε CP *if and only if* X *is an arc.*

Proof. Assume X is an arcwise connected continuum. If X ε CP, then X is irreducible [by (14.73.1)] and, hence, is an arc. Conversely, any arc has the covering property by (14.13.1).

In (14.73.21) we give an interesting and useful connection between the covering property and Class[W]. A continuum X is said to be *in Class[W]*, written $X \in Class[W]$, provided that every mapping of any continuum onto X is weakly confluent. The notion of Class[W] originated in Lelek's topology seminar at the University of Houston on November 8, 1972. Since then, several authors have determined various types of continua which are in Class[W]. For example: Read [274, Theorem 4] showed (using inverse limits) that every chainable continuum is in Class[W], and Feuerbacher [106] obtained a general result which implies every non-planar circle-like continuum is in Class[W]. As we can see, the next theorem implies these results by using (14.13.1) and (14.73.17) respectively. It also implies [by (14.14.1)] that any hereditarily indecomposable continuum in in Class[W] --a stronger result is in (1.207.6). The applications just mentioned are due to Bruce Hughes.

(14.73.21) THEOREM [Bruce Hughes]. *If* X *has the covering property, then* $X \in Class[W]$.

Proof. Assume X has the covering property, let Y be a continuum, and let f be any mapping of Y onto X. To show f is weakly confluent, it suffices by (0.49.1) to show

(*) $\hat{f}$ maps C(Y) onto C(X).

Let μ be a Whitney map for C(X). We will prove (**) below [clearly, (**) implies (*)]:

(**) $\mu^{-1}(t) \subset \hat{f}[C(Y)]$ for any $t \in [0, \mu(X)]$.

To prove (**), let $t \in [0, \mu(X)]$ be fixed. Let

$$\Gamma_1 = \{A \in C(Y): \mu(\hat{f}(A)) \leq t\}$$

and let

$$\Gamma_2 = \{A \in C(Y): \mu(\hat{f}(A)) \geq t\}.$$

Note the following simple facts: $\hat{f}(A) \subset \hat{f}(B)$ if $A, B \in C(Y)$ and $A \subset B$ [by definition of $\hat{f}$ --see (0.49)], $F_1(Y) \subset \Gamma_1$ [since $\mu(\hat{f}(\{y\})) = 0 \leq t$ for each $\{y\} \in F_1(Y)$], $Y \in \Gamma_2$ [since $\mu(\hat{f}(Y)) = \mu(X) \geq t$], and $\Gamma_1 \cup \Gamma_2 = C(Y)$. Using these facts to give an argument similar to the one for (14.2), we obtain:

(1) $\Gamma_1 \cap \Gamma_2 = \{A \in C(Y): \hat{f}(A) \in \mu^{-1}(t)\}$ is a subcontinuum of $C(Y)$.

Note that the proof of (1) did not use the covering property. Now we show, using (1) and the covering property, that $\hat{f}[\Gamma_1 \cap \Gamma_2] = \mu^{-1}(t)$. By (1) and the continuity of $\hat{f}$,

(2) $\hat{f}[\Gamma_1 \cap \Gamma_2]$ is a subcontinuum of $\mu^{-1}(t)$.

We will show that $\cup \hat{f}[\Gamma_1 \cap \Gamma_2] = X$. To do this, let $x \in X$. Then, since $f[Y] = X$, there exists $y \in Y$ such that $f(y) = x$. By (1.8) and (1.11), there is an order arc α in $C(Y)$ from $\{y\}$ to Y. Note that

$$\mu(\hat{f}(\{y\})) = 0 \leq t \leq \mu(X) = \mu(\hat{f}(Y)).$$

Therefore, since $\hat{f}$ and μ are continuous and since $\{y\}, Y \in \alpha$, it follows easily that there exists $K \in \alpha$ such that $\mu(\hat{f}(K)) = t$. Thus, $K \in [\Gamma_1 \cap \Gamma_2]$. Since α is an order arc and $\{y\}, K \in \alpha$, we have that $y \in K$. Hence, since $x = f(y)$, $x \in \hat{f}(K)$. Thus, since $K \in [\Gamma_1 \cap \Gamma_2]$, $x \in \cup \hat{f}[\Gamma_1 \cap \Gamma_2]$. Therefore, since $x \in X$

(14.73.21)

was arbitrary, we have proved that

(3) $\cup \hat{f}[\Gamma_1 \cap \Gamma_2] = X$.

Since $X \in CP$, we have by (2) and (3) that

(4) $\hat{f}[\Gamma_1 \cap \Gamma_2] = \mu^{-1}(t)$.

Since $t \in [0, \mu(X)]$ was arbitrary, it follows from (4) that we have proved (**). Since (**) implies (*), we have proved (14.73.21).

An application of (14.73.21) is in the proof of (14.55).

Recall the applications of (14.73.21) mentioned in the paragraph preceding it. In part, this book has attempted to relate hyperspace theory to the subject of confluent and weakly confluent mappings--for example, see (0.49.1), (1.190), (14.73.25), (16.29) through (16.32), and the proofs of (1.193) and (14.73.21). As can be seen from the references just cited, hyperspaces are useful in investigating confluent and weakly confluent mappings (and the reverse). This is especially evident from the proof of (14.73.21), in which various properties of hyperspaces, such as their arc structure and unicoherence, were utilized to obtain information about mappings between the original continua. Hyperspace techniques should be very helpful in answering some questions in the literature about mappings [see, for example, some questions in papers of Lelek in the Bibliography]. Also, results and techniques about special mappings between continua should be useful in hyperspace theory [for example, in answering (1.200)].

A number of questions about the covering property and CPH

have yet to be answered. The first five questions below are due to Bruce Hughes.

(14.73.22) QUESTION [Bruce Hughes]. If $X \in CP$, then is $\mu^{-1}(t)$ unicoherent for each Whitney map μ for $C(X)$ and each $t \in [0, \mu(X))$? Note that, since X is unicoherent when $X \in CP$ [by the second part of (14.14.1)], the answer to the question above is "yes" when $t = 0$. Let us also note that if $X \in CP$, then by (14.73.3) $\mu^{-1}(t)$ is irreducible for each $t \in [0, \mu(X))$ --however, an irreducible continuum need not be unicoherent [see X in (14.29)].

(14.73.23). QUESTION [Bruce Hughes]. If there is a Whitney map μ for $C(X)$ such that $\mu^{-1}(t_o)$ is irreducible for some $t_o \in [0, \mu(X))$, then must X be irreducible? Partial answers are in (14.55.1).

(14.73.24) QUESTION [Bruce Hughes]. If X is any decomposable unicoherent continuum which is not a triod [in the sense of (0.21)], then does there exist a Whitney map μ for $C(X)$ such that $\mu^{-1}(t_o)$ is irreducible for some $t_o > 0$? Affirmative answers to (14.73.23) and (14.73.24) would give a hyperspace proof of Sorgenfrey's theorem [306 Theorem 3.2].

(14.73.25) QUESTION [Bruce Hughes]. Is the converse of (14.73.21) true? Lelek [194, Problem 1] has asked for a character-

ization of those continua which are in Class[W]. An affirmative
answer to (14.73.25) would give a hyperspace characterization. It
is known that any continuum in Class[W] is unicoherent and not a
triod (thus, by [306, Theorem 3.2], irreducible). The first of
these facts is in the notes (unpublished) from Lelek's seminar at
the University of Houston. The other facts are due to Bruce Hughes
(unpublished). By these facts, some previous results [such as the
second part of (14.14.1) and (14.73.18)] can be viewed as corollaries to (14.73.21). The question above has been answered affirmatively for circle-like continua in the paper cited in Footnote 1
at the end of Chapter XV. For a question related to the one above,
see (14.76.10).

(14.73.26) QUESTION [Bruce Hughes]. What classes of mappings
preserve the covering property? In particular, if X has the covering property, then does every monotone image of X?[3] Note that
open mappings do not preserve CP, or even CPH, since non-planar
solenoids have CPH [by (14.73.17)] and can be openly mapped onto
a circle [by (1.207.5)].

Let us note the following general question.

(14.73.27) QUESTION. What classes of continua have the covering property or have CPH? Some answers are in (14.13.1),
(14.14.1), (14.73.12), (14.73.17), (14.73.19), and (14.73.20).

K.

Whitney Hyperspaces

In this section we will ask some questions and give a few results about when the "tops", $\mu^{-1}[t, \mu(X)]$, of hyperspaces are again hyperspaces.

(14.73.28) DEFINITION. Let X be a continuum. Call $C(X)$ a *Whitney hyperspace* [resp., a *weak Whitney hyperspace*] provided that given any [resp., some] Whitney map μ for $C(X)$ and any $t \in [0, \mu(X))$, there exists a continuum $Y_{t,\mu}$ such that
$$\mu^{-1}([t, \mu(X)]) \cong C(Y_{t,\mu}).$$
If $C(X)$ is a Whitney hyperspace [resp., a weak Whitney hyperspace], then we say that X *has a Whitney hyperspace* [resp., X *has a weak Whitney hyperspace*].

(14.73.29) QUESTION. For what classes of continua X is $C(X)$ a Whitney hyperspace or weak Whitney hyperspace [see (14.73.30)]? Note the following facts: By (14.73.11), $X \in$ CPH implies X has a Whitney hyperspace. Hence, every chainable continuum [see (14.73.12)], every hereditarily indecomposable continuum [see (14.73.12) or (14.73.14)], and every non-planar circle-like continuum [see (14.73.17)] has a Whitney hyperspace. Stronger statements for some subclasses of these continua (and others) are in (14.73.34).

(14.73.30) QUESTION. Is there a continuum X such that C(X) is a weak Whitney hyperspace but not a Whitney hyperspace?

(14.73.31) DEFINITION. Let X be a continuum. Call C(X) an *invariant Whitney hyperspace* provided that for each Whitney map μ for C(X),

$$\mu^{-1}([t, \mu(X)]) \cong C(X)$$

for each $t \in [0, \mu(X))$. If C(X) is an invariant Whitney hyperspace, then we say that X *has an invariant Whitney hyperspace.*

Let us note the following general fact (comp., [249, p.403]):

(14.73.32) THEOREM. *Let* X *be a continuum and let* μ *be a Whitney map for* C(X). *If* $\mu^{-1}(t_o)$ *is locally connected for some* $t_o \in [0, \mu(X)]$, *then* $\mu^{-1}([t_o, \mu(X)])$ *is locally connected and, hence, an absolute retract.*

Proof. Since $\mu^{-1}(t_o) = \{X\}$ if $t_o = \mu(X)$, assume that $0 \leq t_o < \mu(X)$. Then, $\mu^{-1}(t_o)$ is a continuum [by (14.2)] and, by assumption, $\mu^{-1}(t_o)$ is locally connected. Hence, by (1.92), $C(\mu^{-1}(t_o))$ is a locally connected continuum. Therefore, $\mu^{-1}([t_o, \mu(X)])$ is locally connected by (14.73.8) and [186, Theorem 5, p.257]. Hence, by (0.74.1), $\mu^{-1}([t_o, \mu(X)])$ is an absolute retract.

(14.73.33) COROLLARY. *If a chainable* [*resp., circle-like*] *continuum* X *has an invariant Whitney hyperspace, then* X *is an arc* [*resp., circle*] *or* X *is indecomposable.*

Proof. Assume X is any decomposable chainable or circle-like continuum, and let μ be a Whitney map for $C(X)$. Then, by (14.31), there exists $t_0 \in [0, \mu(X))$ such that $\mu^{-1}(t_0)$ is locally connected. Hence, by (14.73.32),

(1) $\mu^{-1}([t_0, \mu(X)])$ is locally connected.

Now, assume X (also) has an invariant Whitney hyperspace. Then, noting $t_0 < \mu(X)$, we have by (1) and (14.73.31) that $C(X)$ is locally connected. Hence, by (1.92), X is locally connected. Therefore, since a locally connected chainable or circle-like continuum must be an arc or a circle, X is an arc or a circle. This proves (14.73.33) [Note: The theorem could also have been proved by using (14.31), part (1) of (14.73.13), and the second part of (1.213.2)].

(14.73.34) QUESTION. For what classes of continua X is $C(X)$ an invariant Whitney hyperspace? Note the following facts: If $X \in CPH$ and if X is Whitney stable [see (14.39.1)], then it follows easily from (14.73.11) that $C(X)$ is an invariant Whitney hyperspace. Hence, an arc, a pseudo-arc [also see (14.73.15)], any particular pseudo-solenoid [also see (14.73.16)], and any non-planar solenoid [also see the paragraph following (14.73.17)] has an invariant Whitney hyperspace. So does a circle, by (1.213.2). The result in (14.73.33) leads to the following question which is more specific than the one above: Which indecomposable chainable or circle-like continua have an invariant Whitney hyperspace?

(14.74) REMARK. I have recently been told that many of Petrus' results in [267], some of which have been mentioned in this chapter, will be appearing in [266] and [268].

(14.75) REMARK. In [249], after the results stated here in (14.8) and (14.9) were obtained, the following question was asked: If $\mu^{-1}(t_o)$ is arcwise connected [resp., locally connected] for some $t_o \varepsilon [0, \mu(X)]$, then is $\mu^{-1}(t)$ arcwise connected [resp., locally connected] whenever $t_o < t \leq \mu(X)$ [249, P.2, p.407]? Petrus ([267] and [268]) has answered this question affirmatively and obtained some generalizations. Also, Eberhart [418] has recently obtained some related results.

L.
Two Hyperspace Functions

In this section we discuss two functions, σ_t [defined in (14.73.7)] and Ψ_t [defined below], and give some relationships between them. We have seen that σ_t is useful in investigating various properties of $C(X)$ and of X. We will see that the same is true of Ψ_t. Beginning with (14.76.6), we give some applications and questions.

(14.76.0) NOTATION. Let X be a continuum and let μ be a Whitney map for $C(X)$. For any given $t \varepsilon [0, \mu(X)]$, let $\Psi_{t,\mu}(A) = C(A) \cap \mu^{-1}(t)$ for each $A \varepsilon \mu^{-1}([t, \mu(X)])$. By part

(2) of (14.11.1), the formula just given defines a function $\Psi_{t,\mu}$ from $\mu^{-1}([t, \mu(X)])$ into $C(\mu^{-1}(t))$. As in the case of $\sigma_{t,\mu}$, we will simply use the symbol Ψ_t instead of $\Psi_{t,\mu}$ when there is no confusion.

The function Ψ_t is not in general continuous, as can be seen by taking $X = S^1$ and any $t \in (0, \mu(S^1))$ [267, p.24]. In (14.76.4), we determine a necessary and sufficient condition in order that Ψ_t be continuous. Note the following result:

(14.76.1) THEOREM [267, Proposition 20]. *Let* X *be a continuum and let* μ *be a Whitney map for* $C(X)$. *If* σ_t *is one-to-one for some* $t \in [0, \mu(X)]$, *then* σ_t *is a homeomorphism from* $C(\mu^{-1}(t))$ *onto* $\mu^{-1}([t, \mu(X)])$ *and* $\Psi_t = \sigma_t^{-1}$ [*hence,* Ψ_t *is continuous*].

Proof. The theorem is a consequence of (14.73.8) and the formulas for σ_t and Ψ_t.

Let us also note the following general fact about Ψ_t:

(14.76.2) THEOREM. *Let* X *be a continuum and let* μ *be a Whitney map for* $C(X)$. *Then, for any* $t \in [0, \mu(X)]$, $\sigma_t \circ \Psi_t$ *is the identity on* $\mu^{-1}([t, \mu(X)])$ *and, hence,* Ψ_t *is one-to-one and* σ_t *is one-to-one on the range of* Ψ_t.

Proof. The theorem follows easily using (1.8), (1.11), and the formula for Ψ_t.

Note the following theorem [comp., (14.76.11)]:

(14.76.3) THEOREM. *Let* X *be a continuum and let* μ *be a Whitney map for* C(X). *If* X ε CPH, *then* Ψ_t *is a homeomorphism from* $\mu^{-1}([t, \mu(X)])$ *onto* $C(\mu^{-1}(t))$ *for each* $t \varepsilon [0, \mu(X)]$.

Proof. The theorem is a simple consequence of (14.73.10) and (14.76.1).

Now, let us make some observations. In (14.76.3), we gave a sufficient condition for Ψ_t to be both continuous and onto $C(\mu^{-1}(t))$. As was mentioned above, Ψ_t is not always continuous. Let us note that Ψ_t is not always onto $C(\mu^{-1}(t))$; in fact, as can be seen by taking X to be a simple triod, Ψ_t can be continuous without mapping onto $C(\mu^{-1}(t))$ [note: Ψ_t is continuous when X is any dendrite by (14.76.4) and (15.11)]. Thus--the following two questions arise: What are necessary and sufficient conditions in order that Ψ_t be continuous? What are necessary and sufficient conditions in order that the range of Ψ_t be all of $C(\mu^{-1}(t))$? The answer to the first question is in (14.76.4), and the answer to the second question, along with other facts, is in (14.76.5). In connection with the observations above, note that, by (14.76.5), Ψ_t being onto all of $C(\mu^{-1}(t))$ implies that Ψ_t is continuous.

For the following theorem, we refer the reader to (15.5) for the definition of C*-smoothness [also, see the first page of Chapter XV for a discussion]. As can be seen, C*-smoothness is a natural geometrical condition.

(14.76.4) THEOREM. *Let* X *be a continuum and let* μ *be a Whitney map for* C(X). *Then,* Ψ_t *is continuous for each* t ε [0, μ(X)] *if and only if* X *is* C*-smooth. In fact: Let* A ε C(X); *then,* X *is* C*-smooth at* A *if and only if* Ψ_t *is continuous at* A *for each* t ε [0, μ(A)].

The proof of (14.76.4) is omitted. The proof of the following theorem is not difficult and is omitted.

(14.76.5) THEOREM. *Let* X *be a continuum and let* μ *be a Whitney map for* C(X). *Let* t ε [0, μ(X)] *be fixed. Of the following seven properties, the first five are equivalent to each other, any of the first five implies the sixth, and the sixth and seventh are equivalent to each other:*

(1) $\Psi_t[\mu^{-1}([t, \mu(X)])] = C(\mu^{-1}(t))$;

(2) σ_t *is one-to-one;*

(3) *no proper subcontinuum of* $C(A) \cap \mu^{-1}(t)$ *covers* A *for any* $A \in \mu^{-1}([t, \mu(X)])$ [*written:* X ε CPH *at* t --*comp.*, (14.73.2)];

(4) σ_t *is a homeomorphism of* $C(\mu^{-1}(t))$ *onto* $\mu^{-1}([t, \mu(X)])$;

(5) Ψ_t *is a homeomorphism of* $\mu^{-1}([t, \mu(X)])$ *onto* $C(\mu^{-1}(t))$;

(6) Ψ_t *is continuous;*

(7) Ψ_t *is a homeomorphism of* $\mu^{-1}([t, \mu(X)])$ *into* $C(\mu^{-1}(t))$.

(14.76.6) **APPLICATION.** By combining (14.76.3) and (14.76.4), we see that a continuum X is C*-smooth if X ∈ CPH. Thus, in particular, the continua in the hypotheses of (14.73.12) and (14.73.17) are C*-smooth. This gives a proof of (1.207.8) and of (15.13). It also gives a partial answer to (15.15).

We will give another application of (14.76.3) in (14.76.8). We also use the following theorem in (14.76.8).

(14.76.7) **THEOREM** [267, Lemma 31]. *Let* X *be a continuum and let* μ *be a Whitney map for* $C(X)$. *If* σ_t *is one-to-one for some fixed* $t \in [0, \mu(X)]$, *then there is a Whitney map* μ_t *for* $C(\mu^{-1}(t))$ *such that:*

(1) *Given any* $r \in [t, \mu(X)]$, *there exists* $s \in [0, \mu_t(\mu^{-1}(t))]$ *such that* $\Psi_t[\mu^{-1}(r)] = \mu_t^{-1}(s)$;

(2) *Given any* $s \in [0, \mu_t(\mu^{-1}(t))]$; *there exists* $r \in [t, \mu(X)]$ *such that* $\Psi_t[\mu^{-1}(r)] = \mu_t^{-1}(s)$.

Proof. Since σ_t is one-to-one, we have by (14.76.5) [or (14.76.1)] that

(*) $\Psi_t[\mu^{-1}([t, \mu(X)])] = C(\mu^{-1}(t))$.

Define $\mu_t: C(\mu^{-1}(t)) \to [0, +\infty)$ by

$$\mu_t(\Gamma) = \mu(\sigma_t(\Gamma)) - t$$

for each $\Gamma \in C(\mu^{-1}(t))$. Using the one-to-oneness of σ_t, it is easy to verify that μ_t is a Whitney map for $C(\mu^{-1}(t))$. The proof of the following formula is easy using the one-to-oneness of σ_t and the formulas for μ_t, Ψ_t, and σ_t:

(**) For any $r \in [t, \mu(X)]$, $\Psi_t[\mu^{-1}(r)] = \mu_t^{-1}(r - t)$.

The fact that (1) and (2) of (14.76.7) hold follows immediately from (*) and (**).

(14.76.8) APPLICATION. It follows from (14.73.5) that:

(*) $X \in$ CPH *if and only if* $\mu^{-1}(t)$ *is hereditarily irreducible for each* $t \in [0, \mu(X))$. Now, assume $X \in$ CPH. Let $t_o \in [0, \mu(X))$. Then, by (14.73.10), the hypothesis of (14.76.7) holds. Hence, by (14.76.7), (14.76.3), and (*) above, $\mu^{-1}(t_o) \in$ CPH. Therefore, we have proved: *CPH is a Whitney property*.

(14.76.9) QUESTION. Is CP a Whitney property?

(14.76.10) QUESTION. Is being in Class[W] a Whitney property? If the answer is "yes," (14.73.5) can then be used to give an affirmative answer to (14.73.25).

Finally, let us note the following characterization of CPH in terms of ontoness of the functions Ψ_t.

(14.76.11) THEOREM. *Let* X *be a continuum and let* μ *be a Whitney map for* C(X). *Then: The equality* $\Psi_t[\mu^{-1}([t, \mu(X)])] =$

$C(\mu^{-1}(t))$ *holds for all* $t \in [0, \mu(X)]$ *if and only if* $X \in$ CPH.

Proof. Assume the equality holds. Then, by (14.76.2), σ_t is one-to-one [on all of $C(\mu^{-1}(t))$]. Hence, by (14.73.10), $X \in$ CPH. The other half of (14.76.11) follows immediately from (14.76.3). We remark that (14.76.11) is also a simple consequence of (14.73.10) and the equivalence of (1) and (2) of (14.76.5).

NOTES

1. For information about the function which assigns to A the continuum Λ in (14.11.1), see section L of this chapter.

2. With respect to (14.57), E. Abo-Zeid has recently shown that each of the following is a strong Whitney-reversible property: Hereditary decomposability, one-dimensionality if $\dim[C(X)] < \infty$, C*-smoothness, and hereditary arcwise connectedness (in which case, X must be an arc or a circle). The above-mentioned results are part of Abo-Zeid's dissertation currently being written at the University of Saskatchewan (Saskatoon, Sask., Canada). I have recently shown that each of the following is a strong Whitney-reversible property: Being of dimension $\leq n$ (and, hence, being one-dimensional), being contractible with respect to any absolute neighborhood retract, being tree-like, and being acyclic. My paper containing these and other results is under preparation at this time.

3. With respect to (14.73.26), E. Abo-Zeid has shown that monotone mappings preserve the covering property.

XV.

CONTINUITY PROPERTIES

OF THE HYPERSPACE OPERATIONS

Let X be a continuum. Define $C^*: C(X) \to C(C(X))$ and $2^*: 2^X \to 2^{2^X}$ by

$$C^*(A) = C(\{x \in X: x \in A\}) \quad , \quad \text{each} \quad A \in C(X)$$

$$2^*(A) = 2^{\{x \in X: x \in A\}} \quad , \quad \text{each} \quad A \in 2^X.$$

It will be convenient to think of each A both as a member of 2^X and as a compact subset of X. From now on we will do this. Thus: $C^*(A) = C(A)$ for each $A \in C(X)$, and $2^*(A) = 2^A$ for each $A \in 2^X$. Note that to say C^* is continuous means [by using (15.2)] that: For any sequence $\{A_i\}_{i=1}^\infty$ of subcontinua A_i of X converging to a subcontinuum A of X, any subcontinuum B of A is a limit of subcontinua B_i of A_i (as $i \to \infty$). This is a natural geometric condition to consider, i.e., it is natural to ask when C^* is continuous. As far as I know, there is only one result about this in the literature [250, 2.4], given below in (15.13). As we will see, though C^* is not always continuous [by (15.1)], C^* is continuous at "most" subcontinua of X [by (15.3)]. We will determine classes of continua X for which C^* is continuous at each subcontinuum of X. For example, this is true when X is a dendrite [by (15.11)] and when X is chainable [by (15.13)]. On the other hand, 2^* is always continuous by (15.4).

Let us begin by observing that C^* is not necessarily continuous:

(15.1) EXAMPLE. If X is a circle, then $C^*: C(X) \to C(C(X))$ is not continuous. This is easily seen by taking X to be the unit circle S^1 in R^2 and letting (in polar coordinates)

$$A_n = \{(1, \theta): 0 \leq \theta \leq 2\pi - [1/n]\} \qquad \pi$$

for each $n = 1, 2, \ldots$. Then, clearly, the sequence $\{A_n\}_{n=1}^{\infty}$ converges to S^1. However, $\{C^*(A_n)\}_{n=1}^{\infty}$ does not converge to $C^*(S^1)$. This last statement is easily verified by noting that if B is any arc in S^1 with $(1, 0)$ in its interior, then $B \notin \lim \sup C^*(A_n)$.

The following theorem shows that C^* and 2^* each have a type of continuity. It will be used, for example, to show that C^* is continuous at "most" places and that 2^* is, in fact, continuous everywhere [see (15.3) and (15.4), respectively].

(15.2) THEOREM. *For any continuum* X, C^* *and* 2^* *are each upper semi-continuous [see (0.47) for definition].*

Proof. We prove the theorem for C^*; the proof for 2^* is similar. Suppose C^* is not upper semi-continuous at $A \in C(X)$. Then there exists $\varepsilon > 0$ such that, for some sequence $\{A_n\}_{n=1}^{\infty}$ converging in $C(X)$ to A,

$$C^*(A_n) \notin N_H[\varepsilon, C^*(A)]$$

(15.3)   CONTINUITY OF HYPERSPACE OPERATIONS   515

for any $n = 1,2,\ldots$. Hence, for each $n = 1,2,\ldots$, there exists $B_n \in C^*(A_n)$ such that

(i) $H(B_n, K) \geq \varepsilon$ for each $K \in C^*(A)$.

Assume without loss of generality (by going to a subsequence if necessary) that the sequence $\{B_n\}_{n=1}^{\infty}$ converges to, say, B (note that B is a subcontinuum of X). Then, by (i),

(ii) $H(B, K) \geq \varepsilon$ for each $K \in C^*(A)$.

Since $B_n \in C^*(A_n)$ for each $n = 1,2,\ldots$, $B_n \subset A_n$ for each $n = 1,2,\ldots$. Therefore, since $\{A_n\}_{n=1}^{\infty}$ converges to A and $\{B_n\}_{n=1}^{\infty}$ converges to B, it follows that $B \subset A$. Thus, since B is a subcontinuum of X, $B \in C^*(A)$ which contradicts (ii). This completes the proof of (15.2).

A subset N of a topological space S is said to be *nowhere dense in* S provided that $cl[N]$ contains no nonempty open subset of S. A subset R of a topological space S is said to be a *residual subset of* S provided that $S \setminus R = \cup_{i=1}^{\infty} N_i$ where N_i is nowhere dense in S for each $i = 1,2,\ldots$. Recall [185, p.414] the Baire Theorem which says that a residual subset of a complete metric space is dense. Hence, since the countable intersection of residual subsets of a space is again a residual subset of the space, a residual subset of a complete metric space comprises, in a sense, "most" of the points of the space. Thus, the second part of the following result says that C^* is continuous at "most" places.

(15.3) COROLLARY. *For any continuum* X, C^* *is a Baire class* one function. Hence, the points of continuity of C^* form a residual G_δ (equivalently, a dense G_δ) subset of $C(X)$.

Proof. Since C^* is upper semi-continuous by (15.2), C^* is a Baire class one function by [186, p.70]. Hence, by [185, p.394], the points of continuity of C^* form a residual subset--this fact also follows from (15.2) and [186, Corollary 1, p.71]. The fact that the points of continuity form a G_δ subset of $C(X)$ is a general fact [185, pp.207-208] for any function.

(15.4) THEOREM. *For any continuum* X, 2^* *is continuous on* 2^X.

Proof. Let $A \in 2^X$. Let $\{A_n\}_{n=1}^\infty$ be any sequence in 2^X converging to A. Let $\{2^*(A_{n(j)})\}_{j=1}^\infty$ denote any convergent subsequence of $\{2^*(A_n)\}_{n=1}^\infty$, $\{2^*(A_{n(j)})\}_{j=1}^\infty$ converging to Λ for some $\Lambda \in 2^{2^X}$. We will show $\Lambda = 2^*(A)$. By (15.2) we have

(i) $\Lambda \subset 2^*(A)$.

To prove the reverse inclusion, let $B \in 2^*(A)$. Let $\varepsilon > 0$. Since $\{A_{n(j)}\}_{j=1}^\infty$ converges to A and since $\{2^*(A_{n(j)})\}_{j=1}^\infty$ converges to Λ, there exists a natural number $k = n(j)$ for some j such that (d denotes the metric for X)

(ii) $A \subset N_d[\varepsilon/3, A_k]$

and

(iii) $2^*(A_k) \subset N_H[\varepsilon/3, \Lambda]$.

Since B is a nonempty compact set, there exists a nonempty finite subset $F = \{x_1, x_2, \ldots, x_m\}$ of B such that $H(B, F) < \varepsilon/3$. Note that $F \subset A$. Thus, by (ii), there exists $Y_i \in A_k$ such that

$d(x_i, y_i) < \varepsilon/3$ for each $i = 1, 2, \ldots, m$. Let $F_k = \{y_1, y_2, \ldots, y_m\}$. It is easy to see that $H(F, F_k) < \varepsilon/3$. Since $F_k \in 2^*(A_k)$, we have by (iii) that there exists $L \in \Lambda$ such that $H(F_k, L) < \varepsilon/3$. Now,

$$H(B, L) \leq H(B, F) + H(F, F_k) + H(F_k, L) < \varepsilon.$$

Therefore, since Λ is a compact subset of 2^X and ε was arbitrary, it now follows that $B \in \Lambda$. This proves $2^*(A) \subset \Lambda$ and so, by (i), $2^*(A) = \Lambda$. It now follows that 2^* is continuous at A.

Thus, we have seen 2^* is always continuous, and C^* is not necessarily continuous. We now give some classes of continua for which C^* is continuous. One result about this has been left as an exercise in (1.207.8). Some other results are in (14.76.6).

For convenience, we introduce the following terminology.

(15.5) DEFINITION. A continuum X is said to be C^*-*smooth at* $A \in C(X)$ provided that C^* is continuous at A. A continuum X is said to be C^*-*smooth* provided that C^* is continuous on $C(X)$, i.e., at each $A \in C(X)$. An equivalent formulation of C^*-smoothness was given in (14.76.4). Also note that, by (15.3), any continuum X is C^*-smooth at each point of a dense G_δ subset of $C(X)$.

Let us observe the following simple but useful fact:

(15.6) THEOREM. *The property of being C^*-smooth is*

hereditary, i.e., if a continuum is C^*-smooth, so are each of its subcontinua.

Thus, combining (15.6) and (15.1), we have:

(15.7) COROLLARY. *If a locally connected continuum is C^*-smooth, then it is a dendrite.*

We will prove that every dendrite is C^*-smooth. This will follow from the next two results. First, recall the definition of 0-regular convergence (see the discussion immediately following the proof of (15.9) for some historical comments).

(15.8) DEFINITION. A sequence $\{Y_n\}_{n=1}^{\infty}$ of subcontinua of a continuum Y is said to *converge 0-regularly to* Y_o, written

$$Y_n \to Y_o \text{ 0-regularly,}$$

provided (a) and (b) below both hold:
  (a) $Y_n \to Y_o$ as $n \to \infty$;
  (b) Given $\varepsilon > 0$, there exist $\delta(\varepsilon) > 0$ and a natural number N such that if $n \geq N$, then any two points of Y_n less than δ apart lie together in a connected subset of Y_n of diameter less than ε.

(15.9) THEOREM [236, 3.5]. *If $\{Y_n\}_{n=1}^{\infty}$ is a sequence of subcontinua of a continuum X such that $Y_n \to Y_o$ 0-regularly, then $\{C(Y_n)\}_{n=1}^{\infty}$ converges to $C(Y_o)$ with respect to the Hausdorff* *metric.*

Proof. We must show

(1) $\lim \sup C(Y_n) = C(Y_0) = \lim \inf C(Y_n)$.

Since $\{Y_n\}_{n=1}^{\infty}$ converges to Y_0, we have by (15.2) that

(2) $\lim \sup C(Y_n) \subset C(Y_0)$.

Hence, since $\lim \inf C(Y_n) \subset \lim \sup C(Y_0)$, to prove (1) it suffices by (2) to show

(3) $C(Y_0) \subset \lim \inf C(Y_n)$.

To verify (3), let $A \in C(Y_0)$. Let $\varepsilon > 0$. Since $Y_n \to Y_0$ 0-regularly, there exist δ, $0 < \delta \le \varepsilon$, and a natural number N_1 such that for each $n \ge N_1$ (d will denote the metric for X):

(4) if $p, q \in Y_n$ such that $d(p, q) < \delta$, then there is a subcontinuum, denoted $B_n(p, q)$, of Y_n such that $p, q \in B_n(p, q)$ and $\text{diam}[B_n(p, q)] < \varepsilon/2$.

Also, since $Y_n \to Y_0$, there exists a natural number N_2 such that if $n \ge N_2$, then

(5) $H(Y_n, Y_0) < \delta/3$.

Let $N = \max\{N_1, N_2\}$. Then,

(6) for each $n \ge N$, (4) and (5) each hold.

Now, choose and fix $n \ge N$. Since A is compact, there is a finite subset $W = \{w_1, w_2, \ldots, w_i\}$ of A such that $H(W, A) < \delta/3$. Since A is connected there exists, for each $j = 1, 2, \ldots, i-1$, a finite subset $F_j = \{z_1^j, z_2^j, \ldots, z_{k(j)}^j\}$ of A such that $z_1^j = w_j$, $z_{k(j)}^j = w_{j+1}$, and $d(z_m^j, z_{m+1}^j) < \delta/3$ for each $m = 1, 2, \ldots, k(j)-1$ (see [338, pp.13-14]). Let $F = \bigcup_{j=1}^{i-1} F_j$. It is easy to enumerate the points of F so that the distance between any two consecutively indexed points is less than $\delta/3$. Let $\{a_1, a_2, \ldots, a_t\}$ be such an enumeration. Thus:

(7) $\quad F = \{a_1, a_2, \ldots, a_t\}$

where

(8) $\quad d(a_s, a_{s+1}) < \delta/3 \quad$ for each $\quad s = 1, 2, \ldots, t-1$.

Note that, since $H(W, A) < \delta/3$ and $W \subset F \subset A$,

(9) $\quad H(F, A) < \delta/3$.

From (5) and (6) it follows that, for each $s = 1, 2, \ldots, t$, there exists a point $p_s \in Y_n$ such that

(10) $\quad d(a_s, p_s) < \delta/3$.

Using (8) and (10) we see that

(11) $\quad d(p_s, p_{s+1}) < \delta \quad$ for each $\quad s = 1, 2, \ldots, t-1$.

Using (6) and (11) there exists, for each $s = 1, 2, \ldots, t-1$, a continuum $B_n(p_s, p_{s+1})$ as in (4). Let

$$B_n = \bigcup_{s=1}^{t-1} B_n(p_s, p_{s+1}).$$

Clearly, $B_n \in C(Y_n)$. Also, the following two facts

(12) $\quad \delta \leq \varepsilon$

(13) $\quad \text{diam}[B_n(p_s, p_{s+1})] < \varepsilon/2 \quad$ for each $\quad s = 1, 2, \ldots, t-1$

together with (9) and (10) can be used to show $H(A, B_n) < \varepsilon$. Therefore, since $A \in C(Y_o)$, $\varepsilon > 0$, and $n \geq N$ were each arbitrary and since $B_n \in C(Y_n)$, it follows that (3) holds and the theorem is proved.

In general, the space $L(X)$ of all locally connected subcontinua of a given continuum X is not a complete space with the

Hausdorff metric H. In fact, it is known that L(X) with H is not even topologically complete for certain continua X [210]. In [211], Mazurkiewicz showed that L(X) can be given a complete metric ρ_T such that convergence with respect to ρ_T implies convergence with respect to H [see (19.36)]. It can be seen that convergence with respect to ρ_T is precisely 0-regular convergence, a notion which appeared later in [341]. In [334], White metrized r-regular convergence for the space of all closed lc^r subsets of any compact metric space. White was primarily interested in giving a metric for Whyburn's notion of r-regular convergence. Unlike the metric ρ_T [211], White's metric is not in general complete, even for the case of L(X) and 0-regular convergence.

One of White's theorems [334, 7.32] implies that if X is a locally connected continuum, then L(X) with 0-regular convergence is compact if and only if X is a dendrite. This fact easily implies the following result. However, we give a proof which is independent of White's work.

(15.10) THEOREM (comp., [334, 7.32]). *Let* X *be a continuum. Then, the following two statements are equivalent:*

(15.10.1) *A sequence of subcontinua of* X *converges 0-regularly to a subcontinuum* Y_o *of* X *if and only if it converges with respect to the Hausdorff metric to* Y_o;

(15.10.2) X *is a dendrite.*

Proof. Assume (15.10.1) holds. Then the sequence

X, X,..., X,... must converge 0-regularly to X. Hence, X is a locally connected continuum. Suppose X contains a circle Y. Then a sequence $\{A_n\}_{n=1}^{\infty}$ in Y, analogous to the one in (15.1), converges with respect to the Hausdorff metric, but not 0-regularly, to Y. This contradicts (15.10.1). Hence, X does not contain a circle. Thus, X is a dendrite. Next, assume (15.10.2) holds. Let $\{Y_n\}_{n=1}^{\infty}$ be a sequence of subcontinua of X such that $\{Y_n\}_{n=1}^{\infty}$ converges with respect to the Hausdorff metric to a subcontinuum Y_o of X. Let $\varepsilon > 0$. Since X is a locally connected continuum [by (15.10.2)], there exists a $\delta = \delta(\varepsilon) > 0$ such that if p and q are any two distinct points of X less than δ apart, then p and q are the end points of an arc $\gamma(p, q)$ in X of diameter less than ε. From properties of dendrites [338, p.88] it is immediate that if p and q are each in some given Y_n, then $\gamma(p, q) \subset Y_n$. Thus, it now follows that $\{Y_n\}_{n=1}^{\infty}$ converges 0-regularly to Y_o.

(15.11) THEOREM. *A locally connected continuum is C^*-smooth if and only if it is a dendrite.*

Proof. Half the theorem is (15.7). The other half is proved by first using the fact that (15.10.2) implies (15.10.1), and then applying (15.9).

Thus, (15.11) precisely determines those locally connected continua which are C^*-smooth. The next main result, (15.13), shows that every chainable continuum is C^*-smooth. First we prove the following lemma [another proof of (15.13) was mentioned in (14.76.6)—the proof given here is independent of material in Chapter XIV and is, in a sense, more geometric than the one in (14.76.6)].

(15.12) LEMMA. *Let* Z *be a continuum. Let* $W = \{W_1, W_2, \ldots, W_n\}$ *be a chain [see (0.22) for definition] of open subsets of* Z *covering* Z. *Then, for any two natural numbers* p *and* q *such that* $1 \leq p \leq q \leq n$, *there is a subcontinuum* M *of* Z *such that* $M \cap cl[W_i] \neq \emptyset$ *if and only if* $p \leq i \leq q$.

Proof. For the purpose of proof, assume $|p - q| \geq 2$ (if $|p - q| \leq 1$, then the lemma is trivial—simply choose M to be a set consisting of just one point of $W_p \cap W_q$). Let M be irreducible with respect to the property of being a subcontinuum of Z which intersects both $cl[W_p]$ and $cl[W_q]$ [see (0.68.7)]. To show M is as desired, we first prove

(*) $M \cap W_p = \emptyset = M \cap W_q$.

To prove (*), suppose $M \cap W_p \neq \emptyset$. Let $A = M \backslash W_p$. Since $M \cap W \neq \emptyset$ and W_p is an open subset of Z, we have that A is a proper closed subset of M. Also, since $cl[W_q] \cap W_p = \emptyset$ (because W is a chain of open sets and $|p - q| \geq 2$) and $M \cap cl[W_q] \neq \emptyset$, we have that $A \cap cl[W_q] \neq \emptyset$. Thus, letting K denote a component of A such that $K \cap cl[W_q] \neq \emptyset$, we see by (20.2) that $K \cap cl[M \backslash A] \neq \emptyset$. Therefore, since $[M \backslash A] \subset W_p$, we have that $K \cap cl[W_p] \neq \emptyset$. Hence, K is a subcontinuum of M such that

$K \cap cl[W_p] \neq \emptyset \neq K \cap cl[W_q]$.

Thus, by the irreducibility of M, we see that K = M. Clearly $K \cap W_p = \emptyset$ and, therefore, $M \cap W_p = \emptyset$. Similarly, it can be shown that $M \cap W_q = \emptyset$. This completes the proof of (*). Now we use (*) to show M is the desired subcontinuum of Z. First note that, since W is a chain of open sets,

(1) $cl[W_j] \subset [W_{j-1} \cup W_j \cup W_{j+1}]$ for any $j = 1, 2, \ldots, n$

(note: For the formula to make sense when $j = 1$ or n, assume $W_0 = \emptyset = W_{n+1}$).

Choose and fix i such that $M \cap cl[W_i] \neq \emptyset$. Suppose $i < p$. By (1), we have that $M \cap W_{i-1} \neq \emptyset$, $M \cap W_i \neq \emptyset$ or $M \cap W_{i+1} \neq \emptyset$. Thus, since $i + 1 \leq p$ and $M \cap W_p = \emptyset$ by (*), it follows easily that

(2) $M \cap [\bigcup_{k=1}^{p-1} W_k] \neq \emptyset$.

Also, since $M \cap cl[W_q] \neq \emptyset$ and since $q \geq p + 2$, it follows from (1) that

(3) $M \cap [\bigcup_{k=p+1}^{n} W_k] \neq \emptyset$.

Finally, since $M \cap W_p = \emptyset$ by (*),

(4) $M \subset [\bigcup_{k=1}^{p-1} W_k] \cup [\bigcup_{k=p+1}^{n} W_k]$.

It follows from (2) through (4) that M is not connected, a contradiction. Hence, $i \geq p$. A similar argument shows that $i \leq q$. We have now proved that if $M \cap cl[W_i] \neq \emptyset$, then $p \leq i \leq q$; we need to prove the converse. To do this, first observe that, from what we have just proved,

(5) $M \cap W_{p-1} = \emptyset = M \cap W_{q+1}$.

Also we have that

(6) $M \cap cl[W_p] \neq \emptyset \neq M \cap cl[W_q]$.

Now, using (5), (6), (*), and (1) with $j = p$ and with $j = q$ we see that

(7) $M \cap W_{p+1} \neq \emptyset \neq M \cap W_{q-1}$.

From (7), the connectedness of M, and the fact that W is a chain of open sets it follows that

(8) $M \cap W_i \neq \emptyset$ whenever $p + 1 \leq i \leq q - 1$.

Hence, by (6) and (8), $M \cap cl[W_i] \neq \emptyset$ whenever $p \leq i \leq q$. This completes the proof of (15.12).

(15.13) THEOREM [250, 2.4]. *Any chainable continuum is C^*-smooth.*[1]

Proof. Let X be a chainable continuum. Let $\{Y_m\}_{m=1}^{\infty}$ be a sequence of subcontinua of X converging to a subcontinuum Y_o of X. By (15.2), it suffices to prove

(1) $C(Y_o) \subset \liminf C(Y_m)$.

To prove (1), let $K \in C(Y_o)$. Let $\varepsilon > 0$. Let $U = \{U_1, U_2, \ldots, U_r\}$ be a chain of open subsets U_j of X covering X such that $diam[U_j] < \varepsilon$ for each $j = 1, 2, \ldots, r$. Since Y_o is connected, there exist natural numbers $s_1 \leq t_1$ such that

(2) $Y_o \cap U_j \neq \emptyset$ if and only if $s_1 \leq j \leq t_1$.

Let [see (0.10.2)]

$V = <U_{s_1}, U_{s_1+1}, \ldots, U_{t_1}> \cap C(X)$.

Then, V is an open subset of $C(X)$ [by (0.13)] and, by (2),

$Y_o \in V$. Hence, since $\{Y_m\}_{m=1}^{\infty}$ converges to Y_o, there exists a natural number N such that

(3) if $m \geq N$, then $Y_m \in V$.

Choose and fix $m \geq N$. For each $i = 1,2,\ldots, n = t_1 - s_1 + 1$, let $W_i = [U_{s_1+i-1}] \cap Y_m$. It follows using connectedness of Y_m and (3) that $W = \{W_1, W_2, \ldots, W_n\}$ is a chain of open subsets of Y_m covering Y_m. Since K is a connected subset of Y_o, it follows using (2) that there exist natural numbers s_2 and t_2 such that $s_1 \leq s_2 \leq t_2 \leq t_1$ and

(4) $K \cap U_j \neq \emptyset$ if and only if $s_2 \leq j \leq t_2$.

Applying (15.12) to the case when $Z = Y_m$, W is as just defined, $p = s_2 - s_1 + 1$, and $q = t_2 - s_1 + 1$, we see that there exists a subcontinuum M of Y_m such that

(5) $M \cap \text{cl}[W_i] \neq \emptyset$ if and only if $p \leq i \leq q$.

Note that (4) is equivalent to

(6) $K \cap U_{s_1+i-1} \neq \emptyset$ if and only if $p \leq i \leq q$.

Also note that

(7) $\text{diam}[\text{cl}(U_{s_1+i-1})] < \varepsilon$ for $p \leq i \leq q$.

It follows, using (5) through (7) and the definition of W_i, that $H(K, M) < \varepsilon$ [use the definition of H as in (0.1)]. Thus, since m was an arbitrary natural number greater than or equal to N, it follows that $K \in \liminf C(Y_m)$. Therefore, we have proved (1), which completes the proof of (15.13).

(15.14) QUESTION. What are classes of C^*-smooth continua

besides those given in (1.207.8), (14.76.6), (15.11), and (15.13)? A few answers are given below--also, see Footnote 1 at the end of this chapter.

A special case of (15.14) which would seem to be of interest is the following question:

(15.15) QUESTION. What are the C^*-smooth circle-like continua [see (15.16)]?[1] A partial answer was given in (14.76.6), i.e., non-planar circle-like continua are C^*-smooth.

Let us observe the following fact concerning (15.15).

(15.16) THEOREM. *If* X *is a C^*-smooth circle-like continuum, then* X *is chainable or* X *is indecomposable.*

Proof. Assume that X is a circle-like continuum and that X is neither chainable nor indecomposable. Thus, since X is decomposable, there exist proper subcontinua A and B of X such that X = A ∪ B. Clearly X\[A ∩ B] is not connected. Burgess has proved [56, Theorem 4] that if a circle-like continuum is separated by one of its subcontinua, then it is chainable. Hence, since X is not chainable, we have that A ∩ B is not connected. Therefore, since X is a-triodic, A ∩ B has exactly two components (comp., [56, Theorem 5]). Denote the two components of A ∩ B by K and L. Let B_o be irreducible with respect to the property of being a subcontinuum of B which intersects both K and L [see (0.68.7)]. By (1.25), there exists a segment $\sigma: [0, 1] \to 2^{B_o}$ from $B_o \cap K$ to B_o. For each $n = 1, 2, \ldots$, let $A_n = A \cup \sigma(1 - [1/n])$.

Let M be a subcontinuum of $A \cup B_o$ such that $L \subset M$, $M \cap [A \setminus L] \neq \emptyset$, $M \cap [B_o \setminus L] \neq \emptyset$, and $M \cap K = \emptyset$ [such a continuum M can be seen to exist by using (20.2)]. It follows that $\{A_n\}_{n=1}^\infty$ is a sequence in $C(X)$ converging to $A \cup B_o$, but that $M \not\subset \lim \sup C^*(A_n)$. Therefore, X is not C^*-smooth. This completes the proof of (15.16).

The following result is easy to prove using ideas from the proof of (15.16).

(15.17) THEOREM. *Assume* X *is a continuum containing two subcontinua* A *and* B *satisfying:*

(1) A ∩ B *is not connected;*

(2) There is a component L *of* A ∩ B *and an open subset* U *of* X *such that* U ∩ [A ∩ B] = L.

Then, X *is not* C^*-*smooth.*

However, I do not know the answer to the following question:

(15.18) QUESTION. Must a C^*-smooth continuum be unicoherent [hence, by (15.6), hereditarily unicoherent]?[1] For arcwise connected continua the answer is "yes," as the following result shows.

(15.19) THEOREM. *An arcwise connected* C^*-*smooth continuum* *must be a dendroid* [see (15.20)].

Proof. Let X be an arcwise connected C^*-smooth continuum. First we prove:

(#) Each subcontinuum of X is arcwise connected.

To prove (#), suppose that there is a subcontinuum B of X such that B is not arcwise connected. Then, since X is arcwise connected, there is an arc A in X such that A ∩ B consists of exactly the two end points of A. Clearly A and B satisfy all the hypotheses of (15.17) and thus, by (15.17), X is not C^*-smooth. This is a contradiction. Therefore, (#) holds. Now, suppose X is not a dendroid. Then, by (0.28), there exist two subcontinua Y and Z of X such that Y ∩ Z is not connected. Thus, since Y and Z are each arcwise connected by (#), it follows easily that Y ∪ Z contains a circle. Hence, by (15.1) and (15.6), X is not C^*-smooth. This is a contradiction. Therefore, X is a dendroid.

(15.20) REMARK. There seems to be little if any relationship between smoothness [defined in (1.68)] and C^*-smoothness for dendroids. For example, the dendroids in (1.66) and (1.69) are C^*-smooth but not smooth. On the other hand, the dendroid in (6.10.2) is smooth but not C^*-smooth. A necessary and sufficient condition in order that a smooth dendroid be C^*-smooth is given in the paper cited in Footnote 1 at the end of this chapter.

Recall from (1.207.8) that any hereditarily indecomposable

continuum is C^*-smooth. I do not know the answer to the following questions. If the answer to the second one is "yes," then we would have a characterization of C^*-smooth continua from among the class of planar homogeneous continua.

(15.21) QUESTIONS. If X is a homogeneous C^*-smooth continuum, then must X be indecomposable? If, in addition, X is planar, then must X be hereditarily indecomposable?

NOTES

1. Recently J. Grispolakis, E. D. Tymchatyn, and I have answered the question in (15.15) and have affirmatively answered the question in (15.18). We also obtain many results relating C^*-smoothness, local separating continua, Class[W], and the covering property. Some known results are obtained as corollaries. For example, we show that if $A \in C(X)$ such that $A \in CP$, then X is C^*-smooth at A. This, together with (14.73.12), gives a new proof of (15.13) [another, somewhat similar proof is in (14.76.6)]. Our paper, entitled "Some properties of hyperspaces with applications to continua theory," is currently under preparation.

XVI.
CONTRACTIBILITY OF HYPERSPACES
AND A PROPERTY OF KELLEY

The first person to investigate contractibility of hyperspaces was Wojdyslawski--his result is stated in (1.93). The next investigation was done by Kelley. He generalized Wojdyslawski's result [see (16.15)] and, in so doing, gave a simpler proof of it [see the proof in (16.18)]. For these purposes, he defined what we will call property[κ]. In the first part of this chapter, we discuss Kelley's work on contractibility of hyperspaces. The rest of the chapter [i.e., after (16.22)] is devoted mainly to a study of peoperty[κ] itself (also, we give what few other results there are about contractibility of hyperspaces). Kelley used property[κ] in connection with contractibility. As we will see, the property is of interest in itself and has some applications outside of hyperspace theory. Near the end of the chapter some unanswered questions are stated.

Let us note the following definitions.

(16.1) DEFINITIONS. Let Z_1 and Z_2 be two topological spaces. A *homotopy* is a mapping h from $Z_1 \times [0, 1]$ into Z_2. If $f: Z_1 \to Z_2$ and $g: Z_1 \to Z_2$ are two mappings and if $h: Z_1 \times [0, 1] \to Z_2$ is a homotopy such that, for each $z_1 \in Z_1$,

$h(z_1, 0) = f(z_1)$ and $h(z_1, 1) = g(z_1)$,
then h is said to be a *homotopy from* f *to* g. Also, when there is a homotopy from f to g, then f and g are said to be *homotopic*.

(16.2) DEFINITIONS. A topological space Z_1 is said to be *contractible in itself*, or simply *contractible*, provided that the identity mapping from Z_1 onto Z_1 is homotopic to a constant mapping from Z_1 into Z_1 [equivalently: provided that every mapping from Z_1 into Z_1 is homotopic to a constant mapping from Z_1 into Z_1]. More generally--Z_1 is said to be *contractible with respect to a topological space* Z_2 provided that every mapping from Z_1 into Z_2 is homotopic to a constant mapping from Z_1 into Z_2. A topological space which is contractible with respect to S^1 is sometimes said to have *property (b)* (this is not the usual definition of property(b), but it is equivalent [338, p.226]). If Z_1 is contractible with respect to Z_2 and if $Z_1 \subset Z_2$, then we say that Z_1 *is contractible in* Z_2 ; as is easy to prove, Z_1 is contractible in Z_2 if and only if the identity mapping from Z_1 into Z_2 is homotopic to a constant mapping from Z_1 into Z_2.

In the following lemma, a homotopy h' is constructed from an arbitrary homotopy h. The homotopy h' is given a name in (16.4), and some of its properties are delineated in (16.5); h' will be used in some of the proofs in this chapter.

(16.3) LEMMA [165, proof of 3.1]. *Let* $\Lambda \subset 2^X$ *and let* $h: \Lambda \times [0, 1] \to 2^X$ *be a homotopy. Then, for each* $(A, t) \in \Lambda \times [0, 1]$,

$$[\cup\{h(A, s): 0 \leq s \leq t\}] \in 2^X.$$

Furthermore: The function $h': \Lambda \times [0, 1] \to 2^X$, *defined by*

$$h'(A, t) = \cup\{h(A, s): 0 \leq s \leq t\}$$

for each $(A, t) \in \Lambda \times [0, 1]$, *is continuous.*

Proof. For each $(A, t) \in \Lambda \times [0, 1]$, let

$$g(A, t) = \{h(A, s): 0 \leq s \leq t\}.$$

Since h is a continuous function into 2^X and since

$$g(A, t) = h(\{A\} \times [0, t])$$

for each $(A, t) \in \Lambda \times [0, 1]$, we have that $g(A, t) \in 2^{2^X}$ for each $(A, t) \in \Lambda \times [0, 1]$. In other words,

(1) g is a function from $\Lambda \times [0, 1]$ into 2^{2^X}.

Since union is a function from 2^{2^X} into 2^X [see (1.48)], we have by (1) that

(2) $[\cup g(A, t)] \in 2^X$ for each $(A, t) \in \Lambda \times [0, 1]$.

By (2) and the definition of g, we have proved the first part of (16.3). To prove the second part, note that

(3) $h': (A, t) = \cup g(A, t)$ for each $(A, t) \in \Lambda \times [0, 1]$.

It is easy to see, using the continuity of h, that

(4) $g: \Lambda \times [0, 1] \to 2^{2^X}$ is continuous.

By (3), (4), and the continuity of union [see (1.48)], h' is continuous. This completes the proof of (16.3).

(16.4) TERMINOLOGY. The homotopy h', defined in terms of

h in (16.3), will be called *the segment homotopy associated with*
h. The reason for this terminology is evident from (16.5.1) of the
next lemma.

(16.5) LEMMA (165, proof of 3.1]. *Let* $\Lambda \subset 2^X$ *and let*
$h: \Lambda \times [0, 1] \to 2^X$ *be a homotopy. Then, the segment homotopy* h'
associated with h *has the following properties:*

(16.5.1) *For each* $A \in \Lambda$, $h'(\{A\} \times [0, 1])$ *is the range
of a segment in* 2^X *from* $h'(A, 0)$ *to* $h'(A, 1)$,
equivalently, $h'(A, t_1) \subset h'(A, t_2)$ *whenever*
$0 \leq t_1 \leq t_2 \leq 1$;

(16.5.2) *For each* $A \in \Lambda$, $h'(A, 0) = h(A, 0)$;

(16.5.3) *For each* $(A, t) \in \Lambda \times [0, 1]$, $h'(A, t) \supset h(A, s)$
for each $s \in [0, t]$;

(16.5.4) *If* $h'(A, t) \in C(X)$ *for some* $(A, t) \in \Lambda \times [0, 1]$,
then $h'(\{A\} \times [t, 1]) \subset C(X)$.

Proof. Both (16.5.2) and (16.5.3) follow immediately from the
definition of h' in (16.3). Also, (16.5.4) follows from (16.5.1)
and (1.26). Hence, it suffices to prove (16.5.1). Let $A \in \Lambda$.
Since h' is a continuous function into 2^X [by (16.3)],

(1) $h'(\{A\} \times [0, 1])$ is a subcontinuum (possibly
degenerate) of 2^X.

Also, from the formula for h' in (16.3),

(2) $h'(A, t_1) \subset h'(A, t_2)$ whenever $0 \leq t_1 \leq t_2 \leq 1$.

By using (1), (2), (1.4), (1.15), and (1.22), (16.5.1) now follows.
This completes the proof of (16.5).

The next lemma defines a special mapping which will be used in the proof of (16.7).

(16.6) LEMMA (comp., [165, proof of 3.1]). *The function* $f_1: 2^X \to 2^{2^X}$ *defined by*

$$f_1(A) = \{\{a\}: a \in A\} \text{ for each } A \in 2^X$$

is continuous. In fact, f_1 *is an isometry, i.e.,*

$$H^2[f_1(A), f_1(B)] = H(A, B) \text{ for each } A, B \in 2^X$$

where H^2 *denotes the Hausdorff metric for* 2^{2^X} *induced by the Hausdorff metric* H *for* 2^X *[see (1.31.1)].*

Proof. The lemma follows easily from the definition of H in (0.1) and of H^2 in (1.31.1).

Now we begin our main results on contractibility. Wojdyslawski [343] stated that 2^X is contractible when X is a locally connected continuum. In the paragraph following (1.93), we remarked that Wojdyslawski's proof [343] of the 2^X part of (1.93) also works for $C(X)$. Now, let us note the following general theorem of Kelley which shows that, for *any* continuum X, 2^X is contractible if and only if $C(X)$ is contractible.

(16.7) THEOREM [165, 3.1]. *For any continuum* X, *the following three statements are equivalent:*

(16.7.1) $F_1(X)$ *is contractible in* 2^X;

(16.7.2) 2^X *is contractible;*

(16.7.3) $C(X)$ *is contractible (in itself).*

Proof. Assume (16.7.1) holds. Then, there is a homotopy
$f: F_1(X) \times [0, 1] \to 2^X$ such that

$f(\{x\}, 0) = \{x\}$ for each $\{x\} \in F_1(X)$

and

$f(\{x\}, 1) = A_0$ for each $\{x\} \in F_1(X)$

where A_0 is some fixed member of 2^X. Let f_1 be the mapping defined in (16.6). For each $(A, t) \in 2^X \times [0, 1]$, let

$g(A, t) = f[f_1(A) \times \{t\}]$.

Since f is continuous and $f_1(A)$ is a nonempty compact subset of $F_1(X)$ for each $A \in 2^X$, we have that $g(A, t) \in 2^{2^X}$ for each $(A, t) \in 2^X \times [0, 1]$, i.e.,

g is a function from $2^X \times [0, 1]$ into 2^{2^X}.

Since f_1 is continuous [by (16.6)] and f is continuous, it follows that

(1) $g: 2^X \times [0, 1] \to 2^{2^X}$ is continuous.

By (1.48), the union function

(2) $\cup: 2^{2^X} \to 2^X$ is continuous.

Define the function $k: 2^X \times [0, 1] \to 2^X$ by

$k(A, t) = \cup g(A, t)$ for each $(A, t) \in 2^X \times [0, 1]$.

Since k is the composition of g and $\cup$, we have by (1) and (2) that k is continuous, i.e., k is a homotopy. Using the formulas, it follows easily that

$k(A, 0) = A$ for each $A \in 2^X$

and

$k(A, 1) = A_0$ for each $A \in 2^X$.

Therefore, 2^X is contractible. We have proved (16.7.1) implies

(16.7.2). Next assume (16.7.2) holds. Then, since 2^X is arcwise connected (by (1.9) or, more simply, by the fact that a contractible space is arcwise connected [186, Theorem 1, p.374]), there is a homotopy $h: 2^X \times [0, 1] \to 2^X$ such that

(3) $h(A, 0) = A$ for each $A \in 2^X$

and

(4) $h(A, 1) = X$ for each $A \in 2^X$.

Let h' be the segment homotopy associated with h [see (16.4)]. By (3) and (16.5.2),

(5) $h'(A, 0) = A$ for each $A \in 2^X$.

By (5) and (16.5.4),

(6) $h'(C(X) \times [0, 1]) \subset C(X)$.

By (4) and (16.5.3),

(7) $h'(A, 1) = X$ for each $A \in 2^X$.

Therefore, by restricting h' to $C(X) \times [0, 1]$, we see that $C(X)$ is contractible. We have proved (16.7.2) implies (16.7.3). The fact that (16.7.3) implies (16.7.1) is obvious. Hence, we have proved (16.7).

Let us note the following simple consequence of (16.7). It can also be proved directly [see (0.67.5)].

(16.8) COROLLARY [see (16.9)]. *If* X *is a contractible continuum, then* 2^X *and* $C(X)$ *are each contractible.*

(16.9) REMARK. Since (16.8) is clearly implied by (16.7),

(16.8) should be credited to Kelley. It is not explicitly stated in [165].

Kelley defined the following property in [165, 3.2], and used it to investigate contractibility of hyperspaces. As we will see, the property is of interest in its own right and has some applications outside of hyperspace theory. I will say more about this after (16.22). For now, let us define the property and discuss what Kelley did with it.

(16.10) DEFINITION [165, 3.2]. A continuum X is said to have *property*$[\kappa]$ (called property 3.2 in [165]), written $X \in prop[\kappa]$, provided that given any $\varepsilon > 0$ there exists $\delta = \delta(\varepsilon) > 0$ such that if

$a, b \in X$, $d(a, b) < \delta$, and $a \in A \in C(X)$,

then there exists $B \in C(X)$ such that $b \in B$ and $H(A, B) < \varepsilon$.

(16.11) EXAMPLE [165, proof of 4.1]. Any locally connected continuum has property$[\kappa]$. This is easy to prove from the fact that any locally connected continuum is uniformly locally arcwise connected [186, Theorem 4, p.257].

Two other classes of continua which have property$[\kappa]$ will be given in (16.26) and (16.27).

The next main result is (16.15). In (16.12), we defined a special function which Kelley used in [165, proof of 3.3]. In

(16.14) it is shown that the function defined in (16.12) is continuous if and only if the continuum has property[κ]. The "if" part is then used in the proof of (16.15). The "only if" part, which was observed by Wardle [329, 2.4], will be used in the proof of (16.27).

(16.12) NOTATION. Let X be a continuum and let μ be any given fixed Whitney map for C(X) [see (0.50)]. Define
$$F_\mu: X \times [0, \mu(X)] \to 2^{2^X} \quad [\text{see (16.13)}]$$
by
$$F_\mu(x, t) = \{A \in C(X): x \in A \in \mu^{-1}(t)\}$$
for each $(x, t) \in X \times [0, \mu(X)]$.

(16.13) REMARK. It is easy to prove $F_\mu(x, t) \in 2^{2^X}$ for each $(x, t) \in X \times [0, \mu(X)]$. Furthermore, though it will not be important here, let us note that, by (14.8.1), $F_\mu(x, t)$ is an arcwise connected subcontinuum of 2^X for each $(x, t) \in X \times [0, \mu(X)]$. As mentioned in (3.1.1), F_μ was first defined by Kelley in [165, proof of 3.3]. In [165], F_μ was denoted $\mathcal{F}$; since Kelley only used normalized Whitney maps [see (1.16)], $\mathcal{F}$ had domain $X \times [0, 1]$.

(16.14) LEMMA (necessity in [165, proof of 3.3]; sufficiency observed in [329, 2.4]). *Let F_μ be as in (16.12). A necessary and sufficient condition that* $X \in \text{prop}[\kappa]$ *is that F_μ be continuous.*

Proof. Assume $X \in \text{prop}[\kappa]$. To show F_μ is continuous, let

$\varepsilon > 0$. First we show two facts, (*) and (**) below. Let D denote the metric for $X \times [0, \mu(X)]$ given by

$$D[(x_1, t_1), (x_2, t_2)] = ([d(x_1, x_2)]^2 + |t_1 - t_2|^2)^{\frac{1}{2}}$$

for each $(x_1, t_1), (x_2, t_2) \in X \times [0, \mu(X)]$. Also, let H^2 denote the Hausdorff metric for 2^{2^X} induced by the Hausdorff metric H for 2^X [see (1.31.1)].

(*) There exists $\delta_1 > 0$ such that if

$$D[(x, t_1), (x, t_2)] < \delta_1, \text{ then}$$

$$H^2[F_\mu(x, t_1), F_\mu(x, t_2)] < \varepsilon.$$

Proof of (*). Choose δ_1 to be $\eta(\varepsilon)$ as in (1.28) [(1.28) may be restated with 2^X replaced by $C(X)$ and ω replaced by μ --the proof is the same as that given for (1.28)]. Assume

$$D[(x, t_1), (x, t_2)] < \delta_1 = \eta(\varepsilon).$$

This means

(1) $|t_1 - t_2| < \eta(\varepsilon).$

For convenience, assume $t_1 < t_2$. Let $A_1 \in F_\mu(x, t_1)$. By (1.8) and (1.11), there exists an order arc $\alpha \subset C(X)$ from A_1 to X. Since μ is continuous [by (0.50)], there exists $A_2 \in \alpha$ such that $\mu(A_2) = t_2$. Since $x \in A_1$ and $A_1 \subset A_2$ [by (1.2) and (1.7)], $x \in A_2$. Since $A_2 \in \alpha \subset C(X)$, $A_2 \in C(X)$. Thus,

(2) $A_2 \in F_\mu(x, t_2)$.

Note that, since $\mu(A_1) = t_1 < t_2 = \mu(A_2)$, we have from (1) that

(3) $[\mu(A_2) - \mu(A_1)] < \eta(\varepsilon).$

Since (3) holds and $A_1 \subset A_2$, we have by (1.28) that

(4) $H(A_1, A_2) < \varepsilon.$

It follows from (2) and (4) that we have proved

(5) $F_\mu(x, t_1) \subset N_H[\varepsilon, F_\mu(x, t_2)]$.

Next, let $B_2 \in F_\mu(x, t_2)$. By (1.8) and (1.11), there exists an order arc $\beta \subset C(X)$ from $\{x\}$ to B_2. Since μ is continuous and

$$\mu(\{x\}) = 0 \leq t_1 < t_2 = \mu(B_2),$$

there exists $B_1 \in \beta$ such that $\mu(B_1) = t_1$. By (1.2) and (1.7), $x \in B_1$. It follows that

(6) $B_1 \in F_\mu(x, t_1)$.

In a manner similar to that done for A_1 and A_2 above, it can be shown that

(7) $H(B_1, B_2) < \varepsilon$.

It follows from (6) and (7) that we have proved

(8) $F_\mu(x, t_2) \subset N_H[\varepsilon, F_\mu(x, t_1)]$.

By (5), (8), and the definition of H^2 we have

(9) $H^2[F_\mu(x, t_1), F_\mu(x, t_2)] < \varepsilon$.

This proves (*).

(**) There exists $\delta_2 > 0$ such that if

$$D[(x_1, t), (x_2, t)] < \delta_2,$$ then
$$H^2[F_\mu(x_1, t), F_\mu(x_2, t)] < \varepsilon.$$

Proof of (**). Let $\eta(\varepsilon/2)$ be as in (1.28) [with $\varepsilon/2$ replacing ε in (1.28)]. Since μ is continuous, there exists γ such that

(10) $0 < \gamma < \varepsilon/2$

and such that

(11) if $K, L \in C(X)$ such that $H(K, L) < \gamma$,

then $|\mu(K) - \mu(L)| < \eta(\varepsilon/2)$.

Since $X \varepsilon \text{prop}[\kappa]$,

(12) there exists $\delta(\gamma)$ as in (16.10) with ε replaced by γ.

Choose δ_2 to be $\delta(\gamma)$ in (12). Now, assume

$$D[(x_1, t), (x_2, t)] < \delta_2.$$

This means

(13) $d(x_1, x_2) < \delta_2 = \delta(\gamma)$.

Let $A \varepsilon F_\mu(x_1, t)$. Then $x_1 \varepsilon A \varepsilon C(X)$. Hence, since (13) holds, we have by (12) that there exists $A_1 \varepsilon C(X)$ such that $x_2 \varepsilon A_1$ and

(14) $H(A, A_1) < \gamma$.

By (14) and the fact that $\mu(A) = t$, we have using (11) that

(15) $|t - \mu(A_1)| < \eta(\varepsilon/2)$.

We will define a set B after considering three cases.

Case I: $\mu(A_1) = t$. Then, $A_1 \varepsilon F_\mu(x_2, t)$.

Case II: $\mu(A_1) < t$. By (1.8) and (1.11), there exists an order arc $\alpha \subset C(X)$ from A_1 to X. Since μ is continuous and $\mu(A_1) < t \leq \mu(X)$, there exists $A_2 \varepsilon \alpha$ such that $\mu(A_2) = t$. Note that, by (1.2) and (1.7),

(16) $A_1 \subset A_2$.

Since $t = \mu(A_2)$, we have by (15) that

(17) $[\mu(A_2) - \mu(A_1)] < \eta(\varepsilon/2)$.

Since (16) and (17) hold, we have by (1.28) that

(18) $H(A_1, A_2) < \varepsilon/2$.

Therefore, since $\gamma < \varepsilon/2$ [by (10)], we have from (14), (18), and the triangle inequality that

(19) $H(A, A_2) < \varepsilon$.

Since $x_2 \in A_1 \subset A_2$ [by (16)], $x_2 \in A_2$. Hence, since $\mu(A_2) = t$ and $A_2 \in C(X)$,

(20) $A_2 \in F_\mu(x_2, t)$.

Case III: $\mu(A_1) > t$. By replacing α in Case II with an order arc β from $\{x_2\}$ to A_1, and by using ideas in the verifications in Case II, we obtain $A_3 \in C(X)$ such that

(21) $H(A, A_3) < \varepsilon$

and

(22) $A_3 \in F_\mu(x_2, t)$.

This completes the three cases. Let $B = A_1$ if $\mu(A_1) = t$, let $B = A_2$ if $\mu(A_1) < t$, and let $B = A_3$ if $\mu(A_1) > t$. Then, in any case,

(23) $B \in F_\mu(x_2, t)$ and $H(A, B) < \varepsilon$.

This proves

(24) $F_\mu(x_1, t) \subset N_H[\varepsilon, F_\mu(x_2, t)]$.

Similarly it can be shown that

(25) $F_\mu(x_2, t) \subset N_H[\varepsilon, F_\mu(x_1, t)]$.

By (24), (25), and the definition of H^2,

(26) $H^2[F_\mu(x_1, t), F_\mu(x_2, t)] < \varepsilon$.

This proves (**).

The continuity of F_μ on $X \times [0, \mu(X)]$ is a simple consequence of (*) and (**). Thus, half the theorem is proved. The sufficiency part of the theorem is easy to prove using the uniform continuity of F_μ. This completes the proof of (16.14).

Now we come to Kelley's result about the relationship of property [K] to contractibility of hyperspaces. An application of (16.15) is given in (5.10).

(16.15) THEOREM [165, 3.3]. *If X is a continuum such that $X \in \text{prop}[K]$, then 2^X and $C(X)$ are each contractible [by (16.21), the converse is false].*

Proof. Assume $X \in \text{prop}[K]$ and let F_μ be as in (16.12). By (16.14),

(1) F_μ is continuous.

Let $j: F_1(X) \to X$ be defined by

$$j(\{x\}) = x \text{ for each } \{x\} \in F_1(X).$$

Clearly

(2) j is continuous.

By (1.48),

(3) union $\cup: 2^{2^X} \to 2^X$ is continuous.

Now, define $h: F_1(X) \times [0, \mu(X)] \to 2^X$ by

$$h(\{x\}, t) = \cup F_\mu(j(\{x\}), t)$$

for each $(\{x\}, t) \in F_1(X) \times [0, \mu(X)]$. As the composition of continuous functions [by (1) through (3)],

(4) h is continuous.

Since $\mu^{-1}(0) = F_1(X)$ by (0.50), it follows easily using the formula for F_μ in (16.12) that $F_\mu(x, 0) = \{\{x\}\}$ for each $x \in X$. Hence, using the formula for h above,

(5) $h(\{x\}, 0) = \{x\}$ for each $\{x\} \in F_1(X)$.

Since $\mu^{-1}[\mu(X)] = \{X\}$ by (0.50), it follows easily using the formulas for F_μ and h that

(6) $h(\{x\}, \mu(X)) = X$ for each $\{x\} \in F_1(X)$.

It follows from (4) through (6) that (16.7.1) holds. Therefore, by (16.7), 2^X and $C(X)$ are each contractible. This proves (16.15).

From (16.11) and (16.15) we see that 2^X and $C(X)$ are each contractible when X is a locally connected continuum. This is part of (16.18). In order to prove the "locally contractible" part of (16.18), we give the following two lemmas.

(16.16) LEMMA [165, proof of 4.2]. *For any continuum* Y, $\text{diam}[C(Y)] = \text{diam}[Y]$ *where* $C(Y)$ *has the Hausdorff metric as defined in (0.1)*.

Proof. The lemma follows easily from the formula for the Hausdorff metric given in (0.1).

The lemma in (16.16) will be useful in the proof of the next lemma.

(16.17) LEMMA [165, 4.2]. *Let* X *be any continuum. If* Λ *is a locally connected subcontinuum of* 2^X *[resp.,* $C(X)$*] with the Hausdorff metric as defined in (0.1), then* Λ *is contractible in a subset* Γ *of* 2^X *[resp.,* $C(X)$*] such that* $\text{diam}[\Gamma] \leq \text{diam}[\Lambda]$.

Proof. By (16.11) and (16.15), $C(\Lambda)$ is contractible. Hence, there is a homotopy $h_1 : F_1(\Lambda) \times [0, 1] \to C(\Lambda)$ such that

$h_1(\{L\}, 0) = \{L\}$ for each $\{L\} \in F_1(\Lambda)$

and

$h_1(\{L\}, 1) = \Lambda_o$ for each $\{L\} \in F_1(\Lambda)$

where Λ_o is some fixed member of $C(\Lambda)$. Now, consider $C(\Lambda)$ as contained in $C(2^X)$ by inclusion. Then, for each $(\{L\}, t) \in F_1(\Lambda) \times [0, 1]$, we have $h_1(\{L\}, t) \in C(2^X)$ and if $\Lambda \subset C(X)$, then $h_1(\{L\}, t) \subset \Lambda \subset C(X)$. For convenience in the proof, let us note that the facts in the last sentence imply (1) and (2) below; as usual, consider $C(X)$ [resp., $C(2^X)$] as contained in 2^X [resp., 2^{2^X}] by inclusion.

(1) $h_1(\{L\}, t) \in 2^{2^X}$ for each $(\{L\}, t) \in F_1(\Lambda) \times [0, 1]$;

(2) if $\Lambda \subset C(X)$, then $h_1(\{L\}, t)$ is a subcontinuum of 2^X such that $h_1(\{L\}, t) \cap C(X) \neq \emptyset$ for each $(\{L\}, t) \in F_1(\Lambda) \times [0, 1]$.

By (1.48),

(3) union $\cup: 2^{2^X} \to 2^X$ is nonexpansive.

By (1) and (3) we may take the composition of h_1 and union, and define a homotopy $h_2: \Lambda \times [0, 1] \to 2^X$ by

$h_2(L, t) = \cup h_1(\{L\}, t)$

for each $(L, t) \in \Lambda \times [0, 1]$. It follows easily that

(4) $h_2(L, 0) = L$ for each $L \in \Lambda$

and

(5) $h_2(L, 1) = \cup \Lambda_o$ for each $L \in \Lambda$.

Furthermore, by (2) and (1.49) applied to $h_1(\{L\}, t)$,

(6) if Λ is a subcontinuum of $C(X)$, then

$h_2(L, t) \in C(X)$ for each $(L, t) \in \Lambda \times [0, 1]$.

Now, let (L_1, t_1) $(L_2, t_2) \in \Lambda \times [0, 1]$. By the definition of h_2, (3), and (16.16),

(7) $H[h_2(L_1, t_1), h_2(L_2, t_2)] =$
$H[\cup h_1(\{L_1\}, t_1), \cup h_1(\{L_2\}, t_2)] \leq$
$H^2[h_1(\{L_1\}, t_1), h_1(\{L_2\}, t_2)] \leq$
$\text{diam}[C(\Lambda)] = \text{diam}[\Lambda].$

The lemma now follows from (4) through (7).

The following result was stated in (1.93).

(16.18) COROLLARY [see (16.19)]. *If* X *is a locally connected continuum, then* 2^X *and* C(X) *are each contractible and locally contractible* [comp., (16.41)].

Proof. Let X be a locally connected continuum. The contractibility of 2^X and C(X) follows immediately from (16.11) and (16.15). The local contractibility of 2^X and C(X) can be proved using (16.17) in conjunction with (1.92) and the fact that any point of a locally connected continuum [i.e., 2^X or C(X) by (1.92)] has arbitrarily small neighborhoods which are locally connected continua (comp., [186, Theorem 3, p.257]).

(16.19) REMARK. In 1938, Wojdyslawski proved (16.18) and, in 1939, he proved 2^X and C(X) are each absolute retracts when X is a locally connected continuum [see (1.96); also, see [131] for a correction to Wojdyslawski's proof]. In 1942, Kelley [165] reproved these results. The main interest in that part of [165] has been the techniques of proof. Also, some new results were obtained which have been of interest [for example, (16.7) and (16.15)]. We have given Kelley's proof of (16.18). A different proof is indicated in (0.65.4).

It is now appropriate to give some examples.

(16.20) EXAMPLE [165, pp.26-27]. Let X_o be the familiar sin[1/x]-continuum defined in (1) of (8.22). Then, it is easy to see that $X_o \in \text{prop}[\kappa]$ and hence, by (16.15), 2^{X_o} and $C(X_o)$ are each contractible. As far as I know, this is the first example in the literature showing the contractibility of a hyperspace of a non-locally connected continuum. Let us note, as Rogers states [280, p.284], $C(X_o) \cong \text{Cone}(X_o)$ [here, see Chapter VIII].

(16.21) EXAMPLE [165, p.27]. Let X_o be as in (16.20) and let

$$X_1 = X_o \cup \{(0, y) \in R^2: 1 \leq y \leq \tfrac{3}{2}\} .$$

Note that $X_1 \notin \text{prop}[\kappa]$, as is easily seen by taking [in (16.10)] $a = (0, 1)$ and $A = \{(0, y) \in R^2: 1 \leq y \leq \tfrac{3}{2}\}$. However, 2^{X_1} and $C(X_1)$ are each contractible. We briefly describe how to prove this. First, there is a homotopy h deforming X_1 to X_o [h leaves each point of X_o fixed throughout the homotopy]. Then, h "lifts up" to a hyperspace homotopy k which at time = 1 only has values in the hyperspace of X_o and is the identity on the hyperspace of X_o. Hence, by (16.20), k may be "followed" by a homotopy which contracts the hyperspace of X_o to a point. Thus, we see that the converse of (16.15) is fales.

Kelley [165] gave the first example of a non-contractible hyperspace. We now give his example. An example for Hausdorff continua which shows sharp contrast to the situation for metric continua is in (16.41).

(16.22) EXAMPLE [165, p.27]. Let X_1 be as in (16.21) and let
$$X_2 = X_1 \cup \{(x, \tfrac{1}{2} + \sin[1/x]) \in R^2: -1 \le x < 0\}.$$
Then, neither 2^{X_2} nor $C(X_2)$ is contractible. To prove this, note that by (16.7) it suffices to show it for $C(X_2)$. Suppose $C(X_2)$ is contractible. Then, since $C(X_2)$ is arcwise connected, there is a homotopy $h: C(X_2) \times [0, 1] \to C(X_2)$ such that

(1) $h(A, 0) = A$ for each $A \in C(X_2)$

and

(2) $h(A, 1) = X_2$ for each $A \in C(X_2)$.

Let h' be the segment homotopy associated with h [see (16.4)]. By (16.5.1),

(3) $h'(A, t_1) \subset h'(A, t_2)$ whenever $0 \le t_1 \le t_2 \le 1$ and $A \in C(X_2)$.

By (1) and (16.5.2),

(4) $h'(A, 0) = A$ for each $A \in C(X_2)$.

By (2) and (16.5.3),

(5) $h'(A, 1) = X_2$ for each $A \in C(X_2)$.

By (4) and (16.5.4),

(6) $h'(A, t) \in C(X_2)$ for each $(A, t) \in C(X_2) \times [0, 1]$.

The facts in (7) and (8) below follow easily from (3) through (6) and the geometry of X_2.

(7) If $z = (z_1, z_2) \in X_2$ such that $z_1 > 0$, then there exists $t(z) \in [0, 1]$ such that $h'(\{z\}, t(z)) \in C(X_o)$, $h'(\{z\}, t(z)) \supset \{(0, y) \in R^2 : -1 \le y \le +1\}$, and $z \in h'(\{z\}, t(z))$.

(8) For any z as in (7) and any $w = (w_1, w_2) \in X_2$ such that $w_1 < 0$, $H[h'(\{z\}, t(z)), h'(\{w\}, t(z))] \ge 1/2$.

Since z and w as in (8) may be chosen arbitrarily close together, we see that h' is not (uniformly) continuous. Since $C(X_2)$ is compact [by (0.8)], this is a contradiction [see (16.3)]. Therefore, $C(X_2)$ is not contractible. Though 2^{X_2} is not contractible, it is g-contractible as is any hyperspace 2^X [see (4.10)].

This completes the discussion of Kelley's results concerning contractibility of hyperspaces. We now give some recent results about property[K].

The only papers which investigate or use property[K], besides [165], are [236], [237], and [329]. In Wardle's paper [329] a somewhat detailed study is made of property[K]. In [236] and [237], some uses are made of property[K] apart from hyperspace theory. Some results in these three papers are given below.

Recall that property[K], as defined by Kelley [see (16.10)], is a global property. Wardle considers a pointwise version of it as follows:

(16.23) DEFINITION [329]. A continuum X is said to have *property*[κ] *at a point* $a \in X$ provided that given any $\varepsilon > 0$ there exists $\delta = \delta(a, \varepsilon) > 0$ such that if

$b \in X$, $d(a, b) < \delta$, and $a \in A \in C(X)$,

then there exists $B \in C(X)$ such that $b \in B$ and $H(A, B) < \varepsilon$. Thus, the definition is the same as (16.10) *except* the point a is "held fixed" and δ depends on both a and ε. It can easily be shown, using the compactness of X and of $C(X)$, that a continuum has property[κ] if and only if it has property[κ] at each of its points.

Wardle obtains the following interesting and surprising fact:

(16.24) THEOREM [329, 2.3]. *Any continuum has property*[κ] *at each point of a dense* G_δ *(equivalently, a residual* G_δ*) subset.*

Proof [see (16.25)]. Let X be a continuum. Define $\alpha: X \to C(C(X))$ by

$\alpha(x) = \{A \in C(X): x \in A\}$

for each $x \in X$. [Note: $\alpha(x) \in C(C(X))$ since $\alpha(x)$ is arcwise connected by (1.8) and compact.] It is not difficult to show that α is upper semi-continuous [in the sense of (0.47) with $Y = X$ and $Z = C(X)$], and that α is continuous at x if and only if X has property[κ] at x. Using these facts, (16.24) follows from the results in [185] and [186] mentioned in the proof of (15.3).

(16.25) REMARK. The proof of (16.24) uses the general

technique of examining a property "pointwise" by using semi-continuous functions. This technique was used, in a different context, in a paper of Fort [108] and in a paper of mine, "Horizontally- and Vertically-fixed points," J. of Nat. Sci. and Math. (Lahore, W. Pakistan), 7(1967), pp.181-189.

Let us note that, by (16.24), every continuum has property[κ] at some point! This fact is utilized to prove the following result.

(16.26) THEOREM [329, 2.7]. *If* X *is a homogeneous continuum, then* X ε prop[κ] *(and, hence,* 2^X *and* C(X) *are each contractible).*

Proof. Assume X is a homogeneous continuum. By (16.24), X has property[κ] at some point a ε X. By the statement at the end of (16.23), we need only show that X has property[κ] at each of its points in order to know X ε prop[κ]. Let p ε X. Let ε > 0. Since X is homogeneous, there is a homeomorphism h: X $\xrightarrow{\text{onto}}$ X such that h(a) = p. Let $\hat{h}$: C(X) $\xrightarrow{\text{onto}}$ C(X) denote the induced map [see (0.49)]. Since $\hat{h}$ is uniformly continuous [by (0.8)], there exists $\delta_1 > 0$ such that

(1) if K, L ε C(X) and H(K, L) < δ_1,
     then H[$\hat{h}$(K), $\hat{h}$(L)] < ε.

Since X has property[κ] at a, there exists $\delta_2 = \delta_2(a, \delta_1)$ as in (16.23) with ε replaced by δ_1. Now, since h is a homeomorphism, there exists δ > 0 such that

(2) {x ε X: d(p, x) < δ} ⊂ h[{x ε X: d(a, x) < δ_2}].

Let $y \in X$ such that $d(p, y) < \delta$. Let $A(p) \in C(X)$ such that $p \in A(p)$. Let $A = h^{-1}[A(p)]$. Then $a \in A \in C(X)$. Let $b = h^{-1}(y)$. By (2), $d(a, b) < \delta_2$. Thus, by the choice made for δ_2, there exists $B \in C(X)$ such that $b \in B$ and $H(A, B) < \delta_1$. Hence, by (1), $H[\hat{h}(A), \hat{h}(B)] < \varepsilon$. Therefore, since $\hat{h}(A) = A(p)$, we have that $H[A(p), \hat{h}(B)] < \varepsilon$. Also, $y \in \hat{h}(B) \in C(X)$. We have proved that X has property[κ] at p, an arbitrarily chosen point of X. This completes the proof of (16.26), the contractibility of 2^X and $C(X)$ being a consequence of (16.15). ∎

Thus, we have two classes of continua which have property[κ]—the class of locally connected continua [see (16.11)] and the class of homogeneous continua [see (16.26)]. In [276, 3.5], Rhee stated that $C(X)$ is contractible when X is the pseudo-arc. His proof showed that for any hereditarily indecomposable continuum X, F_μ [defined in (16.12)] is continuous. Hence, though he did not say so, he obtained the following stronger fact (also obtained, independently, in [329]).

(16.27) THEOREM ([276, proof of 3.5] and [329, 3.1]). *Any hereditarily indecomposable continuum has property*[κ].

Proof. If X is an hereditarily indecomposable continuum, then it is easy to show that F_μ [see (16.12)] is continuous. Hence, (16.27) follows using the sufficiency part of (16.14). ∎

As we can see from comparing (16.20) and (16.21), small changes in a continuum can destroy property[κ]. This is further exemplified by the next result. First, we make a few comments. Let Φ denote the class of all compactifications of a half-line with an arc as the remainder. As is deducible from (0.70.3), there are uncountably many topologically different members of Φ. One member of Φ is the familiar sin[1/x]-continuum S_o, and another member of Φ is the continuum used in (8.29). We have observed in (16.20) that $S_o \in$ prop[κ]. It is easy to see that the continuum in (8.29) does not have property[κ]. We have the following theorem:

(16.28) THEOREM [237, 2.5]. *Let* X *be a compactification of a half-line with an arc as the remainder. Then,* X $\in$ prop[κ] *if and only if* X $\cong S_o$.

The proof of (16.28) is not done here--it uses the characterization in [239, 3.5]. An application of (16.28) will be given in (16.30).

We have already seen some of Wardle's results concerning property[κ]. Wardle also proved the following mapping result which says that property[κ] is an invariant under confluent mappings.

(16.29) THEOREM [329, 4.3]. *If* X *is a continuum such that* X $\in$ prop[κ] *and if* f: X $\xrightarrow{\text{onto}}$ Y *is a confluent mapping, then* Y $\in$ prop[κ].

Several years ago when I was a visitor in Wroclaw, Poland, the

(16.30)   CONTRACTIBILITY OF HYPERSPACES

following question arose in a conversation with Professor Charatonik: What are all the open images of the usual sin[1/x]-continuum S_o [defined in (1) of (8.22)]? Using (16.28) and (16.29), I was able to answer this question by obtaining the following result.

(16.30) THEOREM [237, 2.8]. *If* Y *is a confluent image of* S_o, *then* Y *is homeomorphic to* [0, 1], S_o, *or a one-point space (and conversely).*

The proof of (16.30) uses (16.28), (16.29), and the fact that any confluent image of an hereditarily decomposable chainable continuum is chainable [237, 2.7]. It is worth mentioning that it is not known whether or not the confluent image of a chainable continuum must be chainable [195, Problem 4].

In [236], property[κ] was used to investigate spaces of confluent mappings. A classical theorem of Whyburn [338, 3.11, p.174] says that a uniform limit of a sequence of monotone mappings from a continuum onto a locally connected continuum must be monotone. Since [338, 3.11, p.174] was published, there have been many papers on spaces of monotone and other types of mappings (for example, see [183], [189], [200], [236], and [340]). In such papers, usually the condition of local connectivity is assumed for the range of the mappings. In [236], where the space $C_oFL[X, Y]$ of confluent mappings was first investigated, it was found that many results could be proved for $C_oFL[X, Y]$ by only assuming Y ε prop[κ] (instead of the stronger assumption that Y is locally connected). Two of these results are stated in (16.31) and (16.32). We use the following notation. Let

$$C_o[X, Y] = \{f : f \text{ is a mapping of } X \text{ onto } Y\}$$

where X and Y are continua and the metric for $C_o[X, Y]$ is the "uniform" or "sup" metric, and let

$$C_oFL[X, Y] = \{f \in C_o[X, Y] : f \text{ is confluent}\}.$$

(16.31) THEOREM [236, 3.1]. *If X and Y are continua and $Y \in \text{prop}[\kappa]$, then $C_oFL[X, Y]$ is a closed subspace of $C_o[X, Y]$.*

Recall [55, p.482] that a *near-homeomorphism* of X onto Y is a mapping which is a uniform limit of homeomorphisms of X onto Y. When X and Y are locally connected a near-homeomorphism must be monotone [338, p.174]. However, in general, this is not the case [340, pp.465-466]. We remark that the example in [340, pp.465-466] has property[κ]. Note the following result:

(16.32) THEOREM [236, 3.3]. *Let X and Y be continua such that $Y \in \text{prop}[\kappa]$. Then, any near-homeomorphism of X onto Y is confluent.*

Proof. Use (16.31).

As was mentioned in Chapter I, near-homeomorphism techniques were used in proving (1.97) and (1.98). Thus, it would seem that (16.32) might be useful in investigating the hyperspaces of certain non-locally connected continua.

In (16.31) we saw that $Y \in \text{prop}[\kappa]$ is a sufficient condition for $C_oFL[X, Y]$ to be a closed subspace of $C_o[X, Y]$. Property$[\kappa]$ is not a necessary condition [236, 4.2], nor is it always true that $C_oFL[X, Y]$ is closed [236, 1.4]. However, we have the following general result.

(16.33) THEOREM [236, 2.10]. *For any continua* X *and* Y, $C_oFL[X, Y]$ *is topologically complete.*

The proof in [236] of (16.33) uses semi-continuous mappings into 2^X and some special subsets of 2^X similar to the sets $<U_1, U_2, \ldots, U_n>$ of (0.10.2).

In (16.31) and (16.32), we have seen that property$[\kappa]$ is a condition, weaker than local connectivity, which can be used to determine the nature of a limit mapping. It would perhaps be fruitful to investigate other spaces of mappings where the range space has property$[\kappa]$ rather than the usual condition of local connectivity.

We now give some unanswered questions and some discussion about them.

(16.34) QUESTION. What are necessary and/or sufficient conditions in terms of X in order that 2^X [equivalently, by (16.7), $C(X)$] be contractible? In (16.15) we saw that property$[\kappa]$ is sufficient and, in (16.21), we saw that property$[\kappa]$ is not necessary. Also, we saw in (16.8) that X being contractible is sufficient.

As far as I know, no necessary conditions have ever been given.

Let us note the following example.

(16.35) EXAMPLE [329. 4.7]. Let $X = S^1 \cup \Sigma_1 \cup \Sigma_2$ where:

$S^1 = \{(x, y) \in R^2 : x^2 + y^2 = 1\}$,

$\Sigma_1 = \{[1 + (1/t)] \cdot e^{it} : t \geq +1\}$,

$\Sigma_2 = \{[1 + (1/t)] \cdot e^{it} : t \leq -1\}$.

In other words, X is $(SP)_3$ in (8.22) "before identifying" $2e^{i \cdot 1}$ and $(0, 0)$. Then, $X \in \text{prop}[\kappa]$ but $X \times X \notin \text{prop}[\kappa]$. It seems quite surprising that the cartesian product of two continua, each having property$[\kappa]$, may fail to have property$[\kappa]$. Let us note in passing that [329, 4.6] if a cartesian product of continua has property$[\kappa]$, then so do the factor spaces [use (16.29)--the projections are monotone].

The example in (16.35) suggests the following question.

(16.36) QUESTION. If X and Y are continua whose hyperspaces are contractible, then are the hyperspaces of $X \times Y$ contractible? The converse is true, as can be seen by using the mapping induced by the projection of $X \times Y$ onto a factor [a more general result is in (16.39)].

(16.37) QUESTIONS. If $X \in \text{prop}[\kappa]$, then does $2^X \in \text{prop}[\kappa]$? If $X \in \text{prop}[\kappa]$, then does $C(X) \in \text{prop}[\kappa]$? If $2^X \in \text{prop}[\kappa]$, then does $C(X) \in \text{prop}[\kappa]$? If $C(X) \in \text{prop}[\kappa]$, then does

$2^X \in \text{prop}[\kappa]$? The converse of the second question is true and is stated and proved in [329, 2.8]; the proof shows that the converse of the first question is also true.

(16.38) QUESTION. What kinds of mappings preserve property$[\kappa]$? In (16.29) we saw that confluent mappings preserve property$[\kappa]$; hence, so do monotone [329, 4.4] and open [329, 4.5] mappings. Also, Wardle [329, 2.9] has shown that retractions preserve property$[\kappa]$.

Let us note the following result.

(16.39) THEOREM. *If* X *is a continuum such that* 2^X [*or, equivalently by (16.7),* C(X)] *is contractible and if* Y *is an open image of* X, *then the hyperspaces of* Y *are contractible.*

Proof. Assume X is a continuum such that there is a homotopy $h: 2^X \times [0, 1] \to 2^X$ satisfying:

$$h(A, 0) = A \text{ for each } A \in 2^X$$

and

$$h(A, 1) = A_o \text{ for each } A \in 2^X$$

where A_o is a fixed member of 2^X. Assume Y is a continuum such that there is an open mapping $f: X \xrightarrow{\text{onto}} Y$. Define $k: 2^Y \times [0, 1] \to 2^Y$ by

$$k(B, t) = f[h(f^{-1}(B), t)]$$

for each $(B, t) \in 2^Y \times [0, 1]$. It follows that, since f is an open mapping, k is continuous (see [338, 4.31, p.130]). It follows from the formula for k that

$$k(B, 0) = B \text{ for each } B \in 2^Y$$

and

$$k(B, 1) = f[A_o] \text{ for each } B \in 2^Y.$$

Hence, 2^Y is contractible and so is $C(Y)$ by (16.7).

The result in (16.39) leads to the following question.

(16.40) QUESTION. If X is a continuum whose hyperspaces are contractible, then what kinds of images of X have contractible hyperspaces? For example, open images do by (16.39).

Finally, we mention a result for Hausdorff continua which is in startling and surprising contrast to the metric case. Recall from (1.96) that if X is a locally connected continuum, then $C(X)$ is an absolute retract (therefore, contractible). Now, let L_o be the *extended long line*, i.e., L_o is obtained from the ordinal space $[0, \Omega]$ by placing a copy of the metric open interval $(0, 1)$ between each ordinal $\alpha < \Omega$ and its successor $\alpha + 1$; L_o has the order topology. Let Σ be the "long circle" obtained by indentifying 0 and Ω together. Then, Σ is a locally connected Hausdorff continuum which is connected by metric arcs. However:

(16.41) EXAMPLE ([264, 3.5] or [429, Theorem 5]). Let Σ be the "long circle" as defined above. Then, $C(\Sigma)$ is not "contractible by any Hausdorff continuum," i.e., there is no Hausdorff continuum Z for which there is a mapping $g: C(\Sigma) \times Z \to C(\Sigma)$ such that, for some $p, q \in Z$,

$g(A, p) = A$ for each $A \in C(\Sigma)$

and

$g(A, q) = A_o$ for each $A \in C(\Sigma)$

where A_o is some fixed member of $C(\Sigma)$.

XVII.

HOMOGENEITY OF HYPERSPACES

Recall that a topological space S is said to be *homogeneous* provided that given any two points p and q in S, there is a homeomorphism h of S onto S such that $h(p) = q$. Note that the unit circle $S^1 = Z$ is homogeneous, but $C(Z)$ is not homogeneous since $C(Z)$ is a 2-cell [by (0.55)]. However, since I_∞ is homogeneous [167], 2^Z is homogeneous by (1.97). Thus, the following question arises: For what continua X are $C(X)$ or 2^X homogeneous? By using recent results [(1.97) and (1.98)], this question can be completely answered (remember--continuum means *metric* continuum). We will do so in (17.2) and (17.3). Then, we will briefly discuss the homogeneity question for hyperspaces of Hausdorff continua. As we will see, the situation for Hausdorff continua is quite different than the situation for metric continua. Thus, a number of unanswered questions arise.

We begin with the following simple lemma. It was stated in the metric case in [93, p.1032]. Since we will be dealing with Hausdorff continua after (17.3), we state it in the Hausdorff setting here. Recall that the topology for a hyperspace of a Hausdorff continuum is the Vietoris topology [defined in (0.12)] and that, when the continuum is metric, the Vietoris topology and the

Hausdorff metric topology are the same [by (0.13)].

(17.1) LEMMA (noted in the metric case in [93, p.1032]). *If Y is a Hausdorff continuum such that* $C(Y)$ *or* 2^Y *is homogeneous, then Y is locally connected.*

Proof. Assume Y is a Hausdorff continuum. By (1.136), $C(Y)$ and 2^Y are each locally connected at Y. Hence, if $C(Y)$ [resp., 2^Y] is homogeneous, $C(Y)$ [resp., 2^Y] is locally connected. Since (1.92) holds for Hausdorff continua, (17.1) now follows.

Let P be the pseudo-arc. By [30], P is homogeneous but, by (17.1), neither $C(P)$ nor 2^P is homogeneous. The fact that $C(P)$ is not homogeneous was noted in [93, p.1032].

The next two results completely determine those metric continua X for which $C(X)$ or 2^X is homogeneous. Also, they say what $C(X)$ and 2^X must be when they are homogeneous.

(17.2) THEOREM. *Let X be a (metric) continuum. The following three statements are equivalent:*

(1) $C(X)$ *is homogeneous;*

(2) X *is locally connected and contains no free arc;*

(3) $C(X) \cong I_\infty$.

Proof. Assume (1) holds. Then, by (17.1), X is locally connected. Suppose X contains a free arc A. Let $a \in A$ such that a is not an end point of A. Then, since A is a *free* arc, it follows that $\{a\}$ has arbitrarily small neighborhoods in $C(X)$

which are 2-cells. Hence, by (1), $\dim[C(X)] = 2$. Thus, by (1.107.1) [or (1.109)], it follows that X is an arc or a circle. Hence, by (0.54) or (0.55), $C(X)$ is a 2-cell. This contradicts (1). Thus, X does not contain a free arc. We have proved (1) implies (2). By the second part of (1.98), (2) implies (3). By [167], (3) implies (1). This completes the proof of (17.2).

(17.3) THEOREM. *Let* X *be a (metric) continuum. The following three statements are equivalent:*

(1) 2^X *is homogeneous;*

(2) X *is locally connected;*

(3) $2^X \cong I_\infty$.

Proof. By (17.1), (1) implies (2). By (1.97), (2) implies (3). By [167], (3) implies (1). This proves (17.3).

Thus, having answered the homogeneity question for hyperspaces of *metric* continua, we now discuss the question for hyperspaces of *Hausdorff* continua.

For Hausdorff continua Y, as well as for metric continua, Y is locally connected if $C(Y)$ or 2^Y is homogeneous [by (17.1)]. Here the analogy with the metric situation stops, as the following example shows when compared with the equivalence of (1) and (2) in (17.3).

(17.4) EXAMPLE. Let L_o be the extended long line as defined in the paragraph preceding (16.41). Thus, being a generalized arc, L_o is the "simplest" non-metric Hausdorff locally connected continuum. However, 2^{L_o} is not homogeneous. This is clear since some members of 2^{L_o} have metrizable neighborhoods while others do not. Also, $C(L_o)$ is not homogeneous.

I do not know the answer to any of the following questions.

(17.5) QUESTION. When is 2^Y or $C(Y)$ homogeneous for a (locally connected) Hausdorff continuum Y? I do not know an example of a non-metric Hausdorff continuum Y such that 2^Y or $C(Y)$ is homogeneous [see (17.6)].

(17.6) QUESTION. If Y is a locally connected homogeneous Hausdorff continuum, then is it necessarily true that 2^Y is homogeneous? Since the cartesian product of homogeneous spaces is homogeneous, it follows from [167] that any uncountable product Z of closed real number intervals is a homogeneous Hausdorff continuum. I do not know if 2^Z or if $C(Z)$ is homogeneous for these product spaces Z.

In Chapter XIX, some questions about homogeneity of various spaces of sets are asked [for example, see (19.9)].

XVIII.
SPACES OF COMPACT CONVEX SETS

In 1952, Radstrom [273, Theorem 2] proved an embedding theorem for certain general spaces of convex subsets of a normed linear space. Except for [273], no work was done on spaces of convex sets until [255], [256], and [257]. This chapter is mostly devoted to a discussion of these three papers.

As we will see, results about spaces of compact convex sets and techniques used to obtain them are a common meeting ground for three general areas of mathematics: Functional analysis, convexity theory, and topology. In particular, results and techniques in infinite-dimensional topology are used to obtain many of the results about spaces of compact convex sets. Also, spaces of compact convex subsets of certain finite unions of cells provide a finite-dimensional means of visualizing some phenomena which can occur in infinite-dimensional topology [see, especially, (18.36) and (18.37)]. The roles and inter-play of functional analysis, convexity theory, and topology will become clearer as the chapter progresses.

Let us note the following definition.

(18.1) DEFINITION [255]. Let Z be a subset of a metrizable

real topological vector space L. Throughout this chapter, we will assume 2^Z has the "Hausdorff metric" as defined in (0.63.6). Now, let

$$cc(Z) = \{A \in 2^Z : A \text{ is convex}\}$$

with the (restricted) Hausdorff metric [note: "convex" means convex in the classical sense]. The space cc(Z) will be called the cc-*hyperspace* of Z.

Let us note that if cc(Z) is compact, then Z is compact since $F_1(Z)$ is a closed subset of cc(Z). By using (0.8) and an elementary sequence argument, it can be shown that, cc(Z) is compact if Z is compact. We will use the fact that

(18.1.1) Z is compact if and only if cc(Z) is compact

(usually) without mentioning it. Another general fact is in (1.203.5).

We give some notational conventions: The symbol conv denotes closed convex hull [see (0.63.5)]. If $A \subset L$, L as in (18.1), and if λ is a real number, then

$$\lambda \cdot A = \{\lambda \cdot a : a \in A\}.$$

When the symbols R^n and ℓ_2 are used in this chapter, they are assumed to be as defined in (0.19) and (0.40) and to have all the vector space and norm structure usually associated with them.

We now give a brief summary of how the chapter is structured. The main results about cc-hyperspaces begin with (18.4). Some general theorems about cc(Z) when Z itself is compact and convex are in (18.4) through (18.7). In (18.8), a result about cc(Z) for some fundamental non-compact convex sets Z is given. The rest of the results [except (18.40) and (18.41)] concern cc(Z) when Z is not (necessarily) convex. A theorem with "geometric" flavor is in (18.11). There are some other general results [see, for example, (18.13) and (18.27)]. However, the investigations are in many instances geometrically oriented. In particular, (18.15) and the examples in (18.17) through (18.20) lead us to focus our attention on cc(X) when X is a 2-cell in R^2, and to look for geometric conditions on such an X in order that cc(X) be a Hilbert cube. In (18.33) through (18.35), we examine analogies (and the lack of them) to (18.15) for R^3 and R^4. In (18.36) and (18.37) some very interesting examples due to Quinn and Wong are given. These examples illustrate how cc-hyperspaces can be used to find and visualize Hilbert cubes with interesting intersection properties. Unsolved questions are stated throughout the chapter.

The result in (18.2) determines the topological type of the space $cas(R^n)$ of convex arcs and singletons in R^n for each $n \geq 2$ [when $n = 1$, clearly $cas(R^1) = cc(R^1) \cong R^1 \times [0, +\infty)$]. Note that (18.2) deals with spaces of the "simplest" convex sets. A related result is in (19.13). The fact that $cas(R^2)$ is finite-dimensional will be used in the proof of (18.15).

(18.2) THEOREM [257, Lemma 3.6]. *For* $n \geq 2$, *let* $cas(R^n) = \{A \in cc(R^n) : A \text{ is an arc}\} \cup F_1(R^n)$. *Then*

(18.2.1) $\quad cas(R^n) \cong R^n \times \left[([0, +\infty) \times P^{n-1}) \Big/ (\{0\} \times P^{n-1}) \right]$

where P^{n-1} denotes projective (n-1)-space and where the cartesian factor which is in square brackets is the upper semi-continuous decomposition of $[0, +\infty) \times P^{n-1}$ obtained by "shrinking $\{0\} \times P^{n-1}$ to a point." Hence,

(18.2.2) $\text{cas}(R^2) \cong R^4$.

Proof. A homeomorphism verifying (18.2.1) is obtained by assigning to a convex arc A with end points a and b the point $([a + b]/2, e[a, b])$, where $e[a, b]$ denotes the equivalence class of

$$D = ([0, +\infty) \times P^{n-1}) \big/ (\{0\} \times P^{n-1})$$

determined by the pair $(||a - b||, P_A^{n-1})$, where P_A^{n-1} is the point of P^{n-1} determined by the intersection of S^{n-1} with the translate to the origin of the extended line determined by A. A singleton $\{a\}$ is assigned to (a, s) where $s = \{0\} \times P^{n-1} \in D$.

The following lemma will be used in the proof of the first main result--(18.4).

(18.3) LEMMA [255, proof of Theorem 2.2]. *Let* K *be a compact convex subset of a metrizable locally convex real topological vector space* L. *If* $\dim[K] \geq 2$, *then* $\dim[cc(K)] = \infty$.

Proof. Assume K is as above with $\dim[K] \geq 2$. It is easy to see, by taking the closed convex hull of a maximal convex arc in K and a point in K not on the arc, that K contains a convex 2-cell D. Choose and fix an integer $n \geq 2$. Then there is in D a convex

(18.4) SPACES OF COMPACT CONVEX SETS 571

2n-sided polygon P_{2n} with sides $S_1, S_2, \ldots, S_{2n}$, where the enumeration is chosen so that

$S_i \cap S_j \neq \emptyset$ if and only if $|i - j| \leq 1$ or $i, j \in \{1, 2n\}$.

Let Δ_n be the n-cell given by

$$\Delta_n = \prod_{i=1}^{n} S_{2i-1} .$$

Define $h: \Delta_n \to cc(K)$ by

$$h(x_1, x_2, \ldots, x_n) = \operatorname{conv}\{x_1, x_2, \ldots, x_n\}$$

for each $(x_1, x_2, \ldots, x_n) \in \Delta_n$. It follows easily that h is a homeomorphism from Δ_n into cc(K). Thus, cc(K) contains an n-cell. Therefore, since $n \geq 2$ was arbitrary, $\dim[cc(K)] = \infty$. This proves (18.3).

From the standpoint of theory and motivation, the following theorem is basic to all the results in [255], [256], and [257]. The proof is a modification of a proof used in [166, p.31] to generalize a theorem of Keller [167]. We will state and use Keller's theorem in the proof.

(18.4) THEOREM [255, Theorem 2.2]. *Let* K *be a compact convex subset of a metrizable locally convex real topological vector space* L *such that* $\dim[K] \geq 2$. *Then,* $cc(K) \cong I_\infty$.

Proof. It follows, using a well-known separation theorem [89, Corollary 11, p.418], that there exists a countable family $\{\psi_i : i = 1, 2, \ldots\}$ of continuous (real-valued) linear functionals ψ_i such that

(1) given $A \in cc(K)$ and $x \in [K \setminus A]$, there exists ψ_j such that $\psi_j(x) \notin \psi_j[A]$.

Without loss of generality assume that, for each $i = 1,2,\ldots,$

(2) $\text{l.u.b.}\{|\psi_i(y)| : y \in K\} \leq 1$.

We define an "affine" embedding $h: cc(K) \to \ell_2$ as follows. Let $A \in cc(K)$. For each $i = 1,2,\ldots,$ let

$$[a_i, b_i] = \psi_i[A].$$

Let $h(A)$ be given by

$$h(A) = \left(\frac{a_1}{2^1}, \frac{b_1}{2^2}, \frac{a_2}{2^3}, \frac{b_2}{2^4}, \ldots, \frac{a_n}{2^{2n-1}}, \frac{b_n}{2^{2n}}, \ldots\right).$$

Using (2) we see that $h(A) \in \Pi_{j=1}^{\infty}[-2^{-j}, +2^{-j}]$. Thus, we have defined a function h,

(3) $h: cc(K) \to \prod_{j=1}^{\infty}[-2^{-j}, +2^{-j}]$.

Since the linear functionals $\psi_1, \psi_2, \ldots$ are all continuous, the coordinate functions for h are all continuous. Hence [since the range of h is contained in a cartesian product], h is continuous. It follows easily using (1) that h is one-to-one. Therefore, since $cc(K)$ is compact [see (18.1.1)], h is an embedding of $cc(K)$ into $\Pi_{j=1}^{\infty}[-2^{-j}, +2^{-j}]$. Furthermore, using the linearity of the linear functionals $\psi_1, \psi_2, \ldots,$ it follows that $h[cc(K)]$ is a convex subset of $\Pi_{j=1}^{\infty}[-2^{-j}, +2^{-j}]$. Hence, considering $\Pi_{j=1}^{\infty}[-2^{-j}, +2^{-j}]$ as contained in ℓ_2 by inclusion, we have proved

(4) h is an embedding of $cc(K)$ into ℓ_2 such that $h[cc(K)]$ is a (compact) convex subset of ℓ_2.

By (4) and (18.3), we may apply a theorem of Keller (which says: Any compact convex infinite-dimensional subset of ℓ_2 is a Hilbert cube [167]) to see that $cc(K) \cong I_\infty$. This proves (18.4).

The next result is more general than (18.4). As we will see, it is a convenient result to use for the proof of (18.7). Let us note the following definition.

(18.5) DEFINITION [255, p.556]. Let L be a metrizable locally convex real topological vector space and let $F \subset cc(L)$. Then: F is said to be a *convex family* provided that for any A, B ε F and $\lambda \varepsilon [0, 1]$, $[\lambda \cdot A + (1 - \lambda) \cdot B] \varepsilon F$.

(18.6) THEOREM [255, Theorem 2.3]. *Let* L *be a metrizable locally convex real topological vector space and let* $F \subset cc(L)$. *If* F *is a compact infinite-dimensional convex family, then* $F \cong I_\infty$.

Proof. The theorem follows using (parts of) the proof of (18.4).

Some "relative versions" of (18.4) are given in the following result; also, see (18.40).

(18.7) COROLLARY [255, p.556]. *Let* K *be a compact convex subset of a metrizable locally convex real topological vector space* L *such that* $\dim[K] \geq 2$. *Then:*

(18.7.1) [255, Corollary 2.4]. *If* Q *is a given compact*

subset of K *such that* conv[Q] $\neq$ K, *then*

$$\{A \in cc(K) : Q \subset A\} \cong I_\infty.$$

(18.7.2) [255, Corollary 2.5]. *If* K_o *is a given nonempty compact convex subset of* K, *then*

$$\{A \in cc(K) : A \cap K_o \neq \emptyset\} \cong I_\infty.$$

(18.7.3) [255, Remark 2.6]. *If* p *is a given point of* K, *then*

$$\{A \in cc(K) : p \in K\} \cong I_\infty.$$

Proof. It is easy to verify that the subsets of $cc(K)$ in (18.7.1) and (18.7.2) are compact convex families [in the sense of (18.5)]. Also they are infinite-dimensional, as can be seen by modifying some ideas used in the proof of (18.3). Thus, (18.7.1) and (18.7.2) follow using (18.6). Clearly, (18.7.3) is a consequence of (18.7.1) [or of (18.7.2)]. This completes the proof of (18.7).

For a recent result related to (18.7.2), and which is more general for the setting in which it is stated, see (18.40). Some results related to and generalizing (18.7.1) are in [73, section 6].

By (18.4), $cc(B^n) \cong I_\infty$ for each $n \geq 2$. The next result determines the topological type of $cc(R^n)$ and $cc(B^n \setminus S^{n-1})$, $n \geq 2$, by showing that they are homeomorphic to the Hilbert cube without a point [for the case when $n = 1$, see the paragraph preceding (18.2)]. Recall that the Hilbert cube I_∞ is homogeneous [167] and, thus, $I_\infty \setminus \{p\} \cong I_\infty \setminus \{q\}$ for any points p and q in I_∞.

(18.8) **THEOREM** [256, Theorems 3.1 and 3.2]. *For each* $n \geq 2$
$$cc(B^n \setminus S^{n-1}) \cong I_\infty \setminus \{(0, 0, \ldots, 0, \ldots)\}$$
and
$$cc(R^n) \cong I_\infty \setminus \{(0, 0, \ldots, 0, \ldots)\}.$$

Proof. Let $n \geq 2$. By (18.4),

(1) $cc(B^n) \cong I_\infty$.

Let $\Sigma = \{A \in cc(B^n) : A \cap S^{n-1} \neq \emptyset\}$. For $0 < \varepsilon < 1$, define $f_\varepsilon : cc(B^n) \to cc(B^n) \setminus \Sigma$ by

$$f_\varepsilon(A) = \{\varepsilon \cdot a : a \in A\} \quad \text{for each } A \in cc(B^n).$$

It is easy to see that by picking ε near enough to one, f_ε is as near the identity mapping as we please. Hence, by the definition of Z-set in (1.214.1),

(2) Σ is a Z-set in $cc(B^n)$.

Next we show that Σ is contractible in itself. This is easily seen by defining $g : \Sigma \times [0, 2] \to \Sigma$ by the formula [where d is the usual Euclidean metric for B^n, see (0.19), and K_d is as defined in (0.65.1)]

$$g(A, t) = K_d(t, A)$$

for each $(A, t) \in \Sigma \times [0, 2]$. Since d is a convex metric for B^n, g is continuous by (f) of (0.65.3). Clearly, $g(A, 0) = A$ and $g(A, 2) = B^n$ for each $A \in \Sigma$. Thus,

(3) Σ is contractible (in itself).

Now [by (1)] let $h : cc(B^n) \xrightarrow{\text{onto}} I_\infty$ be a homeomorphism. By (2), it follows easily that, since $h[cc(B^n)] = I_\infty$,

(4) $h[\Sigma]$ is a Z-set in I_∞.

By (3) and [41, 5.5, p.28],

(5) $h[\Sigma]$ has the shape of a point.

As is easy to prove, $\{p\}$ is a Z-set in I_∞ for any point $p \in I_\infty$. Hence,

(6) $\{(0, 0,\ldots, 0,\ldots)\}$ is a Z-set in I_∞.

Chapman has proved the following general result:

(7) If $A \subset I_\infty$ and $B \subset I_\infty$ are Z-sets in I_∞ such that A and B have the same shape, then $I_\infty \backslash A \cong I_\infty \backslash B$ [354, 25.2, p.40].

By (4) through (7), $I_\infty \backslash h[\Sigma] \cong I_\infty \backslash \{(0, 0,\ldots, 0,\ldots)\}$. Hence, $cc(B^n) \backslash \Sigma \cong I_\infty \backslash \{(0, 0,\ldots, 0,\ldots)\}$. Therefore, since (clearly) $cc(B^n) \backslash \Sigma = cc(B^n \backslash S^{n-1})$, we have

(8) $cc(B^n \backslash S^{n-1}) \cong I_\infty \backslash \{(0, 0,\ldots, 0,\ldots)\}$

which proves the first part of (18.8). It can be shown that

(9) $cc(R^n) \cong cc(B^n \backslash S^{n-1})$;

the details are in [257] and will not be done here. By (8) and (9),

(10) $cc(R^n) \cong I_\infty \backslash \{(0, 0,\ldots, 0,\ldots)\}$

which completes the proof of (18.8).

In the proof of (18.8), we showed that Σ (as defined in the proof) is contractible. It follows from a recent result of Doug Curtis that $\Sigma \cong I_\infty$ [see (18.40) and the paragraph preceding it].

The answer to the following question is not known.

(18.9) QUESTION [256, Problem 3.5]. Is $cc(\ell_2) \cong \ell_2$? Since this question was asked in [256], a related result has appeared--

(18.11)   SPACES OF COMPACT CONVEX SETS   577

see (19.40).

So far we have only dealt with cc-hyperspaces when the "base set" itself is convex. The next result shows that $cc(X) \cong I_\infty$ for some (not necessarily) convex sets X. Let us note the following definitions.

(18.10) DEFINITIONS. Let Z be a subset of a real topological vector space V. For x, y ε V, let
$$\widehat{xy} = \{t \cdot x + (1 - t) \cdot y : 0 \leq t \leq 1\}.$$
For x ε Z, let
$$S(x) = \{y \in Z : \widehat{xy} \subset Z\}.$$
Then the *kernel of* Z, written ker(Z), is the subset of Z given by
$$\ker(Z) = \cap \{S(x) : x \in Z\}.$$
The set Z is said to be *starshaped* provided that $\ker(Z) \neq \emptyset$. Thus, a starshaped set is a set which has a point that "sees every point." For some examples which illustrate these concepts and which are related to the next theorem, see (18.17), (18.19), and (18.20).

For a question related to the following theorem, see (18.26).

(18.11) THEOREM [255, Theorem 3.7]. *If* X *is a compact starshaped subset of* R^n, *some* $n \geq 2$, *such that* ker(X) *has nonempty interior in* R^n, *then* $cc(X) \cong I_\infty$.

Proof. Without loss of generality assume $O_n \in U$, where O_n is the zero vector $(0, 0,\ldots, 0)$ of R^n and U is the interior in R^n of $\ker(X)$. Choose $r > 0$ such that $B(r) \subset U$ where $B(r) = \{x \in R^n : ||x|| \leq r\}$.
For each $A \in [cc(X) \setminus \{O_n\}]$, let

$$\alpha(A) = \text{l.u.b.}\{\alpha \in (0, +\infty) : \alpha \cdot A \subset X\},$$

$$\tilde{A} = \alpha(A) \cdot A,$$

$$\lambda(\tilde{A}) = \text{l.u.b.}\{\lambda \in [0, 1] : \lambda \cdot \tilde{A} \subset B(r)\}.$$

Define $f: cc(X) \to cc(B(r))$ by

$$f(A) = \begin{cases} \lambda(\tilde{A}) \cdot A, & \text{if } A \in [cc(X) \setminus \{O_n\}] \\ O_n, & \text{if } A = \{O_n\}. \end{cases}$$

It can be shown that the "dragging in" map f is a homeomorphism of $cc(X)$ onto $cc(B(r))$ --the proof is in [257] and will not be done here. Therefore, since $cc(B(r)) \cong I_\infty$ [by (18.4)--recall that $n \geq 2$], $cc(X) \cong I_\infty$. This completes the proof of (18.11).

(18.12) REMARK. A very general "dragging in " procedure is in [257, proof of Theorem 5.4]. Theorem 5.4 of [257] is a homeomorphism theorem which is applicable in situations more general than the one in (18.11) [see, for example, (18.19)].

In view of results above, the following question arises naturally: For what not-necessarily convex continua X is $cc(X) \cong I_\infty$? One result about this was given in (18.11). Some more are now given, beginning with the following general theorem.

(18.13) THEOREM [255, Lemma 3.2]. *Let* X *be a continuum lying in a Banach space* E *with norm* $||\ ||$. *If* $cc(X) \cong I_\infty$, *then* X *is an absolute retract and* $\dim[X] \geq 2$.

Proof. Let F denote the closed linear span of X. Then, F is a separable Banach space. Hence [36, Theorem 2.1, p.18], F has a strictly convex norm $||\ ||_F$ (giving the same topology on F as did $||\ ||$). Let $p \in F$. Since $||\ ||_F$ is strictly convex, we have the following fact (see [36, Lemma 2.1, p.19]): Given any $A \in cc(X)$, there exists one and only one point a_o of A such that a_o is $||\ ||_F$-nearest p, i.e., such that

$$||p - a_o||_F = \inf\{||p - a||_F : a \in A\}.$$

For each $A \in cc(X)$, let $\psi_p(A)$ denote the unique point of A which is $||\ ||_F$-nearest p. It is easy to see that ψ_p is a (continuous) r-map from $cc(X)$ onto X [actually, ψ_p is a selection--some other properties of ψ_p are in (18.30) and (18.32)]. Thus, since $cc(X) \cong I_\infty$, X is an absolute retract [40, 2.1, p.101]. This proves the first part of (18.13). We sketch a proof of the second part. Suppose $\dim[X] = 1$. Then, by the first part of (18.13), X is a one-dimensional absolute retract. Hence, if $A \in cc(X)$, A is either a convex arc or a one-point set. Therefore [see (18.41)], the barycenter "map" $\beta: cc(X) \to X$ is continuous. It can be shown, using [150, Theorem III2, p.30], that

$$\dim[\beta^{-1}(x)] \leq 1 \text{ for each } x \in X.$$

Hence, by (#) on p.215, $\dim[cc(X)] \leq 2$ which contradicts the assumption that $cc(X) \cong I_\infty$. This completes the proof of (18.13).

(18.14) REMARK. In the proof of (18.13), we used the fact that the barycenter "map" is continuous on the space of convex arcs and singletons. The barycenter "map" is not in general a mapping, i.e., it is not in general *continuous*. For example: Let

$$A = \{(x, 0) \in R^2 : 0 \leq x \leq 1\}$$

and, for each $n = 1, 2, \ldots$, let D_n be the closed convex hull in R^2 of the points $(0, 0)$, $(1, 0)$, and $(0, 1/n)$. Clearly, $D_n \to A$ as $n \to \infty$, the barycenter of D_n is $(1/3, 1/3n)$ for each $n = 1, 2, \ldots$, and the barycenter of A is $(1/2, 0)$. Let us note that the restriction of the barycenter "map" to any collection of compact convex sets, all of the same finite dimension, is continuous. This last fact will be used in the proof of (18.15).

The next theorem is a topological converse to (18.4) for the plane.

(18.15) THEOREM [255, Theorem 3.1]. *Let* X *be a continuum in* R^2. *If* $cc(X) \cong I_\infty$, *then* X *is a 2-cell [converse is false--see (18.18)]*.

Proof. Let $X \subset R^2$ be a continuum such that $cc(X) \cong I_\infty$. By (18.13),

(1) X is a 2-dimensional (compact) absolute retract.

Let U denote the interior in R^2 of X. Suppose there is a point $p \in (X \setminus cl[U])$. Then there is an open subset W of R^2 such that $p \in W$ and $W \cap X$ contains no 2-cell. Let

$\Lambda = \{A \in cc(X) : A \subset W\}$.

Clearly $\Lambda \neq \emptyset$ [since $\{p\} \in \Lambda$] and Λ is an open subset of $cc(X)$ [since $\Lambda = \langle W \rangle \cap cc(X)$; see (0.13)]. Also, since $W \cap X$ contains no 2-cell and $[W \cap X] \subset R^2$, we have using (18.2.2) that $\dim[\Lambda] < \infty$. These properties of Λ contradict the assumption that $cc(X) \cong I_\infty$. Hence, we have proved

(2) $X = cl[U]$.

Next we show that U is connected. Recall the well-known fact that no finite-dimensional subcontinuum of I_∞ can (arcwise) disconnect I_∞ (this can be proved using [150, Example VI 11, p.93]). Hence, since $cc(X) \cong I_\infty$, it follows using (18.2.2) that Γ is arcwise connected, where

$\Gamma = \{A \in cc(X): A \text{ is a 2-cell}\}$.

Now, let $p_1, p_2 \in U$. Then there exist $A_1, A_2 \in \Gamma$ such that p_i is the barycenter of A_i for each $i = 1$ and 2. Since Γ is arcwise connected, there is an arc $\gamma \subset \Gamma$ such that A_1 and A_2 are the end points of γ. Let β denote the restriction of the barycenter "map" to γ. Since $\gamma \subset \Gamma$, β is continuous [see (18.14)] and $\beta[\gamma] \subset U$. Thus, $\beta[\gamma]$ is an arcwise connected subset of U such that $p_1, p_2 \in \beta[\gamma]$. Therefore, since $p_1, p_2 \in U$ were arbitrary, we have proved

(3) U is connected.

By (1) through (3) it follows that X is a 2-cell [see (18.16)]. This proves (18.15).

(18.16) REMARK. The following fact was used at the end of the proof of (18.15): *If Y is a compact absolute retract in the plane* R^2 *such that the interior* V *in* R^2 *of* Y *is connected and* Y = cl[V], *then* Y *is a 2-cell*. This fact follows from [186, Theorem 11, p.534]. A different proof was given in [257].

Let us note the following four examples.

(18.17) EXAMPLE [255, p.558]. Let $X = X_1 \cup X_2$ where
$$X_1 = \{(x, y) \in R^2 : 0 \le y \le x \le 1\}$$
and
$$X_2 = \{(x, y) \in R^2 : 0 \le x \le 1 \text{ and } 0 \le y \le -x+1\}.$$
Then X is a non-convex 2-cell such that, since X satisfies the hypotheses of (18.11), $cc(X) \cong I_\infty$.

(18.18) EXAMPLE [255, Example 3.5]. Let X be the 2-cell in R^2 pictured in Figure (18.18).

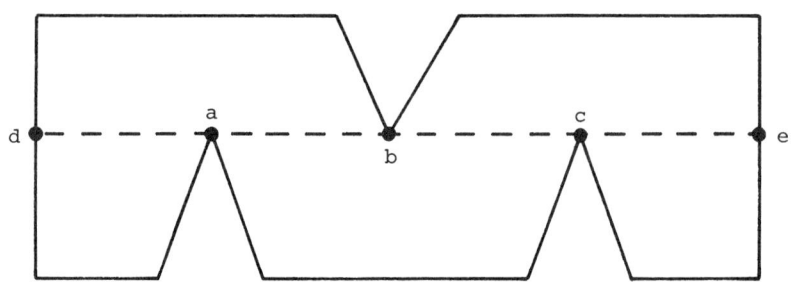

Figure (18.18)

Let A be the convex arc in X from d to e [A is drawn in Figure (18.18) with dashes]. Since a, b, c $\in$ A, it is easy to see that A has a neighborhood N in cc(X) such that each member of N is a subarc of A. Hence, dim[N] ≤ 2 [actually, it can be seen that A has a 2-cell neighborhood in cc(X)]. Therefore, cc(X) is not homeomorphic to I_∞.

(18.19) EXAMPLE [256, Example 2.6(b)]. Let X be the 2-cell in R^2 which is the closure of the bounded complementary domain of $\cup_{i=1}^{4} X_i$, where

$X_1 = \{(x, y) \in R^2: (x - 1)^2 + (y - 1)^2 \leq 1\}$,

$X_2 = \{(x, y) \in R^2: (x - 1)^2 + (y + 1)^2 \leq 1\}$,

$X_3 = \{(x, y) \in R^2: (x + 1)^2 + (y + 1)^2 \leq 1\}$,

$X_4 = \{(x, y) \in R^2: (x + 1)^2 + (y - 1)^2 \leq 1\}$.

Note that X is a starshaped 2-cell in R^2 with ker(X) = {(0,0)}. Hence, (18.11) does not apply. However, it can be shown using the general homeomorphism theorem [257, Theorem 5.4] referred to in (18.12) that cc(X) $\cong$ I_∞ (the details are in [257]). Note that the convex arc B with end points (0, -1) and (0, +1) is a maximal convex subset of X [comp., (18.18); also, see (18.21)].

(18.20) EXAMPLE [256, Example 2.6(a)]. Let X be the 2-cell in R^2 given by $X = \cap_{i=1}^{3} Z_i$ where

$Z_1 = \{(x, y) \in R^2: x \leq -1/2 \text{ and } y \geq 0\}$,

$Z_2 = \{(x, y) \in R^2: (x + 1/2)^2 + y^2 \geq 1/4\}$,

$Z_3 = \{(x, y) \in R^2: x^2 + y^2 \leq 1\}$.

Using inverse limit and near homeomorphism techniques it can be shown that cc(X) $\cong$ I_∞ (the details are in [257]). Note that X is not starshaped and, in fact, that the point (-1, 0) is a maximal convex subset of X.

(18.21) REMARK. We see from results and examples above that there are many 2-cells X in R^2 such that $cc(X) \cong I_\infty$ and that these 2-cells

(1) can have maximal convex subsets which are arcs [see [see (18.19)]];

(2) can have maximal convex subsets which are points and, therefore, do not have to be starshaped [see (18.20)].

Let A be the convex arc in (18.18) [compare with B in (19.19)]. Note that there are three non-end points a, b, and c of A which are points of "local non-convexity" of X; also, b is between a and c, the "wedges" at a and c are on one side of A, and the "wedge" at b is on the other side. The same is true of any subarc of A with one end point strictly to the left of a and the other end point strictly to the right of c. Without getting technical, call such arcs "singular arcs" [75]. In [75] it is shown that for 2-cells with polygonal boundaries, the singular arcs are canonical in the following sense: If X is a 2-cell in R^2 with polygonal boundary, then

$cc(X) \setminus \{A: A$ is a singular arc$\} \cong I_\infty$.

Thus, we have the following characterization of when $cc(X) \cong I_\infty$ provided X is a 2-cell in R^2 with *polygonal* boundary:

(18.21.1) THEOREM [75]. *Let* X *be a 2-cell in* R^2 *such that* X *has polygonal boundary. Then, the following three statements are equivalent:*

(1) $cc(X) \cong I_\infty$;

(2) X *contains no singular arc;*

(3) $cc(X)$ *is locally infinite-dimensional.*

However, the following questions are open:

(18.22) QUESTION. For what 2-cells X in R^2 is $cc(X) \cong I_\infty$? By (18.21.1) we know that if X is a 2-cell in R^2 with polygonal boundary such that

    (1) each member of $cc(X)$ has arbitrarily small
          infinite-dimensional neighborhoods

or

    (2) every maximal convex subset of X is a 2-cell
        [see (18.29)]

holds, then $cc(X) \cong I_\infty$. This is a partial answer to the questions in (18.23) through (18.25).

(18.23) QUESTION [256, Problem 2.4]. If X is a 2-cell in R^2 such that (1) of (18.22) holds, then is $cc(X) \cong I_\infty$? A pertinent example is in (18.20).

An affirmative answer to (18.23) implies an affirmative answer to the next two questions. Also, an affirmative answer to (18.23) would provide a satisfactory characterization, since it would show that the example in (18.18) is typical of what can "go wrong" [as has been shown when the boundary is polygonal--see (18.21.1)].

(18.24) QUESTION [256, Problem 2.5]. If X is a 2-cell in R^2 such that (2) of (18.22) holds, then is $cc(X) \cong I_\infty$?

(18.25) QUESTION [256, Problem 2.4]. If X is a 2-cell in R^2 such that every maximal convex subset of X is a point or a 2-cell, then is $cc(X) \cong I_\infty$? A pertinent example is in (18.20). Also, the converse question for 2-cells in R^2 has a negative answer by (18.19).

In (18.11) we showed that $cc(X) \cong I_\infty$ for certain starshaped sets X. In (18.19) an example was given of a starshaped subset X of R^2 such that $cc(X) \cong I_\infty$ even though X did not satisfy some of the hypotheses in (18.11). The answer to the following question is not known.

(18.26) QUESTION. If X is a starshaped 2-cell in R^2, then is $cc(X) \cong I_\infty$?

The second part of the next result allows us to "localize" the problem of showing that $cc(X) \cong I_\infty$. Let us note that if $cc(X) \cong I_\infty$, then X is an absolute retract by (18.13).

(18.27) THEOREM [256, Theorem 2.1]. *Let* X *be a continuum lying in a Banach space* E.

(18.27.1) *If* X *is contractible, then* cc(X) *is contractible [see (18.31)]*.

(18.27.2) *If* X *is a 2-cell or any continuum which is an absolute retract, then* cc(X) $\cong I_\infty$ *if and only if* cc(X) *is a Hilbert cube manifold*.

Proof. To prove (18.27.1), assume X ⊂ E is a contractible continuum. Let ψ_p: cc(X) $\xrightarrow{\text{onto}}$ X be as in the proof of (18.13), where it was noted that

(1) ψ_p is a (continuous) selection for cc(X).

For each A ε cc(X) and each t ε [0, 1], let

$f(A, t) = t \cdot \{\psi_p(A)\} + (1 - t) \cdot A$.

Note that, by (1), $\psi_p(A)$ ε A for each A ε cc(X). Hence, by the formula for f,

(2) f(A, t) ⊂ A ⊂ X and, thus, f(A, t) ε cc(X) for each A ε cc(X) and t ε [0, 1].

Also, by the formula for f,

(3) f(A, 0) = A and f(A, 1) = $\{\psi_p(A)\}$ ε $F_1(X)$ for each A ε cc(X).

By (2) and the continuity of ψ_p,

(4) f: cc(X) × [0, 1] → cc(X) is continuous.

Since X is contractible and $F_1(X) \cong X$, $F_1(X)$ is contractible. Therefore, it follows using (3) and (4) that cc(X) is contractible. This proves (18.27.1). To prove (18.27.2), assume X is a continuum which is an absolute retract and that cc(X) is a Hilbert cube manifold. Note that cc(X) is compact [see (18.1.1)]. Hence, by (18.27.1) [which applies, since X is an absolute retract],

(5) cc(X) is a compact contractible Hilbert cube manifold.

Chapman has proved the following general result:

(6) Let Y be a compact Hilbert cube manifold. Then:

$Y \cong I_\infty$ if and only if Y is homotopically trivial

[61, Theorem 10].

By (5) and (6), cc(X) $\cong I_\infty$. This completes the proof of (18.27).

Some applications of (18.27) are given in [256] and [257]. We state one of them and give an indication of the proof. The result was mentioned in (18.22).

(18.28) THEOREM [256, Theorem 2.2(a)]. *Let* X *be a 2-cell in* R^2 *such that* X *has polygonal boundary. If every maximal convex subset of* X *is a 2-cell, then* cc(X) $\cong I_\infty$.

Indication of proof [257]. It can be shown that given any M ε cc(X), there is a compact starshaped subset Y of X such that the kernel of Y has nonempty interior in R^2 and such that cc(Y) is a neighborhood in cc(X) of M (the verifications of this are in [257]). By (18.11), cc(Y) $\cong I_\infty$. This proves that cc(X) is a Hilbert cube manifold. Hence, by (18.27.2), cc(X) $\cong I_\infty$.

(18.29) REMARK. The result in (18.28), which was also mentioned in (18.22), originally appeared in [256] where it was a partial answer and motivation for a question--see (18.24)--raised in

[256].

The mapping ψ_p, used in the proofs of (18.13) and (18.27), has some properties which are worth noting:

(18.30) THEOREM (easily deducible from statements in [255] and [256]). *Let* X *be a continuum lying in a Banach space* E. *Let* $\psi_p: cc(X) \xrightarrow{onto} X$ *be as in the proof of (18.13). Then:*

(18.30.1) ψ_p *is a (continuous) selection for* cc(X);

(18.30.2) $\psi_p^{-1}(x)$ *is contractible for each* $x \in X$ [*see (18.32)*];

(18.30.3) *there is a homotopy* f: cc(X) × [0, 1] → cc(X) *such that* f(A, 0) = A *for each* A ∈ cc(X), f(A, 1) = {ψ_p(A)} *for each* A ∈ cc(X), *and* f({x}, t) = {x} *for each* x ∈ X *and each* t ∈ [0, 1];

(18.30.4) $F_1(X)$ *is a strong deformation retract of* cc(X).

Proof. The statement in (18.30.1) was noted in the proof of (18.13). The statement in (18.30.2) is not difficult to prove. The statement in (18.30.3) is easily proved by taking f to be as in the proof of (18.27.1). The statement in (18.30.4) is a consequence of (18.30.3) and the observation that f(A, 1) ∈ $F_1(X)$ for each A ∈ cc(X).

As a consequence of (18.30), note the following relationships between X and cc(X).

(18.31) COROLLARY. *Let* X *be a continuum lying in a Banach space. Then*

(18.31.1) X *and* $cc(X)$ *are of the same homotopy type.*
Hence, in particular:

(18.31.2) X *is contractible if and only if* $cc(X)$ *is contractible;*

(18.31.3) X *and* $cc(X)$ *have the same shape.*

Proof. By (18.30), $F_1(X)$ is a (strong) deformation retract of $cc(X)$. Hence [40, 14.2, p.27], $F_1(X)$ is of the same homotopy type as $cc(X)$. Therefore, since X is homeomorphic to $F_1(X)$, X is of the same homotopy type as $cc(X)$. This proves (18.31.1). The statement in (18.31.2) is an easy-to-prove consequence of (18.31.1), as is (18.31.3) [41, p.28].

(18.32) REMARK. By (18.30.4) [or, by the proof of the first part of (18.13)], it is clear that if $cc(X)$ is an absolute retract [resp., absolute neighborhood retract], then X is an absolute retract [resp., absolute neighborhood retract]. The converses are false, as can be seen from some simple examples. Now, assume $cc(X)$ and (hence) X are absolute neighborhood retracts. Then, since a mapping with contractible point inverses between compact absolute neighborhood retracts is a CE map [354, p.91], we have by (18.30.2) that ψ_p is a CE map. From this fact, some relationships between X and $cc(X)$ can be deduced. For example, by [335], X and $cc(X)$ have the same simple homotopy type. Also, by results of R. D. Edwards and R. J. Miller (see Chapters XIII and

XIV of [354]), it follows that $X \times I_\infty \cong cc(X) \times I_\infty$.

Beginning with (18.15) we have emphasized the situation when X is in the plane. Now we turn our attention briefly to the case when X is in 3-space.

The following example shows that (18.15) does not have a "straightforward" generalization to 3-space.

(18.33) **EXAMPLE** [256, Example 2.8]. Let X be the continuum in R^3 given by $X = X_1 \cup X_2$, where
$$X_1 = \{(x, y, z) \in R^3: x^2 + y^2 + z^2 \leq 1\}$$
and
$$X_2 = \{(x, y, 0) \in R^3: \max\{|x|, |y|\} \leq 1\}.$$
Then X, called the McDonald Cheeseburger, is a three-dimensional continuum in R^3 such that X is not a 3-cell but such that $cc(X) \cong I_\infty$. To prove the last part of this statement, first observe that

(1) $cc(X) = cc(X_1) \cup cc(X_2)$,

(2) $cc(X_1) \cap cc(X_2) = cc(X_1 \cap X_2)$,

and, by (18.4),

(3) $cc(Y) \cong I_\infty$ where $Y = X_1$, X_2, or $X_1 \cap X_2$.

It can be shown that (the details are in [257])

(4) $cc(X_1) \cap cc(X_2)$ has property Z in $cc(X_1)$.

Handel has proved the following general "sum theorem":

(5) If Q_1, Q_2, and $Q_3 = Q_1 \cap Q_2$ are all Hilbert cubes such that Q_3 has property Z in Q_1, then $Q_1 \cup Q_2$

is a Hilbert cube [137].

By taking $Q_1 = cc(X_1)$ and $Q_2 = cc(X_2)$, it follows from (2) through (4) that Q_1 and Q_2 satisfy the hypotheses in (5). Hence, (5) "Handels it" [sic(k)], i.e., by (1) and (5), $cc(X) = Q_1 \cup Q_2 \cong I_\infty$.

In spite of the example in (18.33), there is an analogue to (18.15) for 3-space. We state it without proof--the proof is in [257].

(18.34) THEOREM [255, Theorem 3.4]. *Let* X *be a continuum in* R^3. *If* $cc(X) \cong I_\infty$ *and if* X *is not contained in any two-dimensional hyperplane in* R^3, *then* X *contains a 3-cell.*

The analogue to (18.34) for R^4 is false, as the following example shows.

(18.35) EXAMPLE [255, Example 3.8]. Let X be the continuum in R^4 given by $X = X_1 \cup X_2$, where

$X_1 = \{(x, y, z, 0) \in R^4 : \max\{|x|, |y|, |z|\} \leq 1\}$

and

$X_2 = \{(x, y, 0, w) \in R^4 : \max\{|x|, |y|, |w|\} \leq 1\}$.

Then X is a continuum in R^4 such that $cc(X) \cong I_\infty$, X is not contained in any three-dimensional hyperplane in R^4, but X does not contain a 4-cell. The proof that $cc(X) \cong I_\infty$ is done in a manner similar to what was done in (18.33). In particular, (1) through

(4) in (18.33) hold here also. Thus, (5) in (18.33) may be used—note that $cc(X_1) \cap cc(X_2)$ also has property Z in $cc(X_2)$ and, thus, the "sum theorem" in [332, 5, p.456] may be used in place of (5) in (18.33).

As mentioned at the beginning of the chapter, cc-hyperspaces provide a finite-dimensional means of visualizing some infinite-dimensional phenomena. The next two examples illustrate this.

(18.36) EXAMPLE [272, Example 4.2]. This is an example of four convex Hilbert cubes in ℓ_2 whose pairwise intersections are Hilbert cubes, whose total intersection is a 2-cell, no one of them contains any other, and whose union is a Hilbert cube [no example of *three* such Hilbert cubes is known—see (18.37)]. Let Y_1, Y_2, Y_3 and Y_4 be the subsets of R^2 given by:

$Y_1 = conv\{(0, 0), (0, +1), (-1, +1), (-1, 0)\}$,

$Y_2 = conv\{(0, 0), (+1, 0), (+1, +1), (0, +1)\}$,

$Y_3 = conv\{(-1, +1/2), (-1, 0), (+1, 0), (+1, +1/2)\}$,

$Y_4 = conv\{(-1, 0), (-1, -1), (+1, -1), (+1, 0)\}$.

In R^3, let $X_i = Y_i \times [0, 1]$, i.e.,

$X_i = \{(x, y, t) \in R^3 : (x, y) \in Y_i \text{ and } 0 \le t \le 1\}$

for each i = 1,2,3, and 4. Now, using the embedding h in the proof of (18.4) [take K to be $\cup_{i=1}^{4} X_i$ and apply the proof of (18.4) to embed $cc(K)$ in ℓ_2 by h], we see that the images $K_i = h[cc(X_i)]$, $i \in \{1,2,3,4\}$, satisfy:

(1) K_i is a convex Hilbert cube in ℓ_2 for each $i \in \{1,2,3,4\}$;

(2) $K_i \cap K_j$ is a Hilbert cube whenever $i, j \in \{1,2,3,4\}$ [this follows from (18.4) since each $K_i \cap K_j$ is easily seen to be a cc-hyperspace of a convex 2-cell or of a convex 3-cell];

(3) $\cap_{i=1}^{4} K_i$ is a 2-cell [because $\cap_{i=1}^{4} K_i$ is the cc-hyperspace of the convex arc $\cap_{i=1}^{4} X_i$].

Also (as is clear):

(4) If $i \neq j$, then $K_i \not\subset K_j$.

Now, to verify that $\cup_{i=1}^{4} K_i$ is a Hilbert cube, we first note the following "sum theorem":

(5) If $Q_1, Q_2, \ldots, Q_n$ is a finite number of convex Hilbert cubes in ℓ_2 such that the intersection of every nonempty subcollection of them is a Hilbert cube, then $\cup_{i=1}^{n} Q_i$ is a Hilbert cube [272, Lemma 3.3].

To see that $\cup_{i=1}^{4} K_i$ is a Hilbert cube, first consider $\cup_{i=1}^{3} K_i$. Since $\cap_{i=1}^{3} K_i$ is the cc-hyperspace of a convex 2-cell, we have by (18.4) that

(6) $\cap_{i=1}^{3} K_i$ is a Hilbert cube.

Hence, by (1), (2), and (6), the hypotheses of (5) are satisfied by $Q_1 = K_1$, $Q_2 = K_2$, and $Q_3 = K_3$. Thus, by (5),

(7) $\cup_{i=1}^{3} K_i$ is a Hilbert cube.

Now note that $K_4 \cap [\cup_{i=1}^{3} K_i] = K_4 \cap K_3$. Hence, by (2),

(8) $K_4 \cap [\cup_{i=1}^{3} K_i]$ is a Hilbert cube.

A procedure similar to that used to prove (2) in the proof of (18.8) shows that:

(9) $K_4 \cap [\cup_{i=1}^{3} K_i]$ has property Z in the Hilbert cube K_4.

By (7) through (9), the hypotheses of (5) in the proof of (18.33) are satisfied by $\cup_{i=1}^{3} K_i$ and K_4. Therefore, by (5) in the proof of (18.33),

(10) $[\cup_{i=1}^{3} K_i] \cup K_4 = \cup_{i=1}^{4} K_i$ is a Hilbert cube.

(18.37) **EXAMPLE AND QUESTION** [272, Example 4.3]. This is an example of three convex Hilbert cubes in ℓ_2 whose pairwise intersections are Hilbert cubes, whose total intersection is a 2-cell, and whose union is not a Hilbert cube. Let Y_1, Y_2, and Y_3 be the subsets of R^2 given by:

$Y_1 = \text{conv}\{(0, 0), (+1, 0), (+2, -1), (0, -1)\}$,

$Y_2 = \text{conv}\{(0, 0), (+1, 0), (+2, +1), (0, +1)\}$,

$Y_3 = \text{conv}\{(+1, 0), (+2, -1), (+2, +1)\}$.

In R^3, let $X_i = Y_i \times [0, 1]$, i.e.,

$X_i = \{(x, y, t) \in R^3 : (x, y) \in Y_i \text{ and } 0 \le t \le 1\}$

for each $i = 1, 2,$ and 3. Then, using the procedure in (18.36) and letting

$K_i = h[cc(X_i)]$

for each $i = 1, 2,$ and 3, we obtain three convex Hilbert cubes

K_1, K_2, and K_3 in ℓ_2 whose pairwise intersections are Hilbert cubes and whose total intersection is a 2-cell (namely, $\cap_{i=1}^{3} K_i$ is the cc-hyperspace of the convex arc $\cap_{i=1}^{3} X_i$). It can be shown that $\cup_{i=1}^{3} K_i$ is not a Hilbert cube. The following question is motivated by this example: Do there exist three convex Hilbert cubes in ℓ_2 whose pairwise intersections are Hilbert cubes, whose total intersection is nonempty and finite-dimensional, and whose union is a Hilbert cube?

The following example was of interest to the authors during the preparation of [255] through [257].

(18.38) EXAMPLE [256]. Early in the investigations which led to the work in [255] through [257], Quinn, Stavrakas, and I noted that there are natural copies [see below] of $2^{[0,1]}$ in $cc(B^2)$. We had hoped to use such copies of $2^{[0,1]}$, together with the theory of cc-hyperspaces, to give a proof that $2^{[0,1]} \cong I_\infty$. Attempts in this direction were unsuccessful. However, one illustrative example is included here in the hope that the ideas may be used by someone to succeed where we failed. Let $X = B^2$ and let

$$Y = \{(x, y) \in B^2 : x^2 + y^2 = 1 \text{ and } y \geq 0\}.$$

Use the embedding h in (18.4) to embed $cc(X)$ in ℓ_2. Let $Q = h[cc(X)]$ and note that [by the proof of (18.4)]

(1) Q is a convex Hilbert cube in ℓ_2.

Now, let [the symbol ext denotes the set of extreme points]

$$Z = \{A \in cc(X) : \text{ext}[A] \subset Y\}.$$

Note from the geometry that if $f: 2^Y \xrightarrow{\text{onto}} Z$ is given by $f(A) = \text{conv}[A]$ for each $A \in 2^Y$, then f is a homeomorphism. Hence, by (1.97),

(2) $Z \cong I_\infty$.

Quinn, Stavrakas, and I were not able to prove (2) without using the fact that $2^Y \cong I_\infty$. Note that $h[Z] \cong I_\infty$ and, by using the formula for h in the proof of (18.4),

(3) $U = Q \setminus h[Z]$ is an open convex subset of Q.

In general, it is easy to obtain examples of open convex subsets of I_∞ whose complements are homeomorphic to I_∞. Nonetheless, with the example just given in mind, we have the following general question:

(18.39) QUESTION [256, Problem 3.3]. Let W be an open convex subset of I_∞. What are necessary and sufficient conditions on W in order that $I_\infty \setminus W \cong I_\infty$?

Two of the questions raised in [255] and [256] have recently been answered--in [255, Problem 2.7], the following question was asked:

Is $\{A \in cc(B^2): A \cap S^1 \neq \emptyset\} \cong I_\infty$?

In [256, section 2], also in [4, p.166], the following question was asked:

If K_1 and K_2 are two convex Hilbert cubes in ℓ_2 whose intersection is a Hilbert cube, then is $K_1 \cup K_2$ a Hilbert cube?

At the time that Quinn, Stavrakas, and I wrote [256], we did not know anyone had ever raised the second question (though we did know that it was a special case of a conjecture in [2, p.213]). Actually, the question was a rather famous one and of interest for some time. The first of these two questions is answered affirmatively by (18.40) and the second by (18.41). Although we have already used a more general result than (18.41), it seems to be appropriate in view of the history of the second question to state (18.41).

(18.40) THEOREM [73, Corollary 6.5]. *Let* K *be a convex n-cell,* $n \geq 2$, *lying in some Euclidean space or in* ℓ_2 *and let* A_o *be a given nonempty compact subset of* K.

(18.40.1) *If* K *is not polyhedral, then*
$$\{A \in cc(K): A \cap A_o \neq \emptyset\} \cong I_\infty.$$

(18.40.2) *If* K *is polyhedral with vertex set* V *and if* $A_o \neq V$, *then* $\{A \in cc(K): A \cap A_o \neq \emptyset\} \cong I_\infty$.

(18.40.3) *If* K *is polyhedral with vertex set*
$$V = \{v_1, v_2, \ldots, v_k\} \text{ and if } A_o = V, \text{ then}$$
$$\{A \in cc(K): A \cap A_o \neq \emptyset\} \cong S^{k-2} \times I_\infty.$$

(18.41) THEOREM [272, Corollary 3.6]. *If* K_1 *and* K_2 *are two convex Hilbert cubes in* ℓ_2 *such that* $K_1 \cap K_2$ *is a Hilbert cube, then* $K_1 \cup K_2$ *is a Hilbert cube* [*see (5) in the verifications in (18.36) for a more general result*].

With respect to (18.41), we mention that Sher [296] has recently given an example of two Hilbert cubes whose intersection is a Hilbert cube but whose union is not a Hilbert cube. His example shows that the conjecture in [2, p.213] is false.

Since the initial work was done in [255], [256], and [257], there has been growing interest in cc-hyperspaces. The work in [73, section 6] and [75] is an interesting recent contribution to this subject, and the examples in [272] show the applicability of cc-hyperspaces to infinite-dimensional topology. Some of the material in these last three papers was presented in this chapter, but there is other material in them which has not been discussed. The subject of cc-hyperspaces is new, has many open questions, and should be of interest to mathematicians in a variety of fields.

(18.42) REMARK. Duda ([83] and [84]) has studied connections between convexity, in the sense of Menger, of a continuum X and convexity of its hyperspaces 2^X and $C(X)$. Though we are not discussing convexity of hyperspaces, we mention these papers for the interested reader (in relation to [83] and [84], also see [416], [422], and [430]).

XIX.
OTHER SPACES OF
SPECIAL SETS

Throughout most of the book, we have been primarily concerned with the two spaces 2^X and $C(X)$. In Chapter XVIII, spaces of convex sets were studied. In this chapter, spaces of arcs, circles, pseudo-arcs, chainable continua, etc. will be discussed. As we will see, very little is known about these spaces. In fact, except for a few scattered results to be mentioned below, they have never been singled out and studied. The purpose of this chapter is to isolate various spaces of special sets which would seem to be of interest, to present some of their fundamental properties, and to indicate possible future investigations. Because so little is known, even in the simplest situations, our emphasis for the most part will be on what is *not* known (vis a vis questions). Many of the results, simple in nature, are included primarily to motivate questions.

(19.1) PRELIMINARIES AND NOTATION. Let M be a metric space. Throughout this chapter, we will assume 2^M has the "Hausdorff metric as defined in (0.63.6). This metric will (with no confusion) be denoted by H. The following notation is also adopted:

(19.1.1) $a(M) = \{A \in 2^M : A \text{ is an arc}\}$;

(19.1.2) $s(M) = \{A \in 2^M : A \text{ is a circle}\}$;

(19.1.3) $t(M) = \{A \in 2^M : A \text{ is a simple triod}\}$;

(19.1.4) $\theta(M) = \{A \in 2^M : A \text{ is a } \theta\text{-curve}\}$ (a θ-*curve* is any continuum which is homeomorphic to $S^1 \cup B^1$);

(19.1.5) $p(M) = \{A \in 2^M : A \text{ is a pseudo-arc}\}$;

(19.1.6) $ch(M) = \{A \in 2^M : A \text{ is a chainable continuum}\}$;

(19.1.7) $hi(M) = \{A \in 2^M : A \text{ is an hereditarily indecomposable continuum}\}$;

(19.1.8) $L(M) = \{A \in 2^M : A \text{ is a locally connected continuum}\}$.

Of course, we consider all the spaces above as being subspaces of 2^M with the (restricted) metric obtained from H. Frequently, we will want M to be (simultaneously) any of the spaces R^n, $n \geq 2$, or ℓ_2. Thus, for convenience,

(19.1.9) Ω^* denotes any of the spaces R^n, $n \geq 2$, or ℓ_2.

We begin by investigating the spaces defined in (19.1.1) through (19.1.4) since the elements of these spaces are the "simplest" continua. Our discussion ends with (19.25). Much of it emphasizes spaces of arcs.

(19.2) THEOREM [well-known]. *Each of the spaces* $a(\Omega^*)$, $s(\Omega^*)$, $t(\Omega^*)$, *and* $\theta(\Omega^*)$ *is dense in* $c(\Omega^*)$.

Joe Quinn has proved the following result using the notion of ε-crooked arcs [see (19.27.1)]. For a somewhat similar type of result, see (19.35).

(19.3) THEOREM [Joe Quinn; unpublished]. *Each of the spaces* $a(\Omega^*)$, $s(\Omega^*)$, $t(\Omega^*)$, *and* $\theta(\Omega^*)$ *is of the first category in itself.*

Proof. The proof can be done by using (##) in the proof of (19.27) and by observing that, for each $n = 1, 2, \ldots$, the space of arcs which are not $(1/n)$-crooked is a closed subset of $a(\Omega^*)$.

(19.4) THEOREM. *Each of the spaces* $a(R^2)$, $s(R^2)$, $t(R^2)$, *and* $\theta(R^2)$ *is homogeneous.*

Proof. We will prove the theorem for the space $a(R^2)$ --the proofs for the other spaces are similar. Let A_1, $A_2 \in a(R^2)$. By the Schönflies Theorem [186, Theorem 1, p.535], there is a homeomorphism $h: R^2 \xrightarrow{\text{onto}} R^2$ such that $h[A_1] = A_2$. Then $\hat{h}$, defined by [comp., (0.49)]

$$\hat{h}(A) = \{h(x) : x \in A\} \text{ for each } A \in a(R^2),$$

is a homeomorphism of $a(R^2)$ onto $a(R^2)$ such that $\hat{h}(A_1) = A_2$. This proves $a(R^2)$ is homogeneous [see the definition at the beginning of Chapter XVII].

(19.5) LEMMA. *If* $M \in a(\Omega^*)$ *such that* M *is convex, then there is a Hilbert cube* Δ_∞ *contained in* $a(\Omega^*)$ *such that* $M \in \Delta_\infty$.

Proof. For convenience, we will only prove the lemma for $a(R^2)$ and for $M = \{(x, 0) \in R^2 : 0 \leq x \leq 1\}$. It will then be clear from the proof that the entire lemma is valid. Let $j \in \{1, 2, \ldots\}$; then, for each t such that $0 \leq t \leq 2^{-j}$, let

(1) $\alpha(j, t)$ be the convex arc in R^2 with end points $(2^{1-j}, 0)$ and $([3] \cdot 2^{-1-j}, t)$,

(2) $\beta(j, t)$ be the convex arc in R^2 with end points $([3] \cdot 2^{-j-1}, t)$ and $(2^{-j}, 0)$,

and let

(3) $\gamma(j, t) = \alpha(j, t) \cup \beta(j, t)$.

Thus, for each $j \in \{1,2,\ldots\}$ and any t such that $0 \le t \le 2^{-j}$, we have defined an arc $\gamma(j, t)$ in R^2 with end points $(2^{1-j}, 0)$ and $(2^{-j}, 0)$. For each $(t_1, t_2, \ldots, t_j, \ldots) \in I_\infty$ [see (0.40) for definition of I_∞], it is easy to see, using (1) through (3) above, that $cl[\cup_{j=1}^\infty \gamma(j, t_j)]$ is an arc in R^2 [with end points $(0, 0)$ and $(1, 0)$]. Define $h: I_\infty \to a(R^2)$ by

$$h(t_1, t_2, \ldots, t_j, \ldots) = cl[\bigcup_{j=1}^\infty \gamma(j, t_j)]$$

for each $(t_1, t_2, \ldots, t_j, \ldots) \in I_\infty$. It is easy to see that h is a homeomorphism of I_∞ into $a(R^2)$ and that $h(0, 0, \ldots, 0, \ldots) = M$. This completes the proof of (19.5).

(19.6) THEOREM. *Each of the spaces* $a(\Omega^*)$, $s(\Omega^*)$, $t(\Omega^*)$, *and* $\theta(\Omega^*)$ *contains a Hilbert cube and, hence, is infinite-dimensional.*

Proof. For $a(\Omega^*)$, (19.6) is a trivial consequence of (19.5). For $s(\Omega^*)$, $t(\Omega^*)$, or $\theta(\Omega^*)$, (19.6) follows from simple adaptations of the proof of (19.5).

(19.7) THEOREM. *For any* $K \in a(R^2)$ [*resp.*, $s(R^2)$, $t(R^2)$, $\theta(R^2)$], *there exists a Hilbert cube* Δ_∞ *contained in* $a(R^2)$ [*resp.*, $s(R^2)$, $t(R^2)$, $\theta(R^2)$] *such that* $K \in \Delta_\infty$.

Proof. The theorem follows immediately from (19.4) and (19.6).

The following questions are motivated by results in (19.2) through (19.7).

(19.8) QUESTION. Is $a(R^n)$ homeomorphic to $a(R^m)$ for each n, $m \geq 2$?

Of particular interest in (19.8) is the case when $n = 2$ and $m = 3$. In (19.4), we proved that $a(R^2)$ is homogeneous by using the Schönflies Theorem. Since there are arcs in R^3 which are not R^3-equivalent [186, Remark (i), p.535], the method of proof does not generalize to show that $a(R^3)$ is homogeneous. Moreover, I do not know the answer to the following question [a negative answer, of course, would give a negative answer to (19.8)].

(19.9) QUESTION. Is $a(R^3)$, more generally $a(R^n)$ for $n \geq 3$, homogeneous?

(19.10) QUESTION. Is $a(\ell_2)$ homogeneous?

(19.11) QUESTION. Is $a(\ell_2)$ homeomorphic to $a(R^n)$ for some (or for any) $n \geq 2$?

(19.12) QUESTION. The same questions as (19.8) through (19.11) with a( ) replaced by any of the spaces s( ), t( ), or θ( ). Also, I do not know if all these spaces are mutually homeomorphic. In particular: Is $a(R^2)$ homeomorphic to $s(R^2)$?

It would be interesting if some better-understood topological models for the spaces $a(\Omega^*)$, $s(\Omega^*)$, $t(\Omega^*)$, and $\theta(\Omega^*)$ could be obtained. Such models, depending on what they were, could be used to answer questions above. Let us note the following result which gives a "model" for the space of convex arcs in R^n, $n \geq 2$. It is a derivative of the result in (18.2).

(19.13) THEOREM [257, proof of Lemma 3.6]. *For* $n \geq 2$, *let*
$$ca(R^n) = \{A \in C(R^n) : A \text{ is a convex arc}\}. \text{ Then}$$
$$ca(R^n) \cong R^{n+1} \times P^{n-1}$$
where P^{n-1} *denotes projective* $(n-1)$-*space. Hence,*
$$ca(R^2) \cong R^3 \times S^1.$$

Proof. See the proof of (18.2).

In [199], Lum investigated spaces of arcs in dendroids. Some of his results are stated below. We adopt the following notation.

(19.14) NOTATION. Let D be a dendroid. For $x, y \in D$, let $\overline{xy}$ denote the unique subcontinuum of D irreducible about x, y, i.e.: If $x \neq y$, $\overline{xy}$ is the unique arc in D with end points x and y; if $x = y$, $\overline{xy} = \{x\}$. For $p \in D$, let
$$a(D, p) = \{\overline{px} : x \in D\}$$
[in particular, $\{p\} \in a(D, p)$] and let
$$\psi^p : D \to a(D, p)$$
be the natural function defined by
$$\psi^p(x) = \overline{px} \text{ for each } x \in D.$$

Note that ψ^p is one-to-one and sends D onto a(D, p). Also note that [199, Theorem 1] ψ^p is continuous [as a function into C(D)] if and only if D is smooth at p [see (1.68)], i.e.: D is smooth at p if and only if $\psi^p: D \xrightarrow{\text{onto}} a(D, p)$ is a homeomorphism. An instructive exercise about a(D, p) is in (0.72.1).

Several years ago, I asked Lum the following question [199, p.115]: If a dendroid D is homeomorphic to a(D, p), then is D smooth (at some point)? Lum showed that the answer is "yes," as the next result states.

(19.15) THEOREM [199, Theorem 7]. *Let* D *be a dendroid. Then:* D *is smooth (at some point) if and only if there exists a point* $p \in D$ *such that* D *is homeomorphic to* a(D, p).

The result in (19.15) is a little more sophisticated than the reader may suspect--there is a dendroid D and a point $p \in D$ such that D is homeomorphic to a(D, p) but D is not smooth at p [199, Example 2].

Let D be a dendroid. Note that if D is smooth, then a(D, p) is compact for some $p \in D$ [namely, a point p at which D is smooth]. However, the converse is false, i.e., there is a dendroid D such that a(D, p) is compact for some $p \in D$ but D is not smooth [at any point--see (1.68)]. The dendroid D in (1.69) can be used to see this; another example is in [199, Example 1]. Lum defines the following notion.

(19.16) DEFINITION [199, p.113]. A dendroid D is said to be *weakly smooth at a point* p ε D provided that a(D, p) is compact. A dendroid is said to be *weakly smooth* provided that it is weakly smooth at some point.

Before stating the next result, let us note the following facts. If a dendroid is smooth at a point p, then it is connected im kleinen at p [172, p.680]; hence, a dendroid is a dendrite if and only if it is smooth at each of its points. As can be verified using D and p in (0.72.1), a dendroid D can be weakly smooth at a point p and, yet, not be connected im kleinen at p. However, Lum proves the following "global" result. It characterizes the dendrites from among the dendroids and is analogous to the above-mentioned characterization.

(19.17) THEOREM [199, Theorem 8]. *Let* D *be a dendroid. Then:* D *is a dendrite if and only if* D *is weakly smooth at each of its points.*

Thus we see that compactness of certain "nice" subspaces of C(D) [namely, a(D, p) for each p ε D] is equivalent to local connectedness of the dendroid D [comp., the discussion in the paragraph preceding (15.10)].

The next result is a characterization of dendrites within the class of weakly smooth dendroids.

(19.18) THEOREM [199, Corollary 8.1]. *Let* D *be a weakly smooth dendroid. Then, the following two statements are equivalent:*

(19.18.1) $a(D, x) \cong a(D, y)$ *for each* $x, y \in D$;

(19.18.2) D *is a dendrite*.

Proof. Since D is weakly smooth at some point, (19.18.1) clearly implies D is weakly smooth at every point and hence, by (19.17), D is a dendrite. Conversely, assume (19.18.2) holds. Then, by [66, Corollary 4], D is smooth at every point. Hence [see the equivalence at the end of (19.14)], $D \cong a(D, p)$ for each $p \in D$. Therefore, (19.18.1) holds and we have proved (19.18).

(19.19) REMARK. An example [199, Example 3], due to D. G. Paulowich, shows the necessity for the assumption in (19.18) that D be weakly smooth. It is an example of a dendroid D which satisfies (19.18.1) but which is not a dendrite.

We have only covered a few of the results in [199]. I refer the reader to [199] for the others.

In (19.17), the dendrites are characterized from among the dendroids as follows: A dendroid D is a dendrite if and only if $a(D, p)$ is compact for each $p \in D$. This suggests the following questions.

(19.20) QUESTIONS. For what dendroids D is $a(D) \cup F_1(D)$ compact [as a subspace of $C(D)$]? More generally: For what arcwise connected continua X is $a(X) \cup F_1(X)$ compact? Clearly [see (19.22)], there are many dendroids D such that $a(D) \cup F_1(D)$ is compact but such that D is not a dendrite [hence, by (19.17), not weakly smooth at some point]. Also, there are many arcwise connected continua X such that $a(X) \cup F_1(X)$ is compact but X is not a dendroid; such is the case, for example, for the cone over any hereditarily indecomposable continuum [see (19.22)]. Thus, it does not seem to be possible to obtain a "nice" determination of those arcwise connected continua X such that $a(X) \cup F_1(X)$ is compact. However, "nice" determinations should be possible within some subclasses of the class of all arcwise connected continua. For example, we have the following two results.

(19.21) THEOREM. *Let* X *be a locally connected continuum. Then:* $a(X) \cup F_1(X)$ *is compact if and only if* X *is a dendrite.*

Proof. If X is a dendrite, then the compactness of $a(X) \cup F_1(X)$ is a simple consequence of (0.8), (15.10), and Whyburn's theorem [341, 3.1] which says: If $\{A_n\}_{n=1}^{\infty}$ is a sequence of arcs such that $\{A_n\}_{n=1}^{\infty}$ converges 0-regularly to a nondegenerate subcontinuum K of X, then K is an arc. Conversely, assume X is not a dendrite. Then, since X is locally connected, X contains a circle M [by (0.26)]. Clearly, there is a sequence of arcs in M converging to M. Hence, $a(X) \cup F_1(X)$ is not compact. This completes the proof of (19.21).

In (19.20) we mentioned that if K is the cone over any hereditarily indecomposable continuum, then $a(K) \cup F_1(K)$ is compact.

Let us note the following result (which is not difficult to prove).

(19.22) THEOREM. *Let* M *be any compact metric space and let* K *denote the cone over* M. *Then:* $a(K) \cup F_1(K)$ *is compact if and only if* M *does not contain an arc.*

The next result completely determines those hyperspaces $C(X)$ such that $a(C(X)) \cup F_1(C(X))$ is compact. It implies the result in [176, 2.9]--see (19.25).

(19.23) THEOREM. *Let* X *be a continuum. Then:* $a(C(X)) \cup F_1(C(X))$ *is compact if and only if* X *is hereditarily indecomposable.*

Proof. If X is an hereditarily indecomposable continuum, then it follows using (1.29), (1.30), and (1.62) that $a(C(X)) \cup F_1(C(X))$ is compact. The other half of (19.23) follows using (1.146).

Let us note the following fixed point theorem. It is more general than (7.6)--see (19.25).

(19.24) THEOREM. *If* Y *is an arcwise connected continuum such that* $a(Y) \cup F_1(Y)$ *is compact, then* Y *has the fixed point property.*

Proof. Since $a(Y) \cup F_1(Y)$ is compact, it is clear that

(1) if $A_1 \subset A_2 \subset \cdots \subset A_n \subset \cdots$ are arcs in Y, then $\cup_{n=1}^{\infty} A_n$ is contained in an arc in Y.

Young [349, Theorem 16] has shown that if an arcwise connected continuum Y has the property in (1), then Y has the fixed point property. This proves (19.24).

(19.25) REMARK. Combining (19.24) [with Y replaced by $C(X)$] and (19.23), the result in (7.6) is obtained. As can be seen from the proof of (19.24), the proof just given for (7.6) is in essence the one given by Krasinkiewicz in [176, p.180]. In particular, Krasinkiewicz observed [176, 2.9] that (1) in the proof of (19.24) held for the specific situation when X is an hereditarily indecomposable continuum and $Y = C(X)$. Then he used [349, Theorem 16].

We end our discussion of the spaces in (19.1.1) through (19.1.4) with the observation that there is much left to learn about them.

Now we turn our attention to the spaces in (19.1.5) through (19.1.7). In 1930, Mazurkiewicz [209] proved the following two theorems in the context of $C(B^2)$. However, the proofs in [209] can be modified to obtain the more general versions of them stated below [see (19.28)].

(19.26) THEOREM [209, p.151]. *Let* M *be a metric space. The space* $hi(M)$ *is a* G_δ *subset of* $C(M)$.

Proof. For each $n = 1, 2, \ldots$, let

$\Lambda_n = \{B \in C(M):$ there exists a subcontinuum $B_o = B_1 \cup B_2$ of B, where B_1 and B_2 are subcontinua of B_o such that $B_1 \not\subset N(1/n, B_2)$ and $B_2 \not\subset N(1/n, B_1)\}$.

An elementary sequence argument shows that:

(1) For each $n = 1, 2, \ldots$, Λ_n is a closed subset of $C(M)$.

Clearly [using the definition of $\Lambda_1, \Lambda_2, \ldots$],

(2) $hi(M) = C(M) \setminus [(\bigcup_{n=1}^{\infty} \Lambda_n) \cup F_1(M)]$.

By (1) and (2), $hi(M)$ is the complement in $C(M)$ of an F_σ subset of $C(M)$. Therefore, $hi(M)$ is a G_δ subset of $C(M)$.

(19.27) THEOREM (for the case of $C(B^2)$, [209, p.151]; see (19.28)). *The space* $hi(\Omega^*)$ *is a residual* G_δ *subset of* $C(\Omega^*)$. *Hence,* $hi(\Omega^*)$ *is a dense* G_δ *subset of* $C(\Omega^*)$.

Proof. For each $n = 1, 2, \ldots$, let Λ_n be as in the proof of (19.26), where it was shown that

(1) $hi(\Omega^*) = (\bigcap_{n=1}^{\infty} [C(\Omega^*) \setminus \Lambda_n]) \cap [C(\Omega^*) \setminus F_1(\Omega^*)]$ and $C(\Omega^*) \setminus \Lambda_n$ is an open subset of $C(\Omega^*)$ for each $n = 1, 2, \ldots$.

Note that $C(\Omega^*) \setminus F_1(\Omega^*)$ is a dense open subset of $C(\Omega^*)$ and that, by (0.63.6), $C(\Omega^*)$ is a complete metric space. Hence, to prove (19.27), it suffices by (1) to prove that:

(#) For each $n = 1, 2, \ldots$, $C(\Omega^*) \setminus \Lambda_n$ is a dense subset of $C(\Omega^*)$.

To prove (#), let us note the following definition:

(19.27.1) DEFINITION [29, p.267]. An arc $A \subset \Omega^*$ is said to be ε-*crooked* provided that given any two points $p \neq q$ of A, there exist points $r, s \in A$ between p and q such that r lies between p and s, $d(p, s) < \varepsilon$, and $d(r, q) < \varepsilon$ (where d denotes the metric for Ω^*).

For each $\varepsilon > 0$, let
$$Cr_\varepsilon = \{A \in C(\Omega^*) : A \text{ is an } \varepsilon\text{-crooked arc}\}.$$
It is easy to prove that for each $n = 1, 2, \ldots$,
$$Cr_{\frac{1}{n}} \subset [C(\Omega^*) \setminus \Lambda_n].$$
Hence, to prove (#), it suffices to show that for each $n = 1, 2, \ldots$

(##) $Cr_{\frac{1}{n}}$ is a dense subset of $C(\Omega^*)$.

We will not give the details which verify (##), but we will present the ideas involved. Choose and fix $n = n_o$. Let $K \in C(\Omega^*)$ and let $K_1 \subset \Omega^*$ be a polygonal arc near K (in the Hausdorff metric). Let $I \subset \Omega^*$ be a convex arc. Since I and K_1 are polygonal, there is a homeomorphism $h: \Omega^* \xrightarrow{\text{onto}} \Omega^*$ such that $h[I] = K_1$ (see [130] for an even stronger result). For each $\varepsilon > 0$, an ε-crooked arc can be constructed as near to I (in the Hausdorff metric) as one wishes [see (19.28)]. Thus, by taking a sequence of arcs A_j ($j = 1, 2, \ldots$) such that

(i) A_j is $(1/j)$-crooked for each $j = 1, 2, \ldots$

and

(ii) $A_j \to I$ as $j \to \infty$

(19.28)   OTHER SPACES OF SPECIAL SETS   615

and by using the uniform continuity of h on $[\cup_{j=1}^{\infty} A_j] \cup I$ [which is compact by (ii)], it follows that for j large enough, $h[A_j]$ is $(1/n_o)$-crooked and $h[A_j]$ is as near as one wishes to K_1 (which is near K). Except for some historical comments in (19.28), this is all I will say about the proof of (19.27).

(19.28) REMARK. Some historical comments concerning (19.27) and its proof are in order. In Lemma 6 of [209], Mazurkiewicz showed that given a convex arc $I \subset R^2$ and any $\varepsilon > 0$, there is an ε-crooked arc as near to I (in the Hausdorff metric) as one wishes--at least it seems that the arc he constructs in [209] is ε-crooked in the sense of (19.27.1). Then, to complete his proof of (19.27), Mazurkiewicz does what is outlined in the proof above--*except*: Since he is in the plane, he uses the Schönflies Theorem. However, since one of his arcs is convex and the other one is polygonal, he could have used the result (apparently due to Alexandroff) that any two polygonal arcs in Ω^* are Ω^*-equivalent. This is what was used in the proof above to obtain (19.27) in the generality in which it is stated. As we will see in (19.30), Bing has proved a stronger result than (19.27). However, because of the questions aksed in (19.32) and (19.33), the result in (19.27) is of independent interest.

When Mazurkiewicz proved that $hi(B^2)$ is a residual G_δ subset of $C(B^2)$, mathematicians considered the result as being very surprising. Hereditarily indecomposable continua were considered to be quite pathological and difficult to come by. It was only eight years before that Knaster [169] had shown by means of a specific example that such continua exist [see (19.31)]. Mazurkiewicz's result says that *most* continua in B^2 are hereditarily indecomposable. We will see in (19.30) that even more is true. First, let us prove the following theorem.

(19.29) THEOREM [26, proof of Theorem 2]. *Let* M *be a metric space. The space* ch(M) *is a* G_δ *subset of* C(M).

Proof. For each $n = 1, 2, \ldots$, let [see (0.22) for the definition of chain]

$\Gamma_n = \{B \in C(M):$ there does not exist a chain

$U = \{U_1, U_2, \ldots, U_k\}$ of open subsets U_i of M

such that $\text{diam}[U_i] < 1/n$ for each $i = 1, 2, \ldots, k$

and $B \subset [\cup_{i=1}^{k} U_i]\}$.

It is easy to see that:

(1) For each $n = 1, 2, \ldots$, $C(M) \setminus \Gamma_n$ is an open subset of $C(M)$.

Also, as is clear [see (0.23)],

(2) $\text{ch}(M) = (\bigcap_{n=1}^{\infty} [C(M) \setminus \Gamma_n]) \cap [C(M) \setminus F_1(M)]$.

By (1) and (2), we have proved (19.29).

(19.30) THEOREM [26, Theorem 2]. *The space* $P(\Omega^*)$ *is a residual* G_δ *subset of* $C(\Omega^*)$. *Hence*, $P(\Omega^*)$ *is a dense* G_δ *subset of* $C(\Omega^*)$.

Proof [see (19.31)]. Bing has proved [comp., the definition of pseudo-arc in (0.23)]:

(1) Any two hereditarily indecomposable chainable continua are homeomorphic [26, Theorem 1].

By (1) and the definition of pseudo-arc in (0.23),

(2) $P(\Omega^*) = hi(\Omega^*) \cap ch(\Omega^*)$.

By (19.27),

(3) $hi(\Omega^*)$ is a dense G_δ subset of $C(\Omega^*)$.

Since $a(\Omega^*)$ is a dense subset of $C(\Omega^*)$, we have by (19.29) that

(4) $ch(\Omega^*)$ is a dense G_δ subset of $C(\Omega^*)$.

Since $C(\Omega^*)$ is a complete metric space [by (0.63.6)] and (2) through (4) hold, we have proved (19.30).

(19.31) REMARK. In his proof in [26] of (19.30), Bing used the way pseudo-arcs can be constructed to prove $P(\Omega^*)$ is a dense subset of $C(\Omega^*)$. The proof we have given for (19.30) shows that this detail is not essential if one first modifies Mazurkiewicz's proof in [209], as we have done, in order to obtain (19.27) in the generality stated. Thus, as can be seen from the proofs of (19.27) and (19.30), "everything is done with arcs!" From the process Knaster used in [169] to construct his hereditarily indecomposable continuum, it is easy to see that his continuum is chainable. Hence [by (1) in the proof of (19.30)], Knaster's continuum is a pseudo-arc. Thus, by (19.30), we see that the continuum which was considered in 1922 to be so unusual is really the one which is most plentiful!

The results in (19.26) through (19.30) lead to the following (unanswered) questions.

(19.32) QUESTIONS. The same as (19.8) through (19.12) with the spaces considered there replaced by any of the spaces hi( ), ch( ), or P( ). Also [comp., (19.4) and (19.9)], I do not know if hi(R^2), ch(R^2), or P(R^2) is homogeneous.

(19.33) QUESTION. Is P(R^2) [resp., hi(R^2), ch(R^2)] homeomorphic to ℓ_2? About six years ago I asked this question at a conference on infinite-dimensional topology held at Louisiana State University. As fas as I know, it has not been answered. Note that the spaces P(R^2), hi(R^2), and ch(R^2) are each topologically complete [since they are each a G_δ in C(R^2) and C(R^2) is complete by (0.63.6)]. However [see (19.32)], I do not know if they are homogeneous. I also mention that I do not know the answer to the question at the beginning of (19.33) when R^2 is replaced by any of the spaces $\Omega *$. Note that P$(\Omega *)$, hi$(\Omega *)$, and ch$(\Omega *)$ are each topologically complete.

Next, we will briefly discuss the space L(M). The first theorem about L(M) is the following one due to Mazurkiewicz but proved in [181, pp.269-270].

(19.34) THEOREM (due to Mazurkiewicz; see [181, footnote 1, p.270]). Let M *be a compact metric space. Then,* a(M) *and* L(M) are each at most of class $F_{\sigma\delta}$ (with metric H).

Mazurkiewicz then proved the following result.

(19.35) THEOREM [210]. *For each* $n \geq 2$, $L(B^n)$ *is exactly of class* $F_{\sigma\delta}$ *(with metric* H*)*.

Thus, for any given $n \geq 3$, we see by (19.35) that there is no metric for $L(B^n)$ which is topologically equivalent to the Hausdorff metric [for $L(B^n)$] and which is a complete metric. Observing this fact, Mazurkiewicz obtained a metric ρ_τ whose basic properties are listed in (19.36). Some comments about ρ_τ are in the paragraph following the proof of (15.9).

(19.36) THEOREM [211]. *Let* X *be a locally connected continuum. Then there is a metric* ρ_τ *for* L(X) *such that:*
  (19.36.1) *Convergence with respect to* ρ_τ *implies convergence with respect to the Hausdorff metric (for* L(X)*);*
  (19.36.2) *The metric space* (L(X), ρ_τ) *is a complete separable metric space;*
  (19.36.3) $\{A \in L(X): \dim[A] = 1\}$ *is a dense* G_δ *subspace of* L(X) *with metric* ρ_τ.

(19.37) REMARK. Since [211], other work has been done on metrics different from the Hausdorff metric.[1] Of particular interest is the so-called *homotopy metric* ρ_h, which was introduced and studied by Borsuk [37]. The metric ρ_h is defined for

$\{A \in 2^X : A$ is an absolute neighborhood retract$\}$.

The space obtained is complete and homotopy type is "invariant under the limit operation." A survey of what was known about ρ_h up to 1966 is in [40, pp.220-221]. The papers [14] and [15] have appeared since 1966 and, recently, Joe Quinn has proved the following result:

(19.37.1) THEOREM [Joe Quinn; unpublished]. *Let* R_h^2 *[resp.,* S_h^2*] denote the space of all circles (i.e., simple closed curves) in* R^2 *[resp.,* S^2*] with the homotopy metric. Then,* R_h^2 *is an absolute retract but* S_h^2 *is not an absolute retract.*

Another metric, the so-called *metric of continuity*[2], was defined and studied in [37]. It was also investigated by J. De Groot [81] and T. Ganea [119]. These and related papers are mentioned in [40, p.221]. Finally, we mention the so-called *generalized Hausdorff metric* H_∞ which was introduced in [251]. It is defined on the collection CL(M) of all nonempty closed subsets of a metric space (M, d) as follows: As in (0.1), let

$N(\varepsilon, A) = \{x \in M: d(x, a) < \varepsilon$ for some $a \in A\}$

for each $\varepsilon > 0$ and each $A \in CL(M)$. Then, for $A, B \in CL(M)$, define

$E_{A,B} = \{\varepsilon > 0: A \subset N(\varepsilon, B)$ and $B \subset N(\varepsilon, A)\}$

and define

$$H_\infty(A, B) = \begin{cases} \inf E_{A,B} &, \text{if } E_{A,B} \neq \emptyset \\ \infty &, \text{if } E_{A,B} = \emptyset \end{cases}.$$

(19.37)   OTHER SPACES OF SPECIAL SETS   621

Note that H_∞ divides CL(M) into "components" by the relation $A \sim B$ if and only if $H_\infty(A, B) < \infty$; on these "components", H_∞ is an actual metric. The reader may ask: Why not bound the metric on M so that H_∞ is a metric? From a topological point of view, this is not done because the topology of CL(M) depends on the metric for M. For example: If R^1 has its usual unbounded metric, then $CL(R^1)$ with H_∞ is *not* connected; but, if R^1 has a metric which makes R^1 isometric to the open interval (0, 1), then $CL(R^1)$ with H_∞ *is* connected. However, H_∞ was not introduced in [251] in order to study the topology it induces on CL(M). We give some historical comments. The notion of a multi-valued contraction mapping had been defined and studied for the first time in [246] and [247], where analogues of many classical single-valued contraction mapping principles were proved [for example, in a special case, see (0.73.3)]. One of the results [247, Theorem 7] shows that, under certain conditions, special types of locally expansive single-valued maps have fixed points. The result was proved by showing that the inverse function is a special type of locally contractive set-valued mapping to which a previous theorem [247, Theorem 6] could be applied. In [247], multi-valued contraction mappings were defined for set-valued mappings F such that $F(x) \varepsilon CB(M)$ for each $x \varepsilon M$, where CB(M) denotes all the non-empty closed and bounded subsets of M. The restriction to CB(M) was done for two reasons: First, it was desirable not to change an already existing, possibly unbounded, metric for M because a change in metric (could) result in a change in the class of maps which are Lipschitz. Second, it was (mistakenly) desired to have an actual *metric* for the range space. However, requiring $F(x) \in CB(M)$ necessitated some severe restrictions in [247, Theorems 7 and 8]. The generalized Hausdorff metric H_∞ was introduced in [251] to eliminate these restrictions (compare [251, Theorem 3] with [247, Theorem 7]). Furthermore, the notion of a multi-valued contraction mapping was put in a more general setting in [251]--namely, maps into $CL(M)$ rather than into $CB(M)$. Finally, the basic theory of fixed points of multi-valued contraction mappings was done in a "more appropriate setting" in [72]. The reader is referred to [8], [107], [115], [206], [304], [305], [444], etc. for related matters. We mention that the topology for $CL(M)$ induced by H_∞ has not been studied.

We have now completed our discussion of the spaces given in (19.1.1) through (19.1.8).

Some natural spaces to investigate are 2^{R^n}, $C(R^n)$, 2^{ℓ_2}, and $C(\ell_2)$. The results in (19.38) through (19.40) say what these spaces are. These results are very recent, and the first two, which have not been written up for publication, have been generalized [see (19.44)].

(19.38) THEOREM [R. M. Schori]. *For each* $n = 1, 2, \ldots$, $2^{R^n} \cong I_\infty \setminus \{(0, 0, \ldots, 0, \ldots)\}$.

(19.39) THEOREM [R. M. Schori]. *For each* $n = 2, 3, \ldots$,

$C(R^n) \cong I_\infty \setminus \{(0, 0, \ldots, 0, \ldots)\}$.

(19.40) THEOREM [180, Corollary 2.3]. *Each of the spaces* 2^{ℓ_2} *and* $C(\ell_2)$ *is homeomorphic to* ℓ_2.

Kroonenberg has many other results in [180]. We state three more. The first is a generalization of (19.40).

(19.41) THEOREM [180, Corollary 2.5]. *Let* M *be a connected* ℓ_2-*manifold. Then each of the spaces* 2^M *and* $C(M)$ *is homeomorphic to* ℓ_2.

(19.42) THEOREM [180, Theorem 2.4]. *Let* M *be a compact connected Hilbert cube manifold. Then the space of Z-sets [resp., connected Z-sets] in* M *is a pseudo-interior for* 2^M *[resp.,* $C(M)$*]*.

(19.43) THEOREM [180, Theorem 3.4]. *The space of Cantor sets in* [0, 1], *also the space of compact zero-dimensional subsets of* [0, 1], *is a pseudo-interior for* $2^{[0,1]}$.

Doug Curtis has recently generalized the results in (19.38) and (19.39) as follows:

(19.44) THEOREM [74]. *Let* W *be a connected, locally connected, locally compact, non-compact metric space. Then:*

$$2^W \cong I_\infty \setminus \{(0, 0, \ldots, 0, \ldots)\}$$

and, if W contains no free arc,

$$C(W) \cong I_\infty \setminus \{(0, 0, \ldots, 0, \ldots)\}.$$

NOTES

1. In addition to the material mentioned in (19.37), the reader may be interested in [420], [427], [430], and [437].

2. A metric, related to the metric of continuity, is in [437].

XX.

APPENDIX--

THE BOUNDARY BUMPING THEOREMS

Proofs will be given for what I will call the "Boundary Bumping Theorems"--(20.1), (20.2), and (20.3). These theorems, as well as (20.6) and (20.7), are used throughout the book. The first time they are used is in the proof of (0.51)--the reader should become familiar with the material up to and including (0.8) before reading the proofs in this Appendix. From the point of view of hyperspace theory as it is presented in the book, the first *major* use of the Boundary Bumping Theorems is in the proof of (1.8). In a sense, (1.8) is a "continuous boundary bumping theorem" [as is (1.25)].

The proofs below are organized as follows. The "Cut Wire Theorem", (20.6), is used to prove (20.1). Then, (20.1) is used to prove (20.2). Finally, (20.2) is used to prove (20.3).

The reader will note that (20.3) implies each of (20.1) and (20.2) by easy arguments. The result in (20.3) is due to Janiszewski; the reader is referred to [186, p.172] for historical comments.

The definition of boundary, Bd, is in (0.42).

(20.1) THEOREM [338, 10.1, p.16]. *Let* X *be a continuum and let* U *be a nonempty proper open subset of* X. *If* K *is a compo-*

nent of cl[U], then K ∩ Bd[U] ≠ ∅ *(equivalently, since*
K ⊂ cl[U] *and* U *is open,* K ∩ [X\U] ≠ ∅*)*.

(20.2) THEOREM [186, Theorem 1, p.172]. *Let* X *be a continuum and let* A *be a nonempty proper closed subset of* X. *If* K *is a component of* A, *then* K ∩ Bd[A] ≠ ∅ *(equivalently, since* K ⊂ A, K ∩ cl[X\A] ≠ ∅*)*.

(20.3) THEOREM (186, Theorem 2, p.172]. *Let* X *be a continuum and let* E *be a nonempty proper subset of* X. *If* K *is a component of* E, *then* cl[K] ∩ Bd[E] ≠ ∅ *(equivalently, since* cl[K] ⊂ cl[E], cl[K] ∩ cl[X\E] ≠ ∅*)*.

The following definition and simple lemma will be used in the proof of (20.6).

(20.4) DEFINITION. Let (M, d) be a metric space, let a, b ∈ M, and let ε > 0. A *(d,ε)-chain of points from* a *to* b is a finite subset

$$\{a = x_1, x_2, \ldots, x_n = b\}$$

of M such that $d(x_i, x_{i+1}) < \varepsilon$ for each i = 1, 2, ..., n-1.

(20.5) LEMMA [338, 8.1, p.13]. *Let* (M. d) *be a metric space, let* A ⊂ M, *and let* ε > 0. *Then* {x ∈ M: *there is a (d,ε)-chain of points from a point of* A *to* x}

is both an open and a closed subset of M.

The following theorem will be used in the proof of (20.1).

(20.6) THEOREM [338, 9.3, p.15]. *Let* (M, d) *be a compact metric space. If* A *and* B *are closed subsets of* M *such that no connected subset of* M *intersects both* A *and* B, *then there are disjoint compact subsets* M_1 *and* M_2 *of* M *such that* $A \subset M_1$, $B \subset M_2$, *and* $M = M_1 \cup M_2$.

Proof. First we show:

(*) There exists $\varepsilon > 0$ such that there is no (d,ε)-chain of points from a point of A to a point of B.

Suppose (*) is false. Then, for each $n = 1, 2, \ldots$, there is a $(d, 1/n)$-chain of points Q_n from a point $a_n \in A$ to a point $b_n \in B$. Since 2^M is a compact metric space [as follows from (0.8)], the sequence $\{Q_n\}_{n=1}^{\infty}$ has a subsequence $\{Q_{n(j)}\}_{j=1}^{\infty}$ which converges to some $Q_o \in 2^M$. Since A and B are compact and since

$Q_{n(j)} \cap A \neq \emptyset \neq Q_{n(j)} \cap B$ for each $j = 1, 2, \ldots,$

it follows easily that $Q_o \cap A \neq \emptyset \neq Q_o \cap B$. Also, it is easy to prove that Q_o is connected. These last two facts contradict the hypotheses about A and B in (20.6). Hence, (*) holds. Let $\varepsilon > 0$ be as in (*). Let

$M_1 = \{x \in M:$ there is a (d,ε)-chain of points from a point of A to x} and let

$M_2 = M - M_1$.

By (20.5), M_1 is both an open and a closed subset of M. Hence, M_1 and M_2 are disjoint compact subsets of M. Clearly, $A \subset M_1$. Since ε satisfies (*), $B \subset M_2$. Clearly, $M = M_1 \cup M_2$. Therefore, we have proved (20.6).

Now we prove (20.1) through (20.3).

PROOF OF (20.1). Suppose that $K \cap Bd[U] = \emptyset$. Then $M = cl[U]$, $A = K$, and $B = Bd[U]$ satisfy the hypotheses of (20.6). Hence, by (20.6), there exist disjoint compact subsets M_1 and M_2 of $cl[U]$ such that $K \subset M_1$, $Bd[U] \subset M_2$, and $cl[U] = M_1 \cup M_2$. Let $M_3 = M_2 \cup (X \setminus cl[U])$. Since $K \neq \emptyset$ (because $U \neq \emptyset$) and $K \subset M_1$,

(1) $M_1 \neq \emptyset$.

Since $U \neq X$ and X is connected, $Bd[U] \neq \emptyset$. Hence, since $Bd[U] \subset M_2 \subset M_3$,

(2) $M_3 \neq \emptyset$.

Since $cl[U] = M_1 \cup M_2$ and $M_3 = M_2 \cup (X \setminus cl[U])$,

(3) $M_1 \cup M_3 = X$.

Note that $M_1 \cap M_2 = \emptyset$ and $M_1 \subset cl[U]$. Hence, since $cl[M_1] = M_1$ (because M_1 is compact),

(4) $cl[M_1] \cap M_3 = \emptyset$.

Note that $cl[X \setminus cl(U)] \subset [(X \setminus cl[U]) \cup (Bd[U])]$. Since $M_1 \subset cl[U]$, $M_1 \cap (X \setminus cl[U]) = \emptyset$. Also, since $M_1 \cap M_2 = \emptyset$ and $Bd[U] \subset M_2$, $M_1 \cap Bd[U] = \emptyset$. Thus, we have that $M_1 \cap cl[X \setminus cl(U)] = \emptyset$. Hence,

(20.2)   APPENDIX   629

since $M_1 \cap \text{cl}[M_2] = \emptyset$ (because $M_1 \cap M_2 = \emptyset$ and M_2 is compact),

(5) $M_1 \cap \text{cl}[M_3] = \emptyset$.

By (1) through (5), X is not connected. This is a contradiction. Therefore, $K \cap \text{Bd}[U] \neq \emptyset$ and we have proved (20.1).

PROOF OF (20.2). Since A is a proper closed subset of X, there exists a sequence $\{\varepsilon_n\}_{n=1}^{\infty}$ of numbers ε_n satisfying:

(1) $\varepsilon_n > 0$ for each $n = 1, 2, \ldots$,

(2) $\varepsilon_n > \varepsilon_{n+1}$ for each $n = 1, 2, \ldots$,

(3) $N(\varepsilon_n, A) \neq X$ for any $n = 1, 2, \ldots$,

(4) $\{\varepsilon_n\}_{n=1}^{\infty}$ converges to 0.

Note that [by (1)],

(5) $K \subset A \subset N(\varepsilon_n, A)$ for each $n = 1, 2, \ldots$.

Since $A \neq \emptyset$, $N(\varepsilon_n, A) \neq \emptyset$ for each $n = 1, 2, \ldots$ [by (5)]. Hence,

(6) $N(\varepsilon_n, A)$ is a nonempty proper [by (3)] open subset of X for each $n = 1, 2, \ldots$.

For each $n = 1, 2, \ldots$, let K_n denote the component of $\text{cl}[N(\varepsilon_n, A)]$ such that $K_n \supset K$ [by (5), K_n exists for each $n = 1, 2, \ldots$]. By (6), we may use (20.1) to see that

(7) $K_n \cap \text{Bd}[N(\varepsilon_n, A)] \neq \emptyset$ for each $n = 1, 2, \ldots$.

From the definition of K_n ($n = 1, 2, \ldots$) we have that

(8) K_n is a subcontinuum of X such that $K_n \supset K$ for each $n = 1, 2, \ldots$.

By (2) and the definition of K_n ($n = 1, 2, \ldots$),

(9) $K_n \supset K_{n+1}$ for each $n = 1, 2, \ldots$.

By (8), (9), and [338, 9.4, p.15],

(10) $\bigcap_{n=1}^{\infty} K_n$ is a subcontinuum of X such that $K \subset [\bigcap_{n=1}^{\infty} K_n]$.

Since $A \subset N(\varepsilon_n, A)$ and $N(\varepsilon_n, A)$ is an open subset of X for each $n = 1, 2, \ldots$, we have [by (0.42)] that $A \cap Bd[N(\varepsilon_n, A)] = \emptyset$ for each $n = 1, 2, \ldots$. Hence, by (7) $K_n \cap [X \setminus A] \neq \emptyset$ for each $n = 1, 2, \ldots$. Thus, $K_n \cap cl[X \setminus A] \neq \emptyset$ for each $n = 1, 2, \ldots$. Hence, since $cl[X \setminus A]$ is compact, it follows easily using (9) that

(11) $[\bigcap_{n=1}^{\infty} K_n] \cap cl[X \setminus A] \neq \emptyset$.

Since $K_n \subset cl[N(\varepsilon_n, A)]$ for each $n = 1, 2, \ldots$, it follows using (4) that

(12) $[\bigcap_{n=1}^{\infty} K_n] \subset A$.

Since [by (0.42)] $Bd[A] = cl[A] \cap cl[X \setminus A]$, we have by (11) and (12) that

(13) $[\bigcap_{n=1}^{\infty} K_n] \cap Bd[A] \neq \emptyset$.

Since K is a component of A, we have by (10) and (12) that

(14) $\bigcap_{n=1}^{\infty} K_n = K$.

By (13) and (14),

(15) $K \cap Bd[A] \neq \emptyset$.

Therefore, by (15), we have proved (20.2).

PROOF OF (20.3). Let $x_0 \in K$. If $x_0 \in cl[X \setminus E]$, then

$K \cap \text{cl}[X\setminus E] \neq \emptyset$ and we are done [see the parenthetical statement in the conclusion of (20.3)]. Thus, for the purpose of proof,

(1) assume $x_o \notin \text{cl}[X\setminus E]$.

By (1), there exist $W_1, W_2, \ldots, W_n, \ldots$ such that

(2) W_n is an open subset of X for each $n = 1, 2, \ldots$,

(3) $W_n \supset W_{n+1}$ for each $n = 1, 2, \ldots$,

(4) $W_n \supset [X\setminus E]$ for each $n = 1, 2, \ldots$,

(5) $x_o \notin W_n$ for any $n = 1, 2, \ldots$,

and

(6) $\bigcap_{n=1}^{\infty} \text{cl}[W_n] = \text{cl}[X\setminus E]$.

For each $n = 1, 2, \ldots$, let $A_n = X\setminus W_n$. By (5),

(7) $x_o \in A_n$ for each $n = 1, 2, \ldots$.

By (4),

(8) $A_n \subset E$ for each $n = 1, 2, \ldots$.

By (2),

(9) A_n is a closed subset of X for each $n = 1, 2, \ldots$.

Hence, for each $n = 1, 2, \ldots$, A_n is a nonempty [by (7)] proper [by (8), since $E \neq X$] closed [by (9)] subset of X, i.e.,

(10) A_n satisfies the hypotheses on A in (20.2).

For each $n = 1, 2, \ldots$, let K_n denote the component of A_n such that $x_o \in K_n$ [by (7), K_n exists for each $n = 1, 2, \ldots$]. Then, by (10), we may apply (20.2) to see that

(11) $K_n \cap \text{Bd}[A_n] \neq \emptyset$ for each $n = 1, 2, \ldots$.

Recall that $x_o \in K$. Hence, by (8) and the definitions of K and K_n ($n = 1, 2, \ldots$), it follows easily that $K_n \subset K$ for each

$n = 1,2,\ldots$. Thus, by (11),

(12) $K \cap Bd[A_n] \neq \emptyset$ for each $n = 1,2,\ldots$.

By (0.42), $Bd[A_n] = cl[A_n] \cap cl[X\setminus A_n]$ and, by definition, $A_n = X\setminus W_n$ for each $n = 1,2,\ldots$. Thus, $Bd[A_n] \subset cl[W_n]$ for each $n = 1,2,\ldots$. Hence, by (12), $K \cap cl[W_n] \neq \emptyset$ for each $n = 1,2,\ldots$. Therefore, letting $L_n = cl[K] \cap cl[W_n]$ for each $n = 1,2,\ldots$, we see that

(13) for each $n = 1,2,\ldots$, L_n is a nonempty compact set and, by (3), $L_n \supset L_{n+1}$.

By (13),

(14) $\bigcap_{n=1}^{\infty} L_n \neq \emptyset$.

By (6) and the definition of L_n $(n = 1,2,\ldots)$,

$$\bigcap_{n=1}^{\infty} L_n = cl[K] \cap cl[X\setminus E].$$

Hence, by (14),

(15) $cl[K] \cap cl[X\setminus E] \neq \emptyset$.

Therefore, by (15) we have proved (20.3) [see the parenthetical statement in the conclusion of (20.3)].

The following theorem is called the Continuum of Convergence Theorem. It can be proved using (0.8) and (20.1). The proof is left to the reader as an exercise [in (0.64.6)].

(20.7) THEOREM [338, 12.1 and 12.3, pp.18-19]. *Let* X *be a continuum with metric* d. *If* X *is not connected im kleinen at a point* $p \in X$, *then there exists* $\varepsilon > 0$ *such that there is a sequence* $\{K_n\}_{n=1}^{\infty}$ *of distinct components* K_n *of* $\text{cl}[B_d(\varepsilon, p)]$, $B_d(\varepsilon, p)$ *as defined in (0.65.1), satisfying:* $\{K_n\}_{n=1}^{\infty}$ *converges to a subcontinuum* K *of* X *such that* $p \in K$, $K \cap [\cup_{n=1}^{\infty} K_n] = \emptyset$, *and* $\text{diam}[K] \geq \varepsilon$. *Furthermore*, X *is not connected im kleinen at any point of* K.

Let us note that, by (1.88) through (1.91) and (20.7), a continuum can not fail to be locally connected at only one point. Also, a continuum X can fail to be locally connected at a point p [in the sense of (1.88.1)], and yet p is not a point "where" the conclusion of (20.7) holds--see (1.89), in particular [147, Figure 3-9, p.113].

BIBLIOGRAPHY

1. R. D. Anderson, On topological infinite deficiency, Mich. Math. J., 14(1967), 365-383.

2. _____, Topological properties of the Hilbert cube and the infinite product of open intervals, Trans. Amer. Math. Soc., 126(1967), 200-216.

3. _____ and Gustave Choquet, A plane continuum no two of whose nondegenerate subcontinua are homeomorphic: an application of inverse limits, Proc. Amer. Math. Soc., 10(1959), 347-353.

4. _____ and Nelly Kroonenberg, Open problems in infinite-dimensional topology, Math Center Tracts, 52(1974), 141-175.

5. James J. Andrews, A chainable continuum no two of whose nondegenerate subcontinua are homeomorphic, Proc. Amer. Math. Soc., 12(1961), 333-334.

6. H. A. Antosiewicz and A. Cellina, Continuous selections and differential relations, J. of Diff. Eq., 19(1975), 386-398.

7. N. Artémiadis, A remark on metric spaces, Proc. Kon. Ned. Akad. van Wetensch., A68(1965), 316-318.

8. Nadim A. Assad and W. A. Kirk, Fixed point theorems for set-valued mappings of contractive type, Pac. J. Math., 43(1972), 553-562.

9. Nadim A. Assad, Fixed point theorems for set-valued transformations on compact sets, Boll. Un. Mat. Ital., 8(1973), 1-7.

10. G. Aumann, Über Räume mit Mittelbildungen, Math. Ann., 119(1943), 210-215.

11. Philip Bacon, An acyclic continuum that admits no mean, Fund. Math., 67(1970), 11-13.

12. _____, Compact means in the plane, Proc. Amer. Math. Soc., 22(1969), 242-246.

13. _____, Unicoherence in means, Colloq. Math., 21(1970), 211-215.

14. B. J. Ball and Jo Ford, Spaces of ANR's, Fund. Math., 77(1972), 33-49.

15. _____, Spaces of ANR's, II, Fund. Math., 78(1973), 209-216.

16. B. J. Ball and R. B. Sher, Embedding circle-like continua in E^3, Can. J. Math., 25(1973), 791-805.

17. H. T. Banks and Marc Q. Jacobs, A differential calculus for multifunctions, J. Math. Analysis and Applications, 29(1970), 246-272.

18. Paul Bankston, Clopen sets in hyperspaces, to appear Proc. Amer. Math. Soc.

19. Gerald A. Beer, The Hausdorff metric and convergence in measure, Mich. Math. J., 21(1974), 63-64.

20. _____, Starshaped sets and the Hausdorff metric, Pac. J. Math., 61(1955), 21-27.

21. David P. Bellamy, The cone over the Cantor set--continuous maps from both directions, Proc. Topology Conference (Emory University, Atlanta, Ga., 1970), J. W. Rogers, Jr., ed., 8-25.

22. _____, Mapping continua onto the cone over the Cantor set, Proc. Charlotte Topology Conference (University of North Carolina at Charlotte, 1974), *Studies in Topology*, Academic Press, New York, 1975, Nick M. Stavrakas and Keith R. Allen, Editors, 43-45.

23. Ralph Bennett, Locally connected 2-cell and 2-sphere-like continua, Proc. Amer. Math. Soc., 17(1966), 674-681.

24. _____ and W. R. R. Transue, On embedding cones over circularly chainable continua, Proc. Amer. Math. Soc., 21(1969), 275-276.

25. Louis J. Billera, Topologies for 2^X; set-valued functions and their graphs, preprint.

26. R. H. Bing, Concerning hereditarily indecomposable continua, Pac. J. Math., 1(1951), 43-51.

27. _____, The elusive fixed point property, Amer. Math. Monthly, 76(1969), 119-132.

28. _____, Embedding circle-like continua in the plane, Can. J. Math., 14(1962), 113-128.

29. _____, Higher-dimensional hereditarily indecomposable continua, Trans. Amer. Math. Soc., 71(1951), 267-273.

BIBLIOGRAPHY

30. _____, A homogeneous indecomposable plane continuum, Duke Math. J., 15(1948), 729-742.

31. _____, Partitioning a set, Bull. Amer. Math. Soc., 55(1949), 1101-1110.

32. _____, Snake-like continua, Duke Math. J., 18(1951), 653-663.

33. _____, Some characterizations of arcs and simple closed curves, Amer. J. Math., 70(1948), 497-506.

34. _____ and R. J. Bean, Editors, *Topology Seminar, Wisconsin*, 1965, Annals of Mathematics Studies No. 60, Princeton University Press, Princeton, New Jersey, 1966.

35. J. Blatter, P. D. Morris, and D. E. Wulbert, Continuity of the set-valued metric projection, Math. Annalen, 178(1968), 12-24.

36. F. F. Bonsall, Lectures on some fixed point theorems of functional analysis, Tata Institute of Fundamental Research, Bombay, 1962, Notes by K. B. Vedak.

37. K. Borsuk, On some metrizations of the hyperspace of compact sets, Fund. Math., 41(1955), 168-202.

38. _____, On the third symmetric potency of the circumference, Fund. Math., 36(1949), 235-244.

39. _____, A theorem on fixed points, Bull. Pol. Acad. Sci., 2(1954), 17-20.

40. _____, *Theory of Retracts*, Monografie Matematyczne, vol. 44, Polish Scientific Publishers, Warszawa, Poland, 1967.

41. _____, Theory of Shape, Lectures, Fall, 1970, Lecture Notes Series, No. 28, Aarhus Universitet, Matematisk Institut, 1973.

42. _____ and S. Mazurkiewicz, Posiedzenie, Towarzystwo Nauk. Warszawskie, 24(1931), 149-152.

43. _____, Sur l'hyperespace d'un continu, C. R. Soc. Sc. Varsovie, 24(1931), 149-152.

44. K. Borsuk and S. Ulam, On symmetric products of topological spaces, Bull. Amer. Math. Soc., 37(1931), 875-882.

45. Carlos J. R. Borges, On a question of Wojdyslawski and Strother, Duke Math. J., 35(1968), 829-833.

46. _____, A study of multivalued functions, Pac. J. Math., 23(1967), 451-461.

47. R. Bott, On the third symmetric potency of S_1, Fund. Math., 39(1952), 364-368.

48. D. G. Bourgin, Cones and Vietoris-Begle type theorems, Trans. Amer. Math. Soc., 174(1972), 155-183.

49. _____, Fixed point and min-max theorems, Pac. J. Math., 45(1973), 403-412.

50. _____, A generalization of the mapping degree, Can. J. Math., 26(1974), 1109-1117.

51. _____, Set valued transformations, Bull. Amer. Math. Soc., 78(1972), 597-599.

52. Bruno Brosowski, Fixed point theorems in approximation theory, Proc. NATO Advanced Study Institute (Venice, Italy, 1968), *Theory and Applications of Monotone Operators*, 1969, 1-6.

53. _____, Über eine Fixpunkteigenschaft der metrischen Projektion, *Computing 5*, Springer-Verlag, 1970, 295-302.

54. M. Brown, Sets of constant distance from a planar set, Mich. Math. J., 19(1972), 321-323.

55. _____, Some applications of an approximation theorem for inverse limits, Proc. Amer. Math. Soc., 11(1960), 478-483.

56. C. E. Burgess, Chainable continua and indecomposability, Pac. J. Math., 9(1959), 653-659.

57. C. E. Capel, Inverse limit spaces, Duke Math. J., 21(1954), 233-245.

58. _____ and W. L. Strother, A space of subsets having the fixed point property, Proc. Amer. Math. Soc., 7(1956), 707-708.

59. _____, Multi-valued functions and partial order, Portugaliae Math., 17(1958), 41-47.

60. James Harvey Carruth, On the structure of hyperspace semigroups, Master's Thesis, Louisiana State University, Baton Rouge, La., 1963.

61. T. A. Chapman, Hilbert cube manifolds, Bull. Amer. Math. Soc., 76(1970), 1326-1330.

62. _____, On some applications of infinite-dimensional manifolds to the theory of shape, Fund. Math., 76(1972), 181-193.

63. J. J. Charatonik, Confluent mappings and unicoherence of continua, Fund. Math., 56(1964), 213-220.

BIBLIOGRAPHY

64. _____, On fans, Dissertationes Math., 54(1967).

65. _____, On ramification points in the classical sense, Fund. Math., 51(1962), 229-252.

66. _____ and C. A. Eberhart, On contractible dendroids, Colloq. Math., 25(1972), 89-98.

67. _____, On smooth dendroids, Fund. Math., 67(1970), 297-322.

68. Frank A. Chimenti, Tychonoff's theorem for hyperspaces, Proc. Amer. Math. Soc., 37(1973), 281-286.

69. H. Cook, Concerning three questions of Burgess about homogeneous continua, Colloq. Math., 19(1968), 241-244.

70. _____, Continua which admit only the identity mapping onto non-degenerate subcontinua, Fund. Math., 60(1967), 241-249.

71. _____, Tree-likeness of dendroids and λ-dendroids, Fund. Math., 68(1970), 19-22.

72. H. Covitz and S. B. Nadler, Jr., Multi-valued contraction mappings in generalized metric spaces, Israel J. Math., 8(1970), 5-11.

73. D. W. Curtis, Growth hyperspaces of Peano continua, preprint.

74. _____, Hyperspaces of noncompact metric spaces, preprint.

75. _____, J. Quinn, and R. M. Schori, The hyperspace of compact convex subsets of a polyhedral 2-cell, Houston J. Math., 3(1977), 7-15.

76. D. W. Curtis and R. M. Schori, 2^X and $C(X)$ are homeomorphic to the Hilbert cube, Bull. Amer. Math. Soc., 80(1974), 927-931.

77. _____, Hyperspaces of Peano continua are Hilbert cubes, to appear Fund. Math.

78. _____, Hyperspaces of polyhedra are Hilbert cubes, to appear Fund. Math.

79. _____, Hyperspaces which characterize simple homotopy type, Gen. Top. and its Applications, 6(1976), 153-165.

80. J. M. Day and K. Kuratowski, On the non-existence of a continuous selector for arcs lying in the plane, Proc. Koninkl. Nederl. Akademie Van Wetenschappen (Amsterdam), Series A, 69, and Indag. Math., 28(1966), 131-132.

81. J. De Groot, On some problems of Borsuk concerning a hyperspace of compact sets, Proc. Kon. Nederl. Akad. Wetenschappen, 59(1956), 95-103.

82. V. A. Dold and R. Thom, Quasifaserungen und Unendliche Symmetrische Produkte, Annals of Math., 67(1958), 239-281.

83. R. Duda, On convex metric spaces, III, Fund. Math., 51(1962), 23-33.

84. _____, On convex metric spaces, V, Fund. Math., 68(1970), 87-106.

85. _____, On the hyperspace of subcontinua of a finite graph, I, Fund. Math., 62(1968), 265-286.

86. _____, Correction to the paper "On the hyperspace of subcontinua of a finite graph, I," Fund. Math., 69(1970), 207-211.

87. _____, On the hyperspace of subcontinua of a finite graph, II, Fund. Math., 63(1968), 225-255.

88. James Dugundji, *Topology*, Allyn and Bacon, Inc., Boston, Mass, 1966.

89. Nelson Dunford and Jacob T. Schwartz, *Linear Operators: Part I: General Theory*, Interscience Publishers, New York.

90. E. Dyer and M.-E. Hamstrom, Completely regular mappings, Fund. Math., 45(1958), 103-118.

91. Carl Eberhart, A note on smooth fans, Colloq. Math., 20(1969), 89-90.

92. _____, The semigroup of subcontinua of a solenoid, Proceedings of the University of Houston Point Set Topology Conference (University of Houston, 1971), D. R. Traylor, Ed., 1971, 9-10.

93. _____ and S. B. Nadler, Jr., The dimension of certain hyperspaces, Bull. Pol. Acad. Sci., 19(1971), 1027-1034.

94. B. Eckmann, Räume mit Mittelbildungen, Comment. Math. Helv., 28(1954), 329-340.

95. _____, T. Ganea, and P. J. Hilton, Generalized means, Studies in Mathematical Analysis and Related Topics, Stanford University Press, 1962.

96. S. Eilenberg and D. Montgomery, Fixed point theorems for multi-valued transformations, Amer. J. Math., 68(1946), 214-222.

BIBLIOGRAPHY

97. R. Engelking, Selectors of the first Baire class for semicontinuous set-valued functions, Bull. Pol. Acad. Sci., 16(1963), 277-282.

98. _____, R. W. Heath, and E. Michael, Topological well-ordering and continuous selections, Inventiones Math., 6(1968), 150-158.

99. R. Engelking and A. Lelek, Cartesian products and continuous images, Colloq. Math., 8(1961), 27-29.

100. Henryk Fast, A remark on continuous selectors, Notre Dame J. Formal Logic, 7(1966), 106-107.

101. _____ and S. Swierczkowski, The New Scottish Book, Wroclaw, 1946-1958.

102. Lawrence Fearnley, Characterizations of the continuous images of the pseudo-arc, Trans. Amer. Math. Soc., 3(1964), 380-399.

103. _____, Classification of all hereditarily indecomposable circularly chainable continua, Trans. Amer. Math. Soc., 168(1972), 387-401.

104. _____, The pseudo-circle is unique, Trans. Amer. Math. Soc., 149(1970), 45-63.

105. Steve Ferry, When ε-boundaries are manifolds, Fund. Math., 90(1976), 199-210.

106. Gary Feuerbacher, Weakly chainable circle-like continua, Dissertation, University of Houston, Houston, Texas, 1974; W. T. Ingram, Director of Dissertation.

107. Arthur S. Finbow, The fixed point set of real multi-valued contraction mappings, Can. Math. Bull., 15(1972), 507-511.

108. M. K. Fort, Jr., Essential and non essential fixed points, Amer. J. Math., 72(1950), 315-322.

109. _____ and Jack Segal, Minimal representations of the hyperspace of a continuum, Duke Math. J., 32(1963), 129-138.

110. Linda Falcao Foulis, Subsets of an absolute retract, Proc. Amer. Math. Soc., 8(1957), 365-366.

111. Geoffrey Fox and Pedro Morales, A general Tychonoff theorem for multifunctions, Can. Math. Bull., 17(1974), 519-521.

112. S. P. Franklin, Compactness and semi-continuity, Israel J. Math., 3(1965), 13-14.

113. _____, Open and image-open relations, Colloq. Math., 12(1964), 209-211.

114. _____ and A. D. Wallace, The least element map, Colloq. Math., 15(1966), 217-221.

115. R. B. Fraser, Jr. and S. B. Nadler, Jr., Sequences of contractive maps and fixed points, Pac. J. Math., 31(1969), 659-667.

116. H. Freudenthal, Entwicklungen von Räumen und ihren Gruppen, Compositio Math., 4(1937), 145-234.

117. M. Friedberg and W. S. Mahavier, Semigroups on chainable and circle-like continua, Math. Zeitschr., 106(1968), 159-161.

118. J. B. Fugate, Decomposable chainable continua, Trans. Amer. Math. Soc., 123(1966), 460-468.

119. T. Ganea, Stability of polyhedra and hyperspaces of compact sets, Bull. Pol. Acad. Sci., 5(1957), 975-978.

120. _____, Symmetrische Potenzen topologischer Räume, Math. Nach., 11(1954), 305-316.

121. R. Gariepy and W. D. Pepe, On the level sets of a distance function in a Minkowski space, Proc. Amer. Math. Soc., 31(1972), 255-259.

122. M. Gindifer, On generalized spheres, Fund. Math., 38(1951), 167-178.

123. Jack T. Goodykoontz, Jr., Aposyndetic properties of hyperspaces, Pac. J. Math., 47(1973), 91-98.

124. _____, Connectedness im kleinen and local connectedness in 2^X and $C(X)$, Pac. J. Math., 53(1974), 387-397.

125. _____, Connectivity properties of hyperspaces, Dissertation, University of Kentucky, Lexington, Kentucky, 1971; Carl Eberhart, Director of Dissertation.

126. _____, More on connectedness im kleinen and local connectedness in $C(X)$, Proc. Amer. Math. Soc., 65(1977), 357-364.

127. _____, Some functions on hyperspaces of hereditarily unicoherent continua, Fund. Math., 95(1977), 1-10.

128. G. R. Gordh, Jr., Terminal subcontinua of hereditarily unicoherent continua, Pac. J. Math., 47(1973), 457-464.

BIBLIOGRAPHY

129. G. R. Gordh, Jr. and S. B. Nadler, Jr., Arc components of chainable Hausdorff continua, Gen. Top. and Its Applications, 3(1973), 63-76.

130. W. Graeub, Die semilinearen Abbildungen, Akad. Wiss. Math.--Nat. Kl., 1950, 205-272.

131. Neil Gray, Note on a paper of Wojdyslawski, Fund. Math., 63(1968), 215-216.

132. _____, On the conjecture $2^X \cong I^\omega$, Fund. Math., 66(1969), 45-52.

133. _____, Unstable points in the hyperspace of connected subsets, Pac. J. Math., 23(1967), 515-520.

134. Charles L. Hagopian, A fixed point theorem for hyperspaces of λ connected continua, Proc. Amer. Math. Soc., 53(1975), 231-234.

135. _____, Addendum to: A fixed point theorem for hyperspaces of λ connected continua, preprint.

136. Benjamin Halpern, Fixed point theorems for set-valued maps in infinite-dimensional spaces, Math. Ann., 189(1970), 87-98.

137. Michael Handel, On certain sums of Hilbert cubes, preprint.

138. F. Hausdorff, *Mengenlehre*, 3^{ed} ed., Springer, Berlin, 1927.

139. George W. Henderson, On the hyperspace of subcontinua of an arc-like continuum, Proc. Amer. Math. Soc., 27(1971), 416-417.

140. _____, Proof that every compact decomposable continuum which is topologically equivalent to each of its nondegenerate subcontinua is an arc, Annals of Math., 72(1960), 421-428.

141. Henry Hermes, Existence and properties of solutions of $\dot{x}\varepsilon R(t, x)$, Studies in App. Math., 5(1969), SIAM Pub., 188-193.

142. _____, The generalized differential equation $\dot{x}\varepsilon R(t, x)$, *Advances in Mathematics*, Academic Press, Inc., 4(1970), 149-169.

143. _____, On continuous and measurable selections and the existence of solutions of generalized differential equations, Proc. Amer. Math. Soc., 29(1971), 535-542.

144. Charles J. Himmelberg, M. Q. Jacobs, and F. S. Van Vleck, Measurable multifunctions, selectors, and Filippov's implicit functions lemma, J. Math. Anal. Appl., 25(1969), 276-284.

145. Charles J. Himmelberg, J. R. Porter, and F. S. Van Vleck, Fixed point theorems for condensing multifunctions, Proc. Amer. Math. Soc., 23(1969), 635-641.

146. Charles J. Himmelberg and F. S. Van Vleck, Some selection theorems for measurable functions, Can. J. Math., 21(1969), 394-399.

147. John G. Hocking and Gail S. Young, *Topology*, Addison-Wesley Publishing Co., Inc., Reading, Mass., 1961.

148. C. Bruce Hughes, Some properties of Whitney continua in the hyperspace C(X), Topology Proceedings (Proceedings of 1976 Topology Conference, Auburn University), 1(1976), 187-189.

149. W. N. Hunsaker and S. A. Naimpally, Local compactness of families of continuous point-compact relations, Pac. J. Math., 52(1974), 101-105.

150. Witold Hurewicz and Henry Wallman, *Dimension Theory*, Princeton University Press, Princeton, New Jersey, 1948.

151. W. T. Ingram, Decomposable circle-like continua, Fund. Math., 63(1968), 193-198.

152. J. R. Isbell, Embeddings of inverse limits, Ann. of Math., 70(1959), 73-84.

153. J. W. Jaworowski, Involutions of compact spaces and a generalization of Borsuk's theorem on antipodes, Bull. Pol. Acad. Sci., 3(1955), 289-292.

154. _____, A theorem on antipodal sets on the n-sphere, Bull. Pol. Acad. Sci., 3(1955), 247-250.

155. _____, Theorem on antipodes for multi-valued mappings and a fixed point theorem, Bull. Pol. Acad. Sci., 4(1956), 187-192.

156. F. Burton Jones, Aposyndetic continua and certain boundary problems, Amer. J. Math., 63(1941), 545-553.

157. _____, Concerning aposyndetic and non-aposyndetic continua, Bull. Amer. Math. Soc., 58(1952), 137-151.

158. S. Kakutani, A generalization of Brouwer's fixed point theorem, Duke Math. J., 8(1941), 457-459.

159. James Keesling, The hyperspace of the integers and some applications, preprint.

160. _____, Normality and compactness are equivalent in hyperspaces, Bull. Amer. Math. Soc., 76(1970), 618-619.

161. _____, Normality and properties related to compactness in hyperspaces, Proc. Amer. Math. Soc., 24(1970), 760-766.

162. _____, On the equivalence of normality and compactness in hyperspaces, Pac. J. Math., 33(1970), 657-667.

163. _____, On the Stone-Cech compactifications of the hyperspace, preprint.

164. John L. Kelley, *General Topology*, D. Van Nostrand Co., Inc., Princeton, New Jersey, 1960.

165. _____, Hyperspaces of a continuum, Trans. Amer. Math. Soc., 52(1942), 22-36.

166. V. L. Klee, Jr., Some topological properties of convex sets, Trans. Amer. Math. Soc., 78(1955), 30-45.

167. Ott-Heinrich Keller, Die Homoiomorphie der kompakten konvexen Mengen im Hilbertschen Raum, Math. Ann., 105(1931), 748-758.

168. Bronislaw Knaster, Über rationale Kurven ohne Bögen, Monatshefte für Mathematik und Physik, 42(1935), 37-44.

169. _____, Un continu dont tout sous-continu est indécomposable, Fund. Math., 3(1922), 247-286.

170. R. J. Knill, Cones, products and fixed points, Fund. Math., 60(1967), 35-46.

171. R. J. Koch, Arcs in partially ordered spaces, Pac. J. Math., 9(1959), 723-728.

172. _____ and I. S. Krule, Weak cutpoint ordering on hereditarily unicoherent continua, Proc. Amer. Math. Soc., 11(1960), 679-681.

173. J. Krasinkiewicz, Certain properties of hyperspaces, Bull. Pol. Acad. Sci., 21(1973), 705-710.

174. _____, No 0-dimensional set disconnects the hyperspace of a continuum, Bull. Pol. Acad. Sci., 19(1971), 755-758.

175. _____, On the hyperspaces of certain plane continua, Bull. Pol. Acad. Sci., 23(1975), 981-983.

176. _____, On the hyperspaces of hereditarily indecomposable continua, Fund. Math., 84(1974), 175-186.

177. _____, On the hyperspaces of snake-like and circle-like continua, Fund. Math., 83(1974), 155-164.

178. _____, Shape properties of hyperspaces, preprint.

179. _____ and S. B. Nadler, Jr., Whitney properties, to appear Fund. Math.

180. Nelly Kroonenberg, Pseudo-interiors of hyperspaces, Compositio Math., 32(1976), 113-131.

181. K. Kuratowski, Evaluation de la classe borélienne ou projective d'un ensemble de points à l'aide des symboles logiques, Fund. Math., 17(1931), 249-272.

182. _____, Les fonctions semi-continues dans l'espace des ensembles fermés, Fund. Math., 18(1931), 148-159.

183. _____, On the completeness of the space of monotone mappings and some related problems, Bull. Pol. Acad. Sci., 16(1968), 283-285.

184. _____, Sur une méthode de métrisation complete de certains espaces d'ensembles compacts, Fund. Math., 43(1956), 114-138.

185. _____, *Topology*, *Vol. I*, Academic Press, New York, New York, 1966.

186. _____, *Topology*, *Vol. II*, Academic Press, New York, New York, 1968.

187. _____, S. B. Nadler, Jr. and G. S. Young, Continuous selections on locally compact separable metric spaces, Bull. Pol. Acad. Sci., 18(1970), 5-11.

188. K. Kuratowski and C. Ryll-Nardzewski, A general theorem on selectors, Bull. Pol. Acad. Sci., 13(1965), 397-403.

189. K. Kuratowski and R. C. Lacher, A theorem on the space of monotone mappings, Bull. Pol. Acad. Sci., 17(1969), 797-800.

190. A. Y. W. Lau, Acyclicity and dimension of hyperspace of subcontinua, Bull. Pol. Acad. Sci., 22(1974), 1139-1141.

191. J. D. Lawson, Applications of topological algebra to hyperspace problems, *Topology*, *Proc. of Memphis St. Univ. Top. Conf.*, Marcel Dekker, Inc., 201-206.

192. A. Lelek, A classification of mappings pertinent to curve theory, Proc. Topology Conference, University of Oklahoma (Norman, Oklahoma, 1972), David Kay, John Green, Leonard Rubin, and Li Pi Su, Editors, 97-103.

BIBLIOGRAPHY 647

193. _____, On weakly chainable continua, Fund. Math., 51(1962), 271-282.

194. _____, Report on weakly confluent mappings, Topology Conference (V. P. I. and S. U., 1973), Lecture Notes in Math., vol. 375, Springer-Verlag, New York, 1974, Raymond F. Dickman, Jr. and Peter Fletcher, Editors, 168-170.

195. _____, Some problems concerning curves, Colloq. Math., 23(1971), 93-98.

196. _____ and David R. Read, Compositions of confluent mappings and some other classes of functions, Colloq. Math., 29(1974), 101-112.

197. S. D. Liao, On the topology of cyclic products of spheres, Trans. Amer. Math. Soc., 77(1954), 520-551.

198. Charles Vincent Lovetro, Jr., Symmetric and permutation product spaces, Master's Thesis, Louisiana State University, 1969.

199. Lewis Lum, Weakly smooth dendroids, Fund. Math., 83(1974), 111-120.

200. Louis F. McAuley, Spaces of certain non-alternating mappings, Duke Math. J., 38(1971), 43-50.

201. M. C. McCord, Classifying spaces and infinite symmetric products, Trans. Amer. Math. Soc., 146(1969), 273-298.

202. T. Bruce McLean, Confluent images of tree-like curves are tree-like, Duke Math. J., 39(1972), 465-473.

203. M. M. McWaters, Arcs, semigroups, and hyperspaces, Can. J. Math., 20(1968), 1207-1210.

204. T. Maćkowiak, Continua having the fixed point property for set-valued mappings, Bull. Pol. Acad. Sci., 23(1975), 421-424.

205. Sibe Mardešić and Jack Segal, ε-mappings onto polyhedra, Trans. Amer. Math. Soc., 109(1963), 146-164.

206. J. T. Markin, Continuous dependence of fixed point sets, Proc. Amer. Math. Soc., 38(1973), 545-547.

207. _____, A fixed point stability theorem for nonexpansive set valued mappings, J. Math. Analysis and App., 54(1976), 441-443.

208. _____, A fixed point theorem for set valued mappings, Bull. Amer. Math. Soc., 74(1968), 639-640.

209. Stefan Mazurkiewicz, Sur les continus absolument indécomposables, Fund. Math., 16(1930), 151-159.

210. _____, Sur l'ensemble des continus péaniens, Fund. Math., 17(1931), 273-274.

211. _____, Sur l'espace des continus péaniens, Fund. Math., 24(1935), 118-134.

212. _____, Sur l'existence des continus indécomposables, Fund. Math., 25(1935), 327-328.

213. _____, Sur l'hyperespace d'un continu, Fund. Math., 18(1932), 171-177.

214. _____, Sur les images continues des continus, Proc. Congress of Mathematicians of Slavic Countries (Warsaw, 1929), F. Leja, Ed., 1930, 66-71.

215. _____, Sur le type c de l'hyperespace d'un continu, Fund. Math., 20(1933), 52-53.

216. _____, Sur le type de dimension de l'hyperespace d'un continu, C. R. Soc. Sc. Varsovie, 24(1931), 191-192.

217. Ernest Michael, Continuous selections, I, Annals Math., 63(1956), 361-382.

218. _____, Continuous selections, II, Annals Math., 64(1956), 562-580.

219. _____, Continuous selections, III, Annals Math., 65(1957), 375-390.

220. _____, Continuous selections in Banach spaces, Studia Math., 22(1963), 75-76.

221. _____, On a map from a function space to a hyperspace, Math. Annalen, 162(1965), 87-88.

222. _____, Paraconvex sets, Math. Scand., 7(1959), 372-376.

223. _____, Selected selection theorems, Amer. Math. Monthly, 63(1956), 233-238.

224. _____, Selection theorems for continuous functions, Proc. International Mathematical Congress, Amsterdam, 1954.

225. _____, Topologies on spaces of subsets, Trans. Amer. Math. Soc., 71(1951), 152-182.

226. George J. Michaelides, Means on topological spaces, Dissertation, University of Georgia, Athens, Georgia, 1974; R. Sher, Director of Dissertation.

BIBLIOGRAPHY

227. Edwin W. Miller, On subsets of a continuous curve which lie on an arc of the continuous curve, Amer. J. Math., 54(1932), 397-416.

228. Harlan C. Miller, On unicoherent continua, Trans. Amer. Math. Soc., 69(1950), 179-194.

229. R. Molski, On symmetric products, Fund. Math., 44(1957), 165-170.

230. Pedro Morales, A note on the topology of C-convergence in hyperspaces, Proc. Amer. Math. Soc., 52(1975), 469-470.

231. H. R. Morton, Symmetric products of the circle, Proc. Cambridge Philos. Soc., 63(1967), 349-352.

232. T. B. Muenzenberger and R. E. Smithson, Refluent multifunctions on semitrees, Proc. Amer. Math. Soc., 44(1974), 189-195.

233. Sam B. Nadler, Jr., Arc components of certain chainable continua, Can. Math. Bull., 14(1971), 183-189.

234. _____, Arcwise accessibility in hyperspaces, Dissertationes Math., 138(1976).

235. _____, Arcwise connected locally planar cartesian products, 3(1977), 409-421.

236. _____, Concerning completeness of the space of confluent mappings, Houston J. Math., 2(1976), 561-580.

237. _____, Confluent images of the sinusoidal curve, to appear Houston J. Math.

238. _____, Continua which are a one-to-one continuous image of [0, ∞), Fund. Math., 75(1972), 123-133.

239. _____, Continua whose cone and hyperspace are homeomorphic, Trans. Amer. Math. Soc., 230(1977), 321-345.

240. _____, Continua whose cones and hyperspaces are homeomorphic, Notices Amer. Math. Soc., 19(1972), A718-A719.

241. _____, Continua whose hyperspace is a product, to appear Fund. Math.

242. _____, An embedding theorem for certain spaces with an equidistant property, Proc. Amer. Math. Soc., 59(1976), 179-183.

243. _____, Inverse limits and multicoherence, Bull. Amer. Math. Soc., 76(1970), 411-414.

244. _____, Locating cones and Hilbert cubes in hyperspaces, Fund. Math., 79(1973), 233-250.

245. _____, Multicoherence techniques applied to inverse limits, Trans. Amer. Math. Soc., 157(1971), 227-234.

246. _____, Multi-valued contraction mappings, Notices Amer. Math. Soc., 14(1967), 930.

247. _____, Multi-valued contraction mappings, Pac. J. Math., 30(1969), 475-488.

248. _____, On locally connected plane one-to-one continuous images of $[0, \infty)$, to appear Colloq. Math.

249. _____, Some basic connectivity properties of Whitney map inverses in $C(X)$, Proc. Charlotte Topology Conference (University of North Carolina at Charlotte, 1974), *Studies in Topology*, Academic Press, New York, 1975, Nick M. Stavrakas and Keith R. Allen, Editors, 393-410.

250. _____, Some problems concerning hyperspaces, Topology Conference (V. P. I. and S. U.), Lecture Notes in Math., vol. 375, Springer-Verlag, New York, 1974, Raymond F. Dickman, Jr. and Peter Fletcher, Editors, 190-197.

251. _____, Some results on multi-valued contraction mappings, Topology Conference (SUNY at Buffalo, 1969), Lecture Notes in Math., vol. 171, Springer-Verlag, New York, 1970, W. M. Fleischman, Ed., 64-69.

252. _____ and J. Quinn, Continua whose hyperspace and suspension are homeomorphic, to appear Gen. Top. and Its Applications.

253. _____, Embeddability and structure properties of real curves, Memoirs Amer. Math. Soc., No. 125, 1972.

254. _____, Embedding certain compactifications of a half-ray, Fund. Math., 78(1973), 217-225.

255. S. B. Nadler, Jr.; J. Quinn and N. M. Stavrakas, Hyperspaces of compact convex sets, I, Bull. Pol. Acad. Sci., 23(1975), 555-559.

256. _____, Hyperspaces of compact convex sets, II, Bull. Pol. Acad. Sci., 25(1977), 381-385.

257. _____, Hyperspaces of compact convex sets, to appear Pac. J. Math.

BIBLIOGRAPHY

258. Sam B. Nadler, Jr. and J. T. Rogers, Jr., A note on hyperspaces and the fixed point property, Colloq. Math., 25(1972), 255-257.

259. Sam B. Nadler, Jr. and L. E. Ward, Jr., Concerning continuous selections, Proc. Amer. Math. Soc., 25(1970), 369-374.

260. J. Nagata, *Modern General Topology*, John Wiley and Sons, Inc., New York, New York, 1968.

261. Togo Nishiura and C. J. Rhee, The hyperspace of a pseudo-arc is a Cantor manifold, Proc. Amer. Math. Soc., 31(1972), 550-556.

262. T. Parthasarathy, Selection theorems and their applications, Lecture Notes in Math., vol. 263, Springer-Verlag, New York, New York, 1972.

263. Norman S. Passmore, III, Embedding products into hyperspaces, Dissertation, University of Delaware, Newark, Delaware, 1976; Sam B. Nadler, Jr., Director of Dissertation.

264. D. G. Paulowich, Weak contractibility and hyperspaces, Master's Thesis, Dalhousie University, Halifax, Nova Scotia, 1970.

265. A. Pelczynski, A remark on spaces 2^X for zero-dimensional X, Bull. Pol. Acad. Sci., 13(1965), 85-89.

266. Ann Petrus, Contractibility of Whitney continua in C(X), to appear Gen.Top. and Its Applications.

267. _____, Whitney maps and Whitney properties of C(X), Dissertation, Tulane University, New Orleans, La., 1976; J. T. Rogers, Jr., Director of Dissertation.

268. _____, Whitney maps and Whitney properties of C(X), Topology Proceedings (Proceedings of 1976 Topology Conference, Auburn University), 1(1976), 147-172.

269. R. L. Plunkett, A fixed point theorem for continuous multi-valued transformations, Proc. Amer. Math. Soc., 7(1956), 160-163.

270. V. I. Ponomariov, Novoye prostranstvo zamknutykh mnozhestv i mnogoznachnye otobrazhenya bikompaktov (in Russian), [A new space of closed subsets and multi-valued mappings of bicompact spaces], Mat. Sb. Novaya Seriya, 48(1959), 191-212.

271. V. V. Popov, The topology of the space of closed subsets, Vestnik Moskov. Univ. Ser. I Mat. Meh., 30(1975), 65-70.

272. J. Quinn and Raymond Y. T. Wong, Union of convex Hilbert cubes, Proc. Amer. Math. Soc., 65(1977), 171-176.

273. Hans Radstrom, An embedding theorem for spaces of convex sets, Proc. Amer. Math. Soc., 3(1952), 165-169.

274. David R. Read, Confluent and related mappings, Colloq. Math., 29(1974), 233-239.

275. C. J. Rhee, Homotopy groups for cellular set-valued functions, Proc. Amer. Math. Soc., 19(1968), 874-876.

276. _____, On dimension of hyperspace of a metric continuum, Bull. Soc. Royale des Sci. de Liège, 38(1969), 602-604.

277. M. Richardson, On the homology characters of symmetric products, Duke Math. J., 1(1935), 50-69.

278. _____, The relative connectives of symmetric products, Bull. Amer. Math. Soc., 41(1935), 528-534.

279. James T. Rogers, Jr., Applications of a Vietoris-Begle theorem for multi-valued maps to the cohomology of hyperspaces, Mich. J. Math., 22(1976), 315-319.

280. _____, The cone = hyperspace property, Can. J. Math., 24(1972), 279-285.

281. _____, Continua with cones homeomorphic to hyperspaces, Gen. Top. and Its Applications, 3(1973), 283-289.

282. _____, Dimension and the Whitney subcontinua of C(X), Gen. Top. and Its Applications, 6(1976), 91-100.

283. _____, Dimension of hyperspaces, Bull. Pol. Acad. Sci., 20(1972), 177-179.

284. _____, Embedding the hyperspaces of circle-like plane continua, Proc. Amer. Math. Soc., 29(1971), 165-168.

285. _____, Hyperspaces of arc-like and circle-like continua, Topology Conference (V. P. I. and S. U., 1973), Lecture Notes in Math., vol. 375, Springer-Verlag, New York, New York, 1974, Raymond F. Dickman and Peter Fletcher, Editors, 231-235.

286. _____, Whitney continua in the hyperspace C(X), Pac. J. Math., 58(1975), 569-584.

287. Ronald H. Rosen, On tree-like continua and irreducibility, Duke Math. J., 26(1959), 113-122.

BIBLIOGRAPHY

288. R. M. Schori, Hyperspaces and symmetric products of topological spaces, Fund. Math., 63(1968), 77-88.

289. _____, Inverse limits and near-homeomorphism techniques in hyperspace problems, Proc. of the Second Pittsburgh International Conference on General Topology and its Applications (University of Pittsburgh, 1972), Springer-Verlag, Lecture Notes in Math., vol. 378, 1974, 421-428.

290. _____ and J. E. West, 2^I is homeomorphic to the Hilbert cube, Bull. Amer. Math. Soc., 78(1972), 402-406.

291. _____, The hyperspace of the closed unit interval is a Hilbert cube, Trans. Amer. Math. Soc., 213(1975), 217-235.

292. _____, Hyperspaces of graphs are Hilbert cubes, Pac. J. Math., 53(1974), 239-251.

293. J. Segal, A fixed point theorem for the hyperspace of a snake-like continuum, Fund. Math., 50(1962), 237-248.

294. _____, Hyperspaces of the inverse limit space, Proc. Amer. Math. Soc., 10(1959), 706-709.

295. _____, A note on the singular homology of the hyperspace of a continuum, preprint.

296. R. Sher, The union of two Hilbert cubes meeting in a Hilbert cube need not be a Hilbert cube, Proc. Amer. Math. Soc., 63(1977), 150-152.

297. Kermit Sigmon, Acyclicity of compact means, Mich. Math. J., 16(1969), 111-115.

298. _____, A note on means in Peano continua, Aequationes Math., 1(1968), 85-86.

299. _____, On the existence of a mean on certain continua, Fund. Math., 63(1968), 311-319.

300. S. Sirota, Spectral representation of spaces of closed subsets of bicompacta, Soviet Math. Dokl., 9(1968), 997-1000.

301. P. A. Smith, Periodic and nearly periodic transformations, *Lectures in Topology*, University of Michigan Press, 1941, 159-190.

302. _____, The topology of involutions, Proc. Nat. Acad. Sci., 19(1933), 612-618.

303. R. E. Smithson, First countable hyperspaces, Proc. Amer. Math. Soc., 56(1976), 325-328.

304. _____, Fixed points for contractive multifunctions, Proc. Amer. Math. Soc., 27(1971), 192-194.

305. _____, Multifunctions, Nieuw Archief voor Wiskunde, 20(1972), 31-53.

306. R. H. Sorgenfrey, Concerning triodic continua, Amer. J. Math., 66(1944), 439-460.

307. K. B. Stein, Homology of the symmetric product, (abstract), Bull. Amer. Math. Soc., 58(1952), 207.

308. W. L. Strother, Continuous multi-valued functions, Bol. Soc. Mat. São Paolo, 10(1955), 87-120.

309. _____, Fixed points, fixed sets, and M-retracts, Duke Math. J., 22(1955), 551-556.

310. _____, Multi-homotopy, Duke Math. J., 22(1955), 281-285.

311. _____, On an open question concerning fixed points, Proc. Amer. Math. Soc., 4(1953), 988-993.

312. _____ and L. E. Ward, Jr., Retracts from neighborhood retracts, Duke Math. J., 25(1958), 11-14.

313. K. Sundaresan, Extreme points of the unit cell in Lebesgue-Bochner functions spaces, II, preprint.

314. _____, Extreme points of convex sets and selection theorems, Topology Conference (SUNY at Buffalo, 1969), Lecture Notes in Math., vol. 171, Springer-Verlag, New York, New York, 1970, W. M. Fleischman, Ed., 82-87.

315. W. R. R. Transue, On the hyperspace of subcontinua of the pseudoarc, Proc. Amer. Math. Soc., 18(1967), 1074-1075.

316. E. D. Tymchatyn, Hyperspaces of hereditarily indecomposable plane continua, Proc. Amer. Math. Soc., 56(1976), 300-302.

317. T. Van Der Walt, *Fixed and Almost Fixed Points*, Math Center Tracts, Math. Centrum, Amsterdam, 1963.

318. R. Vasudevan, A note on separation axioms in hyperspaces, Kyungpook Math. J., 15(1975), 223-230.

319. L. Vietoris, Bereiche zweiter Ordnung, Monatshefte für Mathematik und Physik, 32(1922), 258-280.

320. _____, Kontinua zweiter Ordnung, Monatshefte für Mathematik und Physik, 33(1923), 49-62.

BIBLIOGRAPHY

321. A. D. Wallace, Acyclicity of compact connected semigroups, Fund. Math., 50(1961), 99-105.

322. _____, A fixed-point theorem for trees, Bull. Amer. Math. Soc., 47(1941), 757-760.

323. L. E. Ward, Jr., Arcs in hyperspaces which are not compact, Proc. Amer. Math. Soc., 28(1971), 254-258.

324. _____, Characterization of the fixed point property for a class of set-valued mappings, Fund. Math., 50(1961), 159-164.

325. _____, A fixed point theorem, Amer. Math. Monthly, 65(1958), 271-272.

326. _____, A fixed point theorem for multi-valued functions, Pac. J. Math., 8(1958), 921-927.

327. _____, A general fixed point theorem, Colloq. Math., 15(1966), 243-251.

328. _____, Rigid selections and smooth dendroids, Bull. Pol. Acad. Sci., 19(1971), 1041-1044.

329. Roger W. Wardle, On a property of J. L. Kelley, Houston J. Math., 3(1977), 291-299.

330. T. Wazewski, Sur un continu singulier, Fund. Math., 4(1923), 214-235.

331. James E. West, Cartesian factors of infinite-dimensional spaces, Topology Conference (V. P. I. and S. U., 1973), Lecture Notes in Math., vol. 375, Springer-Verlag, New York, New York, 1974, Raymond F. Dickman and Peter Fletcher, Editors, 249-268.

332. _____, Identifying Hilbert cubes: General methods and their application to hyperspaces by Schori and West, Proc. of the Third Prague Topological Symposium, 1971, General Topology and its Relations to Modern Analysis and Algebra, III, 455-461.

333. _____, The subcontinua of a dendron form a Hilbert cube factor, Proc. Amer. Math. Soc., 36(1972), 603-608.

334. Paul A. White, r-regular convergence spaces, Amer. J. Math., 66(1944), 69-96.

335. Hassler Whitney, On regular families of curves, Bull. Amer. Math. Soc., 47(1941), 145-147.

336. _____, Regular families of curves, I, Proc. Nat. Acad. Sci., 18(1932), 275-278.

337. _____, Regular families of curves, Annals Math., 34(1933), 244-270.

338. Gordon Thomas Whyburn, *Analytic Topology*, Amer. Math. Soc. Colloq. Publ., vol. 28, Amer. Math. Soc., Providence, R. I., 1942.

339. _____, A continuum every subcontinuum of which separates the plane, Amer. J. Math., 52(1930), 319-330.

340. _____, Monotoneity of limit mappings, Duke Math. J., 29(1962), 465-470.

341. _____, On sequences and limiting sets, Fund. Math., 25(1935), 408-426.

342. M. Wojdyslawski, Rétractes absolus et hyperespaces des continus, Fund. Math., 32(1939), 184-192.

343. _____, Sur la contractilité des hyperespaces des continus localement connexes, Fund. Math., 30(1938), 247-252.

344. Raymond Y. T. Wong, Some remarks on hyperspaces, Proc. Amer. Math. Soc., 21(1969), 600-602.

345. W. Wu, Note sur les produits essentiels symétriques des espaces topologiques, I, Comptes Rendus des Séances de l'Académie des Sciences, 16(1947), 1139-1141.

346. Daniel E. Wulbert, Subsets of first countable spaces, Proc. Amer. Math. Soc., 19(1968), 1273-1277.

347. Gail S. Young, Continuous selections on locally compact separable metric spaces, Topology Conference (SUNY at Buffalo, 1969), Lecture Notes in Math., vol. 171, Springer-Verlag, New York, New York, 1970, W. M. Fleischman, Ed., 102-110.

348. _____, Fixed-point theorems for arcwise connected continua, Proc. Amer. Math. Soc., 11(1960), 880-884.

349. _____, The introduction of local connectivity by change of topology, Amer. J. Math., 68(1946), 479-494.

SUPPLEMENTARY BIBLIOGRAPHY, I

350. Robert Bantegnie, Convexité des hyperespaces, Arch. Math., 26(1975), 414-420.

351. T. Bânzaru, Multivalued mappings and M-product spaces, Bul. Sti. Tehn. Inst. Politehn. Timisoara--Ser. Mat.-Fiz.-Mec. Teoret. Apl., 17(1972), 17-23.

352. _____, On certain properties of multivalued mappings, Stud. Cerc. Mat., 24(1972), 1503-1510.

353. L. J. Billera, Topologies for 2^X, set-valued functions and their graphs, Amer. Math. Soc. Transl., 155(1971).

354. T. A. Chapman, *Lectures on Hilbert Cube Manifolds*, Conf. Board of the Math. Sci. Regional Conf. Series in Math., no. 28, Amer. Math. Soc., Providence, R. I., 1975.

355. _____, Simple homotopy theory for ANR's, to appear Gen. Top. and Its Applications.

356. J. J. Charatonik, On the fixed point property for set-valued mappings of hereditarily decomposable continua, Gen. Top. Relat. mod. Anal. Algebra III, Proc. 3rd Prague Top. Symp., 1971, 83-84(1972).

357. M. M. Čoban and S. Ĭ. Nedev, Factorization theorems for multivalued mappings, Multivalued cross sections and topological dimension, Math. Balkanica, 4(1974), 457-460.

358. M. M. Čoban, S. Ĭ. Nedev, and V. M. Vylov, A certain theorem of Michael on sections, C. R. Acad. Bulgare Sci., 28(1975), 871-873.

359. S. Czerwik, A fixed point theorem for a system of multivalued transformations, Proc. Amer. Math. Soc., 55(1976), 136-139.

360. Carlos de Olano, On selections, Proc. of the Tenth Conf. of Spanish Mathematicians (La Laguna, 1969), 85-113. Inst. "Jorge Juan" Mat., Madrid, 1972.

361. M. M. Drešević and M. M. Marjanović, A note on Ascoli's theorem for spaces of multifunctions, Publ. Inst. Math. (Beograd), 14(1972), 111-114.

362. R. Duda, One result on inverse limits and hyperspaces, Gen. Top. Relat. mod. Anal. Algebra III, Proc. 3rd Prague Top. Symp., 1971, 99-102(1972).

363. Oskar Feichtinger, More on lower semi-continuity, Amer. Math. Monthly, 83(1975), 39.

364. M. K. Fort, Jr. and Jack Segal, Local connectedness of inverse limit spaces, Duke Math. J., 28(1961), 253-259.

365. Gilles Fournier and Lech Górniewicz, The Lefschetz fixed point theorem for multi-valued maps of non-metrizable spaces, Fund. Math., 92(1976), 213-222.

366. Geoffrey Fox and Pedro Morales, Non-Hausdorff multifunction generalization of the Kelley-Morse Ascoli theorem, Pac. J. Math., 64(1976), 137-143.

367. John Ginsburg, A note on the G_δ-closure and the real-compactness of 2^X, Gen. Top. and Its Applications, 6(1976), 319-326.

368. _____, Contributions to the theory of hyperspaces, Thesis, University of Manitoba, Winnipeg, Manitoba, Canada, 1975; R. Grant Woods, Director of Dissertation.

369. _____, On the Stone-Čech compactification of the space of closed sets, Trans. Amer. Math. Soc., 215(1976), 301-311.

370. _____, Some results on the countable compactness and pseudocompactness of hyperspaces, Can. J. Math., 27(1975), 1392-1399.

371. C. K. Goel and R. Vasudevan, Separation axioms in bitopological hyperspaces, Ann. Soc. Sci. Bruxelles Sér. I, 89(1975), 480-496.

372. Jack T. Goodykoontz, C(X) is not necessarily a retract of 2^X, to appear Proc. Amer. Math. Soc.

373. G. R. Gordh, Jr. and C. B. Hughes, On freely decomposable mappings of continua, preprint.

374. Lech Górniewicz, Homological methods in fixed-point theory of multi-valued maps, Dissertationes Math., 129(1976).

375. Gerhard Grimeisen, The hyperspace of lower semicontinuity and the first power of a topological space, Czechosl. Math. J., 24(1974), 15-25.

376. Kiyoshi, Iseki, Multi-valued contraction mappings in complete metric spaces, Rend. Sem. Mat. Univ. Padova, 53(1975), 15-19.

377. Saroop K. Kaul, Ascoli theorem and the exponential map in set valued function spaces, Rend. Mat., 9(1976), 165-177.

378. N. N. Kaulgud and D. V. Pai, Fixed point theorems for set-valued mappings, Nieuw Arch. Wisk., 23(1975), 49-66.

379. James Keesling, Compactness related properties in the hyperspaces, Topology Conference (SUNY at Buffalo, 1969), Lecture Notes in Math., vol. 171, Springer-Verlag, New York, New York, 1970, W. M. Fleischman, Ed., 40-44.

BIBLIOGRAPHY

380. K. Kuratowski and A. Maitra, Some theorems on selectors and their applications to semi-continuous decompositions, Bull. Pol. Acad. Sci., 22(1974), 877-881.

381. A. Y. W. Lau, A note on monotone maps and hyperspaces, Bull. Pol. Acad. Sci., 24(1976), 121-123.

382. _____ and C. H. Voas, Connectedness of the hyperspace of closed connected subsets, preprint.

383. Ka Sing Lau, A note on lower semicontinuous set-valued maps, Bull. Pol. Acad. Sci., 24(1976), 271-273.

384. Y. F. Lin, Tychonoff's theorem for the space of multifunctions, Amer. Math. Monthly, 74(1967), 399-400.

385. _____ and D. A. Rose, Ascoli's theorem for spaces of multifunctions, Pac. J. Math., 34(1970), 741-747.

386. L. D. Loveland, A metric characterization of a simple closed curve, Gen. Top. and Its Applications, 6(1976), 309-313.

387. V. J. Mancuso, An Ascoli theorem for multi-valued functions, J. Austral. Math. Soc., 12(1971), 466-472.

388. M. M. Marjanović, Spaces homeomorphic to their hyperspaces, Proc. of the International Symp. on Top. and its Applications (Budva, 1972), 167-169.

389. Karl Menger, *Kurventheorie*, Verlag und Druck von B. G. Teubner, Leipzig and Berlin, 1932.

390. _____, Untersuchungen über allgemeine Matrik, Math. Ann., 100(1928), 75-163.

391. G. J. Michaelides, A note on topological m-spaces, Colloq. Math., 32(1975), 193-197, 309.

392. E. E. Moise, Grille decomposition and convexification theorems for compact locally connected continua, Bull. Amer. Math. Soc., 55(1949), 1111-1121.

393. Pedro Morales, A non-Hausdorff multifunction Ascoli theorem for k_3-spaces, Can. J. Math., 27(1975), 893-900.

394. _____, Non-Hausdorff Ascoli theory, Dissertationes Math., 119(1974).

395. Sam B. Nadler, Jr., A characterization of locally connected continua by hyperspace retractions, to appear Proc. Amer. Math. Soc.

396. N. Noble, Ascoli theorems and the exponential map, Amer. Math. Soc. Transl., 143(1969), 393-411.

397. Barada K. Ray, A note on multi-valued contraction mappings, Atti Accad. Naz. Lincei Rend. Cl. Sci. Fis. Mat. Natur., 56(1974), 500-503.

398. _____, Some fixed point theorems, Fund. Math., 92(1976), 79-90.

399. Helga Schirmer, Biconnected multifunctions of trees which have an end point as fixed point of coincidence, Can. J. Math., 23(1971), 461-467.

400. _____, δ-continuous selections of small multifunctions, Can. J. Math., 24(1972), 631-635.

401. R. M. Schori, Survey of hyperspace results using infinite-dimensional topology and a short proof that $2^I \approx Q$, preprint.

402. Jack Segal, Mapping norms and indecomposability, J. London Math. Soc., 39(1964), 598-602.

403. R. E. Smithson, First countable hyperspaces, Proc. Amer. Math. Soc., 56(1976), 325-328.

404. _____, Subcontinuity for multifunctions, Pac. J. Math., 61(1975), 283-288.

405. _____, Uniform convergence for multifunctions, Pac. J. Math., 39(1971), 253-259.

406. U. Tašmetov, On the shapes of hyperspaces, Sibir. Mat. Žurn., 16(1975), 869-872.

407. V. L. Timohovič, Multivalued extensions of closed mappings, Dokl. Akad. Nauk USSR, 19(1975), 14-16, 89.

408. A. A. Tolstonogov, Topological structure of continuous set-valued mappings, Sibir. Mat. Žurn., 16(1975), 837-852, 886.

409. A. van Heemert, De R_n-adische Voortbrenging van Algemeen-Topologische Ruimten met Toepassingen op de Constructie van Niet Splitsbare Continua, Thesis, University of Gronongen, 1943.

410. N. V. Veličko, Über den Raum der abseschlossenen Teilmensen, Sibir. Mat. Žurn., 16(1975), 627-629.

411. T. Ważewski, Sur les courbes de Jordan ne renfermant aucune courbe simple fermée de Jordan, Ann. Soc. Pol. Math., 2(1923), 49-170.

412. James E. West, Infinite products which are Hilbert cubes, Trans. Amer. Math. Soc., 150(1970), 1-25.

413. L. R. J. Westermann, On the hull operator, Nederl. Akad. Wet., Proc., Ser. A, 79(1976), 179-184.

414. P. Zenor, On the near completeness and the realcompactness of the space of closed subsets, preprint.

SUPPLEMENTARY BIBLIOGRAPHY, II

415. H. A. Antosiewicz and A. Cellina, Continuous extensions of multifunctions, Annales Pol. Mat., 34(1977), 107-111.

416. Robert Bantegnie, Convexité des hyperespaces, Archiv der Math., 26(1975), 414-420.

417. F. F. Bonsall, B. E. Cain, and Hans Schneider, The numerical range of a continuous mapping of a normed space, Aequationes Math., 1(1968), 205.

418. Carl Eberhart, Continua with locally connected Whitney continua, preprint.

419. Zdeněk Frolík, Concerning topological convergence of sets, Czechoslovak Math. J., 10(1960), 168-179.

420. S. Gladysz, E. Marczewski, and C. Ryll-Nardzewski, Concerning distances of sets and distances of functions, Colloq. Math., 8(1961), 71-75.

421. Gilbert Helmberg, On convergence classes of sets, Proc. Amer. Math. Soc., 13(1962), 918-921.

422. Lars Hörmander, Sur la fonction d'appui des ensembles convexes dans un espace localement convexe, Arkiv för Matematik, 3(1954), 181-186.

423. K. Kuratowski, On partitions of complete spaces which do not admit analytic selectors and on some consequences of a theorem of Gödel, Instituto Nazionale di Alta Matematica, Symposia Mathematica, 16(1975), 67-74.

424. _____, On the selector problems for the partitions of Polish spaces and for the compact-valued mappings, Annales Pol. Math., 29(1975), 421-427.

425. _____, The σ-algebra generated by Souslin Sets and its applications to set-valued mappings and to selector problems, Bollettino U. M. I., 11(1975), 285-298.

426. A. Y. W. Lau, Whitney continuum in hyperspaces, Topology Proceedings (Proceedings of 1976 Topology Conference, Auburn University), 1(1976), 187-189.

427. E. Marczewski and H. Steinhaus, On a certain distance of sets and the corresponding distance of functions, Colloq. Math., 6(1958), 319-327.

428. S. Mrówka, On the convergence of nets of sets, Fund. Math., 45(1958), 237-246.

429. D. G. Paulowich, Weak contractibility and hyperspaces, Fund. Math., 94(1977), 41-47.

430. G. C. Shephard and R. J. Webster, Metrics for sets of convex bodies, Mathematika, 12(1965), 73-88.

431. H. Toruńczyk, On CE-images of the Hilbert cube and characterization of Q-manifolds, preprint.

432. Daniel H. Wagner, Integral of a convex-hull-valued function, J. Math. Analysis and Applications, 50(1975), 548-559.

433. _____, Integral of a set-valued function with semi-closed values, J. Math. Analysis and Applications, 55(1976), 616-633.

434. _____, Survey of measurable selection theorems, to appear SIAM Journal on Control and Optimization.

435. Daniel H. Wagner and Lawrence D. Stone, Necessity and existence results on constrained optimization of separable functionals by a multiplier rule, SIAM Journal Control, 12(1974), 356-372.

436. P. D. Watson, On the limits of sequences of sets, Quarterly J. Math., 4(1953), 1-3.

SUPPLEMENTARY BIBLIOGRAPHY, III

437. Karol Borsuk, On a metrization of the hyperspace of a metric space, Fund. Math., 94(1977), 191-207.

438. Carl Eberhart, Intervals of continua which are Hilbert cubes, to appear Proc. Amer. Math. Soc.

439. _____ and Sam B. Nadler, Jr., Hyperspaces of cones and fans, preprint.

440. G. Fournier and L. Górniewicz, The Lefschetz fixed point theorem for some non-compact multi-valued maps, Fund. Math., 94(1977), 245-254.

441. J. B. Fugate, Retracting fans onto finite fans, Fund. Math., 71(1971), 113-125.

BIBLIOGRAPHY

442. J. B. Fugate, Small retractions of smooth dendroids onto trees, Fund. Math., 71(1971), 255-262.

443. J. Grispolakis and E. D. Tymchatyn, Embedding smooth dendroids in hyperspaces, preprint.

444. Peter K. F. Kuhfittig, Fixed points of locally contractive and nonexpansive set-valued mappings, Pac. J. Math., 65(1976), 399-403.

445. R. M. Schori, Survey of hyperspace results using infinite-dimensional topology and a short proof that $2^I \approx Q$, Topology Proceedings (Proceedings of 1976 Topology Conference, Auburn University), 1(1976), 129-139.

446. J. E. West, Induced involutions on Hilbert cube hyperspaces, Topology Proceedings (Proceedings of 1976 Topology Conference, Auburn University), 1(1976), 281-293.

INDEX OF SPECIAL SYMBOLS

X, d	p.1, (0.16)	$F_n(X)$, $C_n(X)$	(0.48)
2^X, $C(X)$	p.1	$\hat{f}$	(0.49)
$N_d(\varepsilon, A)$, $N(\varepsilon, A)$	(0.1)	ω, μ	(0.50)
H, H_d	(0.1), (0.63.6), (19.1)	conv	(0.63.5)
lim inf, lim sup,		$B_d(t, a)$, $C_d(t, a)$, K_d	(0.65.1)
lim, $A_i \to A$	(0.5)	$X(A, \mu, t)$	(0.68.5)
CL(S)	(0.10.1)	$\Gamma(X)$	(1.27.1)
$<\ >$	(0.10.2)	$S(\omega)$	(1.27.2)
$[a, b]$	(0.17)	f_ω	(1.30.1)
R^n, $\|\ \|$, B^n, S^n	(0.19)	H^2	(1.31.1)
cl	(0.35)	η_t	(1.79)
$\setminus$	(0.36)	$\mathrm{ord}_A[X]$	(1.104)
$\times$, π_Y, $\Pi_{j=1}^\infty$	(0.37)	E_p	(1.114)
Cone(X), B(X), v, π	(0.38)	$\{Y_n, f_n\}_{n=1}^\infty$	(1.149)
Sus(X)	(0.39)	$\underset{\leftarrow}{\lim}$	(1.150)
ℓ_2, I_∞	(0.40)	$Q_i[Y_n, f_n]$	(1.151)
diam	(0.41)	π_i	(1.154)
Bd	(0.42)	$B_i[Y_n, f_n]$, $B[Y_n, f_n]$	(1.156)
$\cong$	(0.43)	f_n^*, 2_∞^X, $C_\infty(X)$	(1.168)
dim, $\dim_p$, $\emptyset$	(0.44)	$C_M(X)$	(1.208.5)

INDEX OF SPECIAL SYMBOLS

$C^M(X)$	(1.208.6)	$C*, 2*$	p.513
2_A^X	(1.214.3)	$Y_n \to Y_o$ 0-regularly	(15.8)
$S_o, S_1^1, S_2^1, S_3^1, (SP)_1,$		ρ_τ	p.521
$(SP)_2, (SP)_3$	(8.22)	$X \in \text{prop}[\kappa]$	(16.10)
τ_j, J_∞	p.348	F_μ	(16.12)
$f:(A_1, A_2, \ldots, A_n) \xrightarrow{\text{onto}}$		$C_o[X, Y], C_o FL[X, Y]$	p.556
$(B_1, B_2, \ldots, B_n)$	p.348	$cc(Z)$	(18.1)
$\tilde{J}, \psi[X]$	(10.19)	$\lambda \cdot A$	p.568
$AA[X]$	p.388	$\text{cas}(R^n)$	(18.2)
$X \in CP$	(14.14)	$\widehat{xy}, S(x), \ker(Z)$	(18.10)
Σ^n, Σ_s^n	p.429	ψ_p	p.579
$Z(z), Z(z, \mu, t)$	p.430	$a(M), s(M), t(M), \theta(M),$	
$W[C(X)], W[2^X]$	(14.71)	$P(M), ch(M), hi(M),$	
$X \in CP(\mu)$	p.479	$L(M), \Omega*$	(19.1)
$X \in CP(\mu)$ at t_o	(14.73.2)	$ca(R^n)$	(19.13)
$X \in CPH$	p.486	$\overline{xy}, a(D, p), \psi^p$	(19.14)
$\sigma_{t,\mu}, \sigma_t$	(14.73.7)	Cr_ε	p.614
$X \in \text{Class}[W]$	p.497	ρ_h	pp.619-620
$\Psi_{t,\mu}, \Psi_t$	(14.76.0)	H_∞	p.620
$X \in CPH$ at t	(14.76.5)		

SUBJECT INDEX

Hyphenated words are treated alphabetically as one word. Thus, for example, u-SPACE follows UPPER SEMI-CONTINUOUS FUNCTIONS. Symbols are usually at the beginning of the appropriate alphabetical heading, and Greek symbols precede others (for example, see the beginning of the S's). Multiple symbols occur alphabetically without regard to the part in parentheses. For example, hi(M) comes between HEREDITARILY UNICOHERENT CONTINUUM and HILBERT CUBE. When numbers in parentheses and page numbers appear simultaneously, there are facts about the topic in the numbered item *and elsewhere* on the referenced page (even when the numbered item is on the given page).

A

a(M) *(19.1)*
  $\leq F_{\sigma\delta}$ *(19.34)*
  see $a(\Omega^*)\ldots,a(X)\cup F_1(X),L(M)$,
    weakly smooth dendroid
$a(\Omega^*),s(\Omega^*),\theta(\Omega^*),t(\Omega^*)$ *(19.1)*
  $\supset I_\infty$ *(19.5-7)*
  dense in $C(\Omega^*)$ *(19.2)*
  first category (19.3)
  homogeneous if $\Omega^* = R^2$ *(19.4)*
  ∞-*dim. (19.6)*
  questions (19.8-12)
  see $a(M)$, $a(X)\cup F_1(X),ca(R^n)$,
    $cas(R^n),Cr_\varepsilon,\Gamma(X),L(M),S(\omega)$,
    weakly smooth dendroid

$a(X) \cup F_1(X)$
  compactness of (19.20-25)
  fixed point theorem (19.24)
    application (19.25)
  see $a(M),a(\Omega^*)\ldots,cas(R^n)$,
    weakly smooth dendroid
AA[X]; *p.388*
  Borel type? (12.24)
  continuumwise dense in X rational; p.388
  $= X$, *what rational X? (12.23)*
  see arcwise accessible...
ABSOLUTE NEIGHBORHOOD RETRACT,ANR
  $cc(X)$ *is* $\Rightarrow X$ *is,* ψ_p *CE map, etc. (18.32)*

see AR, cc-hyperspace
hyps.contractible w.r.t.
    (1.181, 183)
strong Whitney-rev.? (14.57)
Whitney prop., not (14.43.7)
see AR, cc-hyperspace

ABSOLUTE RETRACT, AR
$C(X)$ is iff X loc.conn. (1.96)
$C(X)$ nested $\cap$ of (1.171)
$C_M(X)$ is (1.208.5)
$C^M(X)$ is if M locally connected (1.208.6)
$cc(X)$ is $\Rightarrow X$ is (18.32)
    see ANR, cc-hyperspace
essentially different Whitney
    maps for? (14.43.12)
exercise (0.74)
$\Gamma(X)$ is if X loc.conn.? (1.27.3)
    see $\Gamma(X)$
Hilbert cube factor (1.214.2)
$\cong I_\infty$, n.a.s.c. (1.214)
loc.conn. subcontinuum of $C(X)$
    is, suff. cond. (0.74.1)
$\mu^{-1}([t_o, \mu(X)])$ is if $\mu^{-1}(t_o)$
    loc.conn. (14.73.32)
strong Whitney-rev.? (14.57)
2^X, nested $\cap$ of (1.171)
2^X is iff X loc.conn. (1.96)
2^X_A is for each A, n.a.s.c.
    (1.214.3)
X is $\Rightarrow cc(X) \cong I_\infty$ iff $cc(X)$
    I_∞-manifold (18.27.2)
Whitney prop., not (14.43.6)
see ANR, cc-hyperspace

ACCESSIBLE
see arcw.acc. ..., continuum-
    wise acc., segmentwise acc.

ACYCLICNESS
of hyperspaces (1.172-180)
of $\mu^{-1}(t)$ (14.37), (14.43.6)
strong Whitney-rev.? (14.57)

ADMITS ESSENTIALLY DIFFERENT
WHITNEY MAPS (14.43)
see essentially different...

APOSYNDESIS (0.29)
of hyperspaces (1.138)
Whitney property (14.19)

ARC (0.18)
characterized by
    map 2^X into $C_1(X)$ (6.11.1)
    selections (5.3)
cone=hyp.prop., has; p.303
ε-crooked (19.27.1)
    space of..., Cr_ε ; p.614
end point of (0.18)
existence in 2^X, $C(X)$ (1.8-14)
    structure in $C(X)$, X h.i.
        (1.61-62)
existence in $\mu^{-1}(t)$ (14.8.1)
    see arcwise connected con-
        tinuum ($\mu^{-1}(t)$ is)
free (0.18)
$\psi[X]$ (0.19), (10.20)
generalized (0.18)
    see generalized arc
hyperspaces of (0.54); p.137
invariant Whitney hyperspace,
    has (14.73.34)
    stronger fact (14.73.13)
$\mu^{-1}(t)$ is (14.6, 28, 31, 50)
    $\Rightarrow \mu^{-1}([t,\mu(X)]) \cong B^2$ (14.73.13)
see arcwise connected con-
    tinuum($\mu^{-1}(t)$ is)

singular (18.21), (18.21.1)
strong Whitney-rev. (14.50)
Whitney maps ω, exs. (14.61-62)
Whitney property (14.6)
see $a(\Omega^*)..., arcw.conn.continuum, order arc, pseudo-arc, Sus(X) \cong C(X)$

ARC COMPONENTS OF

$C(X)\setminus\{X\}, X$ indecomposable(1.52)
$C(X)\setminus\{Y\}, dim[C(X)] < \infty$ and Y indecomposable (8.19)
$2^X\setminus\{E\}$, how $\cap C(X)$ (11.14)
  related results(11.13), (12.29), (12.31)
X if X is C-H continuum (8.9-10),(8.21)

ARCLESS

rational chainable continuum (12.18)
strong Whitney-rev. (14.52)
  related facts (14.55.2)

ARC-LIKE *(0.23)*

see chainable

ARC-PRESERVING *(1.207.7)*

ARCWISE ACCESSIBLE FROM $\Sigma\setminus C_1(X)$; p.373

...beginning with K; p.373

ARCWISE ACCESSIBLE FROM $2^X\setminus C_1(X)$; p.373

and arcwise disconnecting (12.14-17)
cut point theorem (12.11)
  generalizations (12.27-28)
general theorems (12.2,8), (12.27-28)
her.decomposable X (12.14-25)
  questions (12.19-25)

h.i.X (12.9, 30)
iff...$C_2(X)\setminus C_1(X)$? (12.26)
n.a.s.c.$\{x\}$ is? (12.22)
no $\{x\}$ is
  decomposable X,ex. (12.10)
  if X h.i. (12.9)
rational X
  AA[X]; p.388, (12.23-24)
  continuumwise dense $\subset F_1(X)$ is arcwise acc. (12.13)
  is not arcwise accessible; p.381, (12.16-17)
  n.a.s.c. $\{x\}$ arcwise accessible? (12.22)
see continuumwise accessible, segmentwise accessible

ARCWISE CONNECTED CONTINUUM

arcwise disconnected by each point & not uniquely arcw.conn.;pp.371-372
$C(X), 2^X$ are (1.13)
  for Hausdorff continua(1.14)
  uniquely,C(X) is... (1.61)
$C^M(X)$ is (1.208.6)
cc(X) is iff X is (1.203.5)
chainable (0.23.1)
circle-like (0.70.3)
CP, X has, iff X arc (14.73.20)
$\mu^{-1}(t)$ is, all t>0,if X is(14.8)
  converse false (14.48)
  generalization (14.75)
  see arc
$\mu^{-1}(t)$ is, all $t \geq t_a$, if X decomposable (14.27)
Whitney property (14.8)
  related fact (14.75)

Whitney-rev., not (14.48)

see *arcwise connectedness of, weak Whitney stable*

ARCWISE CONNECTED ⊂ 2^X

∩$C(X)$, X *h.i.* *(12.29,31)*

ARCWISE CONNECTEDNESS OF

C *(composant in X)* *(1.52.1)*

$C(X)\setminus\{A_1,A_2\}$ *(9.2),(11.17)*

$C(X)\setminus C(E)$, $2^X\setminus 2^E$ *(11.2)*

$C(X)\setminus\{E\}$ *all $E \neq X$, X a-triodic (11.16)*

hyperspaces(1.9), (1.12-14)

$\mu^{-1}(t)$

see *arc, arcwise connected continuum*

X *if $C(X) \cong Sus(X)$ (9.3)*

see *arcwise disconnecting, arcwise connected continuum, Cone(X) $\cong$ C(X), locally arcwise connected*

ARCWISE DECOMPOSABLE *(0.30)*

circle-like (0.30)

images of hyperspaces are(4.7)

ARCWISE DISCONNECTING; *p.358*

applications to arcwise accessibility (12.14-17)

char.of h.i. (11.15)

related ex.; pp.371-372

chars. of indecomposable(1.51), (11.1, 4)

hyperspaces by no E

if $C(X) \cong$ product(10.6)

if $2^X \cong$ product (10.24)

$\neq X$ a-triodic $\Rightarrow$ each proper subcontinuum unicoherent, etc. (11.16)

see *arcw.connectedness of*

hyperspaces by some E

char.thms. (11.5,7)

related facts, examples (11.8-12)

nowhere denseness of E (11.6, 9)

some $C(Y)$, Y subcontinuum of X chainable? (11.18)

hyperspaces by $X(1.51),(11.1,4)$

2^X *by A*

iff $C(X)$ by A (11.8)

$\Rightarrow A \in C(X)$ *(11.3)*

A-TRIODIC CONTINUUM, X

$C(X) < \infty\text{-}dim.product \Rightarrow X$ *arc or circle (10.12)*

a-triodic necessary?(10.9)

each proper subcontinuum is unicoherent, etc.

suff.conds.(10.11), (11.16)

if $X \in$ CPH (14.73.18)

not if $X \in$ CP,ex.;pp.493,495

strong Whitney-rev. (14.49)

see *triod*

B

B^n *(0.19)*

see *n-cell, 2-cell*

BAIRE CLASS 1

C^* *is (15.3)*

choice function exists (5.1)

BAIRE THEOREM; *p.515*

BALLS, CONTINUITY OF *(0.65)*

BASE

$B(X)$, *of Cone(X) (0.38)*

C-H continuum X, $h[B(X)] = F_1(X)$ some h? (8.34)

related facts (8.7),
 (8.20.7),(8.25-26),
 p.332
 for inverse limit (1.156)
 for hyps.of (1.166-167)
 for Vietoris topology(0.11-12)
BONDING MAPS *(1.149)*
 commute with projs. (1.155)
 see hyper-onto representation
BOUNDARY *(0.42)*
 Bumping Theorems (20.1-3)
BROUWER REDUCTION THEOREM;
 pp.45, 53
BUCKETHANDLE, X *(1.209.3)*
 basic properties (1.209.3)
 $C(X) \not\cong X \times X$ *(1.128)*
 cone=hyp.prop., X has;p.303
 question (14.22)
 open map onto arc (1.207.5)

C

C*, THE FUNCTION; *p.513*
 Baire class 1 (15.3)
 continuous
  means C-smooth (15.5)*
  not, example (15.1)
  on dense G_δ (15.3)
  w.r.t. 0-reg.conv. (15.9)
 u.s.c. (15.2)
 see C-smooth*
C(X); *p.1*
 $\supset, \not\cong$ *cones, products*
 see embedding in C(X)
 $\cong Cone(X)$
  see $Cone(X) \cong C(X)$
 $\cong product$

 see product, C(X) is a
 $\cong Sus(X)$
  see $Sus(X) \cong C(X)$
 $\cong 2^X \Rightarrow X$ *loc.conn.? (10.27)*
$C_M(X)$ *(1.208.5)*
 AR, locally connected (1.208.5)
 order arc, n.a.s.c. (1.56-57)
 see terminal subcontinuum
 see 2^X_A
$C^M(X)$ *(1.208.6)*
ca(R^n) *(19.13)*
 see cas(R^n), cc-hyperspace
CANONICAL CONTRACTION OF CONE;
 p.325
CANTOR FAN
 see cone over Cantor set
CANTOR MANIFOLD; *p.224*
 $C(X)$ *is if* $dim[C(X)] = 2$ *(2.15)*
 see nondisconnecting
CANTOR SET *(0.69.6)*
 characterization (0.69.6)
 continuous images of (0.69.4)
 middle-thirds (0.69.6)
 Pelczynski's Theorem (0.69.7)
  appls.(0.69.8),(1.40.1-2)
 space of Cantor sets in [0,1]
  (19.43)
 2^M *is iff M is (0.69.8)*
 2^Z_p *is, $Z = \{p = 0,1/2, 1/3,...\}$*
  (0.69.8)
 see cone over Cantor set
CARTESIAN PRODUCT
 see product...
cas(R^n); *p.569, (18.2)*
 see ca(R^n), cc-hyperspace

CATEGORY G; p.410
CATEGORY T; p.410
  see Whitney stable
cc-HYPERSPACE, cc(X) (18.1)
  ANR $\Rightarrow$ X $\in$ ANR, etc. (18.32)
  AR $\Rightarrow$ X $\in$ AR, etc. (18.32)
  arcwise conn.iff X is (1.203.5)
  $cc(B^n \setminus S^{n-1}), cc(R^n)$ for $n \geq 2$
    (18.8)
    $cc(R^1)$; p.569
  compact convex Z, dim[Z] ≥ 2
    $cc(Z) \cong I_\infty$ (18.4)
    $dim[cc(Z)] = \infty$ (18.3)
  compactness of (18.1.1)
  contractibility of (18.27.1),
    (18.31.2)
  Hilbert cube geometry (18.36-39)
  Hilbert cube manifold (18.27.2)
  Hilbert space $\ell_2 \cong cc(\ell_2)$? (18.9)
  $\cong I_\infty$
    theorems (18.4-7), (18.11),
      (18.13,15), (18.21.1-2),
      (18.28,34,40)
    examples (18.17-20)
  homotopy type (18.31-32)
  ψ_p, properties of; p.579,
    (18.30,32)
  $R^2 \supset X$ (18.15-26), (18.28)
  $R^3 \supset X$ (18.33-34)
  $R^4 \supset X$ (18.35)
  selection for cc(X), ψ_p (18.30)
  shape (18.31.3)
  simple homotopy type (18.32)
  starshaped X (18.10)
    ker(X) = {point} & $cc(X) \cong$
      I_∞, example (18.19)

    not & $cc(X) \cong I_\infty$, ex.(18.20)
  theorem (18.11)
    example (18.17)
    question (18.26)
  strong deformation retract,
    $F_1(X)$ is (18.30.4)
  topology for (18.1)
  2-cell X $\subset R^2$ with polygonal
    boundary (18.21.1),
    (18.22, 28)
  see $ca(R^n)$, $cas(R^n)$
C-DETERMINED (0.61)
  chainable? circle-like? fans?
    (0.62)
  finite graphs (0.59)
  h.i. (0.60)
  smooth fans (0.62)
CE MAP
  ψ_p is if cc(X) $\in$ ANR (18.32)
CELL
  see n-cell, two-cell
C_oFL[X, Y]; p.556
  closed $\subset C_o[X,Y]$ if Y $\in$ prop[κ]
    (16.31)
  topologically complete (16.33)
  see confluent map
ch(M) (19.1)
  G_δ in C(M) (19.29)
  questions (19.32-33)
  topologically complete if M =
    Ω^* (19.33)
  see chainable continuum, $P(\Omega^*)$
C-H CONTINUUM; p.308
  see Cone(X) $\cong$ C(X)
CHAIN
  (d,ϵ)-chain (20.4)
  in a space (0.22)

CHAINABLE CONTINUUM, X *(0.23)*
  admits ess.diff.Whitney maps
    example (14.43.2)
    question (14.43.11)
  arcw.connected is arc (0.23.1)
  arcw.disconnected C(Y) by E?
    (11.18)
  $C(X) \subset R^3$ *(3.2)*
  $C(X) \supset X \times X \Rightarrow X$ *arc? (1.129)*
  $C(X) \cong inv.lim.$ *2-cells (1.191)*
  $C(X) \cong product \Rightarrow X$ *arc (10.10)*
  C(X) is quasi-complex; p.291
    see quasi-complex
  C-determined? (0.62)
  Class[W], X in; p.497
  confluent images of; p.555
  CP, X has (14.13.1)
  CPH, X has (14.73.12)
  C*-smooth, X is(14.76.6),
    (15.13)
  $dim[C(X)] = 2$ *(2.1)*
  fixed point property
    C(X) has (7.1)
    multi-valued (0.73.2)
    2^X *has? (7.10)*
  hyper-onto rep., X has (1.191)
  hyp.conditions $\Rightarrow$ *X chainable*
    (11.16)
  invariant Whitney hyp., X has
    $\Rightarrow$ *X arc or indecomposable (14.73.33)*
    question (14.73.34)
  $\mu^{-1}(t)$ *is*
    arc if X decomposable
      (14.31)
    chainable (14.4)
    indecomposable if X is
      (14.13)
    necessity of X chainable;
      p.413, (14.12)
  $\mu^{-1}([t,\mu(X)])$ *2-cell (14.73.13)*
  map 2^X *or C(X) onto X iff X arc*
    (4.8)
  rational and $\natural$ *arc, ex. (12.18)*
  retract, $C_1(X)$ *of* 2^X*,what X?*
    (6.10.3)
  strong Whitney-rev.? (14.57)
  Whitney hyp., X has (14.73.29)
  Whitney map selection continuum? (5.21)
  Whitney property (14.4)
  see Buckethandle, ch(M), ε-map,
    $P(\Omega^*)$*, pseudo-arc*

CHEESEBURGER
  see McDonald...

CHOICE FUNCTION
  Baire class 1 exists (5.1)
  see selection

CIRCLE, X *(0.19)*
  cone=hyperspace property, X has; p.303
  C(X) is 2-cell (0.55)
  $\natural$ *u-space; p.264*
  diam map on 2^X
    not monotone,ex.(1.211.3)
    not open,examples (1.211.3),
      (14.67)
    question (14.68)
  $F_2(X)$ *(0.71.1)*
  invariant Whitney hyperspace,X has (14.73.34)
    stronger fact (1.213.2)
  mean, X not (0.71.1)

S_4 space $\neq$ circle; p.260

strong Whitney-rev. (14.51)

Whitney map for 2^X

    not open, example (14.67)

    open exists? (14.69)

    see open map, Whitney map

Whitney property (14.7)

see $a(\Omega^*)\ldots$, circle = $\mu^{-1}(t)$, diam map, essential map, long circle, pseudo-circle, solenoid, Warsaw circle

CIRCLE = $\mu^{-1}(t)$

    all $t > 0$ if X circle (14.7)

        converse (14.51)

    all $t \geq t_o$ if X decomposable proper circle-like (14.31)

    $\mu^{-1}([t,\mu(X)])$ 2-cell (1.213.2)

see circle

CIRCLE-LIKE CONTINUUM, X (0.24)

    admits ess.diff.Whitney maps

        example (14.43.4)

        question (14.43.11)

    arcw.conn.,uncnt.many (0.70.3)

    arcw.decomposable iff S^1 (0.30)

    $C(X) \subset R^3$ (3.3), (3.6)

    $C(X) \subset R^4$; pp.235,239

    $C(X) \supset X \times X$, some X? (1.130)

    $C(X) \cong$ inv.lim.2-cells (1.196)

    $C(X) \cong$ product $\Rightarrow X$ circle (10.10)

    $C(X) \neq X \times X$, any X (1.129)

    C-determined? (0.62)

    Čech coho.,Wh.prop.for (14.18)

chainable if nonessential ε-maps of X into S^1 (1.194)

C^*-smooth

    if non-planar (14.76.6), (15.15)

    $\Rightarrow$ chainable or indecomposable (15.16)

$dim[C(X)] = 2$ (2.1)

fixed point property

    $C(X)$ has (7.5)

    2^X has? (7.10)

hyper-onto rep., X has (1.195)

hyperspace conditions $\Rightarrow X$ circle-like (11.16)

invariant Whitney hyp., X has $\Rightarrow X$ circle or indecomposable (14.73.33)

    question (14.73.34)

$\mu^{-1}(t)$ is

    circle if X decomposable proper circle-like (14.31)

    non-planar if X is (14.18)

    not indecomposable if X is, example (14.12)

    not unicoherent if X is, example (14.12)

    planar if X is (14.18)

    proper circle-like if X is (14.5)

    same shape as X (14.18)

map 2^X or $C(X)$ onto X iff X circle (4.9)

minimal rep.,$C(X)$ has (1.196)

non-planar

    Class[W], X in; p.497

    CPH, X has (14.73.17)

    C^*-smooth, X is (14.76.6), (15.15)

Wh.hyp., X has (14.73. 29)
Whitney prop.for (14.18)
planar
   iff $C(X) \subset R^3$ (3.3, 6)
   Whitney prop.for (14.18)
proper (0.24)
   $\mu^{-1}(t)$ circle if X decomposable (14.31)
   strong Whitney-rev.?(14.57)
   Whitney property (14.5)
retract, $C_1(X)$ of 2^X, what X? (6.10.3)
shape, Whitney prop.for (14.18)
strong Whitney-rev.? (14.57)
Whitney map selection continuum? (5.21)
Whitney prop., not (14.26)
   pp.424-426
see pseudo-solenoid, solenoid
CL(S) (0.10.1)
with H_d if (S,d) is bounded (0.66.2)
with H_∞ ; pp.620-622
with Vietoris top. (0.10-12)
   basic properties (0.66)
   see Vietoris topology
see selection (for CL(Z))
CLASS[W], X ε CLASS[W]; p.497
   if X chainable; p.497
   if X ε CP (14.73.21)
      converse? (14.73.25)
   if X h.i.; p.497
   if X non-planar circle-like; p.497
   irreducible, X is (14.73.25)
   strong Whitney-rev. (14.55)
      for sequences? (14.55.1)

triod, X is not (14.73.25)
unicoherent, X is (14.73.25)
Whitney property? (14.73.10)
COMPACT
a(D,p)
   see weakly smooth dendroid
$a(X) \cup F_1(X)$ (19.20-25)
$C(X)$, 2^X are (0.8)
cc(Z) is iff Z is (18.1.1)
$S(\omega)$ is (1.29, 34)
   see $\Gamma(X)$, $S(\omega)$
$X(A,\mu,t)$ is (0.68.5)
   see $X(A,\mu,t)$
COMPACT CONVEX SETS
see $ca(R^n)$, $cas(R^n)$, cc-hyp.
COMPACTIFICATION
of R^1, end of; p.323
Pelczynski's Theorem (0.69.7)
   applications (0.69.8), (1.40.1-2)
COMPLETE
$C(M)$, 2^M are iff M is (0.63.6)
$C(M)$, 2^M are if $M=\ell_2, R^n$ (0.63.6)
   see Hilbert space, n-space
see topological completeness
COMPOSANT (0.32)
C(composant) (1.52.1)
if X is C-H continuum(8.2,9-10)
number of composants (1.206.2)
CONE, POSITIVE CONVEX
$W[C(X)]$, $W[2^X]$ are (14.71.3)
CONE = HYPERSPACE PROPERTY, X HAS; p.303
   arc has; p.303
   arcwise conn.composants (8.2)

arcwise connected, X is if X decomposable (8.3, 6)
Buckethandle has; p.303
circle has; p.303
$dim[X] < \infty$
    question (8.11)
    structure theorems (8.5, 6)
if $X < \infty$-dim. indecomposable C-H continuum (8.7, 13)
solenoid has; p.303, (14.20)
see Cone(X) $\cong$ C(X)

CONE OVER CANTOR SET
    $\subset C(X)$, *any X (1.72)*
    images of
        $C(X)$, 2^X *are (1.33)*
        X is if X g-contr. (4.13.1)
    preimages of, n.a.s.c. (1.37), p.247
    preimages of, C(X) is
        characterization (1.41)
        if $X \supset$ open set uncountably many components (1.45)
        converse? (1.47)
        not and X not locally connected, example (1.42)
    preimages of, 2^X is
        iff X not loc. conn. (1.39)
    weakly chainable; p.245

CONE OVER X, CONE(X) (0.38)
    base of, B(X) (0.38)
        see base
    canonical contraction of; p.325
    C(X), relations to for X h.i. (1.207.7)
    dim[Cone(X)] (8.0)
    projection π of Cone(X)\{v} (0.38)

vertex, v (0.38)
$C_{\{v\}}(Cone(X)) \cong I_\infty$ *(1.214.3)*
see cone=hyp.prop., cone over Cantor set, Cone(X) $\cong$ C(X), embedding in C(X), her. indecomposable continuum

CONE(X) $\cong$ C(X), X IS C-H CONTINUUM; p.308
arc components of X (8.9-10, 21)
compactification theorem (8.24)
composants of X (8.2, 9-10)
cone = hyperspace property
    X does not have, ex.; p.303
    X has if $X < \infty$-dim. indecomposable (8.7, 13)
    see cone = hyp. property
$dim[X] < \infty$
    arc components of X (8.21)
    arcwise connected composants $\Rightarrow \ldots$ (8.4)
    $\Rightarrow dim[X] = 1$ *(8.17-18)*
    $\Rightarrow X \supset \leq 1$ *indecomposable continuum Y (8.18)*
    discussion of Y (8.20)
examples (7.2), (8.1), p.303, p.331, (10.16), (14.20)
hereditarily decomposable X, exactly eight (8.23)
    discussion of pf.; pp.322-331
homeos., Cone(X) onto C(X)
    behavior of (0.70.2), (8.7-8), (8.20), (8.25-26), p.332, (8.34)
    Rogers homeo.; p.303, (8.7)
    see cone=hyp.property
questions (8.14-16), (8.20), (8.31-37)
see cone=hyp.property, embedding in C(X)

CONTRACTIBILITY   SUBJECT INDEX   677

CONFLUENT MAP *(0.45.3)*
   *between X, 2^X, C(X)? (4.15),
     (6.8)*
   2^X onto [0,1]? (14.70)
   *conditions implying f is
     (0.45.3),(1.207.6),
     (16.32)*
   images of S_o (16.30)
   preserves
     chainable? p.555
     h.i. (1.207.5)
     prop[κ] (16.29)
     tree-like (14.16)
   see $C_oFL[X,Y]$, weakly confluent map

CONNECTED IM KLEINEN
   at a point (1.88.2)
   C(X) (1.141-143)
   *2^X (1.131), (1.133-137),
     (1.139-140)*
   Continuum of Convergence Theorem (20.7)
   relation to loc.conn.(1.89,91)
   see loc.arcw.conn., loc.conn.at a point, loc.conn.continuum

CONNER'S QUESTION; *p.234*

CONTINUITY OF
   balls (0.65)
   C^*; p.513, (15.5)
     see C^*, C^*-smooth
   diam (1.211.1)
   η_t (1.80)
   F_μ (16.14)
   induced map (0.49),(0.67.1)
   mapping (0.45)
   *Ψ_t ;p.506,(14.76.1),
     (14.76.3 - 5)*

   σ_t (14.73.7)
   segment (1.15)
   segment homotopy (16.3-4)
   2^* (15.4)
   union, ∪ (1.48)
   Whitney map (0.50)
   see metric (of continuity)

CONTINUOUS DECOMPOSITION
   see c-space, monotone cts.decomposition, monotone open map

CONTINUOUS FUNCTION
   means mapping (0.45)
   see continuity of, mapping

CONTINUUM; *p.1, (0.16)*
   $\mu^{-1}(t)$ is (14.2)
   $\omega^{-1}(t)$ is not, example (14.61)
   see C-H continuum, Hausdorff continuum, etc.

CONTINUUM OF CONVERGENCE
   theorem (20.7)

CONTINUUMWISE ACCESSIBLE
   from $\Sigma \backslash C_1(X)$; p.395
   theorem (13.1)
   *relation to decompositions;
     pp.396-398*
   see arcw.acc. ..., segmentwise accessible

CONTINUUMWISE DENSE *(12.13)*
   see arcwise acc. ...(rational X)

CONTINUUMWISE DISCONNECT *(0.34)*

CONTRACTIBILITY IN A SPACE *(16.2)*
   locally conn.subcontinua of hyps.in hyps. (16.17)
   see contractibility of

CONTRACTIBILITY OF *(16.2)*
   *arcwise conn.subc.of C(X), X
     h.i. (1.63)*

cc-hyp. (18.27.1), (18.31.2)
C(X) iff 2^X iff $F_1(X)$ in 2^X
  (16.7)
  question (16.34)
C(X), 2^X if
  X contractible (16.8)
  X h.i. (1.75-76)
  X homogeneous (16.26)
  X locally connected(0.65.4),
    (1.93),(16.18, 41)
  X not locally connected,
    examples (16.20-21)
  X ε prop[κ] (16.15)
    see prop[κ]
  X ∉ prop[κ], example (16.21)
  see open map (images preserve)
C(X), 2^X locally (0.65.4),
    (1.93), (16.18)
C(X), 2^X not; exs. (16.22, 41)
C(X), 2^X w.r.t.
  ANR (1.181, 183)
  S^1 (1.176)
C(X × Y) if C(X) and C(Y) are?
    (16.36)
loc.conn.subc.of 2^X (16.17)
$\mu^{-1}(t)$ (14.43.6-10)
  question (14.43.12)
$\psi_p^{-1}(x)$ (18.30.2)
strong Whitney-rev.? (14.57)
W[C(X)], W[2^X] (14.71.3)
Whitney property,not (14.43.6)
see dendrite, dendroid, g-contractible,non-contractible

CONTRACTIBILITY WITH RESPECT TO A SPACE (16.2)
  hyperspaces (1.176, 181, 183)

CONTRACTIBLE (16.2)
  see contractibility of, g-contractible, non-contractible
CONTRACTION
  canonical, of cone; p.325
  multi-valued (0.73.3),pp.621-622
CONVERGENCE OF SETS (0.5), p.6
  basic facts (0.5-13), (0.64)
  see ρ_h, ρ_τ, zero-reg.convergence
CONVEX ARCS, SETS, ETC.
  see ca(R^n), cas(R^n), cc-hyp.
CONVEX CONTINUUM, X
  cc(X) ≅ I_∞ (18.4)
  $\mu^{-1}(t)$ ≅ I_∞? (14.38)
  ω monotone and open (14.66)
  see cc-hyperspace,convex metric
CONVEX FAMILY, F (18.5)
  F cpt. ∞-dim. ⇒ F ≅ I_∞ (18.6)
  applications (18.7)
  see cc-hyperspace
CONVEX HILBERT CUBES (18.36-39),
    pp.597-598, (18.41)
  see cc-hyperspace, Hilbert cube
CONVEX METRIC, ρ (0.65.3)
  basic props., appls. (0.65.3-4)
  hyps.of X with ρ; p.148,(18.42)
  ω not open, example (14.67)
COOK'S CONTINUA; pp.151, 216-217,
    223, 235
COSELECTION (5.22)
  discussion; pp.265-266
  questions (5.23)
COVERING PROPERTY, CP (14.14)
  embedding any X in continuum ε CP; p.495

independent of μ *(14.73.4)*
maps that preserve? (14.73.26)
questions (14.73), (14.73.27)
strong Whitney-rev. (14.55)
    for sequences? (14.55.1)
Whitney property? (14.76.9)
    CPH is (14.76.8)
X ε CP if
    X chainable (14.13.1)
    X h.i. (14.14.1)
    $\mu^{-1}(t)$ *irreducible*
      (14.73.2, 5)
    see covering prop.her.
X ε CP iff
    arc if X arcwise connected (14.73.20)
    $\mu^{-1}(t)$ *irreducible*
      (14.73.3, 5)
    σ_t *1-1 at X (14.73.9)*
      see covering prop.her., homeo., union
    X ε CP(μ), some μ *(14.73.4)*
X ε CP ⇒
    Class[W], X in (14.73.21)
      converse? (14.73.25)
    decomposable, strong Whitney-rev; p.454
    indecomposable, Whitney property (14.14.2)
    irreducible, $\mu^{-1}(t)$ *is (14.73.1, 5)*
    triod, X not (14.73.18)
      X ⊃ *triod, ex.;pp.493,495*
    unicoherent, X is (14.14.1)
    $\mu^{-1}(t)$ *is? (14.73.22)*
    not her.unicoh.,ex.;p.496

see covering prop.her.,...rel. μ,*C*-smooth,irreducible*
COVERING PROPERTY HEREDITARILY, CPH; *p.486*
  at t iff... (14.76.5)
  independent of μ *(14.73.6)*
  open image not, ex., (14.73.26)
  question (14.73.27)
  Whitney property (14.76.8)
    is CP? (14.76.9)
  X ε CPH if
    X chainable (14.73.12)
    X circle-like non-planar (14.73.17)
    X h.i. (14.73.12)
  X ε CPH iff
    X chainable if X her.decomposable (14.73.19)
    X ε CPH(μ), some μ *(14.73.6)*
    $\mu^{-1}(t)$ *her.irr. (14.76.8)*
    Ψ_t *onto (14.76.11)*
    σ_t *homeomorphism (14.73.10)*
    see covering property, homeomorphism, union
  X ε CPH ⇒
    a-triodic X (14.73.18)
    C*-smooth X (14.76.6)
    her.unicoherent X; p.496
    $\mu^{-1}([t,\mu(X)]) \cong C(\mu^{-1}(t))$ *(14.73.11)*
    Ψ_t *homeo.onto (14.76.3)*
    Whitney hyp.,X has(14.73.29)
  see covering property, ...relative μ, *Whitney stable*
COVERING PROPERTY RELATIVE TO μ, **CP(**μ**);** *p.479*

at t_o if $\mu^{-1}(t_o)$ irreducible (14.73.2)
    converse false,ex.; p.481
    iff CP (14.73.4)
    see covering property,...her., irreducible
Cr_ε ; p.614
    ε-crooked arc (19.27.1)
C*-SMOOTH (15.5)
    arcw.conn. $\Rightarrow$ dendroid (15.19)
        examples (15.20)
    at A, X is (15.5)
        if $A \in CP$; p.530
        iff Ψ_t cts.at A (14.76.4)
    at dense G_δ, X is (15.5)
    chainable are (14.76.6),(15.13)
    circle-like (14.76.6), (15.15-16)
    CPH $\Rightarrow$ C*-smooth (14.76.6)
    h.i.are(1.207.8), (14.76.6)
    hereditary (15.6)
    homogeneous(planar) $\Rightarrow$ indecomposable(h.i.)? (15.21)
    iff Ψ_t continuous (14.76.4)
    locally connected (15.7, 11)
    unicoh.if C*-smooth? (15.18)
        partial answers(15.17, 19)
    see C*, the function
c-SPACE (5.18-19)
CUBE
    see convex Hilbert cubes, Hilbert cube, n-cell
CUT POINT (0.33)
    arcw.acc.theorem (12.11)
        generalizations (12.27-28)
    of $\mu^{-1}(t)$; pp.475-477

rational $\supset$ subc.with (12.12)
    her.decomposable$\supset$...? (12.20)
see nondisconnecting
CUT WIRE THEOREM; p.625, (20.6)

D

DECOMPOSABLE CONTINUUM, X (0.30)
    arcw.acc.from $2^X \setminus C_1(X)$
        no $\{x\}$ is, example (12.10)
    see arcwise accessible...
    arcw.conn.if image of $\mu^{-1}(t)$ (14.59)
    $C(X) \not\ni Cone(X)$, example (1.119)
    see embedding in C(X)
    $C(X) \supset$ 2-cell (1.145)
    if $C(X)$ or $2^X \cong$ prod.(10.7, 25)
    see product...
    if $\mu^{-1}(t_o)$ decomposable and irreducible (14.72.3)
        necessity of irr.; p.477
    $\mu^{-1}(t)$ arcwise connected, $t \geq t_a$ (14.27)
        arc if X chainable (14.31)
        arcless, example (14.55.2)
        circle if X proper circle-like (14.31)
    $\mu^{-1}(t)$ decomposable (14.11)
    strong Whitney-rev.for $X \in CP$; p.454
    Whitney property (14.11)
    Whitney-rev.,not; p.454
    see her.decomposable continuum, irreducible,rational continuum,weak Whitney stable
DECOMPOSITION
    $\mu^{-1}(t)$ is if X h.i. (1.78, 80)
    set-theoretic (1.77)

theorem for terminal subcontinua (1.205.2)

topology and hyp.top. (1.77)

u.s.c.

   and ctmwse.acc.;pp.396-398 of $R^3 \cong R^3$ (3.0)

   see c-space, monotone cts.decomposition, u-space

DEFORMATION RETRACT

see strong...

DENDRITE *(0.26)*

characterizations (1.68), (1.74.1),(5.9),(15.10-11),(19.17-19),(19.21)

   as S_4 space? (5.15, 17)

⊂ C(X), any X (1.74.1)

contractible, $\mu^{-1}(t)$ is (14.43.10)

Ψ_t continuous; p.507

see dendroid, Hilbert cube factor

DENDROID, D *(0.28)*

arcw.conn. C*-smooth is(15.19)

contractibility of (1.64-69)

   if ⊂ C(X), X h.i. (1.64, 67)

   ⇒ smooth (1.70)

dendrite, D is iff... (1.68), (19.17-19)

dim[C(D)] < ∞ iff D finite graph (10.5.1)

ε-retraction to finite graph? (7.8.3)

fixed point prop.for hyperspaces(7.8.1-3),(7.10)

rigid selection for C(D) (5.12)

smooth (1.68)

   basic facts, examples (1.68)-(1.68.1),(15.20)

C(D), 2^D have fixed point property (7.8.2)

chars.(5.12), (19.14-15)

if D ⊂ C(X), X h.i. (1.70)

   converse (1.73-74)

selection for C(D) (5.11-12)

tree-like, D is (0.28)

see fan, selection(for C(X)), weakly smooth dendroid

DETERMINED

see C-determined

DIAMETER MAP *(0.41), (1.211)*

continuous (1.211.1)

on C(X)

   monotone,not necessarily open (1.211.2)

on 2^X

   not monotone, ex. (1.211.3)

   not open, examples (1.211.3), (14.67)

   question (14.68)

DIMENSION *(0.44)*

of Cone(X) (8.0)

of hyperspaces

   see below

DIM[C(X)]

and $dim[\mu^{-1}(t)]$ (2.10-13)

at certain points; pp.224-226

   $dim_A[C(X)] < \infty$ all A & $dim[C(X)] = \infty$, example (1.112)

   if X locally conn. (1.107)

≤ n, suff.conds. (2.0), (2.14)

> dim[X] if dim[X] < ∞ (2.6)

≥ 2 (2.1)

= 2 if X

*chainable or circle-like
        (2.1)*
*Cook's continua; pp.216-217,
        223, 235*
h.i. ⊂ R^2 *(2.8)*
   *other conditions (2.2.2),
        (14.17.1)*
h.i. 1-dimensional? (2.7)
h.i. tree-like (14.17.1)
= 2 ⇒ *C(X) Cantor mfld.(2.15)*
= 2 *or* ∞ *if X h.i. (2.2)*
= ∞ *if X*
  ⊃ *product (2.5)*
  = *2-dim.? (2.4)*
  ≥ *2-dim. & arcw.conn. (2.5)*
  ≥ *2-dim. & h.i. (1.85),(2.2)*
  ≥ *3-dim. (1.86), (2.3)*
  locally connected & not finite graph (1.109-111)

DIM[$\mu^{-1}(t)$]
  and dim[C(X)] (2.10-13)
  = ∞ *(1.85-87)*
  related facts
    see embedding in $\mu^{-1}(t)$

DIM[2^X] = ∞ *(1.95)*

DIMENSIONALLY HOMOGENEOUS; p.225
  facts, questions; pp.225-226

DISCONNECTING
  continuumwise (0.34)
  *see arcw.disconnecting,cut pt.,
        noncut pt.,nondisconnecting*

DYER-HAMSTROM THEOREM *(3.0)*

E

ε-CROOKED ARC *(19.27.1)*
  space of..., Cr_ε ; p.614

ε-MAP *(1.161)*
  C(X) into R^2 (3.10-11)
  *dendroid onto finite graph,
        ε-retraction? (7.8.3)*
  n_t *(1.80)*
  projs.of inverse limit (1.162)
  X to S^1 nonessential ⇒ *X chainable (1.194)*

EMBEDDING *(0.46)*
  any X in continuum ε *CP; p.495*
  *C(X)\\$F_1(X)$ by $\hat{f}$ (0.51),
        (0.67.6), (14.34)*
  see embedding in...

EMBEDDING IN C(X)
  *cones (1.71-72), (1.118-121),
        see Cone(X) $\cong$ C(X)*
  dendrite (1.74.1)
    hairy point (1.210.2)
  *dendroid (1.64),(1.67)-(1.74.1),
        (1.207.3)*
  fan (1.73)
  I_∞ *(1.103, 111, 148),(1.210.1),
        (2.5)*
  locally connected X (1.111)
  question (1.148)
  relation to $\dim_Y[C(X)]$;p.225
  see Hilbert cube
  n-cell
    n = 1 (1.8-14)
    n = 2 (1.145-146)
    *n ≥ 2 (1.100-102),(1.107),
        (1.210.1)*
    question (1.147)
    see embedding in $\mu^{-1}(t)$
  *product (1.122-130)
        see product...*
  see embedding in 2^X

EMBEDDING IN ℓ_2
  (hyps., H_d) isometric? (3.12)
    discussion; pp.239-240
EMBEDDING IN $\mu^{-1}(t)$
  arc (14.8.1)
    see arc, arcwise connected continuum
  (n-1)-cell (14.32-33)
EMBEDDING IN R^1
  loc.cpt.sep.metric (5.4-5)
EMBEDDING IN R^2
  C(X) (3.9)
    by ε-map (3.10-11)
  $\mu^{-1}(t)$ (1.213.7), (14.18)
  see planarity
EMBEDDING IN R^3, C(X)
  Conner's question; p.234
  ⇒ X planar? (3.5)
  X chainable (3.2)
  X circle-like (3.3, 6)
    X pseudo-solenoid; p.236
    X solenoid; p.235
  X Cook's continua; p.235
  X h.i. (2.9),(3.1, 4)
EMBEDDING IN R^4, C(X)
  X circle-like; pp.235, 239
    X pseudo-solenoid; p.239
  X h.i. (2.8)
EMBEDDING IN R^n, C(X)
  necessary conditions (3.8)
  see embedding in R^1,..., embedding in R^4
EMBEDDING IN 2^X
  decompositions of X;pp.396-398
  I_∞ (1.95)
    see embedding in C(X)

END OF COMPACTIFICATION
  of R^1 (8.24), p.323
END POINT OF ARC (0.18)
  of order arc (1.6)
  see terminal point
ESSENTIAL MAP (0.67.4)
  onto S^1 ⇒ wkly.confluent(1.193)
    converse false,etc.(0.67.4), (1.202)
  onto figure eight, not weakly confluent (0.67.4)
  see homotopic, nonessential map
ESSENTIALLY DIFFERENT WHITNEY MAPS (14.43)
  examples (14.43.1-7)
    Hilbert cube (14.43.11)
    related ex.for 2^X (14.62)
  for 1-dim.polyhedron (14.43.5)
  questions (14.43.11-12)
EUCLIDEAN n-SPACE, R^n (0.19)
  see n-space
EXTENDED LONG LINE, L_0 ; p.560
  hyps.not homogeneous (17.4)

F

F_μ (16.12)
  cts.iff X ε prop[κ] (16.14)
  comments (3.1.1), (16.13)
f_ω (1.30)
  homeomorphism (1.30)
  nonexp., not isometry (1.31.1)
$F_1(X)$ (0.48)
  exercises (0.66.3-5),(0.66.7)
  ≅ X (0.48)
  isometric to X (0.63.4)

*retract of hyps.iff X r-image
of hyps.? (6.28)*
see $F_2(X)$, $F_n(X)$, *retraction,
strong deformation retract*

$F_2(X)$ *(0.48)*
$F_2(C(X))$ & $C_2(X)$ *(0.71.2)*
$F_2(S^1)$ *(0.71.1)*
selection for (5.2-4)
see $F_1(X)$, $F_n(X)$, *retraction*

$F_n(X)$ *(0.48)*
exercises (0.66.6-7)
see $F_1(X)$, $F_2(X)$

FAN; *p.119*
*Cantor fan
see cone over Cantor set
C-determined? (0.62)
fixed point property
hyperspaces have (7.8.2)
smooth (1.68); also see proof
of (1.73)
C-determined (0.62)
hyperspace models (1.210.2)
iff ⊂ C(X), X h.i. (1.73)
see dendroid*

FINITE GRAPH *(1.108)*
*admits essestially different
Whitney maps(14.43.5)
C(X) polyh.iff X finite graph
(1.113)
polyhedral structure
(1.114-117)
C-determined (0.59)
dim[C(X)] < ∞ iff X finite
graph
if X dendroid (10.5.1)
if X loc.conn.(1.109-111)*

$\mu^{-1}(t)$ *finite graph* ⇒ *X arc or
circle (1.213.8)*
see *inverse limit(of polyhedra)*

FINITE TOPOLOGY *(0.12)*
see *Vietoris topology*

FIXED POINT PROPERTY; *p.291*
*C(X) does not have (7.2-3)
C(X) has if
C(X) quasi-complex; p.294
2^X has (& converse)? (7.12)
X chainable (7.1)
X circle-like (7.5)
X dendroid (7.8.1-3)
X fan (7.8.2)
X h.i. (7.6), (19.25)
arcw.conn.subc.C(X);p.296
X locally connected; p.295
X 1-dim.with f.p.p.? (7.8.1)
X smooth dendroid (7.8.2)
X tree-like? (7.8)*
$\mu^{-1}(t)$ *has if X chainable(7.13)*
2^X *has if
X fan, smooth dendroid(7.8.2)
X locally connected; p.295
questions (7.9-12)
X has, suff.conditions (19.24)
Knaster's question; p.291
answer (7.3)
multi-valued (0.73)
Whitney prop.,not (14.30),
(14.43.6)
question (7.13)*

FREE ARC *(0.18)*
$\psi[X]$ *(10.19-21)*

FUNDAMENTAL RETRACT
  hyperspaces are (1.182, 184)

G

$\Gamma(X)$ (1.27.1)
  compact (1.34),(1.203.2)
  $\cong S(\omega)$ (1.30), (1.31.1)
  $\cong 2^X$? (1.27.3)
  homotopy type =...of 2^X
      (1.27.3)
  locally connected $\Rightarrow$
    $\Gamma(X) \in AR$? $\Gamma(X) \cong I_\infty$?
        (1.27.3)
    $\Gamma(X)$ locally connected &
        converse; p.209
  strong deformation retract of
      $F_1(2^X)$ is (1.203.3)
  see $S(\omega)$
g-CONTRACTIBLE; pp.246-247
  hyperspaces (4.10-12)
  loc.conn.continuum is; p.247
  mapping theorems (4.13),
      (4.13.1), (4.14)
GENERALIZED ARC (0.18)
  char.by selections (5.3)
  hyps.of Hausdorff continua Y
    arcw.connected by (1.14)
    loc.arcw.conn.at Y (1.136)
    see Hausdorff continuum
  see extended long line
GENERALIZED HAUSDORFF METRIC,
H_∞ ; p.620
  discussion; pp.620-622
  related metrics (0.63.6),
      (0.66.2)
GRAPH
  see finite graph

GROUPOID, TOPOLOGICAL; p.285

H

H, H_d (0.1),(0.63.6), (19.1)
  see Hausdorff metric
H_∞ ; pp.620-622
  related metrics(0.63.6),(0.66.2)
HAHN-MAZURKIEWICZ THEOREM (0.69.5)
HAIRY POINT (1.210.2)
HAUSDORFF CONTINUUM, Y (0.16)
  gen.arc iff selec.exists (5.3)
  hyperspaces
    acyclic (1.178, 180)
    aposyndetic (1.138)
    arcwise connected (1.14)
    contractible, C(Y) not & Y
        loc.conn.; ex. (16.41)
    homog. $\Rightarrow$ Y loc.conn.(17.1)
        discussion;pp.565-566
    inverse limits (1.170)
        Y is...;pp.159-160,(1.159),
            (1.165)
    loc.arcw.conn.at Y (1.136)
    loc.conn.iff Y is (1.208.1)
    topology for
        see Vietoris topology
  map onto cone over Cantor set;
      p.247
  see cone over Cantor set
      (preimages of)
  metrizable if $F_1(Y)$ is G_δ
      (0.66.3)
HAUSDORFF METRIC, H (0.1)
  convergence w.r.t. H (0.7)
    and 0-reg.conv.;pp.520-521,
        (15.10), (19.36)
    see convergence of sets,L(M)

exercises (0.63)
for *CL(S), S bdd.metric space*
  topology depends on metric
    (0.66.2)
for 2^M, *M metric space(0.63.6),*
    (19.1)
generalized, H_∞ *;pp.620-622*
proof H is a metric (0.2)
top. = Vietoris top. (0.13)
see Vietoris topology
HEREDITARILY
  see covering prop.her., her.
    decomposable continuum,etc.
HEREDITARILY DECOMPOSABLE CONTINUUM, X *(0.30)*
  arcw.acc.from $2^X \setminus C_1(X)$
    (12.14-25)
    questions (12.19-25)
  $Cone(X) \cong C(X)$ *(8.23),pp.322-331*
  CPH, X has iff X chainable
    (14.73.19)
  hyp.conds.$\Rightarrow$ *X chainable or circle-like (11.16)*
  if $C(X) < \infty\text{-}dim.product$ *(10.8)*
  strong Whitney-rev.? (14.57)
  see rational continuum
HEREDITARILY INDECOMPOSABLE CONTINUUM, X *(0.31)*
  arcs in C(X) (1.59-62)
  arcw.acc.from $2^X \setminus C_1(X)$
    chars.of h.i. (12.30-31)
    no {x} is (12.9-10)
  C-determined (0.60)
  C-H continuum $< \infty\text{-}dim.$, $X \notin$
    (8.17-18)

characterization theorems
  $a(C(X)) \cup F_1(C(X))$ *compact*
    (19.23)
  arcless, $\mu^{-1}(t)$ *all t*
    (14.55.2)
  arcw.acc. (12.30-31)
  arcw.disconnecting (11.15)
  $C(X)\natural...$*(1.118),(1.122-125),*
    (1.146), (1.207.2)
  C(X) u.arcw.conn.(1.61-62)
    u.order arcs (1.59)
  terminal, each subc. (1.58)
  $\cup \Gamma \varepsilon \Gamma$, *arcs* $\Gamma \subset C(X)$
    (1.207.1)
Class[W], X in; p.497
confluent maps (1.207.5-6)
contractible hyps. (1.75-76)
  arcw.conn.subc.C(X) (1.63)
CP, X has (14.14.1)
CPH, X has (14.73.12)
C*-smooth, X is (1.207.8),
    (14.76.6)
  question (15.21)
c-space, X not; p.264
dim[C(X)] (1.85),(2.2),(2.7-8),
    (14.17.1)
  $dim[\mu^{-1}(t)] = \infty$ *(1.85)*
dim[X], related to
  mono.open maps (1.82-83)
emb.C(X) in Cone(X),not
    (1.207.7)
emb.C(X) in R^3 *if* $X \subset R^2$ *(2.9)*
  special cases (3.1, 4)
emb.C(X) in R^4 *if* $X \subset R^2$ *(2.8)*
embedding in C(X)
  cone over Cantor set is
    (1.71-72)

cones over continua, not (1.118)

dendroids (1.64), (1.67)-(1.74.1), (1.207.3)

fan iff smooth (1.73)

products (1.122-125)

exercises (1.207)

existence of

  in (n+1)-dim.continuum (1.53)

  ∞-*dim.h.i.* (1.83-84)

  most continua are (19.27)

    pseudo-arcs (19.30)

fixed point property for

  $C(X)$ (7.6), (19.25)

    arcw.conn.subc.of; p.296

  2^X? (7.10)

monotone &/or open maps (1.80), (1.82-83), (1.207.4-5), (1.207.7), (2.2, 7, 8)

prop[κ], X *has* (16.27)

S_4 *space,* $X \notin$ (5.14)

strong Whitney-rev. (14.54)

union thms. (1.60), (1.207.1)

  see union

u-space, $X \notin$; p.264

Whitney hyp., X *has* (14.73.29)

  stronger fact (14.73.14)

Whitney map selection continuum, $X \notin$; p.264

Whitney property (14.1)

see hi(M), indecomposable continuum, pseudo-arc, pseudo-solenoid, tree-like continuum

HEREDITARILY IRREDUCIBLE CONTINUUM; p.490

  $\mu^{-1}(t)$ *is, some* $t \Rightarrow \sigma_t$ *homeo.* (14.73.13)

  converse false; p.491

  $\mu^{-1}(t)$ *is iff* $X \in CPH$ (14.76.8)

  see irreducible

HEREDITARILY IRREDUCIBLE MAP (1.212.3)

  f *is iff* $\hat{f}$ *light* (1.212.3)

  see induced map

HEREDITARILY UNICOHERENT CONTINUUM, X (0.27)

  if $X \in CPH$; p.496

    not if $X \in CP$, *example;* p.496

  iff lim inf is conn. (0.64.4)

  see unicoherent

hi(M) (19.1)

  G_δ *in* $C(M)$ (19.26)

hi(Ω^*) (19.1.9)

  dense G_δ *in* $C(\Omega^*)$ (19.27-28)

  questions (19.32-33)

  topologically complete (19.33)

  see her.indecomposable continuum, $P(\Omega^*)$

HILBERT CUBE (0.40)

  admits ess.diff.Whitney maps (14.43.11)

    related questions (14.38), (14.42.2-3), (14.43.11-12)

  char. of & appls. (1.214)

  $\subset a(\Omega^*), s(\Omega^*), etc.$ (19.5-7)

$\subset C(X)$

    see embedding in $C(X)$

$\subset 2^X$ (1.95)

$\cong C(X)$, 2^X, etc. (1.94-99),
    (1.214.2-3), (17.2-3)

$\cong C_{\{v\}}(Cone(E))$ (1.214.3)

$\cong cc(X)$

    see cc-hyperspace($\cong I_\infty$)

$\cong \Gamma(X)$ if X loc.conn.?(1.27.3)

nested ∩ of...,hyps.are(1.171)

touching set? (13.3)

Whitney prop., not (14.42.1)

    questions (14.38),(14.42.2)-
        (14.42.3),(14.43.11-12)

Whitney stable, not (14.42.1)

    questions (14.38),(14.42.2)-
        (14.42.3),(14.43.11-12)

Wojdyslawski's question (1.94)

see convex Hilbert cubes, Hilbert cube factor,...geometry,...manifold

HILBERT CUBE FACTOR

AR's, Edwards' Thm.(1.214.2)

$C(X)$ is iff X loc.conn.(1.98)

    relative version (1.99.2)

    special case, X dendrite; p.137

    related idea:$C(X) \times I_\infty \cong 2^X$; pp.355-356

see Hilbert cube,...geometry, ...manifold

HILBERT CUBE GEOMETRY (18.36-39)

see Hilbert cube,...factor, ...manifold

HILBERT CUBE MANIFOLD

$cc(X)$ (18.27.2)

space of Z-sets in (19.42)

see Hilbert cube,...factor, ...geometry, Hilbert space manifold

HILBERT SPACE, ℓ_2 (0.40)

$a(\ell_2), s(\ell_2), t(\ell_2), \theta(\ell_2) \supset I_\infty$ (19.6)

questions (19.10-12)

$C(\ell_2)$, $2^{\ell_2} \cong \ell_2$ (19.40)

generalization (19.41)

$cc(\ell_2) \cong \ell_2$? (18.9)

$ch(\ell_2)$, $hi(\ell_2)$, $P(\ell_2)$

    dense G_δ in $C(\ell_2)$ (19.27-31)

    questions (19.32-33)

    topologically complete(19.33)

see complete, emb.in ℓ_2, Hilbert space manifold

HILBERT SPACE MANIFOLD, M

$C(M)$, $2^M \cong \ell_2$ (19.41)

see Hilbert cube manifold, Hilbert space

HOMEOMORPHISM

$C(X) \cong 2^X \Rightarrow X$ locally connected? (10.27)

$C(X) \times I_\infty \cong 2^X \Rightarrow X$ locally connected? (10.26)

exercises (0.70)

f_ω of $S(\omega)$ onto $\Gamma(X)$ (1.30), (1.31.1)

$\Gamma(X) \cong 2^X$? (1.27.3)

ω on compact chain (1.3)

of n-tuples; p.348

Rogers homeo.; p.303, (8.7)

    see cone=hyp.property

σ_t ,each t or some t

    see σ_t

segment, if not constant (1.18)

$X \cong Y \Rightarrow 2^X \cong 2^Y$ and $C(X) \cong C(Y)$ (0.52)

    converse false (0.58)

    see C-determined

  see near-homeo., Ψ_t, Rogers homeo.,weak Whitney stable, Whitney stable

HOMOGENEOUS; p.563

  $a(\ell_2), s(\ell_2), t(\ell_2), \theta(\ell_2)?$ (19.10, 12)

  $a(R^n), s(R^n), t(R^n), \theta(R^n)$ are for $n=2$ (19.4)

    for $n \geq 3$? (19.9, 12)

  $ch(\Omega^*), hi(\Omega^*), P(\Omega^*)$ for $\Omega^* = \ell_2$ or $R^n(n \geq 2)$? (19.32)

  C^*-smooth($+$ planar) $\Rightarrow$ indec.(h.i.)? (15.21)

  dimensionally homog.;pp.225-226

  hyperspaces; pp.563-566

  $\Rightarrow$ prop[κ] & contractible hyperspaces (16.26)

  Whitney property? (14.36)

HOMOTOPIC (16.1)

  b and e are (1.203.3)

  induced maps (0.67.5)

  see contractibility of,essential map,g-contr.,homotopy metric,homotopy type,nonessential map,segment homotopy

HOMOTOPY METRIC, ρ_h ; pp.619-620

HOMOTOPY TYPE

  and homotopy metric;pp.619-620

  of $cc(X)=...$of X (18.31-32)

  of $\Gamma(X)=...$of 2^X (1.27.3)

HYPER-ONTO REPRESENTATION (1.186)

  example, motivating (1.185)

  iff bonding maps weakly confluent (1.190)

  X has $\Rightarrow C(X)$ has minimal representation (1.189)

  X has if

    bonding maps monotone or open (1.192)

    X any continuum? (1.200)

    X chainable (1.191)

    X circle-like (1.195)

    X Menger univ.curve; p.187

    X Sierpinski universal curve; p.187

  see minimal rep., inverse limit (of polyhedra)

HYPERSPACES (0.3)

  as nested ∩ Hilbert cubes(1.171)

  $C(X) \cong 2^X \Rightarrow X$ locally connected? (10.27)

  of Hausdorff continua

    see Hausdorff continuum

  other spaces of sets

    see cc-hyp.,space,space of

  properties of

    see various listings

  special types of

    see invariant Whitney hyp., weak Whitney hyp., Whitney hyperspace

  see $C(X)$, 2^X

I

I_∞ (0.40)

  see Hilbert cube

INDECOMPOSABLE CONTINUUM,X (0.31)

  Buckethandle is (1.209.3)

    open map onto arc (1.207.5)

characterizations

    $C(X)\setminus\{X\}$ *not arcwise connected (1.51), (11.1)*

      arc components of (1.52)

    $2^X\setminus\{X\}$ *not arcw.conn. (11.4)*

    $X(A,\mu,t) \neq \mu^{-1}(t)$ *(1.213.4)*

conn.im kleinen of hyps.only at X (1.139-142)

existence (1.209.2-4)

    see her.indec.continuum

if homog.C*-smooth? (15.21)

monotone image is (1.207.4)

open image not, ex. (1.207.5)

solenoids (1.209.4)

    open map onto S^1 (1.207.5)

strong Whitney-rev.; p.454, (14.46)

union theorem (1.50)

    see union

Whitney property for X chainable (14.13)

    not for X circle-like;p.413, (14.12)

Whitney prop.for $X \in CP$ (14.14.2)

see her.indecomposable continuum

INDECOMPOSABLE INVERSE SEQUENCE
    (1.209.2)

theorem (1.209.2)

applications (1.209.3-4)

INDUCED MAP, $\hat{f}$ *(0.49)*

continuous (0.49), (0.67.1)

embedding of $C(X)\setminus F_1(X)$ (0.51), (0.67.6), (14.34)

exercises (0.67), (1.212)

homotopy props. (0.67.5)

image not disconnected by 0-dim. set; p.225

isometry iff f is (0.63.7)

light iff

    f her.irreducible (1.212.3)

    $\mu \circ \hat{f}$ *Whitney map (1.213.6)*

      rel.facts (0.51), (0.68.6)

mono.of (1.212.1-2)

nonexpansive iff f is (0.63.7)

onto iff f weakly confluent (0.49.1)

    see essential map,hyper-onto rep.,weakly confluent map, etc.

see nondisconnecting

INVARIANT WHITNEY HYPERSPACE
    (14.73.31)

facts (14.73.31-34)

question (14.73.34)

see weak Whitney hyp., Whitney hyperspace

INVERSE LIMIT *(1.150)*

as nested $\cap$... (1.152-153)

    hyperspaces as nested $\cap$ Hilbert cubes (1.171)

base for (1.156)

    hyperspaces of (1.166-167)

bonding maps (1.149)

    commute with projs. (1.155)

cpt.,conn.,non-$\emptyset$ (1.157-158)

cpt. $\subset$ as inverse limit (1.160)

exercises (1.209)

hyps.of, main thm. (1.169-170)

    dim[C(X)] $\leq$... (2.0)

interior approx.thm. (1.199)

of polyhedra P_n

    any X is (1.164-165)

$C(X)$ is of $C(P_n)$, $dim[P_n]=1$
(1.187)
projections of (1.154)
commute bonding maps (1.155)
ε-maps (1.162)
rationality of (12.17.1)
see hyper-onto rep., inverse sequence, minimal rep.

INVERSE SEQUENCE (1.149)
see indecomposable inverse sequence, inverse limit

IRREDUCIBLE; p.476
about terminal pts. a & b, X is
struc. of $C(X/_{\{a,b\}})$ (14.34)
$\mu^{-1}(t)$ is, each t
see covering property
$\mu^{-1}(t_o)$ is, some t_o
and decomposable $\Rightarrow$ X dec. (14.72.3)
nec. of $\mu^{-1}(t_o)$ irr.; p.477
if X decomposable, unicoh., not triod? (14.73.24)
$\Rightarrow CP(\mu)$ at t_o (14.73.2)
converse false, ex., p.481
$\Rightarrow$ X irr.? (14.73.23)
member of $\Lambda \subset 2^X$ (0.68.7)
non-unicoherent, ex. (14.73.22)
strong Whitney-rev. (14.55)
related facts, questions (14.55.1), (14.73.23)
X is if
$X \in Class[W]$ (14.73.25)
$X \in CP$ (14.73.1)
see covering prop., cov. prop. her., her. irr. continuum, her. irr. map

ISOMETRY (0.63.4)
and continuity of balls(0.65.6)
f is iff $\hat{f}$ is (0.63.7)
f_1 is (16.6)
f_ω not, example (1.31.1)
from(hyps., H_d) into ℓ_2? (3.12)
discussion; pp.239-240
from X onto $F_1(X)$ (0.63.4)
see nonexpansive map

K

K_d (0.65.1)
main theorem (0.65.1)
other facts (0.65.2-6)

KELLEY'S QUESTIONS
$C(X)$ acyclic? p.177
answer (1.173)
X non-Peanian, $dim[C(X)]=?$ p.211
answer, see $dim[C(X)]$

KERNEL OF Z, ker(Z) (18.10)
see cc-hyp. (starshaped X)

KNASTER CONTINUUM (0.31)
see hereditarily indecomposable continuum

KNASTER'S QUESTION; p.291
answer (7.3)

L

λ CONNECTED CONTINUUM, X (0.30)
ε-map $C(X)$ into R^2 (3.10-11)
strong Whitney-rev.? (14.57)
Whitney property? (14.36)

L(M) (19.1)
$\leq F_{\sigma\delta}$ (19.34)
$= F_{\sigma\delta}$ if $M = B^n$, $n \geq 3$ (19.35)
with ρ_τ; pp.520-521, (19.36)

ℓ_2 *(0.40)*
    see Hilbert space
LIGHT MAP *(1.212.3)*
    see induced map
LIM, LIM INF, LIM SUP *(0.5)*
    exercises *(0.64)*
    see convergence of sets, Hausdorff metric
LINEAR GRAPH *(1.108)*
    see finite graph
LINEAR ORDER AND SELECTIONS *(5.2)*
    topological well-ordering;p.254
LINK OF A CHAIN *(0.22)*
LOCALLY ARCWISE CONNECTED *(1.133)*
    at a pt.,hyp.thms.(1.133-136)
    see conn.im kleinen, loc.conn. at a point, locally connected continuum
LOCALLY CONNECTED AT A POINT *(1.88)*
    c.i.k., relations *(1.89, 91)*
    $C(X)$ at M, n.a.s.c.? *(1.144)*
    2^X at M
        n.a.s.c. *(1.132, 134)*
        suff.conditions *(1.135,137)*
    see conn.im kleinen, loc.arcwise connected, locally connected continuum
LOCALLY CONNECTED CONTINUUM, X *(1.90)*
    $a(X) \cup F_1(X)$ compact iff X dendrite *(19.21)*
    AR, hyperspaces are *(1.96)*
    $C(X) \supset Cone(X)$ *(1.120)*
    $C(X) \supset I_\infty$ iff $X \neq$ finite graph *(1.111)*
    $C(X) \supset n$-cell *(1.107.2)*

$C(X) \supset X \times X$ *(1.127)*
$C(X) \cong$ product*(10.1, 19-21)*
    examples *(10.2, 15-18)*
$C(X) \setminus F_1(X)$, $2^X \setminus F_1(X)$ loc.conn. iff X is *(1.208.2)*
continuity of balls *(0.65.2-5)*
contractible, hyps.are *(0.65.4)*, *(1.93)*, *(16.18)*
    false for Hausdorff, example *(16.41)*
convex metric for *(0.65.4)*
dendrite, X is iff
    $a(X) \cup F_1(X)$ is cpt. *(19.21)*
    C^*-smooth *(15.11)*
    dendroid, smooth at each point *(1.68)*
    dendroid, weakly smooth at each point *(19.17)*
    selection for $C(X)$ *(5.9)*
$dim[C(X)]$ *(1.107-111)*
    dimensionally homog. *(2.16)*
exercises *(1.208)*
fxd.pt.prop.,hyps.have; p.295
$\Gamma(X)$ *(1.27.3)*, p.209
    see $\Gamma(X)$
g-contractible, X is; p.247
Hahn-Mazurkiewicz Thm. *(0.69.5)*
Hilbert cube, hyps.are*(1.97-99)*, *(1.214.2-3)*, *(17.2-3)*
    Wojdyslawski's question*(1.94)*
if $C(X) \cong$ product? *(10.21)*
if $C(X) \cong 2^X$? *(10.27)*
if $C(X) \times I_\infty \cong 2^X$? *(10.26)*
if hyps.homogeneous *(17.1)*
loc.conn.hyps.,iff *(1.92)*, *(1.208.1)*

METRIC SUBJECT INDEX 693

locally contractible hyps.
  (0.65.4),(1.93),(16.18)
maps exist, $\mu^{-1}(t)$ *onto* $\mu^{-1}(s)$;
  p.465
$\psi[X]$ *(10.19-21)*
prop[κ], X has (16.11)
retractions & r-maps between X,
  2^X, $C(X)$ *(6.4-8),*
  (6.11-13)
  retraction chars. (6.12-13)
strong Whitney-rev. (14.47)
$2^X_A \in AR$, *iff (1.214.3)*
union of open $\subset C(X)$ *(1.208.7)*
Whitney property (14.9)
  related fact (14.75)
see connected im kleinen,
  $C_M(X)$, $C^M(X)$, *finite graph,*
  *L(M), loc.arcw.conn., loc.
  conn.at a point, order of
  A in X*

LOCALLY CONTRACTIBLE

hyperspaces are if X loc.conn.
  (0.65.4),(1.93),(16.18)
$\mu^{-1}(t_o)$ *not if Y is 2-cell,*
  example (14.43.7)
$W[C(X)]$, $W[2^X]$ *are (14.71.3)*

LONG CIRCLE, Σ; *p.560*

$C(\Sigma)$ *not contractible by Hausdorff continua (16.41)*

LONG LINE

see extended long line

M

$\mu^{-1}(t)$, $\mu^{-1}([t, \mu(X)])$
  see Whitney map

MAPPING *(0.45)*

between X, 2^X, $C(X)$; *special
  types? (4.15), (6.1),
  (6.7-8), (6.28-29)*

2^X *onto [0,1]? (14.63-64)
  (14.68-70)*
  see various types of maps
$F_2(X) \cup \Gamma$ *into* $C_1(X)$ *(6.13-21)*
  see mean
$\mu^{-1}(t)$ *onto* $\mu^{-1}(s)$; *pp.465-466*
  *see weak Whitney stable,
  Whitney stable*
2^X *onto C(X), exists (4.1)*
2^X *onto Y & Y onto* 2^X *simultaneously,n.a.s.c. (4.14)*
2^X *or C(X) onto Y*
  questions (4.4-5)
  theorems (4.7-9, 13)
X onto C(X) (4.6)
Y onto 2^X, *n.a.s.c. (4.2)*
  *Y = X or C(X); pp.242-243,
  (4.3)*
*see cone over Cantor set, con-
  tinuity of, union, various
  types of maps*

McDONALD CHEESEBURGER *(18.33)*

MEAN; *p.285*

circle is not (0.71.1)
hyperspaces as means
  C(X) is for what X? (6.21)
  2^X *is, any X; p.287*
retraction char.of (6.17)
  pseudo-mean (6.20),pp.286-287
$X \subset R^2$ *is* $\Rightarrow R^2 \setminus X$ *conn. (6.18)*
  converse false, ex. (6.19)
X is, what X? (6.15),(6.17.1)

MENGER UNIVERSAL CURVE

has hyper-onto rep.; p.187

METRIC

continuum (0.16)
of continuity; p.620

see *convex metric, generalized Haus.metric, Hausdorff metric, homotopy metric, L(M)*

MICHAEL'S QUESTION; *p.254*

answers (5.2-4)

MILLER'S THEOREM; *p.255*

generalized (5.5)

MINIMAL REPRESENTATION, C(X) HAS *(1.188)*

if bonding maps monotone or open (1.192)

if X any continuum? (1.201)

if X circle-like (1.196)

if X has hyper-onto rep. (1.189)

if X polyhedron (1.197)

see *hyper-onto rep., inverse lim.(of polyhedra)*

MONOTONE CONTINUOUS DECOMPOSITION

of C(X) by μ^{-1} (14.44)

of 2^X by ω^{-1}

see *monotone open map(ω is)*

of X if X h.i.$\subset R^2$ (2.8)

see *monotone open map (dimension theorems)*

of X h.i. by each $\mu^{-1}(t)$ (1.80)

topology & hyp.topology (1.77)

MONOTONE MAP *(0.45.1)*

between X, 2^X, C(X) (4.15), (6.6-8), (6.12)

2^X onto [0,1]? (14.63, 70)

see *monotone opem map*

confluent, monotone is (0.45.3)

continuumwise acc.thm. (13.1)

discussion; pp.396-398

diameter map (1.211.2-3)

see *diameter map*

η_t *is (1.80)*

hyper-onto rep., use in (1.192), p.187

images preserve

CP? (14.73.26)

h.i., indec. (1.207.4)

prop[κ] (16.38)

terminal subc. (1.205.1)

induced map $\hat{f}$ (1.212.1-2)

μ *is (14.2),p.453,(14.44),p.466*

ω *is*

if X convex (14.66)

not, example (14.61)

question (14.63)

space of mono.maps; p.555

see *monotone cts.decomposition, monotone open map*

MONOTONE OPEN MAP

between X, 2^X, C(X) (4.15), (6.6-8)

2^X onto [0,1]? (14.70)

see *retraction, r-image*

Cone(X) onto C(X), X h.i. (1.207.7)

dim.thms. (1.82-83), (2.2, 8)

question (2.7)

η_t *is (1.80)*

image Q

$\subset C(Z)$; *p.214*

Q image of cpt.tot.-disc.set, n.a.s.c. (1.81)

μ *is; p.453, (14.44), p.466*

ω *is*

if X convex (14.66)

not, examples (14.61, 67)

see *diameter map, monotone map, monotone cts.decomposition, open map*

MONOTONE u.s.c. DECOMPOSITION
  see decomposition, monotone map
MOVABLE
  hyperspaces are (1.182, 184)
MULTI-VALUED MAP
  contraction (0.73.3),pp.621-622
  fixed points for (0.73)
  see upper semi-cts.functions

N

n-CELL (0.19)
  $\subset C(X)$
    see embedding in $C(X)$
  $\subset \mu^{-1}(t)$
    see embedding in $\mu^{-1}(t)$
  $\cong C(X)$ (1.208.4)
  see two-cell
NEAR-HOMEOMORPHISM; p.556
  confluent if onto $Y \in prop[\kappa]$
      (16.32)
n-FOLD SYMMETRIC PRODUCT (0.48)
  exercises (0.71)
  see $F_1(X)$, $F_2(X)$, $F_n(X)$
n-ODD (0.21)
  $\subset X \Rightarrow$ n-cell $\subset C(X)$
      (1.100-102)
    converse? (1.147)
    related results (1.103, 107),
      (1.210.1),(14.32-33)
  $\subset X \Rightarrow$ (n-1)-cell $\subset \mu^{-1}(t_o)$
      (14.33)
  see simple triod, triod
NON-ALTERNATING MAP (0.45.5)
  between X, 2^X, $C(X)$? (4.15),
      (6.8)
  2^X onto $[0,1]$? (14.70)

NON-CONTRACTIBLE
  hyperspaces,examples (16.22, 41)
  $\mu^{-1}(t_o)$, X 2-cell (14.43.6)
    question (14.43.12)
  see non-locally contractible
  subcontinuum of contractible
      dendroid,example (1.66)
NONCUT POINT (0.33)
  see nondisconnecting
NONDEGENERATE (0.15)
NONDISCONNECTING
  C(X) by Λ
    $\Lambda \subset \mu^{-1}(t) \neq \Lambda$ (1.204.2)
    Λ point (1.204.1)
    Λ zero-dimensional (2.15)
  $\hat{f}[C(X)]$ by 0-dim.set; p.225
  2^X by any $\subset C_1(X)$ (1.204.3)
  see arcwise disconnecting, cut
      point
NONESSENTIAL MAP (0.67.4)
  ε-map X to $S^1 \Rightarrow$ X chainable
      (1.194)
  f is $\Rightarrow \hat{f}$ is (0.67.5)
  relation to weakly confluent map
  see essential map
NONEXPANSIVE MAP (0.63.5)
  conv is (0.63.5)
  f_ω is (1.31.1)
  $\hat{f}$ is iff f is (0.63.7)
  union, $\cup$, is (1.48)
  see contraction(multi-valued),
      isometry
NON-LOCALLY CONTRACTIBLE
  $\mu^{-1}(t_o)$, X $\in$ ANR & Y 2-cell
      (14.43.7)

NONPLANARITY (1.213.7)
  Whitney property for circle-like (14.18)
  Whitney property, not (1.213.7)
  see circle-like continuum (non-planar), planarity
NON-TRIVIAL COSELECTION (5.22)
  discussion; pp.265-266
NON-UNICOHERENT
  irreducible, example (14.73.22)
  not Whitney property (14.29)
  not Whitney-rev.; p.454
  see unicoherent
NON-WHITNEY PROPERTY
  see Whitney property, not
NON-WHITNEY-REVERSIBLE PROPERTY
  see Whitney-rev.prop.(not)
NOWHERE DENSE; p.515
  & arcw. disconnecting (11.6, 9)
n-SEGPLEX; p.429
  theorem (14.32)
n-SPACE, R^n (0.19)
  $C(R^n)$ & 2^{R^n}
    complete (0.63.6)
    $\cong I_\infty \setminus \{point\}$ (19.38-39)
      generalization (19.44)
  $ca(R^n)$, $n \geq 2$ (19.13)
  $cas(R^n)$; p.569, (18.2)
  $cc(R^n)$, $n \geq 2$ (18.8)
    $n = 1$; p.569
    see cc-hyperspace
  $\cong C(X) \setminus F_1(X)$ (1.208.3)
  see emb.in R^1,...,emb.in R^n, Hilbert space, space

n-SPHERE, S^n (0.19)
  see circle

O

OPEN MAP (0.45.2)
  between X, 2^X, $C(X)$? (4.15), (6.8)
  2^X onto $[0,1]$? (14.64), (14.68-70)
  Buckethandle onto arc (1.207.5)
  confluent, open is (0.45.3)
  diameter map (1.211.2-3), (14.67-68)
  η_t is (1.80)
  hyper-onto rep., use in (1.192), p.187
  images do not preserve
    CP or CPH (14.73.26)
    indecomposable (1.207.5)
  images preserve
    contractible hyps. (16.39)
      question (16.40)
    h.i. (1.207.5)
    prop[κ] (16.38)
  μ is (1.213.1), p.453, (14.44), p.466
  ω is
    not, examples (14.61, 67)
    theorems (14.65-66)
    questions (14.64, 69)
  solenoid onto circle (1.207.5)
  see monotone open map
ORDER ARC (1.2)
  $C_M(X)$ is, n.a.s.c. (1.56-57)
  see terminal subcontinuum

⊂ C(X) if begins in C(X) (1.11)
    segment (1.26)
char.as chain (1.4)
end points of (1.5-7)
    functions, b & e (1.203.3)
example of (1.1)
exercises (1.203)
exists, n.a.s.c. (1.8, 25)
general relations to segments
    (1.19-25),(1.30-31)
    see S(ω)
parameterization by segments
    depends on ω (1.23-24)
    unique for fxd. ω (1.31.2)
uniqueness of (1.57, 59),
    (1.203.1, 4)
    number of (1.203.1)
see Γ(X), S(ω), segment

ORDER OF A IN X (1.104)
and dim[C(X)] (1.107-110)
= order in $X/_A$ (1.105.1)
Menger theorem (1.106)

P

$ψ_p$; p.579
    properties of (18.30, 32)
$Ψ_t$ (14.76.0)
    continuity of; p.506,(14.76.1),
        (14.76.3-5)
    level-preserving thm.(14.76.7)
    one-to-one (14.76.2)
    onto $C(μ^{-1}(t))$ iff X ε CPH
        (14.76.11)
    other properties,appls.;
        pp.506-511
        see $σ_t$

P(Ω*) (19.1)
    dense $G_δ$ in C(Ω*) (19.30-31)
    questions (19.32-33)
    topologically complete (19.33)
    see ch(M), hi(M), pseudo-arc

PERFECT (0.69.6)
M is iff 2^M is (0.69.8)
see Cantor set

PLANARITY (1.213.7)
of C(X) (3.9)
of homogeneous C*-smooth X ⇒
    X h.i.? (15.21)
Whitney prop.for circle-like
    (14.18)
Whitney property, not (1.213.7)
see circle-like(planar),emb.in
    R^2, nonplanarity

POLYHEDRON
see finite graph, hyper-onto
    rep.,inverse lim.(of poly-
    hedra), minimal rep.

PRODUCT, C(X) IS A; p.339
arcw.connected acyclic, factors
    are (10.4)
C(X)\{A} is arcw.conn. (10.6)
decomposable, X is (10.7)
her.decomposable, X is if
    dim[C(X)] < ∞ (10.8)
if dim[C(X)] = 2 (10.13)
if dim[C(X)] < ∞ and
    C(X) ≅ X × Z (10.3, 5)
    X arcw.conn.acyclic (10.14)
    X a-triodic (10.12)
    X loc.conn. (10.1)
if X chainable or circle-like
    (10.10)

if X loc.conn.(10.1),(10.20-21)
   examples(10.2),(10.15-18)
questions (10.9, 21)
see embedding in C(X)
PRODUCT, 2^X IS A; *p.339*
   questions,theorems (10.22-26)
PROPER CIRCLE-LIKE *(0.24)*
   see circle-like (proper)
PROPERTY(b) *(16.2)*
   C(X), 2^X have (1.176)
PROPERTY[κ], X HAS *(16.10)*
   at a point (16.23-24)
   confluent images have (16.29)
      question (16.38)
   contractible hyps. (16.15)
      converse false (16.21)
   if X h.i. (16.27)
   if X homogeneous (16.26)
   if X loc.conn. (16.11)
   iff F_μ continuous (16.14)
   iff hyps.have? (16.37)
   iff $X \cong S_o$ (16.28)
   retracts have (16.38)
   X × X does not, ex. (16.35)
   see $C_oFL[X,Y]$, near-homeomorphism
PROPERTY Z; *pp.209-210*
   see Z-map, Z-set
PSEUDO-ARC, X *(0.23)*
   $C(X) \not\vdash$ product;p.151,(1.124)
      see emb.in C(X) (product)
   Conner's question; p.234
   invariant Whitney hyperspace,X has (14.73.34)
   retract, $C_1(X)$ of 2^X? (6.10.3)

strong Whitney-rev. (14.54)
Whitney property (14.3)
see P(Ω), weakly chainable*
PSEUDO-CIRCLE, X; *p.423*
   Whitney property (14.24)
   see pseudo-solenoid
PSEUDO-INTERIORS
   for hyps. (19.42-43)
PSEUDO-MEAN *(6.20)*
   discussion; pp.286-287
   see mean
PSEUDO-SOLENOID, X; *p.423*
   $C(X) \subset R^3$ iff X pseudo-circle; p.236
   $C(X) \subset R^4$; p.239
   $C(X) \not\vdash$ product; p.151, (1.124)
   invariant Whitney hyperspace, X has (14.73.34)
   strong Whitney-rev. (14.54)
   Whitney property (14.23)
PUNCTIFORM *(12.13)*
PUNS; *p.339, p.398, p.592*

Q

QUASI-COMPLEX
   acyclic $\Rightarrow$ fxd.pt.prop.; p.291
   C(X) is
      if X chainable; p.291
      $\Rightarrow$ C(X) has fixed point property; p.294
      not, examples; pp.294-295
   what X? (7.4)

R

ρ_h ; *pp.619-620*
ρ_τ ; *p.521, (19.36)*
   see L(M)

R^n (0.19)
  see $a(\Omega^*)$, $ch(M)$, Cr_ε, $hi(M)$,
    $L(M)$, n-space, $P(\Omega^*)$
RATIONAL CONTINUUM, X; p.380
  arcless chainable, ex. (12.18)
  arcw.acc. from $2^X \setminus C_1(X)$
    see arcwise accessible...
      (rational X)
  ⊂ any her.dec.continuum?
      (12.21)
  ⊃ subc.with cut point (12.12)
    her.dec.continuum ⊃...?
      (12.20)
  inverse limit is, sufficient
      conditions (12.17.1)
  see hereditarily decomposable
      continuum
RESIDUAL; p.515
RETRACT; p.269
  see ANR, AR, fund.retract, re-
      traction, r-image, r-map,
      strong deform.retract
RETRACTION; p.269
  $C(X)$ onto $F_1(X)$ (6.4, 9)
    questions (6.2, 28)
  $C(C(X))$ onto $F_1(C(X))$ (6.22)
  $cc(X)$ onto $F_1(X)$; p.579,
      (18.30, 32)
  dendroid onto finite graph,
      ε-retraction? (7.8.3)
  $F_2(X)$ onto $F_1(X)$ iff X mean
      (6.17)
    see mean
  $F_2(X) \cup \Gamma$ into $C_1(X)$ (6.13-21)
  monotone open, etc.? (6.7-8)
  prop[κ], preserves (16.38)
  $2^{C(X)}$ onto $F_1(C(X))$? (6.22)

  2^X onto $C_1(X)$
    does not exist, ex. (6.10.2)
    if X loc.conn.(6.5-7, 12)
    into $C_1(X)$, etc. (6.11)
    questions (6.3, 7),(6.10.1),
      (6.10.3), (6.29)
  2^X onto $F_1(X)$ (6.4, 9)
    questions (6.2, 22, 28)
  2^X onto 2^X_A (1.214.3)
  2^{2^X} onto $F_1(2^X)$; p.287
  see r-image, r-map, strong
      deform.retract
r-IMAGE; p.269
  monotone open, etc. (6.6-8)
  of 2^X
    $C(X)$ if X chainable, etc?
      (6.10.3)
    $C(X)$ if X loc.conn. (6.6)
    $C(X)$ iff $C_1(X)$ retract of
      2^X? (6.29)
    X if X loc.conn., n.a.s.c.
      (6.4)
    Y, necessary conditions (6.9)
  of 2^X or $C(X)$
    2^A or $C(A)$, what X? (6.23-27)
    X iff $F_1(X)$ retract? (6.28)
    Y, necessary conditions (6.9)
  see retraction, r-map
r-MAP; p.269
  between X, 2^X, $C(X)$? (6.1)
    confluent, monotone, etc.?
      (6.8)
  ψ_p is, $cc(X)$ onto X; p.579
  see retraction, r-image
ROGERS HOMEOMORPHISM; p.303
  see cone=hyp.prop., $Cone(X) \cong C(X)$
      (homeomorphisms)

S

σ_t (14.73.7)

　continuous (14.73.7)

　　onto $\mu^{-1}([t,\mu(X)])$ (14.73.8)

　$\sigma_t \circ \Psi_t$ = identity (14.76.2)

　homeomorphism, each t, iff
　　$X \in CPH$ (14.73.10)

　　appls.(14.73.11),
　　　(14.73.14-16), p.493
　　related ideas (14.73.28-34)
　　see covering prop.,...her.

　homeomorphism, some t
　　equivalence (14.76.5)
　　if $\mu^{-1}(t)$ her.irreducible
　　　(14.73.13)
　　　converse false; p.491

　Ψ_t level-preserving (14.76.7)
　$\sigma_t^{-1} = \Psi_t$ (14.76.1)
　see covering property,...her.,
　　Ψ_t, union

S_4 SPACE; pp.260-263

　see c-space, u-space

$S(\omega)$ (1.27.2)

　compact (1.29, 34)

　$\cong \Gamma(X)$ (1.30), (1.31.1)

　$S(\omega_1) \cong S(\omega_2)$ (1.32)

　　$S(\omega_1) \neq S(\omega_2), ex.$ (1.23-24)

　see $\Gamma(X)$, segment

$s(\Omega^*)$ (19.1)

　see $a(\Omega^*)$...

SEGMENT (1.15)

　constant function is (1.17)

　continuous (1.15)

　exists, n.a.s.c. (1.25)

　homeo.if non-constant (1.18)

　range of segment σ

　　chars.of (1.19, 22)

　$\sigma(0), \sigma(1)$ (1.21)

　$\sigma(t) \in C(X),$ suff.cond.(1.26)

　with respect to ω (1.15)

　　dependence on ω (1.23-24)

　　normalized ω (1.16)

　　uniqueness for fxd. ω (1.30)

　　formula (1.31.1)

　　see $S(\omega)$

　see $\Gamma(X)$, order arc, $S(\omega)$, segment homotopy

SEGMENT HOMOTOPY (16.4)

　existence of, continuous (16.3)

　properties of (16.5)

SEGMENTWISE ACCESSIBLE

　from $\Sigma \backslash C_1(X)$ beginning with K;
　　p.373

　Stiles' question (12.33)

　theorems (12.2, 4)

　see arcw.acc. ...,continuumwise accessible

SEGPLEX

　see n-segplex

SELECTION; p.253

　for $C(X)$

　　what X? (5.6)

　　$\Rightarrow$ X dendroid (5.7)

　　　converse false (5.10)

　　　if X smooth (5.11-12)

　　　question (5.11)

　　X directed, etc. (5.8)

　　X loc.conn. (5.9)

　for $cc(X)$, ψ_p is (18.30.1)

　　see ψ_p

　for $CL(Z)$

　　Z = some subsets of R^1;p.266
　　see choice fcn.,selec.(for 2^Z)

SPACE   SUBJECT INDEX   701

for $F_2(Z)$ (5.2-4)
for 2^Z, Z Hausdorff space
   char.by linear order (5.2)
   top.well-ordering; p.254
   Z cpt.tot.-disc.? (5.24)
   Z continuum (5.3),pp.254-255
   see selection(for CL(Z))
rigid selection (5.12)
see coselection, c-space, S_4
   space, u-space, Whitney map
   selection continuum

SEMI-LOCALLY CONNECTED *(0.29)*
   see aposyndesis

SET-THEORETIC DECOMPOSITION*(1.77)*
   $\mu^{-1}(t)$ is if X h.i. (1.78)
   see decomposition

SHAPE
   of $cc(X)$ = shape of X (18.31.3)
   of hyps.is trivial (1.182, 184)
   Whitney property
      for circle-like (14.18)
      not, example (14.43.6)

SIERPINSKI UNIVERSAL PLANE CURVE
   has hyper-onto rep.; p.187

SIMPLE HOMOTOPY TYPE
   of $cc(X)$ (18.32)
   see homotopy type

SIMPLE TRIOD, X *(0.20)*
   admits ess.diff.Whitney maps
      (14.43.1)
   $C(X) \neq$ product (10.2)
   $\cong \mu^{-1}(t_o)$, some $t_o > 0$
      (1.213.8), p.481
   see $a(\Omega^*)...$, n-odd, triod

SINGLETONS
   see $F_1(X)$

SINGULAR ARC *(18.21)*

SMOOTH
   see C*-smooth,dendroid(smooth),
      fan(smooth),weakly smooth
      dendroid

SNAKE-LIKE *(0.23)*
   see chainable

SOLENOID, X *(1.209.4)*
   basic properties (1.209.4)
   $C(X) \neq X \times X$ (1.128)
   $C(X) \cong Cone(X)$ (8.1), p.303,
      (14.20)
   $C(X) \neq R^3$ if $X \neq R^2$; p.235
   invariant Whitney hyp., X has
      (14.73.34)
   open map onto circle (1.207.5)
   strong Whitney-rev.? (14.57)
   Whitney map selection continuum;
      p.264
   Whitney property (14.21)
   see circle

SPACE
   see $a(M)$, $a(\Omega^*)...,a(X) \cup F_1(X)$,
      $C(X)$, $ca(R^n)$, $cas(R^n)$,
      cc-hyp., $C_oFL[X,Y]$, ch(M),
      $CL(S)$, Cr_ε, c-space, $F_1(X)$,
      $F_2(X)$, $F_n(X)$, $\Gamma(X)$, hi(M),
      Hilbert space, L(M),n-space,
      $P(\Omega^*)$, S_4 space, $S(\omega),s(\Omega^*)$,
      space of,$\theta(\Omega^*)$, $t(\Omega^*)$, 2^X,
      u-space, $W[C(X)]$, $W[2^X]$

SPACE OF
   Cantor sets in [0,1] (19.43)
   compact 0-dim.sets in [0,1]
      (19.43)

Z-sets in I_∞-manifold (19.42)
see space
SPHERE
see n-sphere
STABLE
see unstable, weak Whitney stable, Whitney stable
STARSHAPED (18.10)
see cc-hyp.(starshaped X)
STRONG DEFORMATION RETRACT
of $cc(X)$, $F_1(X)$ is (18.30.4)
of $\Gamma(X)$, $F_1(2^X)$ is (1.203.3)
STRONG WHITNEY-REVERSIBLE PROPERTY (14.45)
see Whitney-rev.prop., strong
SUBCONTINUUM; p.1
SUSPENSION OF X, SUS(X) (0.39)
Čech cohomology of; p.336
vertices of (0.39)
see $Sus(X) \cong C(X)$
SUS(X) $\cong$ C(X)
$dim[X] < \infty \Rightarrow X$ arc (9.1)
$dim[X] = \infty$
examples (9.4), (10.16)
$\Rightarrow X$ arcw.conn. (9.3)
questions (9.4-5)
for 2^X (9.6-7)
SYMMETRIC PRODUCTS (0.48)
exercises (0.71)
see $F_1(X)$, $F_2(X)$, $F_n(X)$

T

$\theta(\Omega^*)$, $t(\Omega^*)$ (19.1)
see $a(\Omega^*)$...
2*, THE FUNCTION; p.513
continuous (15.4)

u.s.c. (15.2)
see C^*, the function
2^X; p.1
$dim[2^X] = \infty$, $2^X \supset I_\infty$ (1.95)
$\cong C(X) \Rightarrow X$ loc.conn.? (10.27)
$\cong$ cones? (8.36-37)
see $Cone(X) \cong C(X)$
$\cong \Gamma(X)$? (1.27.3)
see $\Gamma(X)$
$\cong$ products (10.22-26)
see product, $C(X)$ is a
$\cong$ suspensions? (9.6-7)
see $Sus(X) \cong C(X)$
see $C(X)$, $CL(S)$, embedding in 2^X
2^X_A (1.214.3)
2^Z_p is Cantor set for $Z = \{p=0, 1/2, 1/3, ...\}$ (0.69.8)
see $C_M(X)$
θ-CURVE (19.1.4)
see $a(\Omega^*)$...
TERMINAL POINT (1.54)
as monotone image (1.205.2)
X irreducible between terminal points a & b
structure of $C(X/_{\{a,b\}})$ (14.34)
see terminal subcontinuum
TERMINAL SUBCONTINUUM (1.54)
chars. (1.55-58), (1.205.2)
monotone images (1.205.1-2)
of $C(X)$ (1.205.3)
$X_1 \cap X_2$ is etc.$\Rightarrow \mu^{-1}(t)$ is arc (14.28)
see terminal point

SUBJECT INDEX 703

THEOREMS
 conventions for credit; foot-
   notes 1 & 2 on p.52
TOPOLOGICAL COMPLETENESS
 of $C_oFL[X,Y]$ (16.33)
 of $ch(\Omega^*),hi(\Omega^*),P(\Omega^*)$ (19.33)
 of $W[C(X)],W[2^X]$? (14.71.4)
 of 0-dim.$Z \Rightarrow$ selection for
   $CL(Z)$; p.266
 see complete
TOPOLOGICAL GROUPOID; *p.285*
TOPOLOGICAL WELL-ORDERING; *p.254*
TOPOLOGY
 for $a(M)$, etc. (19.1)
 for $C(X)$, 2^X (0.3, 13)
  and decomposition top.(1.77)
  Hausdorff hyps.
   see Vietoris topology
 for $cc(Z)$ (18.1)
 for $CL(S)$
  see $CL(S)$
 for $\Gamma(X)$ (1.27.1)
  see $\Gamma(X)$
 for $L(M)$
  see $L(M)$
 for $S(\omega)$ (1.27.2)
  see $S(\omega)$
 see generalized Hausdorff met-
  ric,Hausdorff metric,homo-
  topy metric,Vietoris top.
TORUŃCZYK'S THEOREM *(1.214.2)*
TOTALLY-DISCONNECTED *(0.44)*
 monotone open mapping lemma
  (1.81)
 2^M is iff M is (0.69.8)
  homeo.theorems (1.40)

0-dim.,relation to (0.44)
see Cantor set, zero-dim.
TOUCHING SET; *p.397*
 discussion; pp.390, 397-398
 questions (13.2-3)
TREE-LIKE CONTINUUM, X *(0.28)*
 confluent image of X is (14.16)
 dendroid is (0.28)
 fxd.pt.prop.for $C(X)$? (7.8)
 X h.i. $\Rightarrow$
  $dim[C(X)] = 2$ (14.17.1)
  $\mu^{-1}(t)$ tree-like (14.15)
TRIOD *(0.21)*
 not triod
  if $X \in Class[W]$ (14.73.25)
  if $X \in CP$ (14.73.18)
  strong Whitney-rev.(14.49.1)
 see a-triodic,irreducible,n-odd,
  simple triod
TWO-CELL *(0.19)*
 admits ess.diff.Whitney maps,
  (14.43.6-7)
  related questions (14.38),
   (14.43.12)
 $\subset C(X)$ (1.145-146)
  see embedding in $\mu^{-1}(t)$
 $\mu^{-1}([t_o,\mu(X)])$ is, suff.condi-
  tions(1.213.2),(14.73.13)
 multi-valued maps without fixed
  points (0.73.1)
 see cc-hyperspace, n-cell

U

UNICOHERENT *(0.27)*
 each proper subcontinuum is
  suff.conditions(10.11),(11.16)
 hyperspaces are (1.176)

$\mu^{-1}(t)$ is if $X \in CP$? (14.73.22)
   see *irreducible*
   strong Whitney-rev.; p.454, (14.46)
Whitney property,not (14.12)
X *is if*
   $X \in Class[W]$ (14.73.25)
   $X \in CP$ (14.14.1)
   X C^*-*smooth* (15.17-19)
   X *is* $\Rightarrow$ *irreducibility is str. Whitney-rev. for sequences* (14.55.1)
see *her.unicoherent continuum, irreducible, non-unicoherent*

UNION
   theorems (0.74.1),(1.43), (1.48-50),(1.60), (1.207.1),(1.208.7), (12.7)
     see σ_t

UNIQUELY ARCWISE CONNECTED (1.61)
   $C(X)$ *is iff* X *h.i.* (1.61)
     structure of arc (1.62)
     see *order arc(uniqueness)*

UNSTABLE
   points of 2^X; p.137

UPPER SEMI-CONTINUOUS FUNCTIONS (0.47)
   C^*, 2^* *are* (15.2)
     2^* *continuous* (15.4)
     see C^*, *the function*
   nested $\cap$ *of & appls.* (0.69)
   see *decomposition(u.s.c.), monotone map*

u-SPACE (5.18-19)

V

VERTEX
   of Cone(X), v (0.38)
     see Cone(X) $\cong C(X)$ (*homeos.*)
   of n-segplex; p.429
     theorem (14.32)
   of Sus(X) (0.39)

VIETORIS TOPOLOGY (0.12)
   base for (0.10-12)
   basic properties (0.13, 66)
   = *Hausdorff metric top.* (0.13)
   exercises (0.66)
   locally connected (1.208.1)
   metrization theorem (0.66.3)
   see *Hausdorff continuum, Hausdorff metric*

W

$W[C(X)]$, $W[2^X]$ (14.71)
   facts, questions; pp.473-475

WARSAW CIRCLE, S_2^1 (8.22)
   C-H continuum (8.23), p.331
   has fixed point property & $\mu^{-1}(t)$ *does not, some* t (14.30)

WEAK WHITNEY HYPERSPACE (14.73.28-30)

WEAK WHITNEY STABLE (14.39.1)
   questions (14.39),(14.42.2-3)
   theorems (14.40-41)
   see *Whitney stable*

WEAKLY CHAINABLE (0.25)
   $C(X)$, 2^X *are*; p.245
     images of hyps. are (4.7)
   cone over Cantor set is; p.245
   strong Whitney-rev.? (14.57)
   Whitney property? (14.36)

WEAKLY CONFLUENT MAP *(0.45.4)*
  formulation of hyper-onto rep. with (1.190)
  induced map of (0.49.1)
  onto S^1 *(0.67.4),(1.193)*
    question (1.202)
  see Class[W], confluent map
WEAKLY SMOOTH DENDROID *(19.16-19)*
WHITNEY HYPERSPACE *(14.73.28-30)*
WHITNEY MAP *(0.50)*
  constructions of (0.50.1-3)
  continuity of (0.50)
  exercises (0.68),(1.213)
  extends from $C(X)$ to 2^X? (14.71.5)
  $\mu^{-1}(t)$ *is continuum (14.2)*
    see irreducible
        weak Whitney stable
        Whitney property
        Whitney property,not
        Whitney-rev.property
        Whitney-rev.prop.,strong
  $\mu^{-1}(t) \cap C(A)$ *is continuum (14.11.1)*
    see Ψ_t
  $\mu^{-1}([t,\mu(X)])$ *is*
    AR (14.73.32)
    $\cong C(X)$ *(14.73.31-34)*
    $\cong C(Y)$ *(14.73.28-34)*
    loc.conn. (14.73.32)
    2-cell (1.213.2),(14.73.13)
    3-cell (1.213.2)
  $\mu \circ \hat{f}$ *is Whitney map*
    f from [0,1] to [0,1] (0.68.6)
    iff $\hat{f}$ light (1.213.6)
      iff f her.irr.(1.212.3)
    suff.cond.(0.51),(0.68.6)

  monotoneness,openness of μ, ω
    see monotone map, monotone open map,open map, Ψ_t
  normalized (1.16)
  on chain (1.3)
  segment, formula for using ω *(1.31.2)*
    see segment
  selection continuum (5.20-21)
  see essentially diff.Whitney maps, W[C(X)]
WHITNEY MAP SELECTION CONTINUUM *(5.20-21)*
WHITNEY PROPERTY; p.399
  acyclic for dim[X] = 1 (14.37)
  aposyndesis (14.19)
  arc (14.6)
  arcw.conn.continuum (14.8)
  Čech cohomology for X circle-like (14.18)
  chainable (14.4)
  circle (14.7)
    see circle $= \mu^{-1}(t)$
  Class[W]? (14.76.10)
  continuum (14.2)
  contractibility for X dendrite (14.43.10)
  CP? (14.76.9)
  CPH (14.76.8)
  decomposable (14.11)
  finite aposyndesis (14.19)
  h.i. (14.1)
  h.i.tree-like (14.15)
  homogeneous? (14.36)
  indecomposable chainable (14.13)
  indecomposable for X ∈ CP (14.14.2)

λ connected? *(14.36)*

locally connected *(14.9)*

mutual aposyndesis *(14.19)*

nonplanarity for X circle-like (14.18)

planarity for X circle-like (14.18)

proper circle-like *(14.5)*

pseudo-arc *(14.3)*

pseudo-circle *(14.24)*

pseudo-solenoid, any *(14.23)*

semi-aposyndesis *(14.19)*

shape for X circle-like(14.18)

solenoid, any *(14.21)*

tree-like for X h.i. *(14.15)*

weakly chainable? *(14.36)*

see Whitney property, not Whitney-rev. prop., strong Whitney stable

WHITNEY PROPERTY, NOT

ANR (14.43.7)

AR (14.43.6)

acyclic (14.43.6)

circle-like (14.26),pp.424-426

contractible (14.43.6)

　question (14.43.12)

　suff.conds. (14.43.8-10)

cyclic (14.29)

fxd.pt.prop.(14.30),(14.43.6)

　question (7.13)

Hilbert cube (14.42.1)

　questions (14.38), (14.42.2-3)

indecomposable; p.413, (14.12)

locally contractible(14.43.7)

nonplanarity (1.213.7)

non-unicoherence (14.29)

planarity (1.213.7)

trivial shape (14.43.6)

unicoherence (14.12)

see Whitney property

WHITNEY STABLE, X IS *(14.39.1)*

and X ε CPH ⇒ invariant Whitney hyp. (14.73.34)

examples (14.42)

Hilbert cube is not (14.42.1)

　see Whitney property, not (Hilbert cube)

questions (14.39),(14.42.2-3)

see weak Whitney stable

WHITNEY-REVERSIBLE PROPERTY *(14.45)*

equiv.to strong...? (14.56)

not

　arcw.connected (14.48)

　decomposability; p.454

　non-unicoherence; p.454

　question about various properties (14.57)

see Whitney-rev.prop., strong

WHITNEY-REVERSIBLE PROPERTY, STRONG *(14.45)*

arc (14.50)

arcless (14.52)

a-triodic (14.49)

　not being triod (14.49.1)

circle (14.51)

circleless (14.53)

Class[W] (14.55)

　for sequences? (14.55.1)

CP (14.55)

　for sequences? (14.55.1)

decomposability for X ε CP;
  p.454
finite graph (1.213.8)
h.i. (14.54)
indecomposability; p.454,
  (14.46)
irreducibility (14.55)
  for sequences if X unicoherent (14.55.1)
  questions (14.55.1),
    (14.73.23)
locally connected (14.47)
pseudo-arc (14.54)
pseudo-solenoid, any (14.54)
unicoherence; p.454, (14.46)
Whitney prop.P, "not P" (14.46)
questions (14.55.1),(14.56-57),
  (14.73.23)
see Whitney-reversible prop.

X

X; *p.1, (0.16)*
$\cong \mu^{-1}(t)$, *all* $t \in [0,\mu(X))$
  see weak Whitney stable,
    Whitney stable
  see $F_1(X)$
X(A, μ, t) *(0.68.5)*
  compact (0.68.5)
    structure of (14.11.1)
  indecomposability of X,
    char.by (1.213.4)

Z

ZERO-DIMENSIONAL
  does not disconnect C(X) (2.15)
  $...\hat{f}[C(X)]$; *p.225*
  question (11.17)
  see dimensionally homogeneous
  topologically complete ⇒ *selection for CL(Z);p.266*
  totally-disconnected, relation to (0.44)
  see space of, totally-disconnected

ZERO-REGULAR CONVERGENCE *(15.8)*
  C* continuous w.r.t. (15.9)
  see C*, the function
  equivalent to Hausdorff metric conv.iff dendrite (15.10)
  see ρ_τ

Z-MAP *(1.214.1)*
  char.of Hilbert cube by (1.214.1-2)
  see Z-set

Z-SET *(1.214.1)*
  basic facts (1.214.1)
  property Z; pp.209-210
  space of Z-sets (19.42)
  see Z-map